Defect Engineering in Photocatalysis

Defect Engineering in Photocatalysis

From Fundamentals to Applications

Edited by
Chuanyi Wang, Abdelkader Labidi, and Qiuhui Zhu

WILEY-VCH

Editors

Prof. Chuanyi Wang
Shaanxi University of Science & Technology
Weiyang District, 710021
China

Chulalongkorn University
Bangkok, 10330
Thailand

Dr. Abdelkader Labidi
Shaanxi University of Science & Technology
Weiyang District, 710021
China

Dr. Qiuhui Zhu
Wuyi University
Wuyishan, 354300
China

Cover Design: Wiley
Cover Image: © Science RF/stock.adobe.com

Typesetting Straive, Chennai, India
Printing and Binding: CPI Group (UK) Ltd, Croydon, CR0 4YY

Library of Congress Card No.: applied for

British Library Cataloguing-in-Publication Data A catalogue record for this book is available from the British Library.

Bibliographic information published by the Deutsche Nationalbibliothek
The Deutsche Nationalbibliothek lists this publication in the Deutsche Nationalbibliografie; detailed bibliographic data are available on the Internet at <http://dnb.d-nb.de>.

Print ISBN: 978-3-527-35453-5
ePDF ISBN: 978-3-527-84981-9
ePub ISBN: 978-3-527-84980-2
oBook ISBN: 978-3-527-84982-6

C9783527354535_240226

Contents

Preface

The developments of defective photocatalysts have been considerably increased during the last few years to achieve the goal of new photoactive materials synthesis as an emerging research topic for interesting photocatalytic applications. Defect engineering in photocatalytic materials is a research area of great interest which has led to the preparation of highly effective photocatalysts for energy generation (i.e., water splitting for hydrogen production), environmental remediation (i.e., degradation of organic pollutants), and gaseous pollutants reduction (Figure 1). However, the challenge lies in precisely controlling the defects to achieve the desired modifications without introducing detrimental effects that could impede the overall performance of the utilized photocatalytic feedstocks. To fully harness the potential of defect engineering in photocatalysis, a multidisciplinary approach involving materials science, chemistry, physics, and engineering is required. Additionally, collaboration between theoretical and experimental scientists is a critical step to develop efficient photocatalytic materials and understand the underlying principles that govern their behavior. Nowadays, with continued research and development, the challenges associated with defect engineering should be addressed, which can result in advancements in sustainable energy production technologies and environmental remediation strategies. This book typically provides an overview of the principles and applications of defective photocatalytic materials in a clear and suitable manner. Additionally, this advanced book on defect-engineered materials can be a good reference for researchers, scientists, engineers, academics, etc., in the field of materials science, environmental engineering, and energy storage. The proposed book may provide deeper insights into the theoretical aspects of defect-engineered photocatalytic materials, discussing advanced techniques for their preparation and characterization, and furnishing detailed case studies and research findings in numerous photocatalytic applications. It assumes a certain level of background knowledge in chemistry, materials science, physics, etc. The effectiveness of photocatalysis processes in energy storage and environmental applications could be an excellent reason that this book can also attract researchers and industrialists in the fields of wastewater remediation, sterilization, desalination, gaseous pollutants degradation, as well as health care and energy conversion for useful chemicals production.

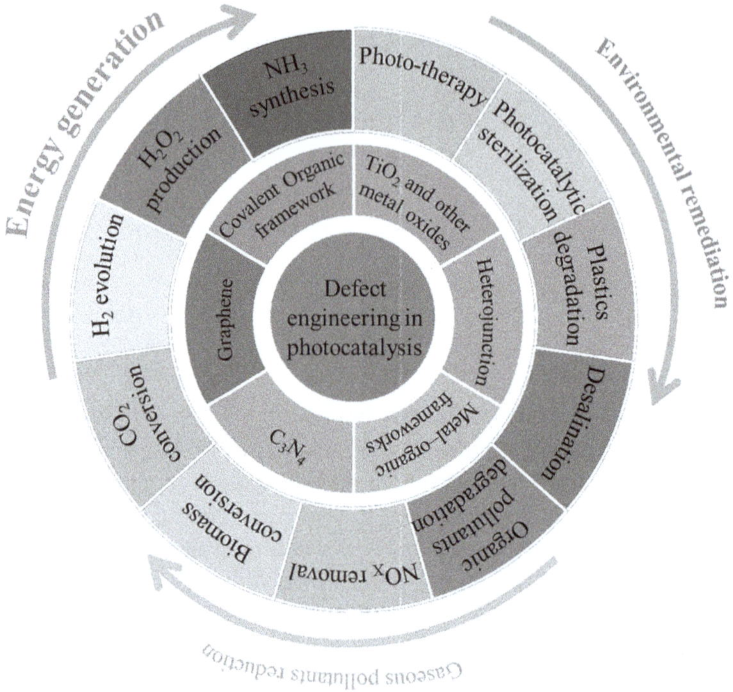

Figure 1 Defective inorganic and organic photocatalysts for catalytic transformations, environmental remediation and green fuels production.

Examples of defective photocatalytic materials are discussed in **Chapter 1** by Labidi et al. They provided an overview of the concept of defects creation and formation in photocatalytic materials and the role of defects for the enhancement of their photocatalytic features. Theoretical modeling and their control utilizing various characterization techniques to study mechanisms of defect formation including band gap engineering, charge carrier separation and transfer, concentration of defects, defect-interface interactions, photostability, etc. **Chapter 2** by Labidi et al. depicts the importance of defective photocatalytic materials synthesis and their usages in several domains. It also highlights the benefits and limitations of defect-engineered photocatalytic materials. Defect engineering as a method to customize the photocatalytic properties of titanium dioxide (TiO_2) and other metal oxides (WO_3, ZnO, etc.) is reviewed by Labidi et al. in **Chapter 3**. They presented the impact of defects on charge carrier dynamics, band structure, and reactivity of photocatalytic materials, highlighting the importance of defect control in enhancing light absorption, charge separation, and photocatalytic activity in numerous applications. **Chapter 4** by Kumar and Rath furnishes an insight into the preparation methods and characterizations of defective heterojunction photocatalysts. It also highlights their photocatalytic applications and future perspectives with a focus on their advantages and disadvantages. **Chapter 5** by Chen and Zhu provides an overview of defects creation and formation in metalorganic frameworks for numerous photocatalytic implementations. Defect engineering strategies in C_3N_4

include chemical modification, thermal treatments, and doping with other elements that enhance their photocatalytic performance by facilitating charge separation and improving their absorption of light for efficient photocatalytic applications are discussed by Gogoi et al. in **Chapter 6**. **Chapter 7** by Elhleli and Moussaoui addresses the properties of defective graphene as photocatalysts with their advantages and disadvantages. They described defects created in graphene which can introduce mid-gap states, enabling the absorption of light across a wider range of wavelengths. This enhanced light absorption increases the efficiency of photocatalytic reactions. They outlined the defects as active sites for several photocatalytic reactions, facilitating the generation and separation of charge carriers, resulting in efficient electron-hole transfer, leading to improved photocatalytic performance of graphene-based photocatalysts. **Chapter 8** by Sudirman highlights the usages of defective photocatalytic materials in various organic transformations, including C–C and C–X bond formation, C–H functionalization, and cross-coupling reactions. **Chapter 9** by Luo et al. depicts the application of defective photocatalytic materials in phototherapy. **Chapter 10** by Ren et al. provides a comprehensive overview of the principles and mechanisms of photocatalysis for sterilization. Subsequently, it presents an in-depth discussion on different types of defects in semiconductor materials and their impact on photocatalytic activity. They also highlight various defect engineering techniques, including doping, surface modification, and defect passivation, aimed at improving the sterilization performance of defective photocatalysts. The future developments of defect-engineered feedstocks to degrade organic matters such as nano- and micro-plastics in wastewater are reviewed in **Chapter 11** by Jalilnejad et al. focusing on the advanced designing of defective photocatalytic materials to improve their efficiency for nano-/micro-plastics photodegradation and their transformation into value-added chemicals and fuels. **Chapter 12** by Wu and Cui details the use of defective photocatalytic materials for enhancing desalination processes highlighting the importance of efficient salt removal and water purification by way of photocatalysis processes. **Chapter 13** by Hu offers knowledge about the potential defective photocatalysts in wastewater treatment technologies. It will also provide information about their synthesis, characterization, and benefits in organic pollutants degradation. Tailoring defective photocatalytic materials to develop highly efficient and selective photoactive feedstocks for NOx remediation is reviewed comprehensively by Zhu et al. in **Chapter 14** focusing on the efficiency of defective photocatalytic materials to achieve NOx reduction purposes and explaining the important role of defect formation in these feedstocks for highly efficient NOx remediation technologies. **Chapter 15** by Pourshirband et al. presents the defect formation/creation in several photocatalytic feedstocks. In addition, it highlights the synthetic protocols and identification of these defective materials for biomass conversion technologies. Understanding the precise role of specific defects and optimizing the photocatalytic systems for practical CO_2 conversion to value-added chemicals and fuels are discussed by Chen et al. in **Chapter 16**, presenting information about usages of defect-engineered materials and their benefits and limitations in CO_2 conversion purposes. Exploring novel photocatalytic materials, optimizing defect structures, and developing new synthesis methods for defective photocatalysts are described by Sial et al. in **Chapter 17**, listing some examples of defective photocatalytic materials and describing their advantages and disadvantages for H_2 evolution. **Chapter 18** by Padervand et al. examines mechanisms involved in H_2O_2 production on defect-engineered photocatalysts and the factors influencing their catalytic performance. It further addresses the challenges and opportunities associated with defect engineering for H_2O_2 production, suggesting future research directions in this area of research interest. The synthesis of ammonia (NH_3) through photocatalytic processes using defect-engineered photocatalytic materials is deeply discussed by Li et al. in **Chapter 19** with understanding the role of defects in enhancing the performance of photocatalysts with the development of new strategies for optimizing NH_3 synthesis processes. **Chapter 20** by Sial et al. provides a comprehensive overview of defect engineering in photocatalytic materials, focusing on their structures, photoelectronic properties, and photocatalytic applications. It explores the role of defects in modifying the electronic structure and optical features of photocatalysts, and how these modifications can enhance their photocatalytic performance in numerous photocatalytic implementations.

Collectively, this cutting-edge book on defect-engineered materials is useful for exploring defective photocatalytic materials in numerous implementations. A deeper understanding of defect-induced phenomena in photocatalytic materials could expand their usage in real-world applications such as sustainable energy production, environmental remediation, and photocatalytic transformations.

Chuanyi Wang
Shaanxi Univ. of Science & Technology
Weiyang District, 710021
China

Chulalongkorn University
Bangkok, 10330
Thailand

Abdelkader Labidi
Shaanxi Univ. of Science & Technology
Weiyang District, 710021
China

Qiuhui Zhu
Wuyi University
Wuyishan, 354300
China

September 29, 2025

About the Book

Defect engineering in photocatalytic materials is an exciting research area, offering opportunities to tailor the properties of photoactive feedstocks for specific photocatalytic applications. This book covers various types of defects such as vacancies, impurities, doping, surface defects, and volume defects and discusses their effects on the charge carrier dynamics and photocatalytic features of photocatalytic materials. It highlights the applications of defect-engineered photocatalysts in several areas including energy conversion and environmental remediation. Collectively, this book offers valuable insights into the design and optimization of photocatalytic materials via defect engineering, along with their challenges and opportunities for driving forward photocatalytic applications at present and in the future.

Key features:

- Provides insights and methodologies for defects formation and creation in photocatalytic materials.
- Offers knowledge about theoretical modeling and detection of defects in photocatalytic feedstocks.
- Covers the applications of defect-engineered materials in photocatalytic remediation and green energy production.

About the Editors

Chuanyi Wang is a distinguished professor at Shaanxi University of Science & Technology (SUST), China. Before moving to SUST in 2017, he was a distinguished professor of Chinese Academy of Sciences (CAS), serving as Director of Laboratory of Environmental Science & Technology of Xinjiang Technical Institute of Physics & Chemistry, CAS (2010–2017). He obtained his Ph.D. degree from Institute of Photographic Chemistry of CAS in 1998, worked in Germany (Institute for Solar Energy Research in Hannover and Free University Berlin) as an Alexander von Humboldt research fellow from 1999 to 2000, and then worked in the United States (Tufts University and Missouri University-Kansas City) as a research faculty from 2000 to 2010. Currently, Prof. Wang also acts as an associate editor or editorial board member for number of international journals. His research mainly focuses on energy and environmental photocatalysis. So far, he has published over 400 peer reviewed papers with an h-index of 94. In addition, Prof. Wang is a recipient of numerous prestigious awards or honors such as Fellow of European Academy of Sciences, Clarivate Highly Cited Researcher, Advanced Materials Laureate, World's Top 2% Scientist (ranked 80th in Physical Chemistry), Highly Ranked Scholar-Prior 5 years (Top 0.05% in all fields; ranked 57th in Environmental Science, ScholarGPS), Fellow of Royal Society of Chemistry, Fellow of International Association of Advanced Materials, Life Fellow of RSA, etc.

Abdelkader Labidi is a Research Associate Professor at Shaanxi University of Science and Technology, China. He received his master's degree in 2012 and Ph.D. degree in 2017 from University of Tunis Elmanar, Tunisia. From 2018 to the end of 2021, he worked as postdoctoral researcher at the Laboratory of Molecular Materials and Applications, University of Carthage, Tunisia, under the supervision of Prof. Manef Abderrabba. From 2022 to the end of 2024, he joined the School of Environmental Science and Engineering, Shaanxi University of Science and Technology (SUST), where he worked as postdoctoral researcher under the supervision of Prof. Chuanyi Wang. At the beginning of 2025, he started working as research associate professor at SUST. His research studies are about new photocatalysts, electrocatalysts, and adsorbents synthesis focusing on the design of photocatalytic materials based on the renewable resources as well as on the control of their morphology, structure, and their efficiency to degrade the environmental pollutants and produce value-added chemical.

Qiuhui Zhu is a Lecturer at Wuyi University, China. He received his Ph.D. degree in July 2023 from Shaanxi University of Science & Technology under the supervision of Prof. Chuanyi Wang. He was a research assistant in the group of Prof. Jianhui Li at Xiamen University from 2018 to 2019. He has studied at Queen's University Belfast as a visiting scholar from 2022 to 2023 under the supervision of Prof. Peter K. J. Robertson. In August 2023, he joined the College of Ecology and Resource Engineering, Wuyi University, where he worked as a Lecturer. His research interest mainly focuses on the design and synthesis of defective bismuth-based photocatalysts for environmental remediation and energy conversion.

Introduction

Dating back to 1972, the controllable production of defect-engineered photocatalytic materials has been widely conducted for tailoring and optimizing the properties and efficiency of photocatalytic feedstocks (i.e., metal, metal oxides, metal organic frameworks [MOFs], covalent-organic frameworks [COFs], hydrogen-bonded organic frameworks [HOFs], nano-composites, semiconductors, conducting polymers, etc.). Furthermore, defects in photocatalytic materials can be sorted either by their atomic structures or by their locations in the solid materials. In consideration of the fact that the functions of defects in the photocatalytic materials are extremely dependent on the types of defects (i.e., point, line, planar, and volume defects), they allow the creation of efficient photocatalysts with improved efficiency, leading to advancements in the photocatalysis processes in several fields, including organic synthesis, phototherapy, environmental applications (air purification, water treatment, photocatalytic sterilization), energy conversion (water dissociation and H_2 production, CO_2 conversion, value-added chemicals, and fuels production), etc. More interestingly, defect engineering refers to irregularities or structural deviations from the ideal lattice structure of a material. These irregularities can include vacancies (missing atoms), interstitials (extra atoms), impurities, dislocations, grain boundaries, or other types of lattice imperfections in photocatalytic feedstocks. Defects in photocatalytic materials can occur naturally or be intentionally created through various techniques such as doping, ion implantation, or annealing processes which are governed by the following key factors.

Defects can significantly influence the properties of photocatalytic feedstocks, especially their photocatalytic efficiency (i.e., light absorption, charge separation, surface reactivity, energy levels, photostability, etc.).

Besides, via the introduction/creation of specific types/concentrations of defects, researchers and industrialists can enhance desired photocatalytic material characteristics for a wide range of photocatalytic applications.

Furthermore, it allows the fine-tuning of material properties by adjusting defect concentrations and types. By strategically introducing and creating defects, photocatalytic materials can be optimized to achieve specific performance requirements such as improved charge separation and light absorption, redox ability, charge transfer channel, and internal electric field which enhance their photocatalytic performance. It is important to note that certain defects can exhibit unique functionalities that are not present in perfect photocatalytic materials. For example, defects like point defects or surface defects can serve as active sites for chemical reactions, leading to an enhanced photocatalytic efficiency of some target feedstocks. Defects can also alter the electronic band structure of photocatalytic materials which is beneficial in the photocatalysis processes. Defect engineering is an integral part of photocatalytic materials' design and synthesis processes.

Additionally, understanding the relationship between defect characteristics (size, distribution, type) and the resulting photocatalytic material features helps guide the development of new photocatalysts with desired attributes. While defect engineering offers exciting opportunities, it also presents some challenges. Interestingly, careful control of defect introduction/creation, optimization, and characterization is necessary to ensure reliability, stability, and reproducibility of photocatalytic materials. Then, the reliability of defect-engineered photocatalytic feedstocks must be evaluated under various conditions to understand their long-term behavior and efficiency.

Collectively, this book provides of overview about the design and characterization of defect engineering in a large family of organic and inorganic feedstocks with a focus on their structures, electronic, and photoelectronic properties. In addition, it summarizes the applications of these materials in the photocatalysis processes providing knowledge about the challenges, limitations, and impacts of the defect-engineered materials on their photocatalytic efficiency in several sectors, especially in the organic chemistry field, sterilization, environmental, and energy implementations.

1

Tailoring Photocatalytic Materials Through Defect Engineering: Control, Mechanism, and Experimental Protocols

Abdelkader Labidi[1], Jiangyushan Liang[1], Haitao Ren[2], Ting Gao[1], Tian Chang[1], Mohsen Padervand[3], Eric Lichtfouse[4], and Chuanyi Wang[5]

[1]Shaanxi University of Science and Technology, School of Environmental Science and Engineering, Weiyang District, Xi'an, 710021, PR China
[2]Xijing University, Technological Institute of Materials & Energy Science (TIMES), Xi'an Key Laboratory of Advanced Photo-Electronics Materials and Energy Conversion Device, Changan District, Xi'an, 710123, PR China
[3]University of Maragheh, Department of Chemistry, Faculty of Science, Amirkabir Highway, Maragheh 55181-83111, Iran
[4]Xi'an Jiaotong University, State Key Laboratory of Multiphase Flow in Power Engineering, Xianning West Road, Xi'an, 710049, PR China
[5]Department of Environmental and Sustainable Engineering, Faculty of Engineering, Chulalongkorn University, 254 Phayathai Road, Pathumwan, Bangkok, 10330, Thailand

1.1 Introduction

Photocatalysis process inspired by natural photosynthesis and dating back more than 100 years is considered a distinct class of chemical transformations [1, 2]. It employs the light energy to activate and drive several reactions that are challenging, sometimes even impossible [1–3]. In this process, electron–hole pairs (e^-/h^+) were formed when catalytic feedstocks were irradiated by UV lamps, xenon lamps, or even sunlight. Then, these generated electrons–hole pairs can migrate to generate oxidizing species, including hydroxyl radical (•OH) or superoxide radical ($\bullet O_2^-$) that display strong oxidation of inorganic and organic matters, converting them into unharmful products and sometimes into high-added chemicals and fuels [4–6]. Most recently, several classes of photocatalytic materials including traditional semiconducting photocatalysts, 2D photocatalysts, molecular photocatalysts, photovoltaic-assisted catalysts, plasmonic metal photocatalysts, and novel emerging photocatalytic feedstocks such as quantum dot-based materials have shown tremendous benefits in several photocatalytic applications (Figure 1.1) [1].

Even though the great progress has been made in the photocatalysis process in many applications including biological sector, organic reactions, energy storage, green fuels production, and environmental implementations, highly reactive photocatalytic feedstocks are still much needed at the present and in the future [7–14]. In order to address this issue, defects can significantly influence the properties of photocatalytic feedstocks, especially their photocatalytic performance

Defect Engineering in Photocatalysis: From Fundamentals to Applications,
First Edition. Edited by Chuanyi Wang, Abdelkader Labidi, and Qiuhui Zhu.

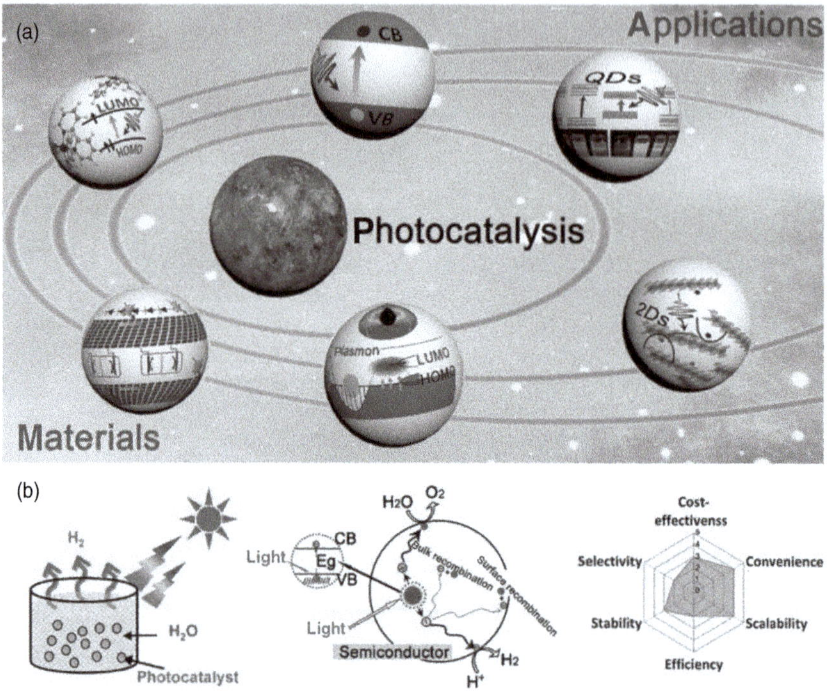

Figure 1.1 (a) Examples of photocatalytic feedstocks. (b) Particles dispersion in single reactor, two side reactions at single particle, benefits of photocatalysis. *Source:* Reproduced from Yang and Wang [1] / with permission of American Chemical Society.

(i.e., light absorption, charge separation, surface reactivity, energy levels, and stability) (Figure 1.2) [15, 16]. Besides, by carefully introducing specific types and concentrations of defects, researchers and industrialists can enhance the desired characteristics of photocatalytic materials for a wide range of photocatalytic applications [17–23]. Moreover, it allows the fine-tuning of material properties by adjusting defect concentrations and types. By strategically introducing and manipulating defects, materials can be optimized to achieve specific performance requirements, such as improved charge separation and light absorption in photocatalytic materials, which enhance their photocatalytic activities [24, 25]. It is important to note that certain defects can exhibit unique functionalities that are not present in perfect photocatalytic materials. For example, defects like point defects or surface defects can serve as active sites for chemical reactions, resulting in enhanced photocatalytic activity of some target feedstocks [15]. Defects can also alter the electronic band structure, which is beneficial in the photocatalysis process. In this account, defect engineering is an integral part of design and synthesis of photocatalytic materials. Additionally, understanding the relationship between defect characteristics (size, distribution, and type) and the resulting photocatalytic material properties helps guide the development of new photocatalysts with desired attributes [15]. While defect engineering offers exciting opportunities, it also presents some challenges [26, 27]. Interestingly, careful control of defect introduction and optimization is

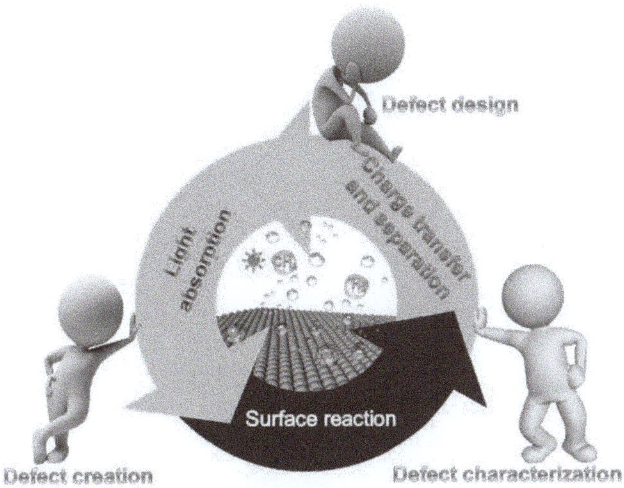

Figure 1.2 Benefits of defects design, creation, and characterization in photocatalytic materials. *Source:* Reproduced from Bai et al. [15] / with permission of Elsevier.

necessary to ensure reliability, stability, and reproducibility of photocatalytic materials [28]. Then, the reliability of defect-engineered photocatalytic feedstocks must be evaluated under various conditions to understand their long-term behavior and efficiency [29]. This chapter provides an overview of the design of defect engineering in organic and inorganic feedstocks with a focus on their structures and electronic and photoelectronic properties. In addition, it summarizes the preparation methods of these feedstocks and provides information about the characterization techniques and theoretical modeling utilized to confirm the successful preparation of defective engineered materials. Finally, benefits and drawbacks of defective photoactive feedstocks usages are presented. To sum up, this chapter provides an excellent platform for the fabrication of defect-engineered materials and their applications in fields such as organic chemistry, phototherapy, and environmental and energy implementations at present and in the future.

1.2 Defects Creation and Applications

In the recent decades, several works have been reported on the defect engineering in photocatalytic feedstocks and excellent results have been discovered so far in many photocatalytic applications (Figure 1.3a) [30, 31]. As a matter of interest, the preparation of defective engineered materials has been regarded as a desired option for highly efficient photocatalytic feedstock synthesis [30, 31]. It is worth mentioning that the control of defect formation in photocatalytic materials is essential for optimizing their efficiency in several applications. Strictly speaking, defect formation or creation (i.e., point defects, volume defects, line defects, surface defects, planar defects, and volume defects) in photocatalytic materials involves the usages of sophisticated strategies (Figure 1.3b) [30, 31].

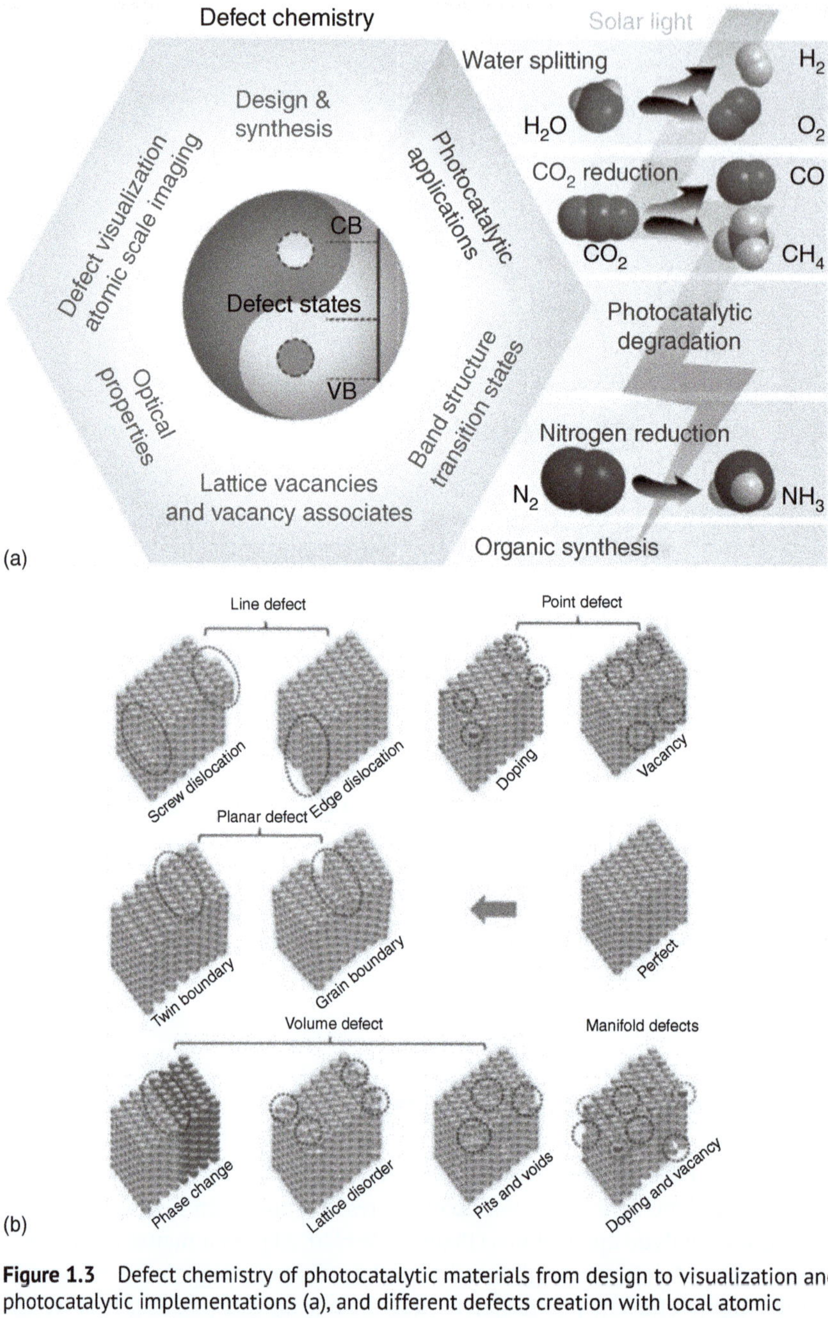

Figure 1.3 Defect chemistry of photocatalytic materials from design to visualization and photocatalytic implementations (a), and different defects creation with local atomic structures in the photocatalytic materials (b). *Source:* Reproduced from Zafar et al. [30] / with permission of John Wiley & Sons.

It should be noted that the creation/formation of defects in photocatalytic feedstocks is the stringent requirement for the defect creation and formation in photocatalysis. The strategies of creating or forming defects in photoactive materials are extremely dependent on the types of defects. Till now, several synthetic strategies have been reported to prepare photocatalytic materials containing defects, which can usually be divided into two classes: introduction of defects into prefabricated photocatalytic feedstocks and creation of defective sites during the preparation of photocatalytic materials [15]. Generally speaking, dopants, vacancies, voids, and lattice disorder in the surface or bulk of photocatalysts have been widely utilized in the preparation of highly efficient photoactive materials.

1.2.1 Dopants

To date, the controlled introduction of specific impurities (dopants), including cationic dopants, anionic dopants, and dual dopants, into the structure of photocatalytic material to modify its photocatalytic features has emerged as one of the excellent options in the synthesis of highly photoactive feedstocks [15]. Doping can be considered a kind of interior alterations or surface modification to rebuild the site configuration of photocatalytic materials, which can be realized at elevated temperature with liquid, gas, or solid precursors [15, 32, 33]. For example, in 2021, Helmy et al. developed S-doped TiO_2 (GST) nanoparticles (NPs) utilizing the sol–gel method, and an aqueous solution of *Malva parviflora* plant is employed as the green and adaptable medium with good reducing and capping features to promote the S-doping and crystal growth of TiO_2 [34]. The authors utilized X-ray powder diffraction (XRD), ATR–Fourier-transform infrared spectroscopy (FTIR), scanning electron microscopy (SEM), electrochemical impedance spectra, energy dispersive X-ray analysis (EDX), X-ray photoelectron spectroscopy (XPS), etc., to confirm the successful doping of sulfur into TiO_2. Using the FTIR technique, they observed absorption peaks at 1139 and 1311 cm^{-1}, which can be attributed to symmetric and asymmetric stretching of S=O bond, respectively; meanwhile, the peak appearing at 1056 cm^{-1} can be attributed to the symmetric S=O bond (Figure 1.4a).

Based on XPS results, the S2p XPS peaks at the binding energy of 168.3 eV for CST and GST materials can be attributed to the higher oxidation state of S^{2-} ions usually present in binding energy over 168 eV (Figure 1.4b–d). They found that S^{2-} ions very likely substituted the O^{2-} ions in TiO_2. Besides, some sulfate entities (SO_4^{2-}) could be produced on the TiO_2 surface. They concluded that the obtained S-doped TiO_2 nanoparticles display excellent antimicrobial activity against *Streptococcus pyogenes, Candida albicans, and Salmonella typhimurium*, and high antioxidant features that reach more than 90% inhibitions (Figure 1.4e).

In 2016, Dai et al. reported Ce-doped Bi_2MoO_6 (Ce-BMO) with distinct Ce/Mo molar ratios (5%, 10%, 20%, and 40%) via a facile one-step hydrothermal strategy with greatly enhanced visible-light-driven photocatalytic efficiency for the degradation of highly dangerous nerve agent simulants and dyes, as well as bacterial photodestruction [35]. They utilized XRD to reveal the composition and crystallinity of the fabricated photocatalysts and inductively coupled plasma mass spectrometry

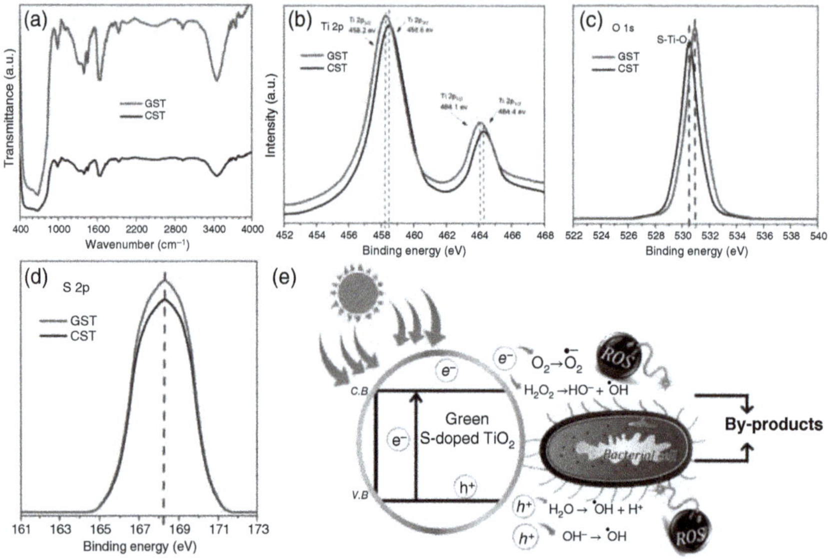

Figure 1.4 (a) ATR–FTIR spectra of S-doped TiO_2 NPs employing ethanol as the solvent (labeled as CST) and S-doped TiO_2 NPs (denoted as GST) employing $TiCl_4$ and thiourea as Ti and S sources; (b–d) XPS spectra for Ti 2p, O 1s, and S 2p core levels for CST and GST NPs; and (e) photodestruction mechanism of bacteria over S-doped TiO_2 NPs. FTIR: Fourier transform infrared spectroscopy, XPS: X-ray photoelectron spectroscopy. *Source:* Reproduced from Helmy et al. [34] / with permission of Elsevier.

(ICP-MS) to assess the relative content of cerium atoms in Ce-doped Bi_2MoO_6. Then, they applied SEM, transmission electron microscopy (TEM), UV-vis diffuse reflectance spectra (UV–Vis-DRS), Brunauer–Emmett–Teller (BET), XPS, photoluminescence (PL), time-resolved photoluminescence spectra (TR-PL), and Raman spectra to get an insight about morphology, chemical structure, and optical features of the as-prepared photocatalysts. Through the aforementioned techniques, Dai's group noticed that Ce-doped BMO displays different distortion and symmetry breaking of the MoO_6 octahedron, especially for the material Ce-BMO (40%). Referring to ICP-MS measurements, they revealed that the replacement of a Mo atom by a Ce element only occurred in Ce-BMO materials with high doping amount, which is different from the doping step of Ce-BMO with insufficient doping amount (Figure 1.5). They also observed that when cerium ions were included into Bi_2MoO_6, the doping step undergoes two stages: in the first step, the substitution of Ce^{3+} cation (ionic radius equal to 0.102 nm) for Bi^{3+} cation (ionic radius = 0.103 nm); in the second step, the substitution of Ce^{4+} cation (ionic radius equal to 0.087 nm) for Mo^{4+} cation (ionic radius = 0.065 nm). More importantly, they evidenced that the cerium doping process can lead to distinct types of crystal defects (i.e., symmetry disturbance, redox couple, and oxygen vacancy).

Interestingly, they concluded that the redox couples formed in the Ce-doping step such as Ce^{3+}/Ce^{4+} and Mo^{4+}/Mo^{6+} could obviously boost the ROS generation by one-electron and two-electron reactions, separately, which is conductive for efficient

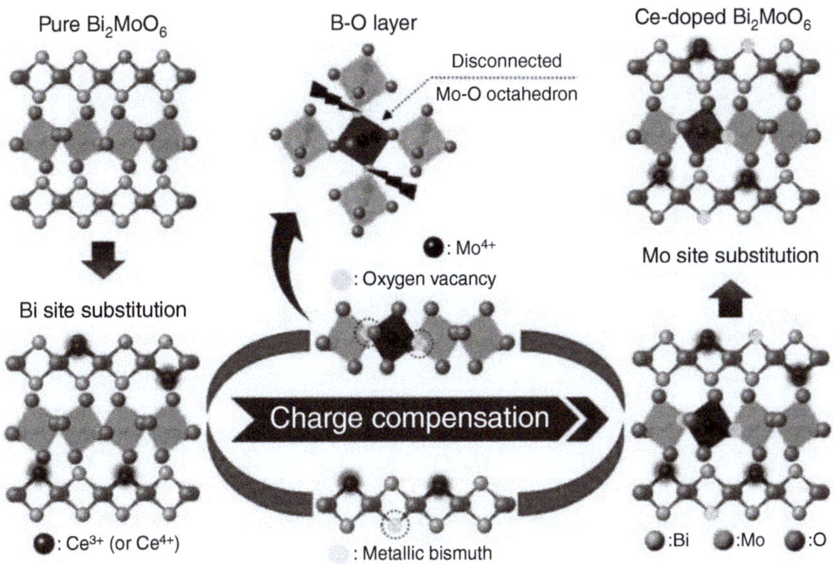

Figure 1.5 Formation pathway of crystal defects in a Ce-doped Bi_2MoO_6 structure. *Source:* Reproduced from Dai et al. [35] / with permission of American Chemical Society.

bacterial inactivation and other applications. The formation of crystal defects by Ce-doping not only brings new advantages for improving the photocatalytic efficiency of Bi_2MoO_6-based feedstocks, but also brings an excellent route to synthesize high-efficient photocatalysts by controlling the type and amount of crystal defects.

In summary, in addition to the above-mentioned photocatalysts, various doped photoactive feedstocks have been reported in the literature, highlighting their photocatalytic features enhancement, including bandgap engineering, charge carrier separation, surface and optical properties, resulting in an improvement of their stability and interaction with reactants. Additionally, doped photocatalysts have shown great promises in various applications such as green energy production, valuable chemical synthesis, and environmental implementations [36–40]. In the future, advanced methods should be applied in the synthesis of doped photocatalysts such as finding new and more efficient dopants and understanding the mechanisms behind doping effects to predict and manipulate photocatalytic features more effectively for large-scale applications of the photocatalysis processes.

1.2.2 Vacancies

Vacancies including cation/anion vacancies and interstitial sites in photocatalytic materials are critical to improve their photocatalytic performance. Besides, by fabricating photocatalysts with a controlled number of vacancies, researchers and industrialists can control their properties for specific applications such as environmental remediation, energy conversion, and other areas requiring photocatalytic processes. For example, in 2021, Kang et al. developed $Mo_{1-x}S_2$ ($Mo_{1-x}S_2/TiO_2$)

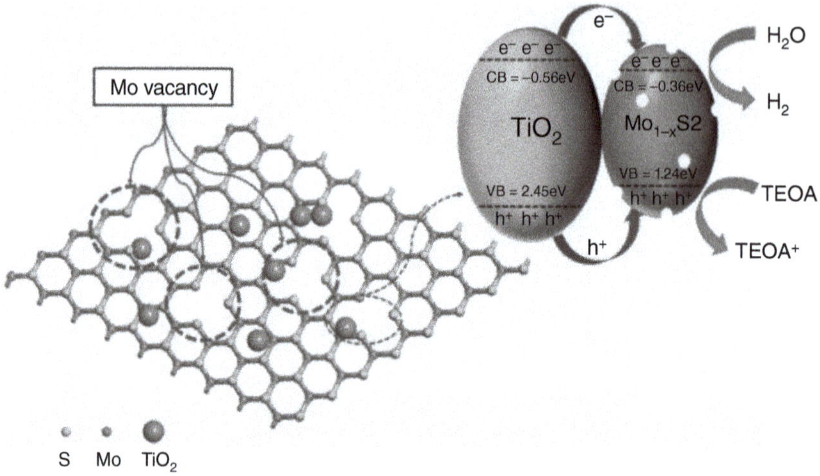

Figure 1.6 Photocatalytic hydrogen production mechanism over the $Mo_{1-x}S_2/TiO_2$ photocatalyst. *Source:* Adapted from Dai et al. [35].

with controllable Mo vacancies using the hydrothermal strategy for the hydrogen evolution reaction (HER) (Figure 1.6) [41].

They utilized XRD, SEM, TEM, high-resolution transmission electron microscopy (HRTEM), UV–Vis-DRS, XPS, Raman spectrometer, PL, and photoelectrochemical measurements in order to confirm the chemical composition, crystalline structure, morphology, and photoelectric properties of the obtained photocatalyst. Besides, the content of the Mo and S elements in the as-prepared photocatalyst was measured employing ICP-OES. They demonstrated that the 2.5:1 ratio of $Mo_{1-x}S_2$ material displays the highest content of Mo vacancies. On the other hand, when the amount of $NaBH_4$ raises to a 6:1 ratio, the reduced amount of Mo vacancies can be explained by the fact that excessive $NaBH_4$ damages the structure of MoS_2, leading to the release of S atoms. More importantly, they proved that the rapid electron transfer hindered the recombination of photogenerated charge carriers. The $Mo_{1-x}S_2$ nanosheets acted as co-catalysts, which exposed more active sites in the presence of Mo vacancies, resulting in enhanced HER activity. They revealed that the rapid charge separation and transportation resulted in good HER activity. As an achievement of their work, they indicated that at an optimal molar ratio of 2.5:1 $Mo_{1-x}S_2/TiO_2$, the H_2 production rate was 1561 $\mu mol\,g^{-1}\,h^{-1}$ in triethanolamine medium, which is nearly 5.3 and 2.7 times higher than that of pure TiO_2 and MoS_2/TiO_2, respectively.

In another report, Liu et al. (2021) successfully prepared Bi atom and oxygen vacancy (OV)-rich $Bi_{12}TiO_{20}$ nanofiber (Bi/BTO) materials using the electrospinning method and subsequent chemical reduction strategy for NO removal [42]. During this study, they employed XRD, SEM, TEM, PL, surface photovoltage (SPV) spectra, and photoelectrochemical properties to get an insight about the composition, structure, and photocatalytic features of the as-fabricated photocatalyst. They concluded that the Bi/BTO photocatalytic feedstock displays significantly improved photocatalytic NO removal efficiency compared to pristine BTO under visible light exposure. The enhancement of photocatalytic efficiency is attributed to the integrative effect of OVs and surface plasmon resonance effect of Bi atoms. The presence of OVs is evidenced by XPS and electron paramagnetic resonance (EPR) spectroscopy. Besides, the density functional theory (DFT) studies indicate that the OVs in BTO lead to the generation of a defect level, which is conductive to increase light absorption. In addition, based on EPR measurements, they discovered that Bi could act as an electron sink, improving the separation and transfer of photoinduced electron–hole pairs (Figure 1.7).

Generally speaking, the introduction or creation of vacancies in defective photocatalytic materials is a desired option for optimizing their photostability and enhancing their photocatalytic performance. In the recent years, vacancies (i.e., anion and cation vacancies) have shown several benefits such as improved charge carrier dynamics, increased surface reactivity, tuned electronic properties of photocatalytic feedstocks, and also enhanced production of reactive oxygen species, which are responsible for the improvement of defective photoactive materials' performance in wide range of implementations such as environmental remediation, valuable products synthesis, and energy conversion technologies [43–47]. Even though great advancements have been made in defective photocatalytic materials with several vacancies, some issues still limit their applications [48, 49]. Among them, careful consideration is much needed regarding the control over defect concentration, reduced structural stability of defective photocatalysts, influence on their electrical properties, and recombination and aggregation issues. More importantly, the existence and distribution of vacancies in defective photocatalytic materials can be complicated to characterize accurately. Thus, advanced techniques are needed urgently to probe their existence and further understanding of how these vacancies influence photocatalytic mechanisms of photoactive feedstocks.

1.2.3 Lattice Disorder

Like element doping and vacancy creation, lattice disorder plays a critical role in improving the photocatalytic features of photocatalytic materials, but it can have both beneficial and negative effects [50, 51]. More precisely, it can enhance the surface reactivity of photocatalytic feedstocks by creating defect sites that can serve as active sites for photocatalytic reactions, trap charge carriers (electrons and holes), and facilitate the absorption of light. However, lattice disorder in defective photoactive materials often contributes to a loss of their crystallinity, and the excess of disorder can introduce recombination centers that trap charge carriers in the defective photocatalytic materials, thereby impacting their stability and efficiency in several applications. Even though there are aforementioned benefits and limitations of lattice disorder in photocatalytic materials, several photoactive feedstocks have been developed, recently, and further utilized in numerous implementations. For example, in 2019, Yang et al. reported a polyol process to form anatase TiO_2 NCs employing distinct surface ligands, suggesting that the proper choice of polyol as a solvent can be utilized to control both the surface disorder and crystalline structure of NCs. In order to evidence the lattice disorder in the obtained photocatalyst,

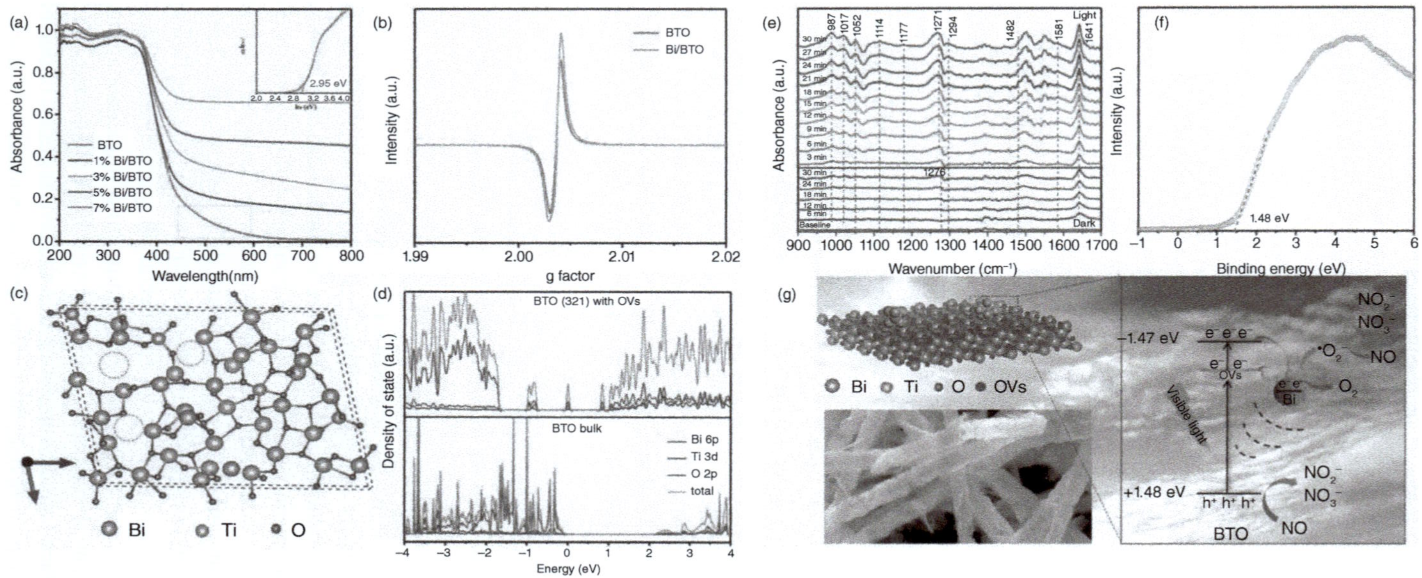

Figure 1.7 (a) UV–vis absorption spectra of the raw BTO and modified X% Bi/BTO (X = 1, 3, 5, and 7); (b) EPR spectra of raw BTO and modified 3% Bi/BTO; (c) crystal structure of BTO (321) plane along with OV formation (red circles denote OVs); (d) the density of states of bulk BTO and BTO (321)-rich OVs; (e) in-situ FTIR spectra of photocatalytic removal of NO gas over modified 3% Bi-BTO catalyst under dark and visible light irradiation; (f) VB-XPS spectra of raw BTO; and (g) plausible photocatalysis mechanism for NO removal over Bi/BTO catalyst. EPR: electron paramagnetic resonance, FTIR: Fourier transform infrared spectroscopy, VB: valence band, XPS: X-ray photoelectron spectroscopy. *Source:* Reproduced from Liu et al. [42] / with permission of Elsevier.

they employed XRD, Raman spectra, TEM, FTIR, UV-visible absorption, XPS, fluorescence spectra, ζ-Potential, electrochemical measurements, etc.; they indicated that Sn^{2+} cation is kinetically stable compared to Ti^{4+} and can occupy only the interstitial sites in $[TiO_4]$ and $[TiO_6]$, leading to a decrease in the lattice distortion energy. Besides, they demonstrated that the decrease in NC size appears with lattice expansion rising, with rising Sn^{2+} cation doping concentration with a rise in bulk defects, involving an optimal balance between them for simultaneous retention of the photoreduction efficiency and visible light absorption behavior of the TiO_2 NCs. Regardless of the as-fabricated photocatalyst application, they observed that under UV exposure, Sn^{2+}-doped TiO_2 NCs displayed higher photoreduction activity compared to undoped TiO_2, suggesting the existence of advantageous carrier trapping in Sn^{2+}-doped TiO_2 NCs. Besides, Sn^{2+} ions doping forms oxygen vacancies into TiO_2 NCs and the oxygen vacancies possess a pivotal role in the color switching of methylene blue [52].

In another research study, Li et al. reported an excellent example for this approach; they investigated structural disorder and its impacts on the optical features, electronic state, and photocatalytic efficiency of $ZnFe_2O_4$ (abbreviated as ZFO) obtained via the conventional solid-state reaction under different annealing temperatures (Figure 1.8).

In this study, Li's group utilized XRD, X-ray absorption spectroscopy (XAS), XPS, EPR, UV–Vis DRS, PL, and also DFT in order to get a deep insight about the impact of structural disordering and OV in the ZFO material on its band structure and oxygen reactive production for enhancing its photocatalytic behavior. They observed that a greater annealing temperature condition heavily leads to Fe^{III} transfers to the

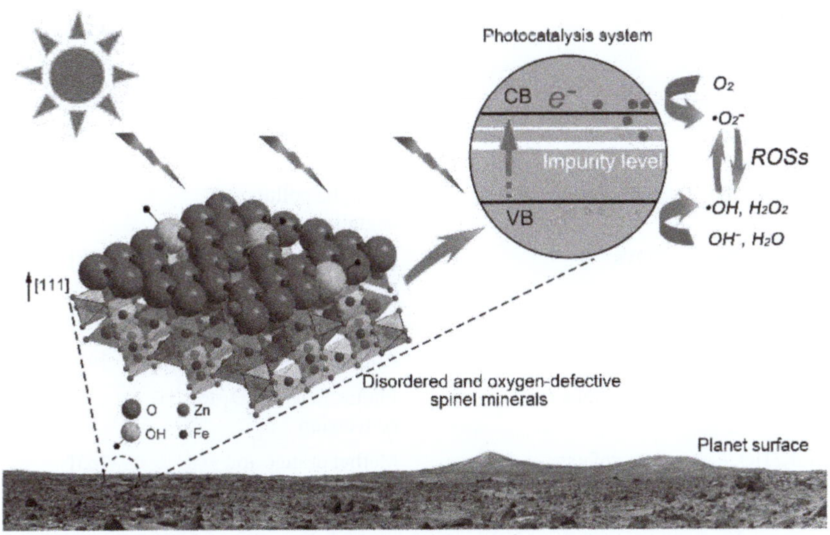

Figure 1.8 Disordered and oxygen-defective spinel minerals for ROSs photocatalytic production on the surface of a sunlit planet. ROSs: reactive oxygen species.
Source: Reproduced from Li et al. [53] / with permission of Elsevier.

tetrahedral site step by step and ZFO turns disordered [53]. They concluded that disordered and OV in $ZnFe_2O_4$ produces more hydroxyl radicals after irradiation compared to ordinary and defect-free ones, resulting in its narrower bandgap, accursed impurity levels, higher trapping e^- centers (VO^+), and a higher adsorption site of hydroxyl species. Besides, they mentioned that $ZnFe_2O_4$ is a model case to examine the effect of cation distribution on the semiconducting feature and photocatalytic efficiency and, therefore, pin down the important role of complicated spinel minerals in taking part in some vital redox chemical reactions in practical geological environment on the solid planet. In conclusion, lattice disorder in photocatalytic feedstocks can enhance their photocatalytic features like reactivity, charge carrier separation, and light absorption. However, it can lead to structural instability, increased charge recombination, and reduced selectivity of the modified photocatalysts. It should be noted that understanding and controlling lattice disorder is essential for optimizing photocatalyst structure for efficient applications, especially in environmental remediation, chemical production, and energy conversion [54–56]. Thus, more efforts in photocatalytic materials with lattice disorder, including computational modeling and comprehensive characterization techniques, should be applied to get insights into how lattice disorder affects photocatalytic efficiency, thereby facilitating the design of more effective photocatalytic feedstocks. Table 1.1

Table 1.1 Examples of photocatalysts along with their applications and type of defects.

Photocatalysts	Defect types	Applications	References
Ag-BiOI	Oxygen vacancies (OVs)	Trimethoprim degradation	[57]
$Bi_2O_2CO_3$	OVs	NO removal	[58]
Bi_2MoO_6	OVs	NO removal	[59]
$Bi_4Ti_3O_{12}$	OVs	NO removal	[43]
Bi@$SrTiO_3$	OVs	Photocatalytic Cr (VI) reduction	[60]
TiO_{2-x}@S-A500	OVs	Phenolic pollutants degradation	[37]
$CoMoO_4$/Carbon Nitride S-scheme	OVs	Photocatalytic H_2 production	[38]
TiO_2/CdS	Ti^{3+} defects	CO_2 reduction	[61]
In_2O_3@CN	N defects	Atrazine degradation	[45]
WO_3	Line defects	Photocatalytic CO_2-to-CH_4 conversion	[62]
$ZnIn_2S_4/TiO_2$/MS-SiO_2	Zn defects	Methyl orange and hexavalent chromium reduction	[63]
Au/ZnO	OVs	Rhodamine B and levofloxacin	[64]
N-deficient g-C_3N_4	N defects	Microcystin-LR degradation	[65]

defines examples of the last reported defective photocatalytic feedstocks, especially with oxygen vacancies, and demonstrates the increase of interest in these defective materials for various photocatalytic applications [37, 43, 57–65].

1.3 Characterization of Defects

It should be noted that defect characterization is important for understanding the relationship between defect formation and photocatalytic material properties. Besides, through the combination of experimental techniques and computational methods, researchers and industrialists can gain comprehensive insights about the types of defects, concentrations, and distributions and their pivotal role in photocatalytic efficiency, stability, and reactivity of photocatalytic feedstocks. In addition to the above-discussed techniques (i.e., XRD, SEM, FTIR, TEM, HRTEM, PL, SPV, EPR, TR-PL, ICP-MS, and Raman spectra) that have been widely utilized in defect characterization, other advanced techniques are also employed for the same purpose. Among them, microscopic characterization including atomic force microscope (AFM), scanning transmission electron microscopy (STEM), aberration corrected transmission electron microscope, scanning tunneling microscopy (STM), and high-angle annular dark-field scanning transmission electron microscopy (HAADF-STEM) has been recently utilized in defect understanding [15]. Besides, spectroscopic characterizations such as X-ray absorption fine structure (XAFS), positron annihilation spectroscopy (PAS), and X-ray spectroscopy (EDX) are employed in defect characterization [15]. Overall, despite the substantial efforts made in characterization techniques of defects, more sophisticated experimental techniques and computational methods are still much needed in order to enhance our ability to engineer photocatalytic materials with tailored defect structures for interesting applications.

1.4 Benefits and Limitations

It was proven that defects can provide a wide range of beneficial effects in photocatalytic materials, further enhancing their performance in photocatalytic applications [56, 57, 66–71]; their benefits are presented as follows:

a) Bulk or surface defects in photocatalytic materials can create new active sites for several chemical reactions, which can enhance their photocatalytic activity. Besides, types of defects such as element doping, vacancies, and lattice disorder can serve as reactive sites for adsorbing liquid and gaseous reactants, enhancing the performance of photocatalytic materials.
b) Some defects can act as traps for photogenerated charge carriers (electrons/holes). Then, by stabilizing these carriers, defects can also reduce their recombination rates and extend their lifetime, thus improving the photocatalytic efficiency of photoactive feedstocks.

c) It is noteworthy that defects could facilitate the separation and transfer of photogenerated charge carriers through providing pathways for charge transfer and their transportation, resulting in improved photocatalytic efficiency of photocatalytic materials.
d) The introduction of various defects can also modify the electronic band structure of photocatalytic materials, leading to the engineering of new bandgaps, enhancing light absorption, and further improving efficiency of the photocatalytic materials under visible light exposure.
e) The existence of defects can alter the surface energy of photocatalytic materials, thereby increasing their reactivity. This is particularly beneficial in specific applications, where surface interactions play an important role, especially in photocatalysis processes.
f) Specifically, defects including oxygen vacancies in different feedstocks can facilitate the formation of reactive oxygen species under light irradiation, which are crucial for the oxidative degradation of organic and inorganic contaminants employing the photocatalytic processes.
g) Some defects can enhance the structural integrity of photocatalytic materials by preventing the movement of dislocations and other defects under thermal stress, thereby improving their thermal and mechanical stabilities.

Even though defect engineering in photocatalytic materials provides a powerful option for optimizing their photocatalytic properties for a large range of applications, it presents some drawbacks which are stated as follows:

a) Unfortunately, certain defects can trap charge carriers (electrons/holes), resulting in increased recombination rates and low photocatalytic efficiency.
b) Defects may result in instability in photocatalysts, especially under prolonged irradiation or harsh environmental conditions. Thus, photocatalytic materials with excessive or uncontrolled defects can lose their photocatalytic activity within the time.
c) The introduction of defects might create new active sites that can improve unwanted side reactions, resulting in reduced selectivity of photocatalysts for the production of desired products.
d) On the one hand, surface defects may result in the accumulation of reaction by-products, especially blocking active sites and thus hindering the photocatalytic activity. Besides, bulk defects can also negatively impact the electronic features of the photocatalytic materials.
e) On the other hand, defects that alter photocatalytic features can make the photocatalytic materials more susceptible to chemical attack, influencing their lifetime and operational viability.
f) Defects can create electric fields that influence the movement of charge carriers. While this can be beneficial, it can also result in unfavorable drift mechanisms that decrease charge transfer and photocatalytic performance of the reported photocatalysts.
g) In some cases, while defects can improve photocatalytic performance, a high density of defects can overwhelm the advantageous effects, potentially

impairing electron transfer and the photocatalytic performance of photocatalytic feedstocks.

h) Some defects can generate new electronic states within the bandgap, which may negatively impact the band structure of the utilized photocatalytic material. This modification can shift the energy levels required for efficient charge separation.
i) If defects in photocatalytic feedstocks are not uniformly distributed, it can result in heterogeneous photocatalytic performance within the photocatalytic material. This nonuniformity can convolute the reaction dynamics, leading to inconsistent performance of photocatalysts.
j) Characterizing defects in photocatalytic feedstocks can be complex. Misunderstanding of defect types or concentrations in defective photocatalysts can lead to inappropriate photocatalytic material selections or processing conditions, hampering the development of desired photocatalysts.
k) The existence of defects might make photocatalytic materials more sensitive to changes in environmental conditions, including temperature or pH, thereby influencing their performance.
l) Developing photocatalytic materials with controlled defects often involves complex synthesis methods, eventually leading to the high cost of producing these classes of photocatalytic materials, further limiting their scalability for practical industrial applications.

1.5 Conclusion

In recent years, several efforts have been devoted to examine the surface-defect-engineered photocatalysts for numerous applications (i.e., environmental applications, biology, and organic synthesis), with substantial progress being made. It should be noted that, in the recent years, the use of defective photocatalytic feedstocks has increased significantly, thanks to their unique features including high reactivity and versatility. This chapter aims to offer a comprehensive overview of recent progress achieved in the usage of defective photocatalytic materials in numerous implementations. Even though defects in photocatalytic materials cannot be easily detected and characterized, several techniques have shown some efficiency to characterize defects including AFM, STEM, STM, HAADF-STEM, XAFS, PAS, EDX, etc. However, studies have shown that certain defects cannot be easily characterized. In the future, newer approaches utilizing advanced research and modern technology should be applied in the synthesis and characterization of defective photocatalysts for highly effective photocatalyst preparation. Besides, photocatalysis processes are limited to academic laboratory settings with existing challenges because of their limited conversion performance, and the creation of practical applications that takes advantage of the unique properties of defective photocatalytic feedstocks will lead to a bright future in a fruitful field of research work, especially for materials science, chemistry, and physics Consequently, theoretical, experimental, and mechanistic challenges in the usage of defective photocatalytic materials should be dealt with urgently, thanks to their substantial features (i.e., effective light absorption, charge separation/transfer, and synergetic effects) for several applications at present and in the future.

Acknowledgments

This work was supported by the Foreign Expert Program of SAFEA of China (Nos DL2023041004L and G2023041021L) and Shaanxi Provincial Foreign Expert Service Plan (Project No. 2025WZ-YBXM44).

Declarations

Funding: This work was financially supported by the National Natural Science Foundation of China (52161145409 and 52306165) and SAFEA of China ("Belt and Road" Innovative Exchange Foreign Expert Project, DL2021041001L).

Conflict of Interest

The authors declare that they have no conflict of interest.

References

1 Yang, X. and Wang, D. (2018). Photocatalysis: from fundamental principles to materials and applications. *ACS Appl. Energy Mater.* 1: 6657–6693.
2 Luo, L., Hou, C., Wang, L. et al. (2024). Layered double hydroxide-based photocatalysts for the removal of emerging contaminants: progress in past ten years. *Catalysts* 14: 252.
3 Li, W., Gao, F., Liao, G. et al. (2024). Synergistically interface networked organic–inorganic photocatalytic membrane for highly stable Cr (VI) reduction and hydrogen production. *Chem. Eng. J.* 492: 152330.
4 Prasad, M.N.V. (2020). *Disinfection By-Products in Drinking Water: Detection and Treatment*. Butterworth-Heinemann.
5 Kuo, H.-H., Vo, T.-G., and Hsu, Y.-J. (2024). From sunlight to valuable molecules: a journey through photocatalytic and photoelectrochemical glycerol oxidation towards valuable chemical products. *J. Photochem. Photobiol. C* 58: 100649.
6 Wimberger, L., Ng, G., and Boyer, C. (2024). Light-driven polymer recycling to monomers and small molecules. *Nat. Commun.* 15: 2510.
7 Liang, X., Gao, T., Cui, Y. et al. (2024). Photoreforming of poly (ethylene-terephthalate) plastic into valuable chemicals and hydrogen over $BiVO_4$/MoOx: synergistic promotion of oxidation and reduction processes. *Appl. Catal. B: Environ.* 357: 124326.

8 Liang, J.-Y.-S., Ren, H.-T., Labidi, A. et al. (2024). Boosted photocatalytic removal of NO using direct Z-scheme UiO-66-NH_2/Bi_2MoO_6 nanoflowers heterojunction: mechanism insight and humidity effect. *Rare Metals* 43: 1–13.

9 Cui, Y., Labidi, A., Liang, X. et al. (2024). Pivotal impact factors in photocatalytic reduction of CO_2 to value-added C_1 and C_2 products. *ChemSusChem* 17: e202400551.

10 Dong, Q., Li, X., Sun, J. et al. (2024). Regulating concentration of surface oxygen vacancies in Bi_2MoO_6/Bi-MOF for boosting photocatalytic ammonia synthesis. *J. Catal.* 433: 115489.

11 Labidi, A., Ren, H., Sial, A. et al. (2024). Photocatalyst materials formed using organic radicals for water splitting. In: *Organic Radicals* (ed. C. Wang, A. Labidi, and E. Lichtfouse), 211–230. Elsevier.

12 Labidi, A., Ren, H., Zhu, Q. et al. (2024). Coal fly ash and bottom ash low-cost feedstocks for CO_2 reduction using the adsorption and catalysis processes. *Sci. Total Environ.* 912: 169179.

13 Atif, S., Afzal, D.A., Abdelkader, L. et al. (2023). Self-regulated interfacial-enhanced piezo-phototronic effect of BNT@ Mn-SnO_2 heterostructures for superlative tetracycline degradation. *J. Catal.* 428: 115179.

14 Zhou, Q., Guo, Y., and Zhu, Y. (2023). Photocatalytic sacrificial H_2 evolution dominated by micropore-confined exciton transfer in hydrogen-bonded organic frameworks. *Nat. Catal.* 6: 574–584.

15 Bai, S., Zhang, N., Gao, C., and Xiong, Y. (2018). Defect engineering in photocatalytic materials. *Nano Energy* 53: 296–336.

16 Chen, M., Sun, M., Cao, X. et al. (2024). Progress in preparation, identification and photocatalytic application of defective g-C_3N_4. *Coord. Chem. Rev.* 510: 215849.

17 Wu, H., Yan, X., and Gu, Q. (2024). Low-content atomically dispersed Mo on defective TiO_2 for significantly improved hydrogen production. *Int. J. Hydrog. Energy* 71: 674–682.

18 He, Y., Lei, Q., Li, C. et al. (2021). Defect engineering of photocatalysts for solar-driven conversion of CO_2 into valuable fuels. *Mater. Today* 50: 358–384.

19 Wu, X., Zhang, Y., Wang, K. et al. (2020). In-situ construction of Bi/defective Bi_4NbO_8Cl for non-noble metal based Mott-Schottky photocatalysts towards organic pollutants removal. *J. Hazard. Mater.* 393: 122408.

20 Wang, M., Wang, S., Li, Z. et al. (2024). Anatase TiO_2/defective UiO-66 porous octahedral S-scheme heterojunctions toward optimized photocatalytic performance. *J. Photochem. Photobiol., A* 455: 115754.

21 Zhang, K., Wei, Y., Xie, L. et al. (2024). Amorphous titanium carbide on N-defective g-C_3N_5 for high-efficiency photocatalytic NO removal. *Chin. Chem. Lett.* 36: 110086.

22 Hamad, H., Samy, M., Bailón-García, E. et al. (2024). Cellulose-based materials in tailoring a novel defective titanium-carbon-phosphorus hybrid composites for highly efficient photocatalytic activity. *Int. J. Biol. Macromol.* 270: 132304.

23 Jadhav, S.R., Mohite, S.V., An, K.C. et al. (2024). Crafting defective ZnO nanoparticles: a green synthesis for enhanced photocatalytic degradation of organic pollutants. *Ceram. Int.* 50: 22783–22794.

24 Hunsom, M., Kunthakudee, N., Sangkhanak, S., and Serivalsatit, K. (2024). Photocatalytic application of defective WO_3 nanoparticles for precious metal recovery from plating effluent. *J. Taiwan Inst. Chem. Eng.* 155: 105301.

25 Dai, Z., Chen, L., Zhang, W. et al. (2024). K doped defective g-C_3N_4 decorated with cyano group for boosting visible-light driven photocatalytic removal of uranium. *Sep. Purif. Technol.* 338: 126537.

26 Mishra, R.K. and Verma, K. (2024). Defect engineering in nanomaterials: impact, challenges, and applications. *Smart Mater. Manuf.* 2: 100052.

27 Huang, Y., Meng, L., Xu, W., and Li, L. (2023). The positive versus negative effects of defect engineering for solar water splitting: a review. *Adv. Funct. Mater.* 33: 2305940.

28 Maarisetty, D. and Baral, S.S. (2020). Defect engineering in photocatalysis: formation, chemistry, optoelectronics, and interface studies. *J. Mater. Chem. A* 8: 18560–18604.

29 Sun, X., Zhang, X., and Xie, Y. (2020). Surface defects in two-dimensional photocatalysts for efficient organic synthesis. *Matter* 2: 842–861.

30 Zafar, Z., Yi, S., Li, J. et al. (2022). Recent development in defects engineered photocatalysts: an overview of the experimental and theoretical strategies. *Energy Environ. Mater.* 5: 68–114.

31 Ding, Y., Maitra, S., Wang, C. et al. (2022). Vacancy defect engineering in semiconductors for solar light-driven environmental remediation and sustainable energy production. *Interdiscip. Mater.* 1: 213–255.

32 Bie, C., Yu, H., Cheng, B. et al. (2021). Design, fabrication, and mechanism of nitrogen-doped graphene-based photocatalyst. *Adv. Mater.* 33: 2003521.

33 Zhang, S., Yi, X., Hu, G. et al. (2023). Configuration regulation of active sites by accurate doping inducing self-adapting defect for enhanced photocatalytic applications: a review. *Coord. Chem. Rev.* 478: 214970.

34 Helmy, E.T., Abouellef, E.M., Soliman, U.A., and Pan, J.H. (2021). Novel green synthesis of S-doped TiO_2 nanoparticles using *Malva parviflora* plant extract and their photocatalytic, antimicrobial and antioxidant activities under sunlight illumination. *Chemosphere* 271: 129524.

35 Dai, Z., Qin, F., Zhao, H. et al. (2016). Crystal defect engineering of aurivillius Bi_2MoO_6 by Ce doping for increased reactive species production in photocatalysis. *ACS Catal.* 6: 3180–3192.

36 Wang, L., Wang, Y., Zhang, W. et al. (2024). Fabrication of N-doped graphitic carbon encapsulated Fe_3C/Fe nanoparticles with dual-reaction centers for highly effective Fenton-like degradation of organic pollutants. *Surf. Interfaces* 49: 104388.

37 Yu, T., Zhang, W., Hou, C. et al. (2024). Confined growth of oxygen-vacancy rich TiO_{2-x} nanoparticles with heterophase junction for high-efficiency photocatalytic water purification. *Surf. Interfaces* 48: 104323.

38 Qin, J., Dong, Y., Lai, X. et al. (2024). Oxygen vacancy-rich $CoMoO_4$/carbon nitride S-scheme heterojunction for boosted photocatalytic H_2 production: microstructure regulation and charge transfer mechanism. *J. Mater. Sci. Technol.* 198: 176–185.

39 Zhang, Y., Sial, A., Shi, M. et al. (2024). Enhancing the catalytic degradation of tetracycline over Bi_2SiO_5 under visible light: synergistic interaction of dopants with oxygen vacancies. *Surf. Interfaces* 46: 104109.

40 Du, D., Shi, M., Guo, Q. et al. (2023). Metal Bi loaded $Bi_2Ti_2O_7/CaTiO_3$ for enhanced photocatalytic efficiency for NO removal under visible light. *Catalysts* 13: 1169.

41 Kong, C., Zhang, F., Wang, Y., and Huang, J. (2021). Synthesis and photocatalytic hydrogen activity of $Mo_{1-x}S_2$ nanosheets with controllable Mo vacancies. *J. Alloys Compd.* 876: 160165.

42 Liu, H., Mei, H., Miao, N. et al. (2021). Synergistic photocatalytic NO removal of oxygen vacancies and metallic bismuth on $Bi_{12}TiO_{20}$ nanofibers under visible light irradiation. *Chem. Eng. J.* 414: 128748.

43 Zhu, Q., Wang, Y., Wang, J. et al. (2024). Synergistic polarization and oxygen vacancies engineering for enhancing photocatalytic NO removal over $Bi_4Ti_3O_{12}$ nanowires. *Appl. Catal. B: Environ.* 346: 123734.

44 Gao, T., Lin, J., Zhang, K. et al. (2022). Porous defective Bi/Bi_3NbO_7 nanosheets for efficient photocatalytic NO removal under visible light. *Processes* 11: 115.

45 Li, W., Luo, J., Liu, J. et al. (2024). Construction of In_2O_3@ CN S-type heterojunctions with defect engineering for highly efficient photocatalytic degrade atrazine: efficacy and mechanism evaluation. *Surf. Interfaces* 46: 104144.

46 Liu, G., Tang, Z., Gu, X. et al. (2022). Boosting photocatalytic nitrogen reduction to ammonia by dual defective-CN and K-doping sites on graphitic carbon nitride nanorod arrays. *Appl. Catal. B: Environ.* 317: 121752.

47 Gao, T., Zhang, K., Zhu, Q. et al. (2022). One step synthesis of oxygen defective $Bi@Ba_2TiO_4/BaBi_4Ti_4O_{15}$ microsheet with efficient photocatalytic activity for NO removal. *Catalysts* 12: 1455.

48 Rezaei, M., Nezamzadeh-Ejhieh, A., and Massah, A.R. (2024). A comprehensive review on the boosted effects of anion vacancy in the heterogeneous photocatalytic degradation, part II: focus on oxygen vacancy. *ACS Omega* 9: 6093–6127.

49 Zhu, Q., Hailili, R., Xin, Y. et al. (2022). Efficient full spectrum responsive photocatalytic NO conversion at $Bi_2Ti_2O_7$: co-effect of plasmonic Bi and oxygen vacancies. *Appl. Catal. B: Environ.* 319: 121888.

50 Piler, K., Bahrim, C., Twagirayezu, S., and Benson, T.J. (2020). Lattice disorders of TiO_2 and their significance in the photocatalytic conversion of CO_2. In: *Advances in Catalysis* (ed. M. Biosca), 109–233. Elsevier.

51 Xiong, J., Di, J., Xia, J. et al. (2018). Surface defect engineering in 2D nanomaterials for photocatalysis. *Adv. Funct. Mater.* 28: 1801983.

52 Yang, K., Chen, X., Zheng, Z. et al. (2019). Solvent-induced surface disorder and doping-induced lattice distortion in anatase TiO_2 nanocrystals for enhanced photoreversible color switching. *J. Mater. Chem. A* 7: 3863–3873.

53 Li, Y., Li, Y., Xu, X. et al. (2019). Structural disorder controlled oxygen vacancy and photocatalytic activity of spinel-type minerals: a case study of $ZnFe_2O_4$. *Chem. Geol.* 504: 276–287.

54 Huang, S., Lang, J., Du, C. et al. (2017). Enhanced driving force and charge separation efficiency in disordered SnNbxOy: boosting photocatalytic activity toward water reduction. *Chem. Eng. J.* 309: 313–320.

55 Xu, Y., Wu, S., Wan, P. et al. (2017). Introducing Ti^{3+} defects based on lattice distortion for enhanced visible light photoreactivity in TiO_2 microspheres. *RSC Adv.* 7: 32461–32467.

56 Li, X., Dong, Q., Li, F. et al. (2024). Defective Bi@BiOBr/C microrods derived from Bi-MOF for efficient photocatalytic NO abatement: directional regulation of interfacial charge transfer via carbon–loading. *Appl. Catal. B: Environ.* 340: 123238.

57 Sial, A., Gao, T., Li, F. et al. (2024). Multifaceted synergism of dual active sites in oxygen vacancies enriched plasmonic Ag-BiOI nanosheets for enhanced piezo-photocatalytic degradation of trimethoprim. *ACS ES&T Eng.* 4: 2923–2937.

58 Wei, S., Zhu, Q., Ren, H. et al. Enhanced photocatalytic activity of nickel (II)-doped $Bi_2O_2CO_3$ nanosheets for efficient nitric oxide removal. *ChemPhotoChem* 9: e202400244.

59 Wang, Y., Wang, J., Zhu, Q. et al. (2024). Spherical Bi_2MoO_6 with oxygen vacancies via solvent manipulation for enhanced photocatalytic NO removal. *J. Environ. Chem. Eng.* 13: 115046.

60 Liu, B., Zhou, N., Liu, J. et al. (2024). Full-spectrum Bi@$SrTiO_3$ for bi-directional promotion effects on photocatalytic redox reaction: insight into intermediates and mechanism. *J. Rare Earths* 42: 917–929.

61 Bai, Y., Li, M., Liu, X. et al. (2022). Ti^{3+} defective TiO_2/CdS Z-scheme photocatalyst for enhancing photocatalytic CO_2 reduction to C_1–C_3 products. *Ind. Eng. Chem. Res.* 61: 8724–8737.

62 Xue, S., Wei, C., Shen, M. et al. (2024). Enriching surface-ordered defects on WO_3 for photocatalytic CO_2-to-CH_4 conversion by water. *Proc. Am. Acad. Arts Sci.* 121: e2319751121.

63 Bai, X., Zhang, H., Sun, S. et al. (2024). Construction of defective $ZnIn_2S_4$ and immobilized TiO_2 heterostructures: synergistic regulation of $ZnIn_2S_4/TiO_2$/MS-SiO_2 composite photocatalyst performance. *Mater. Sci. Semicond. Process.* 171: 108035.

64 Zheng, X., Yu, T., Liu, S. et al. (2024). High-efficiency photocatalyst based on Au nanoparticles loaded on defective ZnO nanorods. *Vacuum* 225: 113236.

65 Li, X., Li, G., Li, M. et al. (2023). Developing self-floating N-defective graphitic carbon nitride photocatalyst for efficient photodegradation of microcystin-LR under visible light. *Sci. Total Environ.* 895: 165171.

66 Liu, G., Huang, Y., Lv, H. et al. (2021). Confining single-atom Pd on g-C_3N_4 with carbon vacancies towards enhanced photocatalytic NO conversion. *Appl. Catal. B: Environ.* 284: 119683.

67 Wang, W., Huang, Y., Rao, Y. et al. (2023). The role of cationic defects in boosted lattice oxygen activation during toluene total oxidation over nano-structured CoMnOx spinel. *Environ. Sci. Nano* 10: 812–823.

68 Li, N., Shi, M., Xin, Y. et al. (2022). Oxygen vacancies-modified S-scheme $Bi_2Ti_2O_7/CaTiO_3$ heterojunction for highly efficient photocatalytic NO removal under visible light. *J. Environ. Chem. Eng.* 10: 107420.

69 Xin, Y., Zhu, Q., Gao, T. et al. (2023). Photocatalytic NO removal over defective Bi/BiOBr nanoflowers: the inhibition of toxic NO_2 intermediate via high humidity. *Appl. Catal. B: Environ.* 324: 122238.

70 Shi, M., Rhimi, B., Zhang, K. et al. (2021). Visible light-driven novel $Bi_2Ti_2O_7/CaTiO_3$ composite photocatalyst with enhanced photocatalytic activity towards NO removal. *Chemosphere* 275: 130083.

71 Wang, T., Zhu, Y., Luo, Z. et al. (2021). Oxygen vacancy confining effect on photocatalytic efficiency of Pt 1-black TiO_2 single-atom photocatalysts for hydrogen generation and phenol decomposition. *Environ. Chem. Lett.* 19: 1815–1821.

2

Defect Engineering in Photocatalysis: Opportunities and Challenges

Abdelkader Labidi[1], Xiaoqian Xu[1], Jiangyushan Liang[1], Ting Gao[1], Haitao Ren[2], Tian Chang[1], Mohsen Padervand[3], Eric Lichtfouse[4], and Chuanyi Wang[5]

[1] *Shaanxi University of Science and Technology, School of Environmental Science and Engineering, Weiyang District, Xi'an, 710021, PR China*
[2] *Xijing University, Technological Institute of Materials & Energy Science (TIMES), Shaanxi Key Laboratory of Liquid Crystal Polymer Intelligent Display, Changan District, Xi'an, 710123, PR China*
[3] *University of Maragheh, Department of Chemistry, Faculty of Science, Amirkabir Highway, Maragheh, 55181-83111, Iran*
[4] *Xi'an Jiaotong University, State Key Laboratory of Multiphase Flow in Power Engineering, Xianning West Road, Xi'an, 710049, PR China*
[5] *Department of Environmental and Sustainable Engineering, Faculty of Engineering, Chulalongkorn University, 254 Phayathai Road, Pathumwan, Bangkok, 10330, Thailand*

2.1 Introduction

The term photocatalyst is a combination of two words: photo, attributed to photon, and catalyst, which is a substance that alters the reaction rate in its presence [1–3]. As far as is known, since the discovery of photocatalytic water splitting on TiO_2-based metal oxide by Fujishima and Honda in 1972, photocatalysis is widely utilized as a strategy that uses light to activate photocatalytic feedstocks to produce reactive oxygen species (i.e., $\bullet OH$, $\bullet O_2-/HO_2\bullet$, $^1O_2$, and H_2O_2), which are the main oxidants that drive chemical reactions. Besides, the photocatalysis process can be divided into two categories based on the appearance of the physical state of the utilized reactants [1].

- Homogeneous photocatalysis: When the semiconductor and reactant are in the same phase including gas, solid, or liquid, in this regard, photocatalytic reactions are termed as homogeneous photocatalysis.
- Heterogeneous photocatalysis: When both the semiconductor and reactant are in distinct phases, in this way, photocatalytic reactions are named as heterogeneous photocatalysis.

Additionally, the photocatalysis process has been widely utilized in several fields (i.e., antifouling, deodorization, antifogging, water splitting, organic synthesis, conservation and storage of energy, air purification, sterilization, valuable fuels and chemicals production, self-cleaning, wastewater treatment, etc.) [1, 4–10] (Figure 2.1).

Defect Engineering in Photocatalysis: From Fundamentals to Applications,
First Edition. Edited by Chuanyi Wang, Abdelkader Labidi, and Qiuhui Zhu.

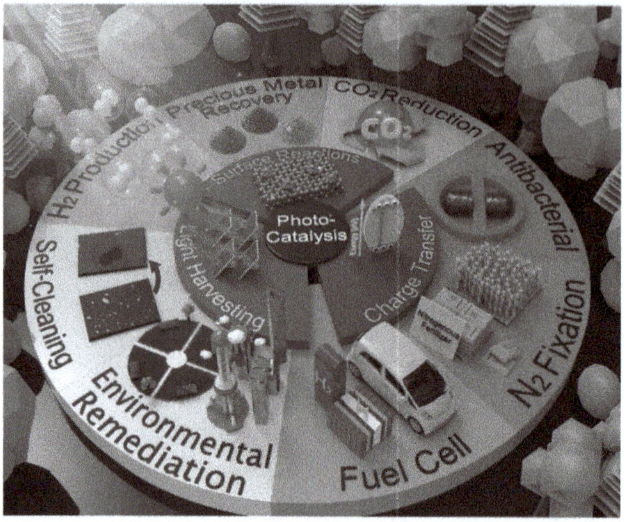

Figure 2.1 Graphic of the different applications of the photocatalysis process. *Source:* Reproduced from Li et al. [10] / with permission of Elsevier.

Generally speaking, heterogeneous and homogeneous photocatalysis are considered to be sustainable and environmentally friendly methods that can be utilized in several fields without producing harmful by-products and operating at mild conditions compared to traditional methods (e.g., adsorption, membrane separation, biological degradation, and electro-catalysis) [8, 9]. As presented in Table 2.1, the photocatalysis process has been recognized as a promising technology in various applications with numerous benefits. Thus, many efforts continue to explore and develop new photocatalytic feedstocks with the improvement of their selectivity, efficiency, and applicability in photocatalysis strategies. Although research works are heavily focused on photocatalysis utilization, this method still suffers from some drawbacks that limit its practical applications (e.g., environment, energy, photosynthesis, medicine, and health) [10–15].

As the main point of the photocatalysis method, photocatalytic materials determine the final conversion performance from solar; consequently, their composition and photo-electronic behaviors need to be carefully addressed [16]. In light of this, several methods have been effectively utilized in enhancing the photocatalytic efficiency of photoactive feedstocks such as morphology controlling, nonmetal doping, vacancy donor–acceptor system construction, and interface engineering [16]. Furthermore, the atoms doping contributes to the valence charge density and thereby reduces the bandgap of photocatalysts considerably [16]. Besides, a vacant state emerges into the bandgap over the Fermi level, and then the valence band of the photocatalyst is changed to a high-energy region [16].

Up until now, a large number of reported works have been introduced in the last few decades that examine different photocatalysts based on semiconductor systems including metal and metal oxide-based photocatalysts, silver-based photocatalysts, and bismuth-based photocatalysts [17–19]. Among various designs, TiO_2, ZnO, NiO,

Table 2.1 Advantages and limitations of the photocatalysis process.

Advantages	Limitations	References
Minimizes the environmental impact of various applications, such as air and water purification, without generating harmful by-products.	Many photocatalysts exhibit low quantum efficiency, meaning that only a small fraction of the absorbed light energy is converted into chemical reactions.	[10]
Photocatalysis is a versatile strategy with diverse applications (i.e., environmental remediation, energy production, materials synthesis, and healthcare).	A close contact between organic/inorganic pollutants and photocatalytic materials required for radicals (i.e., with an average lifetime of few nanoseconds) to apply their oxidation process.	[11]
It can present high selectivity for specific reactions, resulting in precise control over chemical transformations. The optimization of photocatalysts design and irradiation conditions, photocatalysis can improve reaction performance, diminish side reactions, and enhance product yields. This selectivity leads to the development of cleaner and more practical chemical processes.	Photocatalytic processes have drawbacks in terms of reactant specificity, as certain photocatalysts are selective toward specific types of substances or molecules. Some photocatalysts could degrade or lose their activity over time because of the rapid charge recombination, and low migration capability of the photogenerated electron/hole pairs.	[12]
Photocatalytic materials surfaces encompass self-cleaning capabilities, as they could actively decompose organic/inorganic contaminants under light exposure.	Limited visible-light absorption and wide bandgap of most commercial photocatalyst. In some photocatalytic reactions, the generation of undesirable by-products or intermediates may occur, leading to environmental concerns and secondary pollution.	[13]
Photocatalysis plays an important role in energy storage technologies by converting solar energy into useful chemicals and fuels. Photocatalytic systems can produce hydrogen especially in water splitting processes.	Understanding the complex mechanisms required in photocatalytic processes such as charge carrier dynamics, surface reactions, and mass transport phenomena, presents a big obstacle for researchers and industrialists.	[14]
In various applications, photocatalysis offers long-term cost savings because of the durable nature of some photocatalytic materials.	Photocatalysis is always influenced by environmental factors such as light intensity, temperature, humidity, and reactant concentrations, which could highly impact reaction rates and efficiency of the reported photocatalysts.	[15]

$BiVO_4$, CdS, Fe_2O_3, BiOBr, WO_3, g-C_3N_4, Ag_2S, SnO_2, Bi_5O_7I, Bi_2MoO_6, TaON, CuO, etc., have been widely designed and utilized as robust photocatalytic performance in numerous implementations [13, 20–23].

Additionally, the organic semiconductor materials with porous structure, such as covalent organic frameworks (COFs), covalent triazine frameworks (CTFs), graphitic carbon nitride (g-C_3N_4), and conducting polymers, have attracted particular attention in tremendous photocatalytic usages employing non-dangerous or poisonous ingredients, less costly and easy methods [16]. It is worth mentioning that traditional heterojunctions (i.e., type I, type II, and *p-n* type) as well as Z-/S-scheme and even multicomponent heterojunctions with their combination with organic and metal oxide semiconductor systems have played a key role in solving global energy demand and environmental pollution issues [24–26].

As a result, green synthesis of photocatalysts with control of their shape and size in the nanoscale range has led to the creation and fabrication of highly effective photoactive materials for use in innovative applications. Defects in photocatalytic feedstocks (e.g., surface vacancies, surface substitution impurity atoms (dopants), interstitial impurity atoms, vacancies, self-interstitial atoms, and impurity atom (dopant)) have attracted substantial interest in the last few years [27–30]. They are of important research worth in the efficient photocatalysts synthesis with broad light response, adjustable bandgap, excellent electron transfer/storage ability, unusual electronic structure, and good structure/thermal stability in energy and environmental applications. Among them, pollutant degradation, water splitting and H_2 evolution, CO_2 reduction into useful chemicals and fuels, N_2 fixation and NH_3 synthesis, heavy metal reduction, NO removal, and microbial disinfection have been intensively studied [31–36]. As far as we are concerned, various excellent published research works about defective photocatalytic materials and their applications have been intensively reported. Herein, this chapter presents different types of defects, their origin, formation mechanism, synthesis strategies, and the role of defects in enhancing the aforementioned various applications of photocatalytic processes, followed by the description of some recent progress of defective photocatalytic feedstocks in several sectors. Furthermore, unlike previous literature surveys, challenges limiting the commercialization and widespread adoption of engineered defective materials used in photocatalysis technology are discussed in depth. Overall, we believe that this chapter would bring insight for researchers, engineers, academicians, and policymakers into the defective photocatalytic feedstocks' uses and step into the void in the structure of these defective photoactive materials for better photocatalytic performance in several applications.

2.2 Defects in Photocatalytic Feedstocks

2.2.1 Creation and Classifications

Over the past few decades, numerous synthetic strategies have been conducted in order to obtain defect-rich photocatalytic feedstocks, which can be principally

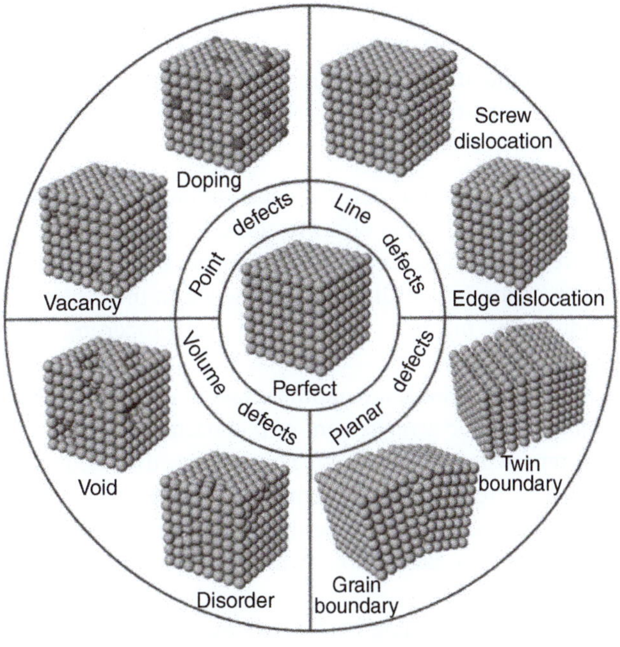

Figure 2.2 Schematic presentation of different defects with distinct atomic arrangement structures in photocatalytic materials. *Source:* Reproduced from Bai et al. [37] / with permission of Elsevier.

divided into two classes: (i) introduction of defects into as-fabricated photocatalytic feedstocks and (ii) creation of different defects under the synthesis processes of photoactive materials [37]. Defects can be classified into two major groups: (i) defects with distinct atomic structures in photocatalytic materials. In this class, all structural irregularities in solid feedstocks are simply classified into four main groups: 0D point defects (i.e., vacancy and doping), 1D line defects (i.e., screw dislocation and edge dislocation), 2D planar defects (i.e., grain boundary and twin boundary), and 3D volume defects (i.e., lattice disorder and void) (Figure 2.2).

Moreover, the aforementioned classifications can be subdivided into more types of defects including oxygen vacancies in various metal oxide semiconductors such as TiO_2, WO_3, Bi_2WO_6, BiO_2, $SrTiO_3$, ZnO, BiOCl, and $BiPO_4$; sulfur vacancies in ZnS, Zn_xCd_yS, and $ZnIn_2S_4$; nitrogen vacancies in Ta_3N_5; and iodine vacancies in $Bi_7O_9I_3$ [37]. In addition, self-doped interstitials, foreign elements doping (anion or cation dopants), impurity doping, and substitutions have been reported for creating point defects in photocatalytic materials. Besides, point defects, planar defects, line defects, and volume defects (voids (or pits) and lattice disorders) in photocatalytic materials, such as CdS, $Cd_{1-x}Zn_xS$, $AgBr_{1-x}I_x$, TiO_2, ZnO, $BiVO_4$, $ZnSnO_3$, Zn_2SnO_4, C_3N_4, In_2O_3, ZrO_2, Fe_2O_3, SnO_2, $BiOCl_xI_{1-x}$, and $SnNb_xO_y$, have been intensively studied in the literature (Figure 2.2). (ii) Defects in distinct locations of photocatalytic materials can be situated either in the bulk phase or on the surface of photocatalytic materials, which are labeled as bulk and surface defects, respectively. They can also be classified as follows: (i) point defects such as vacancies and dopants, which can act as both bulk and surface defects, or (ii) through dislocations and boundaries, which are bulk defects in the same semiconductor materials or in mono-component photocatalytic feedstocks [37]. The different defects are described as follows [37]:

i) Point defects (i.e., vacancies, interstitial defects, substitutional defects, and impurity atoms) have been reported as an effective route to modify the photocatalytic features of several materials. They can also impact the diffusion of atoms within photocatalytic materials, leading to an influence on their thermal stability, electrical conductivity, magnetic behavior, optical properties, and further their photocatalytic efficiency.
ii) Line defects, also known as dislocations, are another class of crystallographic defects that can occur in the atomic arrangement of photocatalytic materials, which have also been reported for highly efficient photoactive feedstock synthesis that can be used in tremendous implementation. They encompass edge dislocations and screw dislocations that highly influence photocatalysts' deformation, symmetry, and light absorption, which also improves their photocatalytic performance.
iii) Planar defects: It was proven that designing photoactive materials with planar defects that occur along a plane or boundary within the crystal lattice of the tailored photocatalysts has shown numerous benefits in the photocatalytic implementation. They consist of grain boundaries, twin boundaries, stacking faults, and surface defects, which have been regarded as a desired option for designing photoactive materials with high efficiency in specific applications.
iv) Volume defects are also known as bulk defects, which are three-dimensional imperfections that occur within the interior of photocatalysts. Volume defects including voids, precipitates, inclusions, and dislocation loops, have been considered as an effective method for tailoring highly efficient photocatalytic feedstocks with modified structure and photocatalytic features.
v) Lattice disorders including vacancies, interstitials, substitutional impurities, and crystallographic defects, are irregularities or distortions in the crystal lattice structure of photocatalysts. Thus, tailoring photoactive feedstocks with lattice disorders can also influence their photocatalytic properties, which is conducive in their use for energy storage and environmental remediation.

2.2.2 Applications

Recently, the utilization of defective photocatalytic materials in several applications has shown great promises for advancing green energy production (i.e., solar fuel generation, ammonia production), organic synthesis, energy storage (photocatalytic hydrogen generation and rechargeable batteries), environmental remediation (organic and inorganic contaminants degradation in aqueous solution, gaseous pollutants remediation), healthcare technologies (antibacterial coatings), etc. [38–44]. In this section, we present some examples of defective photocatalytic

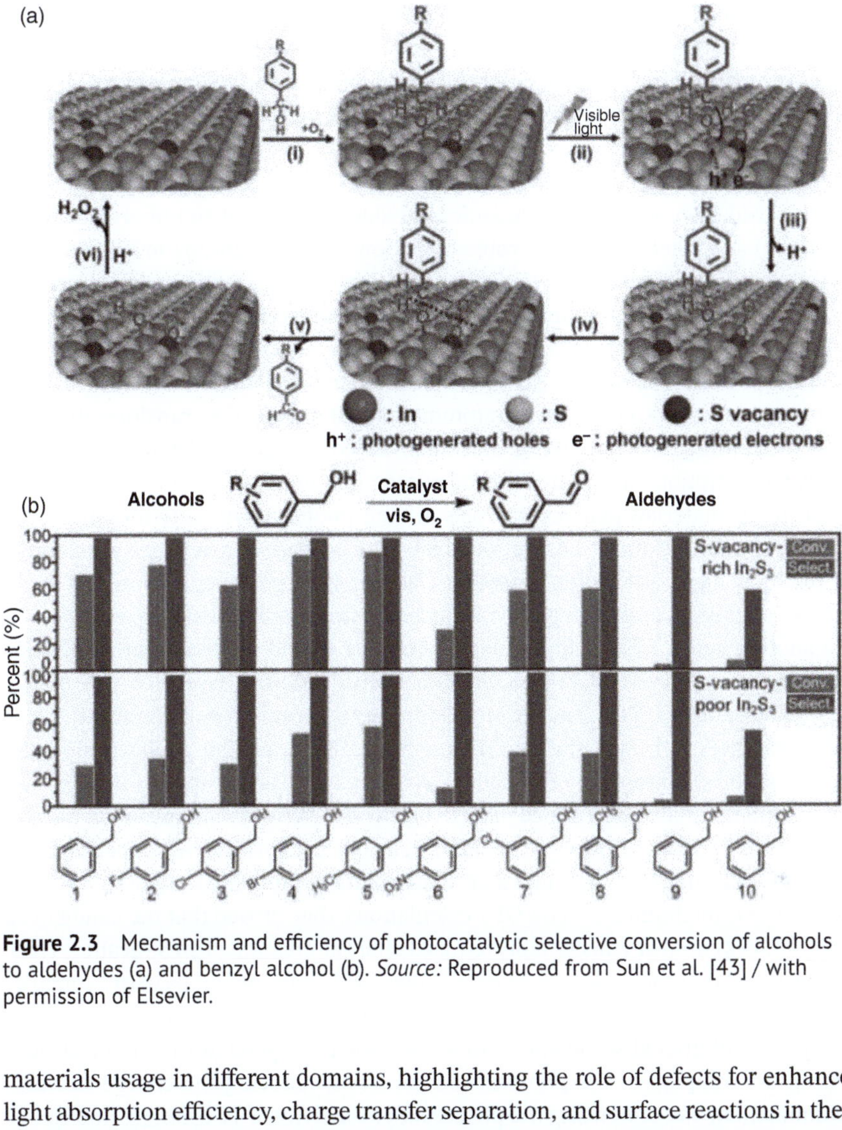

Figure 2.3 Mechanism and efficiency of photocatalytic selective conversion of alcohols to aldehydes (a) and benzyl alcohol (b). *Source:* Reproduced from Sun et al. [43] / with permission of Elsevier.

materials usage in different domains, highlighting the role of defects for enhanced light absorption efficiency, charge transfer separation, and surface reactions in these photocatalytic materials. For instance, in 2020, Sun et al. reported a comprehensive review discussing the defective 2D semiconductor-based photocatalysts for selective organic reactions (Figure 2.3) [43]. Moreover, they discussed the roles played by defects in the improved photocatalytic performance of 2D semiconductor-based photocatalysts, followed by their technical drawbacks and prospects for their future developments toward practical organic synthesis. They indicated that the surface vacancy (cation and anion vacancies) in 2D structures created more coordinately unsaturated sites, which is an encouraging way for switching on photocatalytic materials with more extended light absorption, realizing high selectivity, boosting electron–hole transfer, and separation efficiency [43]. Regarding their applications for fine chemicals production under green conditions, they reported S-vacancy-rich

In_2S_3 nanosheets could efficiently photo-oxidize alcohols to aldehydes because of their superior radicals $\bullet O_2-$ generation and H_2O_2 formation. Additionally, they demonstrated that the S-vacancy-rich In_2S_3 exhibited evidently greater conversion performance (71%) under visible light irradiation compared to the S-vacancy-poor sample for different benzyl alcohols' transformation to aldehydes.

Other photocatalytic reactions were reported by Sun's coworkers, including selective oxidation of sulfides to organic sulfoxides (medicinal and pharmaceutical chemistry) and photocatalytic aerobic oxidation of amine groups to imines (biologically active compounds), employing C_3N_4 nanosheets with carbonyl groups (C=O), vacancy-rich WO_3 nanosheets, and orthorhombic-$BiVO_4$ atomic layers with high content of vanadium defects, respectively. They also described the photocatalytic transformations of CO_2 gas to useful methane utilizing S-deficient $CuIn_5S_8$ single-unit-cell layers, etc. In all the aforementioned photoactive transformations of organic and inorganic compounds into value-added products which decrease fossil energy consumption and environmental contamination, the authors indicated that defects in 2D materials play pivotal roles in their photocatalytic activities improvement (i.e., creating more adsorption sites and activation processes, expanding the region of light absorption, boosting electron–hole pairs separation and their effective migration, and tuning photoinduced excitonic processes) (Figure 2.4).

In another study, Li's group developed a unique hybrid photocatalytic feedstock consisting of twinned $Cd_xZn_{1-x}S$ materials with different layers of defect-rich MoS_2 photocatalysts ($TCd_xZn_{1-x}S/MoS_2$) for the H_2 evolution reaction (HER) [45]. The researchers indicated that T-$Cd_xZn_{1-x}S$ and MoS_2 exhibit good visible-light responses, and the presence of abundant $S_2{}^{2-}$ and Mo^{5+} active species in defective MoS_2 nanomaterials contributed effectively to the activity enhancement of the as-prepared $TCd_xZn_{1-x}S/MoS_2$. The latter defective catalyst showed superior HER rates of 37.22 mmol h^{-1} g^{-1} in lactic acid solution and 69.25 mmol h^{-1} g^{-1} in Na_2S/Na_2SO_3 medium. Based on DFT calculations, they proved that the bandgap of metal sulfides could be considerably reduced after the creation of an S vacancy. They demonstrated that the S vacancies could endow MoS_2 nanomaterials with a remarkably narrowed bandgap, resulting in an enhanced conductivity and broadened light absorption. Additionally, the sulfur vacancies possess a good affinity to electrons, helping the trapping of the photoexcited electrons, reducing the charge carriers' recombination, and enhancing the photocatalytic activity of $TCd_xZn_{1-x}S/MoS_2$.

Gu et al. (2020) developed tungsten trioxide nanosheets (WO_{3-x}) rich-oxygen vacancies employing the microwave-assisted route for *E. coli K-12* inactivation (Figure 2.5). Through XPS analysis, they revealed that the existence of W^{5+} cations on the surface of the WO_{3-x} nanomaterial with an increased surface area of 30.23 m^2 g^{-1}, confirms the successful formation of oxygen vacancies on the photocatalysts' surface. Additionally, Gu's group indicated that the hydroxyl groups formed on the as-fabricated photocatalyst's surface favor the trapping of photogenerated e^--h^+ pairs, thereby inhibiting the charge recombination. The above findings were found to be conducive to bacterial cell inactivation [46].

During this process, the authors indicated that the as-fabricated WO_{3-x} photocatalysts depicted a good photocatalytic activity, achieving a complete inactivation of

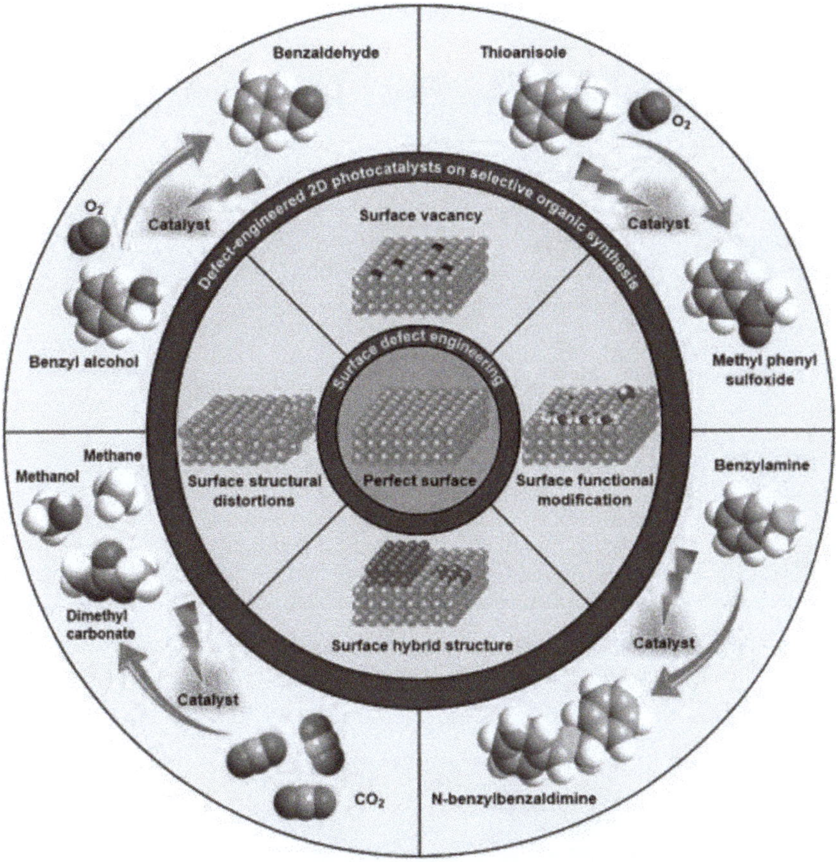

Figure 2.4 Schematic presentation of 2D photocatalysts with surface defect engineering for effective and selective organic synthesis. *Source:* Reproduced from Sun et al. [43] / with permission of Elsevier.

bacterial cells via a direct h^+ oxidation pathway within 150 minutes of contact time, higher than WO_3 without oxygen vacancies (Eqs. (2.1)–(2.3)):

$$WO_{3-x} + h\upsilon \rightarrow WO_{3-x}\left(e^- + h^+\right) \tag{2.1}$$

$$\text{Cell} + h^+ \rightarrow \text{Damaged cell} \tag{2.2}$$

$$OH^-_{ads} + h^+ \rightarrow \bullet OH_{ads} \tag{2.3}$$

In a recent study, Guan and his coworkers successfully synthesized iodine vacancy $BiOI_{1-x}$ photocatalyst via the calcination method for gaseous mercury removal under the visible light illumination (Figure 2.6) [47].

Based on XPS and BET analysis, they demonstrated the existence of iodine vacancy in the obtained material due to the presence of two peaks of I $3d_{3/2}$ and I $3d_{5/2}$ situated at 630.6 and 619.1 eV, respectively, and the Bi-410 °C photocatalytic material exhibited a larger surface because of the presence of iodine vacancies on its structure after employing the calcination strategy. The authors discovered that iodine

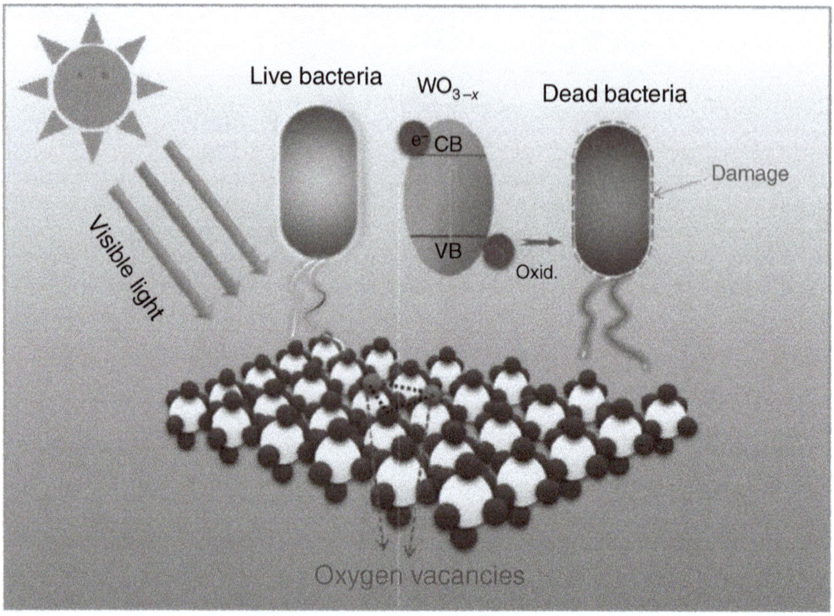

Figure 2.5 Proposed photocatalytic bacterial inactivation mechanism over WO_{3-x} under visible light illumination. *Source:* Reproduced from Gu et al. [46] / with permission of Elsevier.

Figure 2.6 Iodine vacancy flower-like $BiOI_{1-x}$ photocatalyst for gaseous mercury removal. *Source:* Reproduced from Guan et al. [47] / with permission of Elsevier.

vacancies in the as-obtained photocatalyst could enhance its photocatalytic features, further improving its capacity toward mercury removal. They observed that under visible light illumination, the photogenerated electrons migrate from the conduction band of Bi_5O_7I material to the conduction band of BiOI feedstock, and the holes migrate from the valence band of Bi_5O_7I nanosheets to the valence band of BiOI semiconductor. The electrons in the conduction band of BiOI could effectively be combined with the O_2 and H^+ to produce H_2O. Thus, hydroxyl ions (OH^-) can be easily oxidized into hydroxyl radical (•OH) by holes of BiOI, which is conductive for Hg^0 oxidization into HgO (Eqs. (2.4)–(2.9)):

$$BiOI/Bi_5O_7I + h\upsilon \rightarrow BiOI/Bi_5O_7I + e^- + h^+ \tag{2.4}$$

$$H_2O \leftrightarrow h^+ + OH^- \tag{2.5}$$

$$O_2 + 4H^+ + 4e^- \rightarrow 2H_2O \tag{2.6}$$

$$Hg^0\,(g) \rightarrow Hg^0\,(ad) \tag{2.7}$$

$$Hg^0\,(ad) + h^+ + OH^- + OH_{(ad)} \rightarrow HgO + H_2O \tag{2.8}$$

$$Hg^0\,(ad) + 2OH \rightarrow HgO + H_2O \tag{2.9}$$

In 2022, Zhu et al. developed $SrBi_4Ti_4O_{15}$ with sulfur (SVs) and oxygen vacancies (OVs) via a simple molten salt method route for the piezo-photocatalytic degradation of tetracycline pharmaceutical pollutant in aqueous medium. Through experimental investigations and theoretical calculations, Zhu's group demonstrated that oxygen vacancies in the [001]-dominated facets can enhance the internal polarization of the obtained photocatalyst, further improving its photocatalytic efficiency by inhibiting the recombination of photogenerated charge carriers [48]. The authors concluded that the oxygen vacancies were also created in the titanium oxide polyhedra of the as-fabricated material, causing a lattice distortion in its structure and an improvement of its polarization, which is conducive to higher tetracycline degradation (Figure 2.7). Based on DFT simulations, they observed that sulfur vacancies play an unfavorable role in enhancing polarization of $SrBi_4Ti_4O_{15}$, and their enhancement of polarization along the [001] facets is nearly contributed to the generation of OVs instead of SVs. They concluded that OVs contribute to the favorable separation of photogenerated charge carriers and the effective formation of reactive oxygen species derived from the piezoelectric field.

Zhu et al. continued exploring the benefits of the defect-engineered photocatalytic materials in environmental remediation. In the same year, they reported a full-spectrum $Bi@Bi_2Ti_2O_7$ containing oxygen vacancies (OVs) by way of the hydrothermal method for NO conversion [49]. During this process, the charge difference density (CDD) mappings and electronic localization function (ELF) were investigated to study the effect of OVs on the structure and interfacial charge separation/transfer of Bi@BT photocatalyst. They observed that the CDD mappings from top and side views of Bi@BT material and the defective Bi@BT-OV

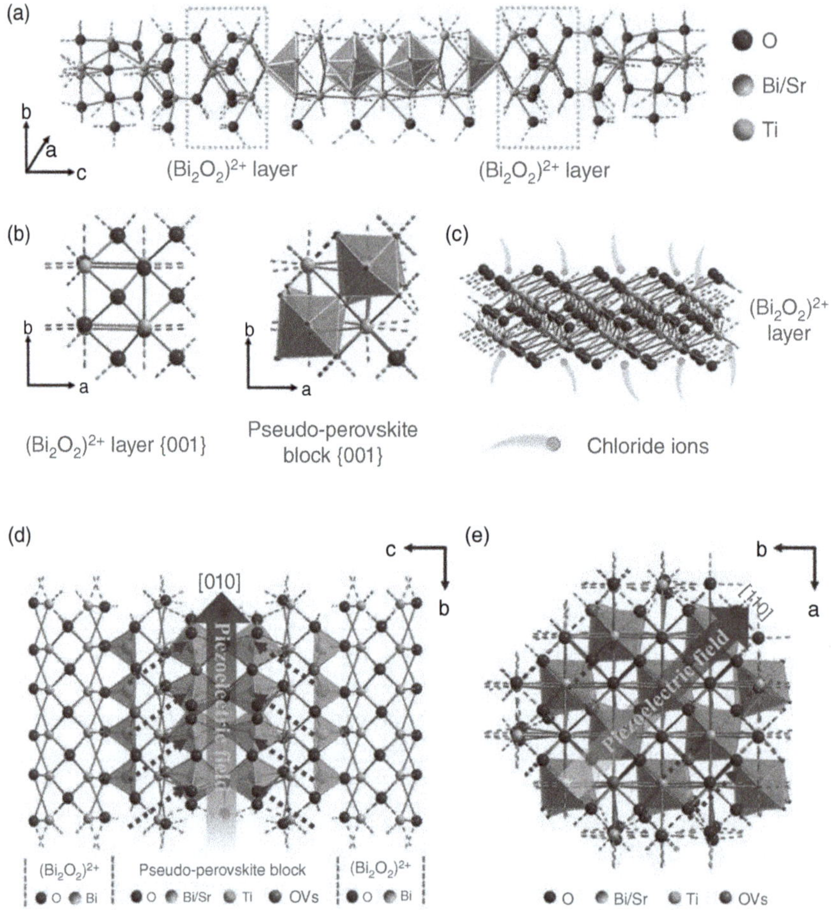

Figure 2.7 Crystal structure of $SrBi_4Ti_4O_{15}$ (a) along b–c and (b) a–b planes and (c) schematic illustration of the growth of $SrBi_4Ti_4O_{15}$ in a molten salt, mechanism of the piezoelectric effect along (d) [010] and (e) [110] directions on SBT-850. *Source:* Reproduced from Zhu et al. [48] / with permission of American Chemical Society.

presented in the Figures 2.8a,b reveal that the irregularly distributed electrons are poor around the metallic Bi elements and rich in the Ti atoms, resulting in a localized region of low electron density of Bi atoms and high electron density of Ti atoms (charge accumulation is in yellow and depletion in cyan). Additionally, they concluded that the adsorption energy ($E_{ads} = -4.25$ eV) and electron transfer Δq of 0.68 of NO on Bi@BT-OV photocatalyst is higher than that of Bi@BT ($E_{ads} = -3.87$ eV, Δq equal to 0.52), suggesting that the oxygen vacancies could drive both the adsorption/activation of NO gas resulting in high NO conversion of 79%. They demonstrated that the enhanced photocatalytic performance of the obtained photocatalyst was assigned to the co-effect of Bi and OVs, referring to control experimental protocol and theoretical studies, which not only enhanced

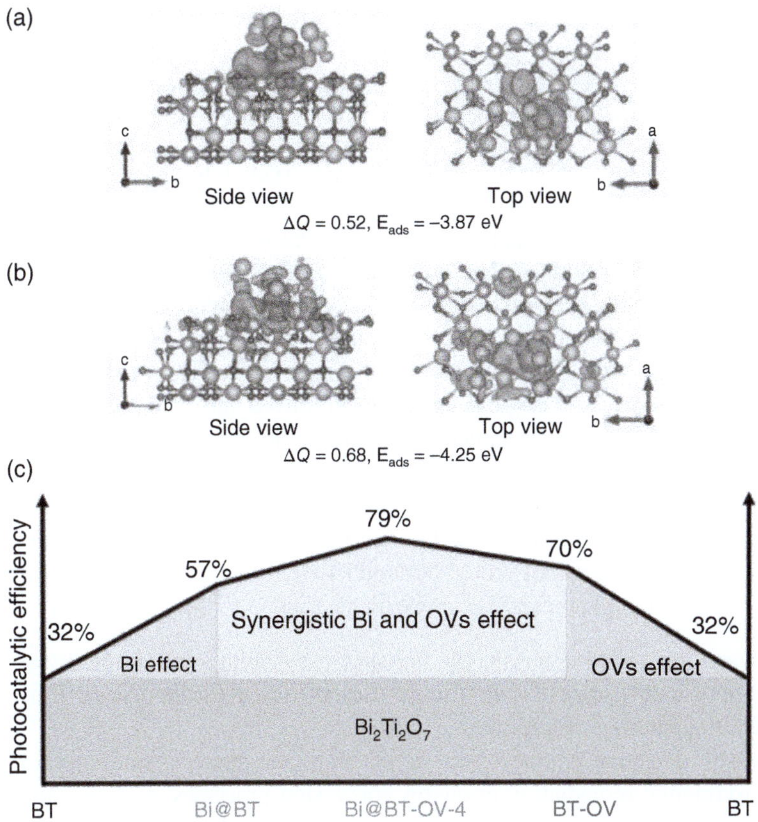

Figure 2.8 Charge-accumulation and charge-depletion regions of (a) Bi@BT material and (b) Bi@BT-OV photocatalyst (green, gray, and red spheres present Bi, Ti, and O atoms) and (c) co-effect of OVs and Bi on the photocatalytic efficiency of Bi@BT-OV-4. *Source:* Reproduced from Zhu et al. [48] / with permission of American Chemical Society.

the adsorption and photocatalytic removal of NO but extended light absorption efficiency to the near-infrared region (Figure 2.8c).

2.3 Advances and Challenges

The literature survey demonstrates that defective photoactive materials are one of the most effective and efficient feedstocks utilized in several applications (i.e., organic synthesis, biological processes, environmental remediation, and photo-electronics). Compared to other reported photocatalytic materials (i.e., traditional semiconductors, heterojunction photocatalysts, photo-electrocatalysts, and plasmon photocatalysts), defective photoactive materials have their own benefits and limitations [50–56]. It is worth mentioning that the construction and optimization of defective photocatalysts involve careful consideration of the type, concentration, and distribution of defects to achieve the desired performance efficiency. Researchers and industrialists continue to explore the substantial advantages and implementation of defect engineering in photocatalysis to harness the benefits offered by imperfect photocatalyst structures. Their benefits are stated below:

a) In most cases, defect engineering in a photocatalyst can generate additional active sites on the photocatalyst's surface, which could enhance its performance toward certain chemical reactions and applications. Additionally, this enhanced reactivity could facilitate the degradation of liquid and gaseous pollutants in the environment or enhance the performance of solar energy conversion processes.
b) Defect engineering in the crystal lattice structure of the reported photocatalyst could lead to transformations in its electronic band structure, encouraging the material to absorb a broader range of light. This improved light absorption capability could enhance the performance of photocatalytic transformations under natural sunlight.
c) It should be noted that the careful introduction or creation of specific defects in the photocatalyst's structure leads to the tailoring of its properties to suit several applications. This flexibility could be conducive to enhanced efficiency and selectivity in different photocatalytic applications.

In some cases, controlled defects could enhance the stability and durability of the reported photocatalysts, resulting in more resistance to degradation over time or under difficult operating conditions.

In some cases, defective photocatalysts could offer cost-effective solutions for various environmental remediation or energy conversion processes, as they could be fabricated employing easier techniques or more readily available raw feedstocks compared to defect-free counterparts. Thus, this can contribute to the scalability and practicality of defective photocatalytic materials in real applications.

Despite the aforementioned benefits, the uses of defective photocatalysts in numerous applications present some limitations. Besides, addressing these drawbacks often requires suitable strategies for defect control (i.e., point defect, planar defect, volume defects, and elements doping), material engineering, and techniques optimization to hinder the negative impacts of defects on the photocatalytic efficiency of the prepared defective materials. Continued research work and advanced developments aim to defeat these limitations and unlock the excellent potential of defective photocatalytic materials for sustainable energy, useful product generation, environmental remediation, and other implementations. Limitations in defective photocatalyst use are mentioned below:

Defects in the crystal structure of the reported photocatalysts could introduce energy levels within the bandgap of the catalysts, resulting in non-radiative recombination of photogenerated charge carriers. Consequently, this can reduce the overall performance of photocatalytic reactions by diminishing the lifetime of the produced reactive species or limiting charge transfer processes.

a) In some cases, defect engineering could serve as an active site for unintended side reactions, leading to the formation of undesired by-products or the consumption of the generated reactive species that are crucial for the main photocatalytic strategies. Thus, these side reactions could hinder the selectivity and performance of the reported photocatalysts.

b) Introducing or creating defects in a photocatalyst usually requires a degree of randomness or uncertainty, resulting in challenges to precisely control the type, distribution, and concentration of defects in the developed defective materials. Moreover, this lack of control could hinder the reproducibility and consistency in the efficiency of defect-engineered photocatalysts.
c) Certain surface defects can promote surface oxidation, photo-corrosion, or structural degradation of the reported photocatalyst over time, further reducing its long-term stability and efficiency. Hence, managing the stability of defect-engineered photocatalysts under prolonged exposure time to light, environmental conditions, or reactive species is important for maintaining their efficiency and scalability.
d) In some cases, analyzing and characterizing defects in photocatalytic feedstocks involve sophisticated techniques and expensive instruments, adding complexity to photocatalytic materials design and optimization processes. More interestingly, understanding the interplay between defects, surface properties, and photocatalytic efficiency requires a comprehensive strategy that could require computational modelling, expensive spectroscopic methods, and sophisticated photocatalytic materials characterization tools.
e) The introduction/creation of defects to enhance certain features of photoactive materials may require other desirable characteristics (i.e., photocatalytic activity, selectivity, or light absorption capability). Thus, balancing these characteristics and optimizing the overall efficiency of defective photoactive materials could pose challenges in practical applications of defect-engineered materials.

2.4 Conclusion

In summary, defective photoactive materials represent a research topic of great interest with a vast potential for revolutionizing photocatalysis and advancing sustainable energy production and environmental remediation technologies. It is worth pointing out that defective photoactive feedstocks encompass structural imperfections (i.e., vacancies, dislocations, and grain boundaries), leading to localized energy states that improve charge separation and migration at the surface or bulk of defective photocatalysts. Consequently, this improved surface reactivity with newly introduced adsorption/activation sites results in an improved photocatalytic efficiency in redox reactions. On the other hand, it is important to note that by exploring the important features of defects in photocatalytic materials, researchers and industrialists could gain new opportunities for addressing tremendous limitations related to water/air purification and sustainable energy production. Even though with the successful implementation of advanced defective photocatalysts in various applications, there are still multiple drawbacks and difficulties that need to be addressed in the future. Taking into account that, under certain conditions, defects could act as possible recombination centers for charge carriers during photocatalytic reactions. To sum up, drawbacks such as defect control, stability, and reproducibility remain challenging in the practical implementation of defective photocatalytic feedstocks. More importantly, addressing these limitations involves more advanced synthetic strategies, in-depth characterization techniques, and theoretical simulation in order to gain insight into the defect-mediated mechanisms in defect-engineered materials at present and in the future.

Acknowledgments

This work was supported by the Foreign Expert Program of SAFEA of China (Nos DL2023041004L and G2023041021L) and Shaanxi Provincial Foreign Expert Service Plan (Project No. 2025WZ-YBXM44).

Declaration of Competing Interest

The authors declare that they have no known competing financial interests or personal relationships that could have appeared to influence the work reported in this chapter.

References

1 Ameta, R., Solanki, M.S., Benjamin, S., and Ameta, S.C. (2018). Photocatalysis. In: *Advanced oxidation processes for waste water treatment* (ed. S.C. Ameta and R. Ameta), 135–175. Elsevier.

2 Khan, M.M. (2021). Principles and mechanisms of photocatalysis. In: *Photocatalytic Systems by Design* (ed. M. Sakar, R.G. Balakrishna, and T.-o. Do), 1–22. Elsevier.

3 Yang, X. and Wang, D. (2018). Photocatalysis: from fundamental principles to materials and applications. *ACS Appl Energy Mater* 1: 6657–6693.

4 Hassaan, M.A., El-Nemr, M.A., Elkatory, M.R. et al. (2023). Principles of photocatalysts and their different applications: a review. *Top. Curr. Chem.* 381: 31.

5 Cui, Y., Labidi, A., Liang, X. et al. (2024). Pivotal impact factors in photocatalytic reduction of CO_2 to value-added C_1 and C_2 products. *ChemSusChem* 17: e202400551.

6 Labidi, A., Ren, H., Sial, A. et al. (2024). Photocatalyst materials formed using organic radicals for water splitting. In: *Organic Radicals* (ed. C. Wang, A. Labidi, and E. Lichtfouse), 211–230. Elsevier.

7 Labidi, A., Ren, H., Sial, A. et al. (2023). Coal ash for removing toxic metals and phenolic contaminants from wastewater: a brief review. *Critical Crit. Rev. Environ. Sci. Technol.* 53: 2006–2029.

8 Labidi, A., Ren, H., Zhu, Q. et al. (2024). Coal fly ash and bottom ash low-cost feedstocks for CO_2 reduction using the adsorption and catalysis processes. *Sci. Total Environ.* 912: 169179.

9 He, F., Jeon, W., and Choi, W. (2021). Photocatalytic air purification mimicking the self-cleaning process of the atmosphere. *Nat. Commun.* 12: 2528.

10 Li, X., Chen, Y., and Tao, Y. (2022). Challenges of photocatalysis and their coping strategies. *Chem. Catal.* 2: 1315–1345.

11 Iervolino, G., Zammit, I., Vaiano, V., and Rizzo, L. (2020). Limitations and prospects for wastewater treatment by UV and visible-light-active heterogeneous photocatalysis: a critical review. In: *Heterogeneous Photocatalysis: Recent Advances* (ed. M.J. Muñoz-Batista, A.N. Muñoz, and R. Luque), 225–264. Cham: Springer.

12 Yang, Y., Guo, L., Wang, X. et al. (2024). Halogen modified organic porous semiconductors in photocatalysis: mechanism, synthesis and application. *Adv. Powder Mater.* 3: 100178.

13 Muscetta, M., Ganguly, P., and Clarizia, L. (2024). Solar-powered photocatalysis in water purification: applications and commercialization challenges. *J. Environ. Chem. Eng.* 12: 113073.

14 Xiong, H.-Q., Bao, H.-R., Long, F. et al. (2024). From lab to nature: overcoming challenges in applying in-situ photocatalysis to water bodies. *J. Environ. Chem. Eng.* 12: 112656.

15 Huang, F., Zhang, F., Wang, Y., and Lang, X. (2024). Semiconductor cooperative photocatalysis with TEMPO. *Trends Chem.* 6: 115–127.

16 Wood, D., Shaw, S., Cawte, T. et al. (2020). An overview of photocatalyst immobilization methods for air pollution remediation. *Chem. Eng. J.* 391: 123490.

17 Wakjira, T.L., Gemta, A.B., Kassahun, G.B. et al. (2024). Bismuth-based Z-scheme heterojunction photocatalysts for remediation of contaminated water. *ACS Omega* 9: 8709–8729.

18 Belessiotis, G.V. and Kontos, A.G. (2022). Plasmonic silver (Ag)-based photocatalysts for H_2 production and CO_2 conversion: review, analysis and perspectives. *Renew. Energy* 195: 497–515.

19 Demarema, S., Nasr, M., Ookawara, S., and Abdelhaleem, A. (2024). New insights into green synthesis of metal oxide based photocatalysts for photodegradation of organic pollutants: a bibliometric analysis and techno-economic evaluation. *J. Clean. Prod.* 463: 142679.

20 Jabbar, Z.H., Graimed, B.H., Ammar, S.H. et al. (2024). The latest progress in the design and application of semiconductor photocatalysis systems for degradation of environmental pollutants in wastewater: mechanism insight and theoretical calculations. *Mater. Sci. Semicond. Process.* 173: 108153.

21 Pastor, E., Sachs, M., Selim, S. et al. (2022). Electronic defects in metal oxide photocatalysts. *Nat. Rev. Mater.* 7: 503–521.

22 Riffat, M., Ali, H., Qayyum, H. et al. (2023). Enhanced solar-driven water splitting by ZnO/CdTe heterostructure thin films-based photocatalysts. *Int. J. Hydrog. Energy* 48: 22069–22078.

23 Makhado, K.P., Mphahlele-Makgwane, M.M., Kumar, N. et al. (2024). Current updates on p-type nickel oxide (NiO) based photocatalysts towards decontamination of organic pollutants from wastewater. *Mater. Today Sustain.* 25: 100664.

24 Tai, R., Wu, R., Zhang, M. et al. (2024). Metal–organic framework–based heterojunctions for photocatalysis. *Curr. Opin. Chem. Eng.* 45: 101033.

25 Ren, H., Qi, F., Labidi, A. et al. (2023). Chemically bonded carbon quantum dots/Bi_2WO_6 S-scheme heterojunction for boosted photocatalytic antibiotic degradation: interfacial engineering and mechanism insight. *Appl. Catal., B Environ.* 330: 122587.

26 Ren, H., Labidi, A., Sial, A. et al. (2024). Carbon quantum dots modified Z and S–Scheme heterojunctions for pharmaceutical contaminants photodegradation: state–of–the–art, benefits, and limitations. *Sep. Purif. Technol.* 346: 127508.

27 Li, Z., Wang, S., Wu, J., and Zhou, W. (2022). Recent progress in defective TiO_2 photocatalysts for energy and environmental applications. *Renew. Sust. Energ. Rev.* 156: 111980.

28 Sun, H., Zou, C., and Liang, H. (2022). Designing dual-defective photocatalyst of Z-scheme H-$BiVO_4$/D-NG composite with hollow structures for efficient visible-light photocatalysis of organic pollutants. *Sep. Purif. Technol.* 297: 121476.

29 Wang, T., Qu, S., Wang, J. et al. (2023). W/Mo-heteropoly blue-modified defective $W_{18}O_{49}$ as Z-Scheme heterostructure photocatalysts for efficient N_2 fixation. *J. Alloys Compd.* 938: 168631.

30 Sulaeman, U., Larasati, R., Putri, D.A.R.W. et al. (2024). Design of defective silver phosphate photocatalyst using *Nigella sativa* seed aqueous extract for enhanced photocatalytic activity. *Inorg. Chem. Commun.* 163: 112368.

31 Du, J., Shen, Y., Yang, F. et al. (2023). In-situ topology synthesis of defective MoN nanosheets/g-C_3N_4 2D/2D heterojunction photocatalyst for efficient H_2 production. *Appl. Surf. Sci.* 608: 155199.

32 Akrami, S., Murakami, Y., Watanabe, M. et al. (2022). Defective high-entropy oxide photocatalyst with high activity for CO_2 conversion. *Appl. Catal., B Environ.* 303: 120896.

33 Mateo, D., Garcia-Mulero, A., Albero, J., and Garcia, H. (2019). N-doped defective graphene decorated by strontium titanate as efficient photocatalyst for overall water splitting. *Appl. Catal., B Environ.* 252: 111–119.

34 Pan, R., Liu, J., and Zhang, J. (2021). Defect engineering in 2D photocatalytic materials for CO_2 reduction. *ChemNanoMat* 7: 737–747.

35 Chen, Y., Yang, W., Gao, S. et al. (2018). Synthesis of Bi_2MoO_6 nanosheets with rich oxygen vacancies by postsynthesis etching treatment for enhanced photocatalytic performance. *ACS Appl. Nano Mater.* 1: 3565–3578.

36 Gao, T., Zhang, K., Zhu, Q. et al. (2022). One step synthesis of oxygen defective Bi@Ba_2TiO_4/$BaBi_4Ti_4O_{15}$ microsheet with efficient photocatalytic activity for NO removal. *Catalysts* 12: 1455.

37 Bai, S., Zhang, N., Gao, C., and Xiong, Y. (2018). Defect engineering in photocatalytic materials. *Nano Energy* 53: 296–336.

38 Xin, Y., Zhu, Q., Gao, T. et al. (2023). Photocatalytic NO removal over defective Bi/BiOBr nanoflowers: the inhibition of toxic NO_2 intermediate via high humidity. *Appl. Catal.* B 324: 122238.

39 Qin, J., Dong, Y., Lai, X. et al. (2024). Oxygen vacancy-rich $CoMoO_4$/carbon nitride S-scheme heterojunction for boosted photocatalytic H_2 production: microstructure regulation and charge transfer mechanism. *J. Mater. Sci. Technol.* 198: 176–185.

40 Qin, J., Wu, H., Dong, Y. et al. (2024). Polymerization synthesis of semi-coke activated carbon modified carbon nitride nanoscrolls with boosted photocatalytic H_2 production. *Mol. Catal.* 556: 113918.

41 Li, H., Wang, Y., Guo, R. et al. (2022). Encapsulating UiO-66-NH_2@Pt with defective PCN-222 as an active armor to fabricate a sandwich-type nanocatalyst for the tandem synthesis via hydrogenation of nitroarenes. *J. Catal.* 407: 253–264.

42 Bucureşteanu, R., Apetrei, R., Ioniţă, M. et al. (2019). Photocatalytic techniques to prevent and combat healthcare associated infections. In: *E3S Web of Conferences*, 04046. EDP Sciences, 111.

43 Sun, X., Zhang, X., and Xie, Y. (2020). Surface defects in two-dimensional photocatalysts for efficient organic synthesis. *Matter* 2: 842–861.

44 Xu, Y., Han, X., Zhou, Z. et al. (2024). Rational construction of defective Z-Scheme tandem heterojunction interfacial charge transfer channels for efficient uranium collection and resistance to biological contamination. *Chem. Eng. J.* 494: 152768.

45 Li, Y., Sun, B., Lin, H. et al. (2020). Efficient visible-light induced H_2 evolution from T-$Cd_xZn_{1-x}S$/defective MoS_2 nano-hybrid with both bulk twinning homojunctions and interfacial heterostructures. *Appl. Catal., B Environ.* 267: 118702.

46 Gu, W., Wang, W., Li, G. et al. (2020). Microwave-assisted synthesis of defective tungsten trioxide for photocatalytic bacterial inactivation: role of the oxygen vacancy. *Chin. J. Catal.* 41: 1488–1497.

47 Guan, Y., Wu, J., Man, X. et al. (2020). Rational fabrication of flower-like $BiOI_{1-x}$ photocatalyst by modulating efficient iodine vacancies for mercury removal and DFT study. *Chem. Eng. J.* 396: 125234.

48 Zhu, Q., Dar, A.A., Zhou, Y. et al. (2022). Oxygen vacancies promoted piezoelectricity toward piezo-photocatalytic decomposition of tetracycline over $SrBi_4Ti_4O_{15}$. *ACS ES&T Eng.* 2: 1365–1375.

49 Zhu, Q., Hailili, R., Xin, Y. et al. (2022). Efficient full spectrum responsive photocatalytic NO conversion at $Bi_2Ti_2O_7$: co-effect of plasmonic Bi and oxygen vacancies. *Appl. Catal., B Environ.* 319: 121888.

50 Zafar, Z., Yi, S., Li, J. et al. (2022). Recent development in defects engineered photocatalysts: an overview of the experimental and theoretical strategies. *Energy Environ. Mater.* 5: 68–114.

51 Wang, Z., Xiao, M., You, J. et al. (2022). Defect engineering in photocatalysts and photoelectrodes: from small to big. *Acc. Mater. Res.* 3: 1127–1136.

52 Zhang, S., Yi, X., Hu, G. et al. (2023). Configuration regulation of active sites by accurate doping inducing self-adapting defect for enhanced photocatalytic applications: a review. *Coord. Chem. Rev.* 478: 214970.

53 Liu, C., Kong, C., Zhang, F.-J. et al. (2021). Research progress of defective MoS_2 for photocatalytic hydrogen evolution. *J. Korean Ceram. Soc.* 58: 135–147.

54 Huang, Y., Meng, L., Xu, W., and Li, L. (2023). The positive versus negative effects of defect engineering for solar water splitting: a review. *Adv. Funct. Mater.* 33: 2305940.

55 Li, X., Dong, Q., Li, F. et al. (2024). Defective Bi@BiOBr/C microrods derived from Bi-MOF for efficient photocatalytic NO abatement: directional regulation of interfacial charge transfer via carbon–loading. *Appl. Catal., B Environ.* 340: 123238.

56 Sial, A., Gao, T., Li, F. et al. (2024). Multifaceted synergism of dual active sites in oxygen vacancies enriched plasmonic Ag-BiOI nanosheets for enhanced piezo-photocatalytic degradation of trimethoprim. *ACS ES&T Eng.* 4: 2923–2937.

3

Tailoring TiO_2 and Other Metal Oxide (WO_3 and ZnO) Photocatalysts with Defect Engineering

Abdelkader Labidi[1], Haitao Ren[2], Jiangyushan Liang[1], Ting Gao[1], Xiaoqian Xu[1], Tian Chang[1], Mohsen Padervand[3], Eric Lichtfouse[4], and Chuanyi Wang[5]

[1] *Shaanxi University of Science and Technology, School of Environmental Science and Engineering, Weiyang District, Xi'an 710021, PR China*
[2] *Xijing University, Technological Institute of Materials & Energy Science (TIMES), Xi'an Key Laboratory of Advanced Photo-Electronics Materials and Energy Conversion Device, Changan District, Xi'an 710123, PR China*
[3] *University of Maragheh, Department of Chemistry, Faculty of Science, Amirkabir Highway, Maragheh 55181-83111, Iran*
[4] *Xi'an Jiaotong University, State Key Laboratory of Multiphase Flow in Power Engineering, Xianning West Road, Xi'an 710049, PR China*
[5] *Department of Environmental and Sustainable Engineering, Faculty of Engineering, Chulalongkorn University, 254 Phayathai Road, Pathumwan, Bangkok, 10330, Thailand*

3.1 Introduction

Photocatalysis process is usually referred to as the boosting of a photoreaction through the assistance of semiconductor catalysts [1, 2]. In this process, photocatalytic feedstocks can be fabricated employing several strategies (i.e., hydrothermal and solvothermal, calcination, sol-gel process, microwave-assisted synthesis, solution processing, and electrochemical) [3–7]. These photoactive feedstocks have been utilized in several implementations (i.e., decontamination, disinfection, gaseous pollutants reduction, self-cleaning, and so on), thanks to their desirable properties and high efficiency in the aforementioned applications [8–12]. Dating back to more than one century, metal oxide-based photocatalysts (i.e., TiO_2, WO_3, ZnO, and ZrO_2) have been widely studied and utilized, thanks to their ability to facilitate chemical reactions upon exposure to light, particularly ultraviolet light [13–18]. Besides, metal oxide-based photocatalytic feedstocks are primarily recognized for their efficient implementations in environmental remediation (air purification and wastewater treatment), energy conversion (solar energy conversion into chemical energy [e.g., hydrogen production through water splitting]), antimicrobial applications, etc. [19–24]. For example, in 2024, Ali et al. developed AgCl/ZIF-8/C–TiO_2 heterojunction photocatalysts for improved levofloxacin antibiotic photodegradation (97.3%). They found that the fabricated photocatalyst possessed remarkable photostability during successive photocatalysis, which can be attributed to the good

Defect Engineering in Photocatalysis: From Fundamentals to Applications,
First Edition. Edited by Chuanyi Wang, Abdelkader Labidi, and Qiuhui Zhu.

separation efficiency of photogenerated carriers from the combined heterojunction between C–TiO_2 and AgCl/ZIF-8 [25]. In another study, Feng et al. (2023) reported Mo-modified (T-M)-ZrO_2 with a narrowed band gap and strong ability toward photocatalytic CO_2 reduction. In their work, they demonstrated that the optimized photocatalyst displayed a 27.3 $\mu mol\ g^{-1}$ of the CO yield with ~100% selectivity, which is 13 times higher than that of bare ZrO_2 semiconductor [26]. Recently, toluene total oxidation by nano-structured $CoMnO_x$ spinel was comprehensively studied by Wang et al. [27]. During this process, the authors found that manganese and cobalt defects in $CoMnO_x$ oxides can enhance the lattice oxygen activation due to the raised cation valance state and shortened metal–oxygen bond, thereby improving toluene oxidation. Recently, Pan et al. [17] successfully synthesized cost-effective, efficient, and durable non-precious metal co-catalysts with an improved CO_2 photocatalytic reduction efficiency [28]. Based on the above-discussed cases, modification strategies of metal oxides can have different effects on their photocatalytic efficiency, including light absorption, charge carrier separation, stability, and practical applications [25–28]. Besides, several modification approaches have been employed for enhancing the photocatalytic features of metal oxides (i.e., doping, composite formation, surface modification, morphology control, hybrid systems, and modification through loading). Nowadays, defect engineering in metal oxides has shown numerous benefits, especially for improving their electronic properties, optical characteristics, and further their photocatalytic activities [29]. From this point of view, very recently, defects including vacancies, interstitials, substitutional defects, grain boundaries, doping, etc., have been extensively announced in the past few years due to their pivotal role in the enhancement of photocatalytic features of metal oxide-based photocatalysts. As the focus of this chapter, herein, we addressed defect engineering in different metal oxides. Besides, various techniques, including doping, surface modification, and annealing processes, are discussed for manipulating the defect concentration and distribution within the materials. The impact of defects on charge carrier dynamics, band structure, and reactivity is examined in detail, highlighting the importance of defect control in enhancing light absorption, charge separation, and catalytic activity. Thus, understanding defects in these photocatalytic feedstocks enables the designing of photoactive materials with improved functionality for diverse applications, such as environmental remediation, solar energy conversion, and other applications (Figure 3.1).

3.2 Defective Metal Oxide Photocatalysts

3.2.1 Examples and Applications

Undoubtedly, defect creation/formation in metal oxides can introduce localized states within the band gap of modified metal oxides, affecting their electronic conduction and further creating new pathways for electron–hole transfer [29, 30]. Additionally, it should be noted that vacancies and other point defects can alter the optical absorption features of photocatalytic materials, thereby affecting their light absorption properties. Defects can serve as active sites for redox reactions,

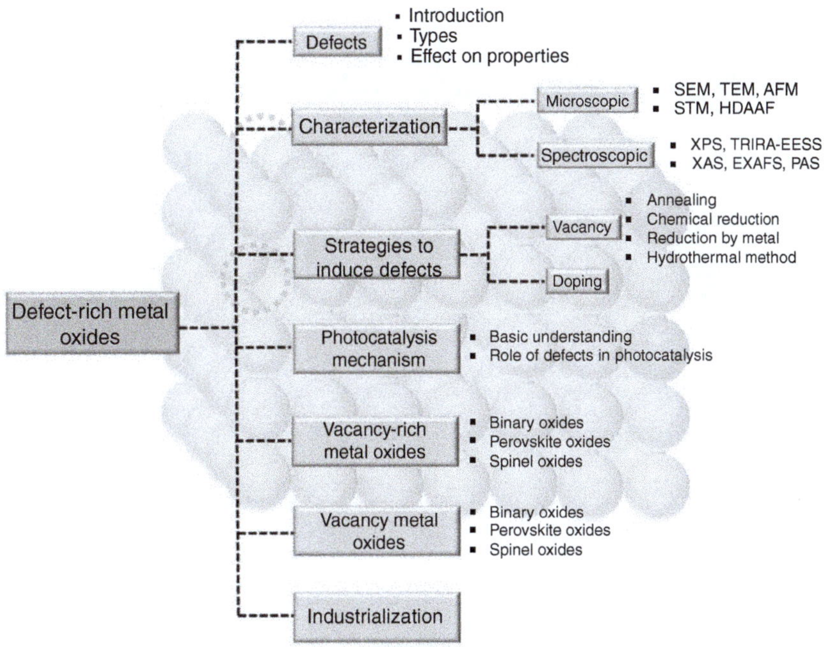

Figure 3.1 Defect controlling formation strategies in metal oxide photocatalysts, along with their types, preparation methods, and characterization. *Source:* Reproduced from Sharma et al. [29] / with permission of John Wiley & Sons.

influencing the metal oxides' photocatalytic efficiency. Furthermore, they can facilitate the adsorption of reactants and desorption of products in the photocatalysis processes [31, 32]. Consequently, defective metal oxides have attracted substantial attention from researchers and industrialists in the past few decades, especially in the photocatalytic applications. Dating back to more than one century, defect creation in metal oxides (i.e., point defects, planar defects, line defects, and volume defects) has been widely reported employing several preparation methods, including annealing, heating, solvothermal, calcination, and ultrasonication (Figure 3.2) [29].

For example, Shi et al. developed $Bi_2Ti_2O_7/CaTiO_3$ heterojunction composites utilizing a facile in situ hydrothermal method. They utilized X-ray diffraction (XRD), field emission scanning electron microscopy (FE-SEM), X-ray photoelectron spectroscopy (XPS), N_2 adsorption/desorption, energy dispersive X-ray spectroscopy (EDX), transmission electron microscopy (TEM), high-resolution transmission electron microscopy (HR-TEM), Raman spectra, etc., to demonstrate the successful preparation of photocatalysts (Figure 3.3). They found that $Bi_2Ti_2O_7/CaTiO_3$ materials exhibited higher visible light photocatalytic performance compared to raw $CaTiO_3$ and $Bi_2Ti_2O_7$ for photocatalytic NO removal at 600 ppb level in air.

They also observed that $Bi_2Ti_2O_7/CaTiO_3$ comprising 20 wt% $Bi_2Ti_2O_7$ displayed greater photocatalytic efficiency, attaining a maximum NO removal performance of 59%. More importantly, they demonstrated that the enhanced photocatalytic

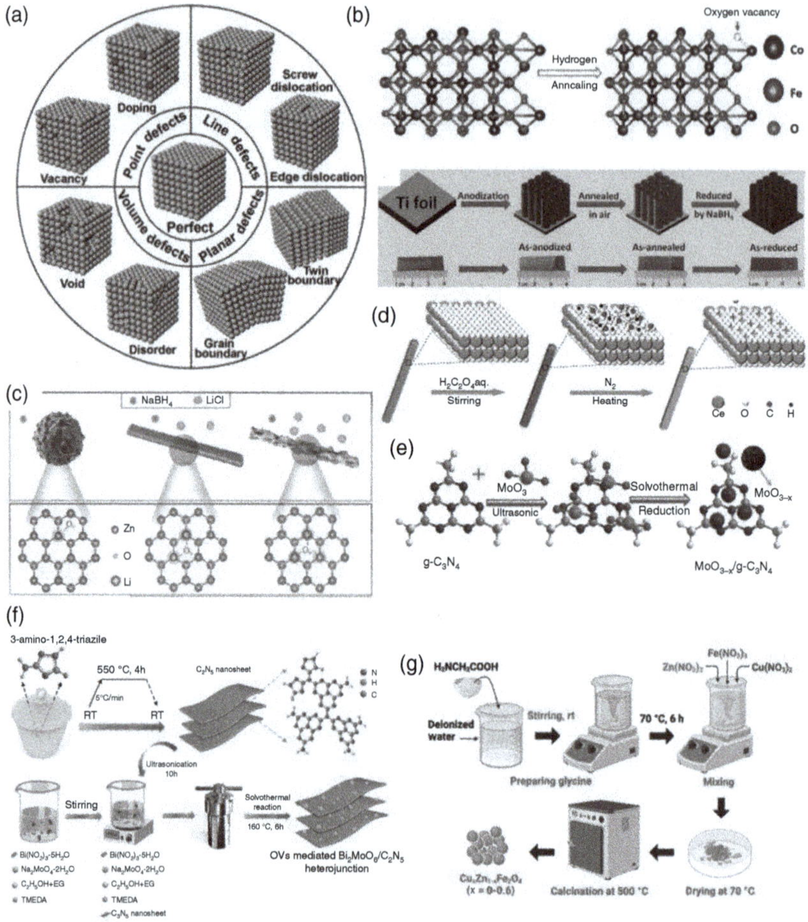

Figure 3.2 (a) Different types of defects in metal oxides; (b–g) preparation methods of defective metal oxide photocatalysts. *Source:* Reproduced from Sharma et al. [29] / with permission of John Wiley & Sons.

efficiency is mainly assigned to the good visible-light-absorbing capability of the obtained photocatalysts, existence of oxygen vacancy defects, and generation of a heterojunction between $CaTiO_3$ and $Bi_2Ti_2O_7$, leading to a good charge carrier separation at the interface of photocatalysts, as proved by photoluminescence (PL) and photocurrent measurements. The authors revealed that the oxidation of NO molecules is due to the synergy between h^+ and $\cdot O_2^-$ entities. Under visible light illumination, with the energy of photons equalling or exceeding the band gap of $Bi_2Ti_2O_7$, electron–hole pairs could be formed. Based on the aforementioned results, because of the reformed band structure of the BTO/CTO-2 catalyst, the photoinduced e^- on the conduction band (CB) of $Bi_2Ti_2O_7$ can efficiently migrate to the CB of $CaTiO_3$ material, while holes are remained on the valence band (VB) of $Bi_2Ti_2O_7$ feedstock resulting in effective separation of the photo-induced charge carriers [33]. During NO removal over the as-prepared photocatalyst, they indicated

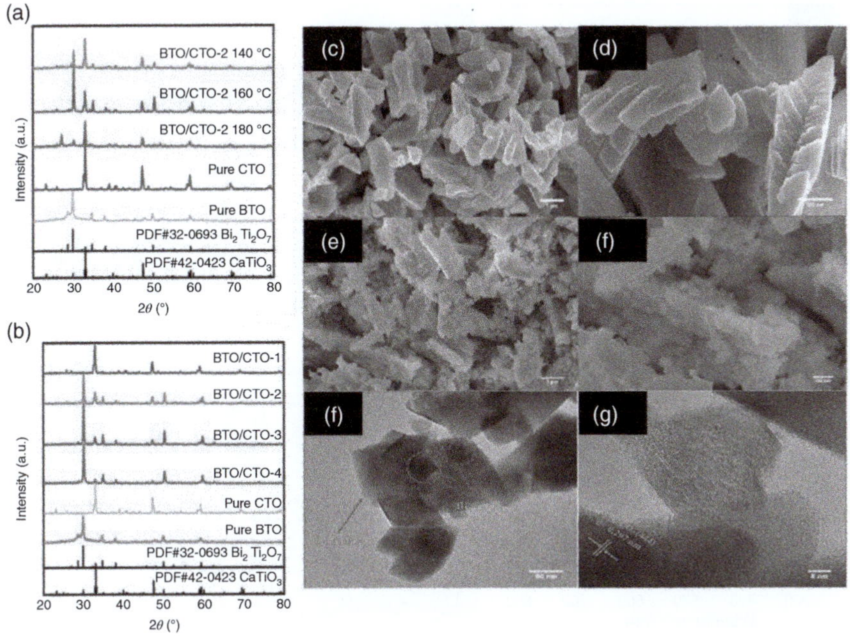

Figure 3.3 XRD patterns of (a) BTO/TCO-*x* photocatalysts fabricated at different synthesis temperatures, (b) BTO/TCO-*x* synthesized with distinct BTO content, (c, d) SEM micrographs of CTO catalyst, (e, f) FE-SEM, (g) TEM, and (h) HRTEM of BTO/CTO-2 photocatalyst. SEM: scanning electron microscopy, FE-SEM: field emission scanning electron microscopy, TEM: transmission electron microscopy, HRTEM: high-resolution transmission electron microscopy. *Source:* Reproduced from Zhu et al. [31] / with permission of Elsevier

that the formed $\bullet O_2^-$ species can directly oxidize NO to NO_3^- and h^+ can effectively oxidize NO to generate NO_2 and NO_3^- intermediates, thanks to its strong oxidation capability:

$$\mathrm{BTO/CTO-2+visible-light} \rightarrow h^+_{VB} + e^-_{CB} \tag{3.1}$$

$$e^-_{CB} + O_{2ad} \rightarrow \bullet O_{2ad}- \tag{3.2}$$

$$NO(g) + \bullet O_{2ad}- \rightarrow NO^-_{3ad} \tag{3.3}$$

$$2h^+_{VB} + NO + H_2O \rightarrow NO_2 + 2H^+ \tag{3.4}$$

$$NO_2 + h^+_{VB} + H_2O \rightarrow NO_3^- + 2H^+ \tag{3.5}$$

In another report, and regardless of defective creation in WO_3, Gu et al. (2020) developed tungsten trioxide nanosheets (WO_{3-x}) with oxygen vacancies via a microwave-assisted pathway from WO_3. Gu's group utilized XRD, SEM, TEM, UV-Vis, and XPS to analyze the crystalline structure, morphology, and chemical compositions of the defective WO_{3-x} (Figure 3.4).

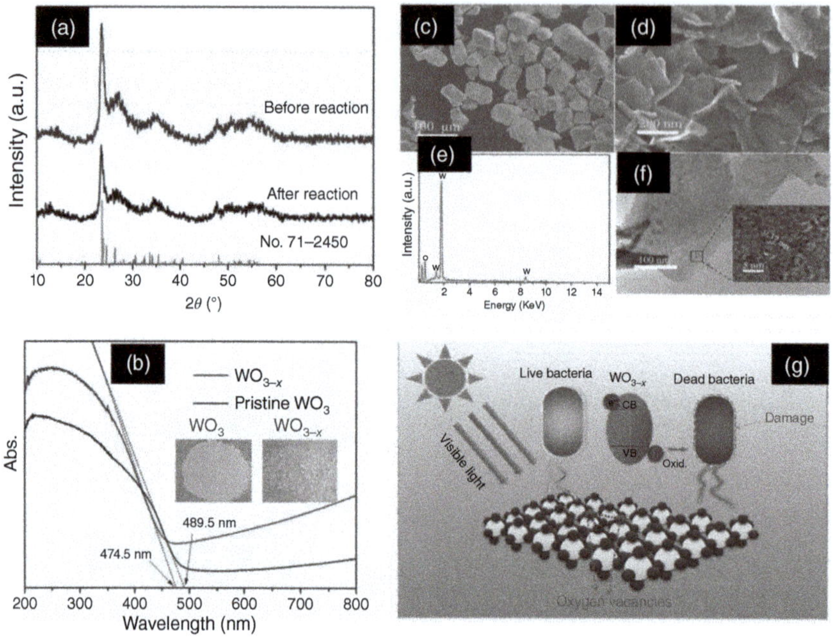

Figure 3.4 (a) X-ray diffraction of WO_{3-x} material; (b) UV-Vis DRS spectra of WO_{3-x} nanosheets and WO_3; SEM micrographs of (c) WO_{3-x} photocatalyst; (d) pristine WO_3; (e) EDX spectrum of WO_{3-x}; (f) TEM and HRTEM micrographs of WO_{3-x} photocatalyst; and (g) photocatalytic bacterial destruction mechanism over WO_{3-x} photocatalyst under VL irradiation. *Source:* Reproduced from Gu et al. [34] / with permission of Elsevier.

Gu et al. indicated that the XRD spectrum can be clearly indexed as monoclinic $W_{18}O_{49}$ ($WO_{2.72}$) (JCPDS No. 71-2450, space group: *P2m*). Besides, peaks situated at 23°, 26°, 35°, and 47° can be assigned to (010), (-104), (113), and (020) crystal planes, respectively, indicating that the WO_{3-x} nanosheet can be effectively prepared by the microwave–solvothermal route at 150 °C for 4 hours [34]. They observed that WO_3 displays an absorption edge at 489.5 nm, which can be assigned to the band gap energy of 2.53 eV. However, the as-synthesized WO_{3-x} nanosheets display an absorption edge of 474.5 nm, resulting in a band gap energy equal to 2.61 eV. Through morphological studies, they revealed that WO_3 material displays large square morphologies along with wide size distributions ranging from 20 to 100 μm. Meanwhile, WO_3 nanosheet fabricated using the microwave–solvothermal route displayed morphology along with a thickness of less than 20 nm. Referring to the XPS results, they demonstrated that W 4f orbit reveals two distinct peaks, which can be attributed to W $4f_{5/2}$ and W $4f_{7/2}$ with binding energies of 38.0 and 35.8 eV, respectively. More importantly, they demonstrated that the existence of W^{5+} cations on the surface of WO_{3-x} nanosheets, suggesting that the oxygen defect vacancies are generated on the photocatalyst surface, which is beneficial for their photocatalytic performance. In this regard, they observed that the as-prepared WO_{3-x} exhibited an improved photocatalytic antibacterial performance against *E. coli*, which could be totally inactivated after 150 minutes compared with pure WO_3. They indicated that the bacterial destruction rate of WO_{3-x} was 15.2 times greater than that of pristine WO_3 in the absence of oxygen vacancies, indicating that the surface oxygen vacancies could effectively boost the bacterial inactivation performance. Additionally, they noticed that the photogenerated h^+ was considered as the main reactive species, which instantaneously impacted the antibacterial activity. They concluded that the creation of oxygen vacancies in WO_{3-x} not only lead to the positive shifts of its VB and CB band positions, leading to high oxidation ability of the photogenerated h^+, but also to an enhanced interfacial charge transfer performance on the WO_{3-x} photocatalyst (Eqs. (3.6)–(3.8)):

$$WO_{3-x} + h\upsilon \rightarrow WO_{3-x}(e^- + h^+) \tag{3.6}$$

$$\text{Cell} + h^+ \rightarrow \text{Damged cell} \tag{3.7}$$

$$OH^-_{ads} + h^+ \rightarrow \bullet OH_{ads} \tag{3.8}$$

In 2022, Zhang et al. prepared a series of Ti/TiO_2 defective photocatalysts using varying amounts of interfacial Ti^{3+} sites via an epitaxial growth strategy under hydrothermal conditions employing commercial titanium powder and titanium (IV) butoxide (TBOT) [35]. Zhang and his co-workers utilized several techniques, including XRD, TEM, BET, UV-vis absorption, XPS, time-resolved photoluminescence (TRPL), etc., to demonstrate the fabrication of TiO_2-based defective photocatalysts for rhodamine B (RhB) degradation in wastewater (Figure 3.5).

They observed the presence of anatase TiO_2 phase Ti metallic phase in the as-prepared Ti/TiO_2 photocatalytic materials and TiO_2 particles possesses similar particle size. Based on UV-vis diffuse reflectance spectroscopy (UV-Vis-DRS) results, they mentioned that the band gap of the as-prepared TiO_2 was found to be 3.2 eV and the obtained Ti/TiO_2 photocatalysts displayed a much higher absorbance in visible light ranging from 400 to 800 nm compared to raw TiO_2, which can be assigned to the presence of black Ti powder. Additionally, the spherical morphology of the Ti phase and the rhombic grain of TiO_2 can be observed with an efficient aggregation of Ti and TiO_2 entities. They noticed that the crystalline phases of hexagonal metallic Ti and tetragonal TiO_2 particles were clearly found with *d*-spacings of 0.224 and 0.234 nm, which can be attributed to the (101) and (002) facets of metallic Ti, respectively, and lattices with *d*-spacing of 0.352 and 0.247 nm can be attributed to the (101) and (10-3) facets of TiO_2 particles, respectively, further proving the preparation of photocatalysts. The photocatalytic efficiency of Ti/TiO_2 photocatalysts was tested with different ratios according the following sequence: Ti/TiO_2 (1:16) ≈ Ti/TiO_2 (1:4) < Ti/TiO_2 (1:8) < Ti/TiO_2 (1:10). More interestingly, they concluded that the CB potential of bare TiO_2 was estimated to be 0.6 eV greater than Fermi energy. They observed that the VB potential of Ti/TiO_2 (1:10) was ~2.4 eV lower than the Fermi energy, which was 0.2 eV below than that of bare TiO_2. The narrowed bandgap and reduced VB potential of Ti/TiO_2 (1:10) material compared to original TiO_2 can be assigned to the introduction of impurity and defects in Ti/TiO_2 photocatalysts, leading to high degradation rate of RhB pollutants. Regardless of defective ZnO usages, Xu et al. (2019) developed

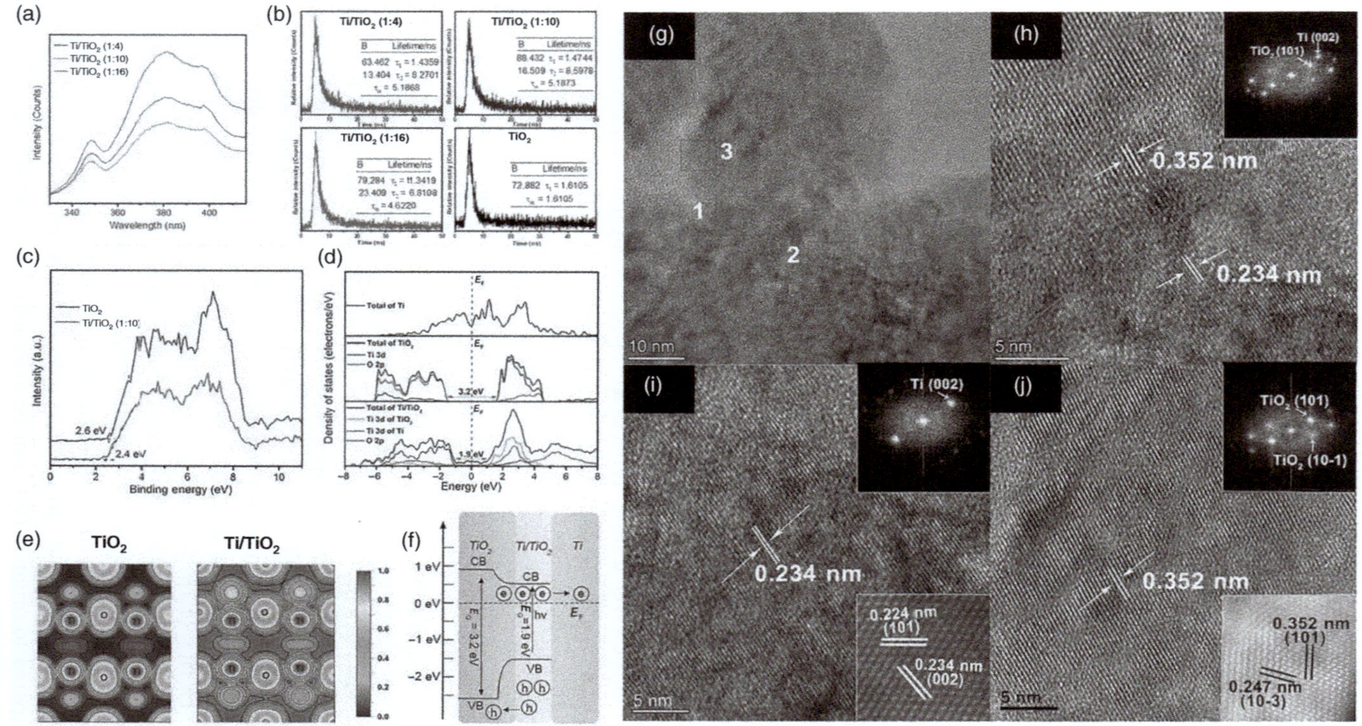

Figure 3.5 (a) Photoluminescence spectra of Ti/TiO_2 with different ratio: Ti/TiO_2 (1/16), Ti/TiO_2 (1/4), and Ti/TiO_2 (1/10); (b) TRPL spectra of the photocatalyst at excitation wavelength equal to 250 nm. (c) XPS spectra of TiO_2 and Ti/TiO_2 (1:10) materials. (d) DOS plots of Ti, TiO_2, and Ti/TiO_2 photocatalysts. (e) 2D-ELF plots of TiO_2 and Ti/TiO_2 interface sliced along (101) plane of TiO_2 material. (f) Charge transfer pathway on Ti/TiO_2 photocatalysts; (g) HRTEM micrographs of Ti/TiO_2 (1:10) photocatalyst with marked *d*-spacing along with the fast Fourier transform (FFT) electron diffraction patterns (insets). The red squares 1, 2, and 3 refer to HRTEM of interface (h), HRTEM of Ti (i), and HRTEM of TiO_2(j) material, respectively. TPRL: time-resolved photoluminescence, XPS: X-ray photoelectron spectroscopy, DOS: density of states, ELF: electron localization function, HRTEM: high-resolution transmission electron microscopy, *Source:* Reproduced from Zhang et al. [35] / with permission of Elsevier.

porous defective ZnO cellular hexagonal plates with surface oxygen vacancies (OVZCHPs), employing the hydrothermal and high-temperature $NaBH_4$ reduction strategies. For the characterization of photocatalysts, they employed XRD, SEM, TEM, XPS, UV–Vis-DRS BET, photoluminescence (PL), etc., to confirm the defect creation in OVZCHPs (Figure 3.6).

They specifically indicated that the XRD intensity of OVZCHPs is raised due to its good crystallinity at high temperature/hydrogenation time, and the as-fabricated OVZCHP displayed a surface area, pore size, and pore volume of 26.7 m^2 g^{-1}, 53.5 nm, and 0.13 cm^3 g^{-1}, respectively. More importantly, they demonstrated that the oxygen vacancies were formed on the surface of the photocatalyst at high temperature, reducing its bandgap from 2.48 eV for the ZCHPs (without oxygen vacancies) to 2.19 eV for the OVZCHPs. Besides, the TEM image of OVZCHPs indicates that ZnO hexagonal plates retain a unique cellular structure after the high reduction of $NaBH_4$ at an elevated temperature. They indicated that the solar-driven photocatalytic removal rate of tetracycline by OVZCHPs is higher than 99.9%, which is around three times greater than that of ZnO without oxygen vacancies (ZCHPs) [36]. Table 3.1 defines other reported defective metal oxides and demonstrates the increase of interest in these photocatalytic materials for various photocatalytic applications [37–50].

3.2.2 Defect Characterization

Defect control and characterization in defective metal oxides are critical aspects of material engineering, particularly in the study of metal oxide combination (heterostructure and heterojunctions) [51, 52]. More importantly, getting insights into the nature, type, and density of defects within metal oxide or combined metal oxides can provide information about their properties and efficiency for effective photocatalytic applications. As well known, various microscopic techniques have been utilized for defect control and characterization in defective metal oxides. Among them, SEM has been regarded to provide high-resolution images of the surface morphology of metal oxides. It can be utilized to observe surface defects, including cracks, pits, and particles. Besides, this technique also allows for the analysis of defect density and distribution in defective metal oxides [53, 54]. Additionally, TEM is an invaluable technique for defect characterization in metal oxide photocatalysts. It can visualize defects including vacancies, dislocations, and grain boundaries. Interestingly, high-angle annular dark field (HAADF) imaging and electron diffraction techniques can further observe the nature of defects. Atomic force microscopy (AFM) is widely utilized to investigate the surface topography at the nanoscale. It can detect surface defects by measuring variations in height and roughness of defective metal oxides [55]. Furthermore, XPS is utilized to analyze the chemical states and elemental composition of metal oxide photocatalysts (Figure 3.7a). It can provide information about surface defects by identifying the changes in chemical bonding and the presence of functional groups introduced by defects in these photocatalytic feedstocks [56]. Raman spectroscopy is often employed to characterize defects in several metal oxide photocatalysts (Figure 3.7b).

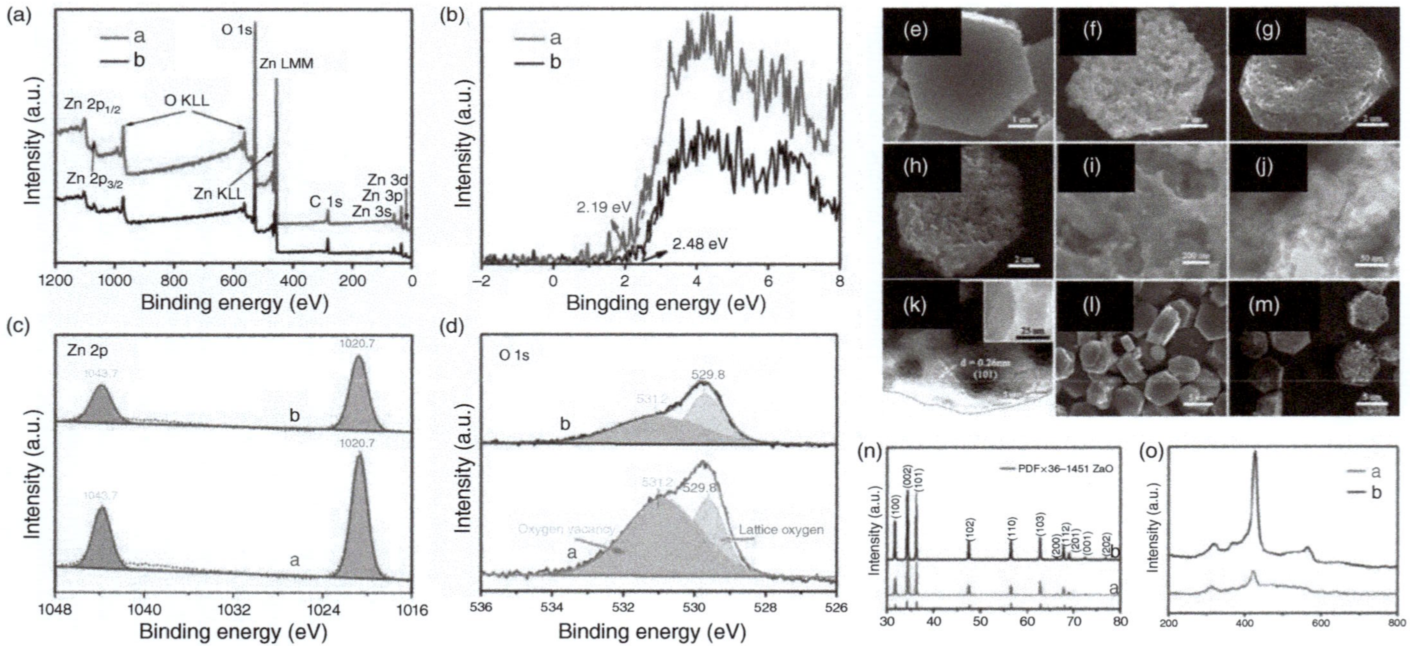

Figure 3.6 XPS spectra for survey spectra (a), valence band XPS (b), Zn 2p (c), O 1s (d) of OVZCHPs catalyst (a) and ZCHPs material (b), respectively; SEM micrographs of ZCHPs (e), OVZCHPs (f, i), OVZCHPs-300 (g), OVZCHPs-500 (h), TEM (j), and HRTEM micrographs (k) for OVZCHPs, large field of vision (l, m) for ZCHPs (e) and OVZCHPs, respectively; X-ray spectra (n) and Raman spectra (o) of OVZCHPs (a) and ZCHPs (b). *Source:* Reproduced from Xu et al. [36] / with permission of Elsevier. OVZCHPs: ZnO cellular hexagonal plates with surface oxygen vacancies, ZCHPs: ZnO cellular hexagonal plates without oxygen vacancies.

Table 3.1 Examples of defective metal oxide photocatalysts for various photocatalytic applications.

Photocatalysts	Type of defects	Applications	Efficiency	References
Spongy defective zinc oxide nanoparticles (ZnONPs)	Oxygen vacancies	Flumequine antibiotic degradation	97.6%	[37]
Au–CuO_{1-x}	Oxygen vacancies	Catalytic CO oxidation	100% at 115.1 °C in dry conditions	[38]
CuO-rich $CuFe_2O_4$ nanofibers	Oxygen vacancies	Ammonia production	NH_3 yield of 9296.21 μg h^{-1} cm^{-2}	[39]
Defective 3D structured CeAl-Co_3O_4	Structure distortion, oxygen vacancies, etc.	CH_4 oxidation	7.6 μmol $g_{cat.}{}^{-1}$ s^{-1}	[40]
Pt/BiOBr-OVs	Oxygen vacancies	Phenol degradation	82.40%	[41]
Defective rTiO_2/GO	Oxygen vacancies	CH_4 conversion	7.65 mmol g^{-1} h^{-1}	[42]
Defective titanium dioxide (TiO_2)	Ti^{3+} defects and oxygen vacancies (Ov)	H_2 production	162 μmol g^{-1} h^{-1}	[43]
BTiO_2–ZnO	Oxygen vacancies	Tetracycline degradation	63%	[44]
$MCoO_2$ (M = Li or Na)	Intrinsic point defects	CO_2 reduction	6.2 and 5.3 μmol g^{-1} h^{-1}	[45]
Zn-Cd-S	Sulfur vacancies, interstitial metal, and Zn and Cd in low valence state on the top surface of nanostructure	H_2 production	571 μmol h^{-1}	[46]
SnO_2 with oxygen vacancies	Oxygen vacancies	CO_2 reduction	17.27 μmol g^{-1} h^{-1}	[47]
N-doped black-V_2O_5	Oxygen-vacancy defects	Green H_2 production	2.369 mmol g^{-1} h^{-1} in methanol	[48]
Fe_3O_4/Co_3O_4–TiO_2	Oxygen defect sites	Hydrogen production	2890 μmol g^{-1}	[49]
WO_3@N/S–Fe_3O_4	Oxygen defects	Degradation of doxycycline and Cr (VI) reduction	99.8% and 98.7%	[50]

GO, graphene oxide; CO, carbon monoxide; B, black.

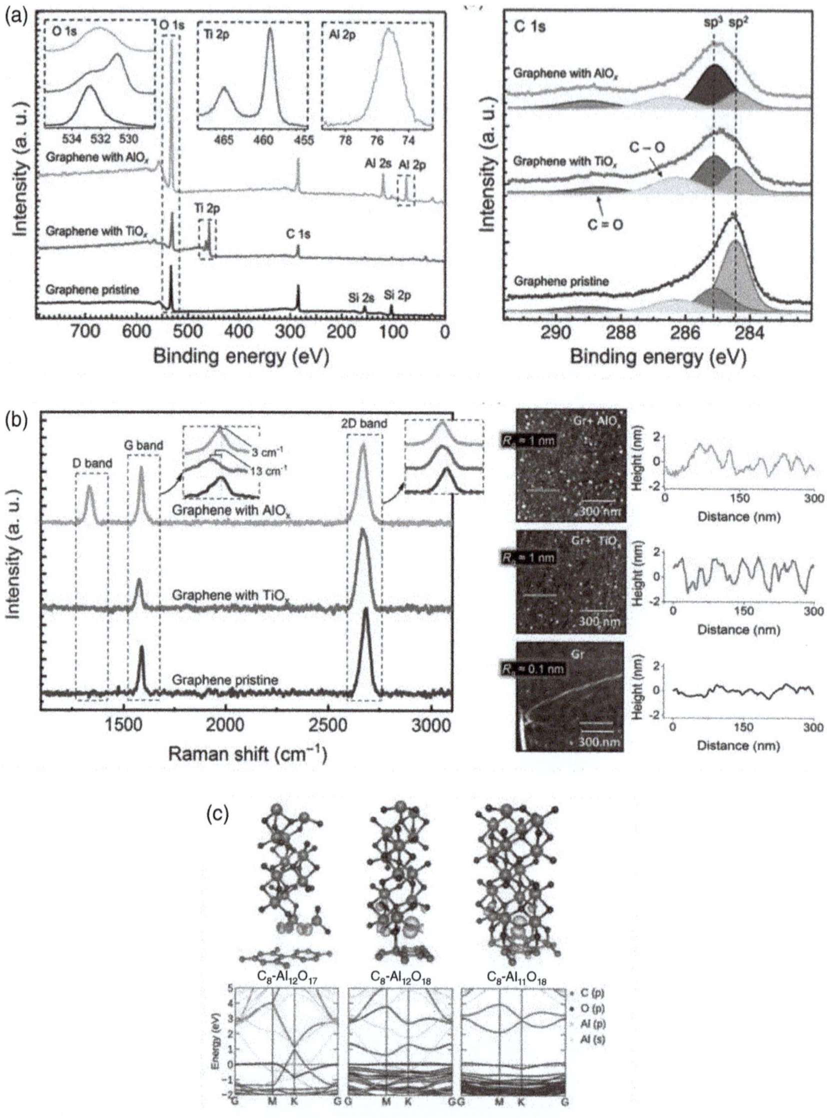

Figure 3.7 (a) XPS characterization, (b) AFM images, and (c) simulated system's geometry of defective graphene and graphene with deposited TiO_x and AlO_x. XPS: X-Ray photoelectron spectroscopy, AFM: atomic force microscopy. *Source:* Reproduced from Belotcerkovtceva et al. [56] / with permission of American Chemical Society.

It should be noted that other methods, such as deep level transient spectroscopy (DLTS), charge pumping (CP), PL, single-photon detection, electron spin resonance (ESR), edge transient-current technique (edge-TCT), and theoretical tools, like density functional theory (DFT) calculations, have been employed to analyze and control defects in metal oxides (Figure 3.7c) [56–58]. In the last few years, several preparations to introduce specific defects in metal oxides have gained particular attention for their efficiency in the photocatalytic applications. Thus, different types of defects are found in metal oxide photocatalysts. Importantly, the creation/formation of defects in metal oxides have led to the generation of new active sites, introduction of strain in metal oxides, and alteration of their bandgaps by forming new defect states, thereby leading to an enhanced photocatalytic efficiency. Furthermore, defects in metal oxides play a critical role in the alteration of their photocatalytic features by changing their band positions. Therefore, the control of defects in metals oxides still need some advanced techniques to effectively tame these materials for efficient photocatalytic applications.

3.3 Benefits and Limitations

Defective metal oxide-based photocatalytic feedstocks have gained particular attention in the field of photocatalysis, especially in environmental remediation and energy storage, thanks to their unique features and efficiency. They have shown various benefits, which are stated below. Defects in metal oxides can create localized energy states within the bandgap of photocatalysts, trapping electrons and holes, thereby reducing recombination rates of charge carriers, and consequently leading to enhanced photocatalytic efficiency of theses photocatalytic materials (Figure 3.8). Besides, defects could increase the surface area and provide more active sites in metal oxides for efficient catalytic reactions, resulting in improved overall reaction rates in different processes, especially in the environmental remediation. In addition, defect creation in metal oxides can modify their electronic

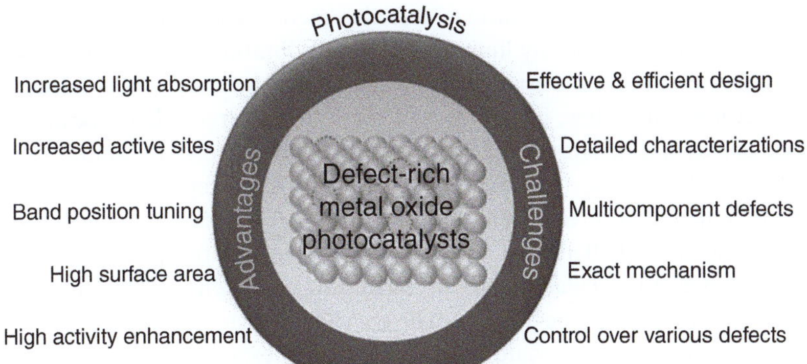

Figure 3.8 Benefits and limitations associated with defect-rich metal oxides. *Source:* Reproduced from Sharma et al. [29] / with permission of John Wiley & Sons.

structure, allowing for tuning of the bandgap energy, resulting in the absorption of a broader spectrum of light by metal oxides like visible light, which is conductive for solar-driven photocatalytic implementations. Moreover, defective metal oxide photocatalysts can exhibit enhanced stability under various operational conditions. The existence of defects can help mitigate the degradation of photocatalysts during photocatalytic reactions. Formation of defects in metal oxides can enhance the migration of charge carriers, facilitating the charge transfer, which is beneficial for specific photocatalytic applications, including water splitting for H_2 production, CO_2 reduction, bacteria disinfection, and organic reaction.

Regardless of their cost, some metal oxides are abundant and relatively inexpensive, eventually leading to the fabrication of defective metal oxide photocatalysts with cost-effective, large-scale fabrication. In case of their reactivity, defective metal oxides can be combined with other organic and inorganic materials to form composite photocatalysts and heterojunctions, resulting in enhanced photocatalytic efficiency due to synergistic effects. For the ecological environment, the usage of defective metal oxide photocatalysts for pollution degradation and energy conversion can contribute to environmental sustainability by reducing the impact of harmful pollutants on the environment and producing clean energy for a circular economy. To sum up, the creation/formation of defects in metal oxide photocatalysts could be considered as a promising strategy to improve their efficiency and broaden their applicability in various applications. Even though defects engineering in metal oxides has shown various benefits, this strategy still suffers from some drawbacks, which are as follows: Defects in metal oxides can sometimes lead to structural instability, causing them to degrade over some photocatalytic reactions. Even though defects can enhance charge separation in metal oxide photocatalysts, they can also introduce additional recombination pathways, reducing their photocatalytic efficiency. Additionally, it is challenging to precisely control the formation and distribution of defects in defective metal oxides, leading to inconsistent efficiency. Regardless of their preparation, the methods of synthesis of defective metal oxides requires introducing defects with complex and multi-step processes, increasing the cost and complexity of production of defective metal oxides. Even though there is potential to tune the bandgap, various defective metal oxides still suffer from narrow absorption range, typically limited to UV light, limiting their applicability in solar-driven photocatalysis. Despite defect creation, in some cases, the efficiency of light utilization can be low, including visible and near-infrared light. Besides, defects can act as active sites for the adsorption of inhibitors, poisoning the photocatalysts and reducing their activities. Based on some reported literature, the processes required to create and control defects can be expensive, making large-scale fabrication challenging. Furthermore, scaling up the production of defective metal oxide photocatalysts while maintaining consistent quality and efficiency is difficult. Defects can be difficult to detect and characterize using standard analytical techniques, making it hard to understand their exact nature and impact on catalytic efficiency of metal oxides. The mechanisms by which defects in metal oxides enhance or degrade photocatalytic efficiency are not always well understood, complicating the optimization of defective metal oxides. Some defective metal oxides can generate toxic substances during their operation, causing environmental and safety risks. The use of certain defective metal oxides may be subject to regulatory restrictions due to their harmful environmental impacts. The efficency of defective metal oxide photocatalysts can vary considerably between applications due to inconsistencies in defect formation, creation, and distribution. It was proven that the performance of defective metal oxides can be sensitive to operational conditions, such as pH solution, temperature, and light absorption, making them difficult to achieve consistent efficiency. On the one hand, defective metal oxide photocatalysts can degrade pollutants under prolonged exposure to light, further reducing their long-term stability and durability. On the other hand, the formation of defects in metal oxides can sometimes lead to photochemical instability, affecting their efficiency in several applications. It is worth mentioning that when metal oxides are combined with other inorganic and organic materials, defects can result in interfacial issues that affect efficiency of the obtained photocatalysts. Lastly, it is still challenging to ensure compatibility between defective metal oxide materials and other feedstocks in the same system.

3.4 Future Perspectives

Despite the numerous reported photocatalytic materials for several applications [59–70], the demand for efficient photocatalysts will increase considerably, in the coming years, especially in the large-scale applications of photocatalysis processes. In this perspective, we have indicated that the usage of defective metal oxides is a promising research area in photocatalytic feedstock synthesis. Thus, the application of defective metal oxides is much needed for several photocatalytic applications. Actually, photocatalyst cost is a global issue due to the high cost and toxicity of some metal oxides. Thus, low-cost metal oxides should be considered for the synthesis of eco-friendly, scalable, and nontoxic defective photocatalysts. The future prospects of defective metal oxide photocatalysts are promising, especially in the photocatalysis processes. In the future, as expected, defective metal oxides will play a critical role in the photocatalysis processes. Therefore, the presence of defects in metal oxides can create active sites that enhance their photocatalytic efficiency, making them valuable for several photocatalytic applications in green chemicals formation, water splitting, sustainable energy production, and liquid/gaseous pollutants degradation. It important to note that defect creation/formation in metal oxides can improve light absorption and charge separation efficiency, eventually leading to an enhanced photocatalytic performance. The future of defective metal oxide photocatalysts may require the development of nanostructured photocatalytic materials and engineered composites. Also, combining defective metal oxides with other organic and inorganic feedstocks can result in an internal electric field and synergistic effects, which can enhance their performance in various photocatalytic applications. As there is a demand for photocatalytic materials in practical applications, the recycling and reusability of defective metal oxide photocatalysts will become more important. Thus, developing sophisticated preparation methods to recycle these inorganic feedstocks while maintaining their outstanding photocatalytic features can contribute to more sustainable practices in photoactive material science.

3.5 Conclusion

This chapter summarizes the application of defective metal oxides using low-cost materials in the photocatalysis processes, including environmental remediation and production of valuable chemicals. It should be noted that exploring novel synthetic methods (i.e., annealing, solvothermal route, calcination, and ultrasonication) and combining them with other techniques allows creating defects in metal oxides, which are widely applied in various photocatalytic implementations. The implementation of metal oxide-based photocatalysts requires the introduction of appropriate defects (e.g., point defects, planar defects, line defects, and volume defects). In fact, defective metal oxides form various compounds are made up of effective photocatalysts utilized for organic and inorganic pollutants degradation, CO_2 conversion, H_2 production, and value-added chemicals production. It can be noted that photocatalysis processes exhibit some benefits and limitations employing defective metal oxides. Furthermore, defects in metal oxides were well confirmed by numerous techniques of analysis (i.e., Ramon spectra, XPS, AFM, TEM, SEM, HAADF, and edge-TCT). Mechanisms and types of defects in metal oxides were described and examples of them were also outlined for different photocatalytic applications. Taking into account the advantages and disadvantages defective metal oxide photocatalysts, metal oxides still present an attractive option for high-efficient photocatalytic feedstocks preparation and their applications in environmental remediation and energy storage.

Acknowledgements

We gratefully acknowledge the support of this research by the Foreign Expert Program of SAFEA of China (Nos DL2023041004L and G2023041021L) and Shaanxi Provincial Foreign Expert Service Plan (Project No. 2025WZ-YBXM44).

Declaration of Competing Interest

The authors declare that they have no known competing financial interests or personal relationships that could have appeared to influence the work reported in this chapter.

References

1 Luo, L., Hou, C., Wang, L. et al. (2024). Layered double hydroxide-based photocatalysts for the removal of emerging contaminants: progress in past ten years. *Catalysts* 14: 252.

2 Bahnemann, D., Robertson, P., Wang, C. et al. (2023). 2023 roadmap on photocatalytic water splitting. *J. Phys.: Energy* 5: 012004.

3 Luo, J., Li, W., Wang, X. et al. (2024). Ytterbium induced surface polarization of graphitic carbon nitride with N vacancies towards enhanced photocatalytic decomposition of tetracycline. *Appl. Surf. Sci.* 644: 158743.

4 Li, X., Dong, Q., Li, F. et al. (2024). Defective Bi@BiOBr/C microrods derived from Bi-MOF for efficient photocatalytic NO abatement: directional regulation of interfacial charge transfer via carbon–loading. *Appl. Catal. B: Environ.* 340: 123238.

5 Li, W. and Wang, C. (2023). Development of highly efficient CdS-based photocatalysts for hydrogen production: structural modification, durability, and mechanism. In: *UV-Visible Photocatalysis for Clean Energy Production and Pollution Remediation: Materials, Reaction Mechanisms, and Applications* (ed. X. Wang, M. Anpo, and X. Fu), 153–169. Wiley.

6 Ren, H., Qi, F., Labidi, A. et al. (2023). Chemically bonded carbon quantum dots/Bi_2WO_6 S-scheme heterojunction for boosted photocatalytic antibiotic degradation: interfacial engineering and mechanism insight. *Appl. Catal. B: Environ.* 330: 122587.

7 Wang, Z., Shi, X., Chen, M. et al. (2023). Polymeric carbon nitride-based photocatalysts for the removal of nitrogen oxides: a review. *Environ. Chem. Lett.* 21: 2913–2952.

8 Padervand, M., Bargahi, A., Eftekhari-Sis, B. et al. (2024). In-situ synthesis of $SnO/CuSnO_3$ nanostructures to catalyze azo dye degradation, CO_2 reduction, and amines direct alkylation reactions under visible light. *Results Eng.* 23: 102515.

9 Liang, X., Gao, T., Cui, Y. et al. (2024). Photoreforming of poly(ethylene-terephthalate) plastic into valuable chemicals and hydrogen over $BiVO_4$/MoOx: Synergistic promotion of oxidation and reduction processes. *Appl. Catal. B: Environ.* 357: 124326.

10 Liang, J.-Y.-S., Ren, H.-T., Labidi, A. et al. (2024). Boosted photocatalytic removal of NO using direct Z-scheme UiO-66-NH_2/Bi_2MoO_6 nanoflowers heterojunction: mechanism insight and humidity effect. *Rare Metals* 43: 1–13.

11 Cui, Y., Labidi, A., Liang, X. et al. (2024). Pivotal impact factors in photocatalytic reduction of CO_2 to value-added C_1 and C_2 products. *ChemSusChem* 17: e202400551.

12 Dong, Q., Li, X., Sun, J. et al. (2024). Regulating concentration of surface oxygen vacancies in Bi_2MoO_6/Bi-MOF for boosting photocatalytic ammonia synthesis. *J. Catal.* 433: 115489.

13 Li, W., Liao, G., Duan, W. et al. (2024). Synergistically electronic interacted PVDF/CdS/TiO_2 organic-inorganic photocatalytic membrane for multi-field driven panel wastewater purification. *Appl. Catal. B: Environ.* 354: 124108.

14 Yu, T., Zhang, W., Hou, C. et al. (2024). Confined growth of oxygen-vacancy rich TiO_{2-x} nanoparticles with heterophase junction for high-efficiency photocatalytic water purification. *Surf Interfaces* 48: 104323.

15 Xu, X., Wang, H., Gao, T. et al. (2024). 2D copper–porphyrin metal–organic framework nanosheet-photosensitized TiO_2 for efficiently broadband light-driven conversion of CO_2 to CH_4. *Sol. RRL* 8: 2400081.

16 Hou, C., Wang, L., Zhang, W. et al. (2023). Construction of TiO_{2-x} confined by layered iron silicate toward efficient visible-light-driven photocatalysis–Fenton synergistic removal of organic pollutants. *ACS Appl. Mater. Interfaces* 15: 23124–23135.

17 Pan, B., Zhou, L., Qin, J. et al. (2022). Modulating $CoFeO_X$ nanosheets towards enhanced CO_2 photoreduction to syngas: effect of calcination temperature and mixed-valence multi-metals. *Chem. Eur. J.* 28: e202201992.

18 Wang, C., Shang, H., Li, J. et al. (2021). Ultralow Ru doping induced interface engineering in MOF derived ruthenium-cobalt oxide hollow nanobox for efficient water oxidation electrocatalysis. *Chem. Eng. J.* 420: 129805.

19 Wang, T., Zhu, Y., Luo, Z. et al. (2021). Oxygen vacancy confining effect on photocatalytic efficiency of Pt 1-black TiO_2 single-atom photocatalysts for hydrogen generation and phenol decomposition. *Environ. Chem. Lett.* 19: 1815–1821.

20 Zhu, P., Yin, X., Gao, X. et al. (2021). Enhanced photocatalytic NO removal and toxic NO_2 production inhibition over ZIF-8-derived ZnO nanoparticles with controllable amount of oxygen vacancies. *Chin. J. Catal.* 42: 175–183.

21 Han, X., Ma, H., Wang, C., and Li, Y. (2020). Enhanced photocatalytic hydrogen production by loading histidine on TiO_2. *J. Phys.: Energy* 3: 014001.

22 Orudzhev, F., Ramazanov, S., Sobola, D. et al. (2020). Atomic layer deposition of mixed-layered aurivillius phase on TiO_2 nanotubes: synthesis, characterization and photoelectrocatalytic properties. *Nano* 10: 2183.

23 Lv, H., Duan, Y., Zhou, X. et al. (2020). Visible-light-driven Ag/AgCl@ In_2 O_3: a ternary photocatalyst for the degradation of tetracycline antibiotics. *Cat. Sci. Technol.* 10: 8230–8239.

24 Azizi-Lalabadi, M., Ehsani, A., Divband, B., and Alizadeh-Sani, M. (2019). Antimicrobial activity of titanium dioxide and zinc oxide nanoparticles supported in 4A zeolite and evaluation the morphological characteristic. *Sci. Rep.* 9: 17439.

25 Ali, F.D., Ammar, S.H., Ali, N.D. et al. (2024). Synthesis of AgCl/ZIF-8/C–TiO_2 heterojunction photocatalysts for enhanced degradation of levofloxacin under visible-light. *Mater. Sci. Semicond. Process.* 172: 108100.

26 Feng, S., Zhao, J., Liang, X. et al. (2023). Mo-modified (tm)-ZrO_2 with narrowed band gap and strong ability to activate reactants for photocatalytic CO_2 reduction. *Mol. Catal.* 544: 113205.

27 Wang, W., Huang, Y., Rao, Y. et al. (2023). The role of cationic defects in boosted lattice oxygen activation during toluene total oxidation over nano-structured $CoMnO_x$ spinel. *Environ. Sci. Nano* 10: 812–823.

28 Li, W., Duan, W., Liao, G. et al. (2024). 0.68% of solar-to-hydrogen efficiency and high photostability of organic-inorganic membrane catalyst. *Nat. Commun.* 15: 6763.

29 Sharma, M., Sajwan, D., Gouda, A. et al. (2024). Recent progress in defect-engineered metal oxides for photocatalytic environmental remediation. *J. Photochem. Photobiol.* 100: 830–896.

30 Kuila, A., Routu, S., Saravanan, P. et al. (2022). Thermo-photodynamic perspective of the simultaneous S-Scheme ternary heterostructure through Ag_3VO_4 shuttle for the increased photo-redox ability. *Appl. Mater. Today* 27: 101435.

31 Zhu, Q., Hailili, R., Xin, Y. et al. (2022). Efficient full spectrum responsive photocatalytic NO conversion at $Bi_2Ti_2O_7$: co-effect of plasmonic Bi and oxygen vacancies. *Appl. Catal. B: Environ.* 319: 121888.

32 Li, N., Shi, M., Xin, Y. et al. (2022). Oxygen vacancies-modified S-scheme $Bi_2Ti_2O_7/CaTiO_3$ heterojunction for highly efficient photocatalytic NO removal under visible light. *J. Environ. Chem. Eng.* 10: 107420.

33 Shi, M., Rhimi, B., Zhang, K. et al. (2021). Visible light-driven novel $Bi_2Ti_2O_7/CaTiO_3$ composite photocatalyst with enhanced photocatalytic activity towards NO removal. *Chemosphere* 275: 130083.

34 Gu, W., Wang, W., Li, G. et al. (2020). Microwave-assisted synthesis of defective tungsten trioxide for photocatalytic bacterial inactivation: role of the oxygen vacancy. *Chin. J. Catal.* 41: 1488–1497.

35 Zhang, Y., Li, Y., Yu, H. et al. (2022). Interfacial defective Ti^{3+} on Ti/TiO_2 as visible-light responsive sites with promoted charge transfer and photocatalytic performance. *J. Mater. Sci. Technol.* 106: 139–146.

36 Xu, Y., Li, H., Sun, B. et al. (2020). Surface oxygen vacancy defect-promoted electron-hole separation for porous defective ZnO hexagonal plates and enhanced solar-driven photocatalytic performance. *Chem. Eng. J.* 379: 122295.

37 Essawy, A.A., Alsohaimi, I.H., Alhumaimess, M.S. et al. (2020). Green synthesis of spongy Nano-ZnO productive of hydroxyl radicals for unconventional solar-driven photocatalytic remediation of antibiotic enriched wastewater. *J. Environ. Manag.* 271: 110961.

38 Wang, Y., Zhu, L., Li, J. et al. (2023). Metal support interaction of defective-rich CuO and Au with enhanced CO low-temperature catalytic oxidation and moisture resistance. *Adv. Powder Mater.* 2: 100119.

39 Zhu, H., Dong, S., Du, X. et al. (2022). Defective CuO-rich $CuFe_2\ O_4$ nanofibers enable the efficient synergistic electrochemical reduction of nitrate to ammonia. *Cat. Sci. Technol.* 12: 4998–5002.

40 Meng, W., Sun, S., Xie, D. et al. (2024). Engineering defective Co_3O_4 containing both metal doping and vacancy in octahedral cobalt site as high performance catalyst for methane oxidation. *Mol. Catal.* 553: 113768.

41 Shi, M., Ren, G., Zhang, Z. et al. (2023). Strong metal-carrier interactions via modulation of Pt oxidation states on defective BiOBr with greatly improved photocatalytic activity. *Sol. Energy* 261: 33–42.

42 Hao, Y., Mao, F., Zhao, Y. et al. (2022). Selective oxidation of CH_4 to valuable HCHO over a defective rTiO_2/GO metal-free photocatalyst. *Cat. Sci. Technol.* 12: 5869–5878.

43 Kunthakudee, N., Puangpetch, T., Ramakul, P. et al. (2024). Ultra-fast green synthesis of a defective TiO_2 photocatalyst towards hydrogen production. *RSC Adv.* 14: 24213–24225.

44 Sawunyama, L., Oyewo, O., Onwudiwe, D.C., and Makgato, S.S. (2023). Photocatalytic degradation of tetracycline using surface defective black TiO_2–ZnO heterojunction photocatalyst under visible light. *Heliyon* 9: e21423.

45 Chen, H.-L., Liu, F.-Y., Xiao, X. et al. (2022). Photoreduction of carbon dioxide and photodegradation of organic pollutants using alkali cobalt oxides $MCoO_2$ (M= Li or Na) as catalysts. *J. Environ. Manag.* 313: 114930.

46 Zhang, X., Zhao, Z., Zhang, W. et al. (2016). Surface defects enhanced visible light photocatalytic H_2 production for Zn-Cd-S solid solution. *Small* 12: 793–801.

47 Huo, Y., Zhang, P., Chi, J. et al. (2024). Surface functionalization and defect construction of SnO_2 with amine group for enhanced visible-light-driven photocatalytic CO_2 reduction. *Adv. Energy Mater.* 14: 2304282.

48 Vijayarangan, R., Mohan, S., Selvaraj, M. et al. (2024). Defect engineered N-doped black-V_2O_5 nanostructures for hydrogen production via solar-catalytic water splitting in the presence of different sacrificial agents. *Surf Interfaces* 45: 103838.

49 Abutalib, M., Alghamdi, H.M., Rajeh, A. et al. (2022). Fe_3O_4/Co_3O_4–TiO_2 S-scheme photocatalyst for degradation of organic pollutants and H_2 production under natural sunlight. *J. Mater. Res. Technol.* 20: 1043–1056.

50 Harikumar, B., Okla, M.K., Kokilavani, S. et al. (2023). Insights into oxygen defect enriched and non-metal dopant co modulated Fe_3O_4 nanospheres embedded WO_3 nanorods for ameliorated photodegradation of doxycycline, Cr (VI) reduction and its genotoxicity. *J. Clean. Prod.* 398: 136549.

51 Chen, A., Su, Q., Han, H. et al. (2019). Metal oxide nanocomposites: a perspective from strain, defect, and interface. *Adv. Mater.* 31: 1803241.

52 Garg, S., Gautam, S., Singh, J.P. et al. (2021). Characterizing the defects and ferromagnetism in metal oxides: the case of magnesium oxide. *Mater. Charact.* 179: 111366.

53 Kaya, S., Yilmaz, E., Aktag, A., and Seidel, J. (2015). Characterization of interface defects in $BiFeO_3$ metal–oxide–semiconductor capacitors deposited by radio frequency magnetron sputtering. *J. Mater. Sci. Mater. Electron.* 26: 5987–5993.

54 Berger, T. and Diwald, O. (2015). Defects in metal oxide nanoparticle powders. In: *Defects at Oxide Surfaces* (ed. J. Jupille and G. Thornton), 273–301. Springer International Publishing.

55 Setvín, M., Wagner, M., Schmid, M. et al. (2017). Surface point defects on bulk oxides: atomically-resolved scanning probe microscopy. *Chem. Soc. Rev.* 46: 1772–1784.

56 Belotcerkovtceva, D., Maciel, R.P., Berggren, E. et al. (2022). Insights and implications of intricate surface charge transfer and sp^3-defects in graphene/metal oxide interfaces. *ACS Appl. Mater. Interfaces* 14: 36209–36216.

57 Bathen, M., Lew, C.-K., Woerle, J. et al. (2022). Characterization methods for defects and devices in silicon carbide. *J. Appl. Phys.* 131: 140903.

58 Cao, D., Li, W., Zhang, X. et al. (2022). Current state-of-the-art characterization methods for probing defect passivation towards efficient perovskite solar cells. *J. Mater. Chem. A* 10: 19278–19303.

59 Labidi, A., Ren, H., Liang, X. et al. (2024). Visible–light–driven photocatalytic degradation of ciprofloxacin antibiotic by novel heterostructured coal fly ash waste: mechanism insight, toxicity pathway and DFT calculation. *Mater. Today Chem.* 42: 102388.

60 Wang, H., Labidi, A., Ren, M. et al. (2024). Recent progress of microstructure-regulated g-C_3N_4 in photocatalytic NO conversion: the pivotal roles of adsorption/activation sites. *Acta Phys. -Chim. Sin.* 41: 100039.

61 Ghasemi, S., Parastesh, A., Padervand, M. et al. (2025). Recent progress on Z-and S-scheme photocatalysis: mechanistic understanding toward green applications. *Curr. Opin. Chem. Eng.* 47: 101059.

62 Sial, A., Gao, T., Li, F. et al. (2024). Multifaceted synergism of dual active sites in oxygen vacancies enriched plasmonic Ag-BiOI nanosheets for enhanced piezo-photocatalytic degradation of trimethoprim. *ACS ES&T Eng.* 4: 2923–2937.

63 Liu, J., Du, S., Zhu, Q. et al. (2024). Recent progress on bismuth-based materials for full-spectrum-driven photocatalysis. *J. Environ. Chem. Eng.* 12: 114790.

64 Ren, H., Labidi, A., Sial, A. et al. (2024). Carbon quantum dots modified Z and S–Scheme heterojunctions for pharmaceutical contaminants photodegradation: state–of–the–art, benefits, and limitations. *Sep. Purif. Technol.* 346: 127508.

65 Labidi, A., Ren, H., Sial, A. et al. (2024). Organic radicals used to modify materials surface and their application to remove pharmaceutical residues from wastewater. In: *Organic Radicals* (ed. C. Wang, A. Labidi, and E. Lichtfouse), 321–339. Elsevier.

66 Labidi, A., Ren, H., Sial, A. et al. (2024). Photocatalyst materials formed using organic radicals for water splitting. In: *Organic Radicals* (ed. C. Wang, A. Labidi, and E. Lichtfouse), 211–230. Elsevier.

67 Labidi, A., Ren, H., Zhu, Q. et al. (2024). Coal fly ash and bottom ash low-cost feedstocks for CO_2 reduction using the adsorption and catalysis processes. *Sci. Total Environ.* 912: 169179.

68 Atif, S., Afzal, D.A., Abdelkader, L. et al. (2023). Self-regulated interfacial-enhanced piezo-phototronic effect of BNT@Mn-SnO_2 heterostructures for superlative tetracycline degradation. *J. Catal.* 428: 115179.

69 Labidi, A., Ren, H., Sial, A. et al. (2023). Coal ash for removing toxic metals and phenolic contaminants from wastewater: a brief review. *Crit. Rev. Environ. Sci. Technol.* 53: 2006–2029.

70 Pan, B., Chen, W., Zhou, L. et al. (2024). $CoFe_2O_4$/carbon nitride Z-scheme heterojunction photocatalytic PMS activation for efficient tetracycline degradation: accelerated electron transfer. *Process Saf. Environ. Prot.* 191: 2522–2532.

4

Defective Heterojunction Photocatalysts: Ample Opportunity for Photocatalytic Implementations

Keshav Kumar and Chandana Rath

Indian Institute of Technology (Banaras Hindu University), School of Materials Science and Technology, Varanasi, 221005, India

4.1 Introduction

During the past decades, the photocatalysis process has been considered the most appealing and emerging process to address the environmental concerns about water pollution by harnessing renewable energy sources [1–7]. The leading cause of water pollution is the direct discharge of waste from industries, agriculture, and other sources. In 1972, Fujishima and Honda introduced TiO_2 as the first photocatalyst for the photolysis of water [8]. Many studies have been done on TiO_2 to investigate its role as a photocatalyst for wastewater treatment. It has a limited photocatalytic efficiency because of a wider band gap (3.2 eV) and faster charge carrier recombination [9]. These factors hinder the ability to absorb the solar spectrum, making them less efficient photocatalysts. Bharti et al. studied the photocatalytic activity of sol–gel synthesized TiO_2 nanoparticles for the degradation of Congo red (CR), methylene blue (MB), and Rhodamine B (Rh B) dyes over ultraviolet (UV) and sunlight. It has been reported that under UV light, the degradation percentage reached up to 100% for MB and Rh B in 180 minutes, while for CR dye, it took 120 minutes [10]. However, under sunlight irradiation, the degradation of MB, CR, and Rh B reached to its maximum degradation up to 100% in 95, 45, and 180 minutes, respectively. In addition to TiO_2, various classes of photocatalysts have been explored, including metal oxides such as ZnO, SnO_2 and CeO_2 [11–14], metal sulfides (CdS, CDS/In_2O_3, Pt-PdS/CdS, and CdS/MoS_2) [15–19], metal-organic frameworks (MOFs) [20–28], and perovskite oxides ($CaTiO_3$, $SrTiO_3$, $LaNiO_3$, $LaCoO_3$, and $GdCoO_3$) [29–33]. The photocatalytic reactions mainly occupy broad application areas, including pollutant degradation, hydrogen production, CO_2 reduction, agriculture, chemical synthesis, the medical sector, and antibacterial inhibition, etc., as shown in Figure 4.1.

In general, the photocatalytic reaction involves four essential steps: (i) the absorption of the light, (ii) electron–hole pair generation, (iii) their subsequent migration from bulk to surface, and (iv) an adsorption or desorption/redox reaction

Defect Engineering in Photocatalysis: From Fundamentals to Applications,
First Edition. Edited by Chuanyi Wang, Abdelkader Labidi, and Qiuhui Zhu.

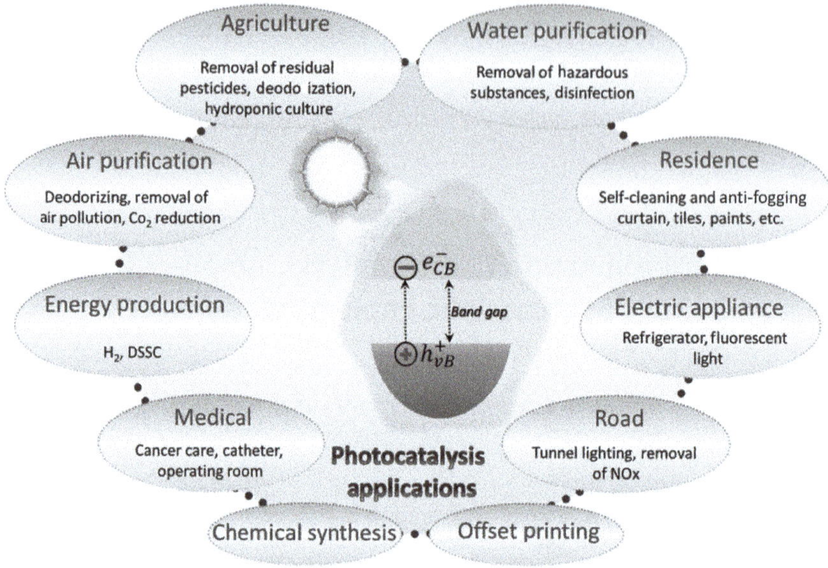

Figure 4.1 Various applications of the photocatalysis process. *Source:* Reproduced from Kumari et al. [18] / with permission of Royal Society of Chemistry.

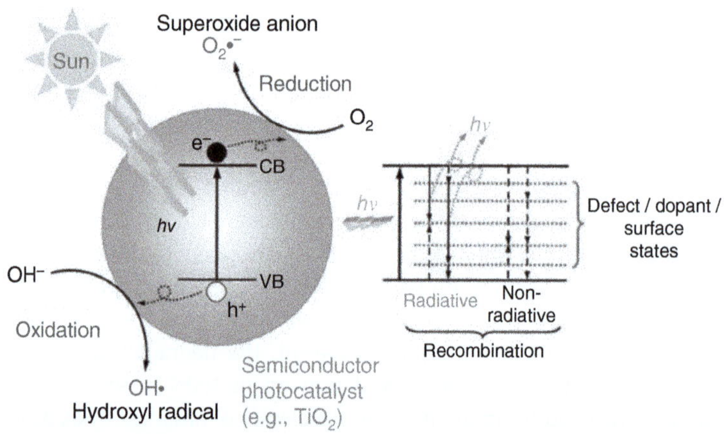

Figure 4.2 Schematic representation of the mechanism pathway for photocatalytic processes for reactive species formation. *Source:* Reproduced from Djurišić et al. [34] / with permission of Royal Society of Chemistry.

on the surface-active sites, as depicted in Figure 4.2. Over the many investigated single photocatalytic materials, even the highly efficient materials still have some limitations, such as a low range of visible light absorption (reducing the redox capacity), the lifetime of photogenerated electron–holes, and their stability. At present, photocatalytic efficiency remains a challenge due to the faster recombination of electron–hole pairs, which makes photocatalytic reactions unfavourable [34]. To address the aforementioned limitations, extensive research has been focused on structural engineering to increase the surface area and active sites for photocatalytic reactions. Heterojunction photocatalysts overcome these drawbacks by combining different semiconductors with complementary characteristics [35]. These systems are able to boost the segregation of photogenerated electron–hole pairs, enhance light absorption, and allow for better utilization of renewable energy sources.

The introduction of defects in these heterojunctions has become a good option to improve the efficiencies in various photocatalytic applications [36]. Defects like oxygen vacancies (OVs) or imperfections are internal flaws in a material's lattice that significantly impact its electronic and structural characteristics [37]. These imperfections may be responsible for modifying the surface of the catalyst, electrical properties, efficient migration of charge carriers, and enhancing the ability to attract molecules, resulting in tuned photocatalytic activity [38]. However, the alignment of the band structure in heterojunction photocatalysts generates some states in the middle of the band gap for increased light absorption. It increases the segregation of charges and transfers them at the interface. Therefore, the material's interfacial contacts, electric field, and surface redox processes are affected.

In this chapter, we have thoroughly discussed the influence of OVs on heterojunction photocatalysts, focusing on their synthesis and the role of key factors such as band alignment, segregation of charge carriers, and their ability to improve the photocatalytic reactions. This chapter offers a basic overview of recent oxygen vacancy engineering, challenges, and advancements and thus predicting future opportunities related to oxygen vacancy engineering for designing more efficient heterojunction photocatalysts.

4.2 Types of Heterojunctions

Heterojunction photocatalysts are usually formed by combining two semiconductor photocatalysts with complementary characteristics, including their relative positions of energy bands and differences in fermi levels to establish carrier pathways between them. This approach facilitates the segregation of electron–hole pairs, broadens the absorbable-spectral range, and enhances the redox ability to overcome the limitations of individual semiconductors [39–41]. Traditional heterojunctions are divided into three categories based on their fermi level alignments. Figure 4.3 depicts the type-I, II, and III heterojunctions, which are classified as straddling, staggered, and broken gaps, respectively. When two semiconductor materials are interfaced with each other in such a way that the conduction band and valence band of the lower band gap semiconductor are enclosed entirely within the wider band gap semiconductor material, it is known as a type-I heterojunction. When suitable light energy is irradiated on the semiconductor (higher band gap), the photogenerated electron–hole pairs are migrated toward the conduction band and valence band of the other semiconductor (lower band gap). This process is unfavourable for the photocatalytic reaction because the charge accumulation occurs across the lower band gap materials.

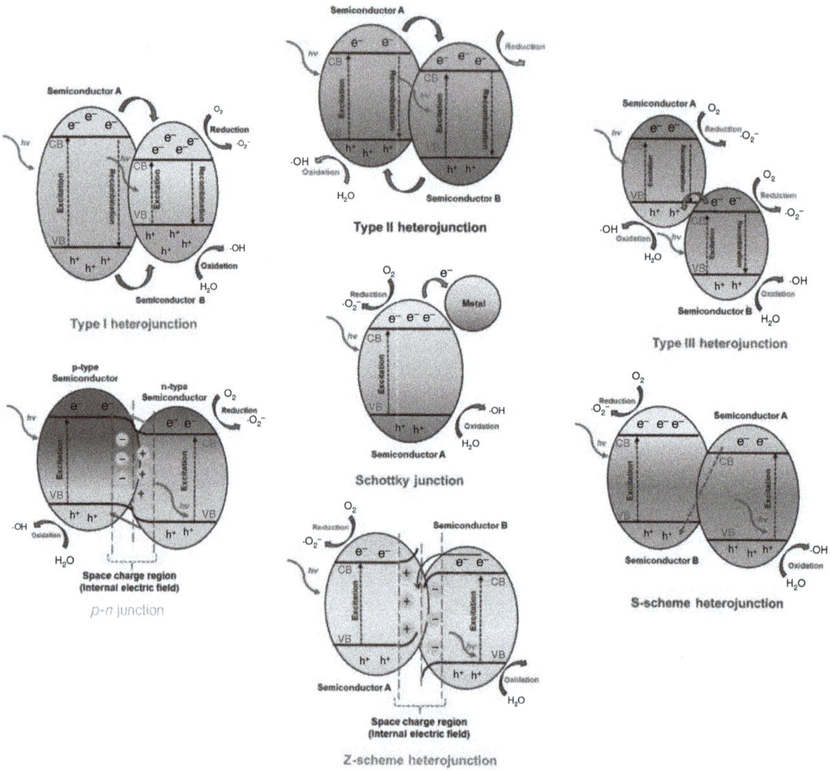

Figure 4.3 Different types of heterojunctions on the basis of the junction formed between semiconductors. *Source:* Reproduced from Gnanaguru et al. [39] / with permission of Springer Nature.

In the type-II heterojunction materials, two semiconductor materials with different band gaps are combined together, resulting in the alignment where the conduction band minima and valence band maxima of one semiconductor are aligned above the conduction band and valence band of the other semiconductor material. When light is irradiated, the photogenerated charge carriers in one semiconductor material move toward the lower conduction band and higher valence band of the other semiconductor material, respectively. Hence, these types of heterojunction photocatalysts are considered efficient photocatalysts due to sufficient charge segregation, which results in the reduction of electron–hole pair recombination. Further, type-II heterojunctions are classified into two categories, Z-Scheme and S-Scheme, based on their band alignment and the migration mechanism of electron–hole pairs. In 1979, Bard proposed the Z-Scheme to enhance the charge carrier segregation and redox potential. This type of heterojunction allows the photogenerated electrons to move from the conduction band of one semiconductor and recombine with the holes of the valence band of the other semiconductor through the interface because of strong electrostatic attraction. As a result, electrons with strong reducibility remain in the conduction band of the second semiconductor, while holes with strong oxidizing ability remain in the valence band of the first semiconductor. These separated charge carriers follow the "Z" pathway movement and participate in redox reactions on the semiconductor surface, enabling the highly efficient photocatalysis process. Till date, many heterojunction photocatalyst have been discovered such as In_2S_3/In_2O_3 [42], $Bi_2O_2CO_3/ZnFe_2O_4$ [43], SnO_2/MoO_3 [44], $FeVO_4/Fe_2TiO_5$ [45], $Mn_2O_3/Mn_3O_4/MnO_2$ [46], $SnFe_2O_4/\alpha\text{-}Fe_2O_3$ [47], $NiTiO_3/Ag_3VO_4$ [48], and Ag_2WO_4/Sb_2WO_6 [49].

The S-scheme mechanism is an advanced model that provides the idea of charge carrier behavior for enhanced photocatalytic performance. This scheme consists of two semiconductor materials, a reducing photocatalyst with a higher Fermi level or more negative conduction band and an oxidizing photocatalyst with a more positive valence band. When two semiconductors come into contact, electrons spontaneously transfer from one semiconductor to another due to differences in their work function. This transfer of electrons continues until the fermi level reaches equilibrium. The built-in electric field is developed at the interface to prevent the reverse movement of the electrons. When light is irradiated at the interface, the photogenerated electrons from the oxidizing photocatalyst conduction band recombine with holes in the reducing photocatalyst valence band due to the electrostatic interactions and inherent electric field. As a result of this recombination, the valence band with oxidizing potential has holes with greater oxidizing ability, while the conduction band with reducing potential has electrons with greater reducibility. The S-scheme heterojunction mechanism uses the complementary effects of the integrated electric field, carrier recombination, and band bending at the interface. It provides efficient spatial charge segregation and energy-efficient catalytic performance. Most of the widely used heterojunctions based photocatalyst have been synthesized such as Bi_2S_3/porous $g\text{-}C_3N_4$ [50], $g\text{-}C_3N_4/Ag_3VO_4$ [51], $TiO_2/W_{18}O_{49}$ [52], AgI/I-BiOAc [53], $g\text{-}C_3N_4/g\text{-}C_3N_4$ [54], SnO_2/SnS_2 [55], $SnFe_2O_4/ZnFe_2O_4$ [56], and WO_3/TiO_2 [57]. When two semiconductors with different bandgap energies are interfaced, their valence band maximum and conduction band minimum overlap, creating a broken gap in the band structure – this is known as type-III heterojunctions (also called broken gap heterojunction materials). There is a significant discontinuity between the bands in this arrangement since the conduction band of one material is lower than the valence band of the other material [58].

4.3 Synthesis of Defective Heterojunctions and Applications

The synthesis of heterojunction nanomaterials plays a pivotal role in achieving efficient charge segregation and light utilization in photocatalytic degradation. Various methods have been developed to fabricate defective heterojunctions with precise control over composition, structure, and interfacial properties. This section discusses the most widely used chemical synthesis techniques for defective heterojunction materials, their principles, advantages, and applications in photocatalysis.

4.3.1 Hydrothermal/Solvothermal Methods

The hydrothermal/solvothermal process is a versatile and widely used technique to synthesize heterojunction photocatalysts with controlled morphology and well-defined interfaces. In this method, the reaction is carried out in a closed vessel under high temperature and pressure above 1 atm. The adjusting parameters like pH, reaction temperature, time, and precursor concentrations allow the formation of an interface between different semiconductor materials, which in turn enhances the charge segregation and transfer efficiency. Several well-defined steps are involved in the synthesis of high-quality materials. Many researchers have developed defective heterojunction photocatalysts by using the hydrothermal method. For example, Li et al. have synthesized oxygen-vacancy-rich CeO_2-based heterojunctions for different photocatalytic applications including tetracycline degradation and the *E. coli.* bacteria inactivation [59]. They indicated that plenty of OVs in AgI/CeO_2 heterojunction can be utilized as oxygen adsorption sites to boost the chemisorption of O_2, thus facilitating the generation of ROS (i.e., H_2O_2, O_2^- and ·OH) (Figure 4.4). Thus, due to the great formation of ROS, the

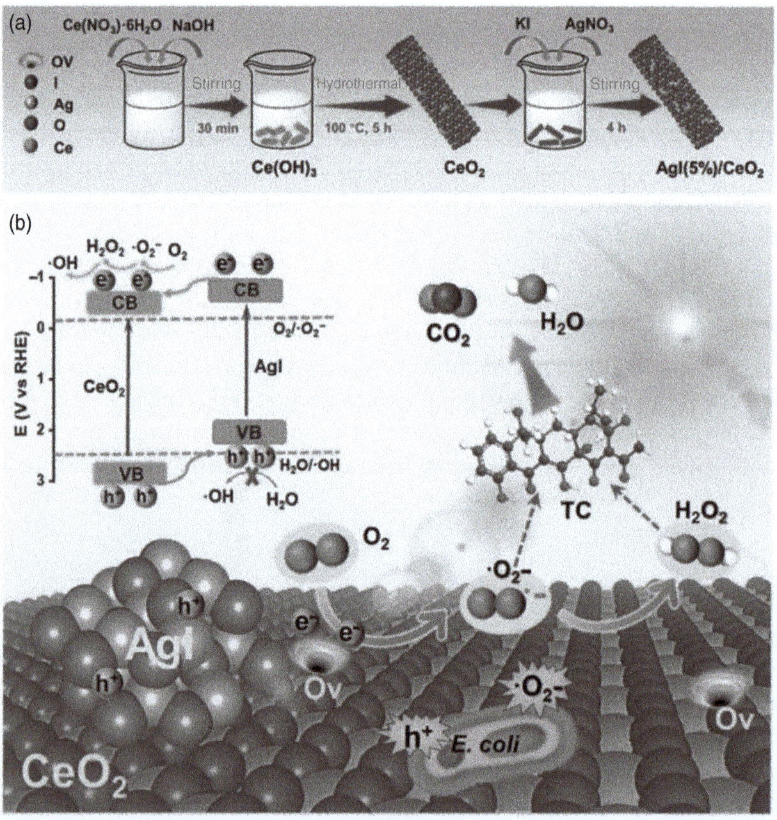

Figure 4.4 (a) Preparation of oxygen-vacancy-rich CeO_2-based heterojunctions and (b) applications of defective heterojunction for tetracycline degradation and bacteria inactivation. *Source:* Reproduced from Wang et al. [59] / with permission of Elsevier.

AgI/CeO_2 heterojunction system displayed high efficiency for photocatalytic degradation and disinfection. They mentioned that this heterojunction with abundant OVs synergistically boosts photocatalytic O_2 activation and may have broad implementation prospects in water purification and other fields. Other prepared defective heterojunctions by the hydrothermal method have been reported, such as PTO@BiOCl-OVs, which displayed an excellent photocatalytic activity for RhB degradation under visible light and boosted charge separation because of the existence of OVs [60]. Zu et al. synthesized WO_3-Ov/In_2S_3 S-Scheme heterojunction photocatalyst with an enhanced photocatalytic activity [61]. In 2021, Wu et al. developed the $BiOCl/CuBi_2O_4$ S-scheme heterojunction with OVs via a hydrothermal route and studied its photocatalytic activity toward the diclofenac degradation and nitric oxide removal [62]. A Z-scheme-based $CeO_2/Bi_7O_9I_3$ heterojunction with OVs synthesized via the hydrothermal method is reported by Liu et al. for the photocatalytic reduction of Hg^0 [63]. Nie et al. developed β-$Bi_2O_3/CeO_{2-\delta}$ *p-n* heterojunction with oxygen vacancy defects via the hydrothermal method [64] and found a deeper oxidation of NO gaseous pollutant under light irradiation.

4.3.2 Sol-gel Method

Researchers widely use the sol–gel method for synthesizing heterojunction photocatalyst nanomaterial, offering precise control over the material, such as particle size, morphology, and surface area. This method involves both physical and chemical processes, including gelation, hydrolysis, condensation, polymerization, and drying. For example, oxygen vacancy defects over TiO_2/WO_3 are synthesized by Huang et al. using the sol–gel method [37] (Figure 4.5). They indicated that the existence of oxygen vacancy and heterostructure improves photocatalytic performance synergistically, achieving the H_2 evolution rate of 509.39 $\mu mol\ g^{-1}\ h^{-1}$ 0.1414 min^{-1}. In another report, Xie et al. prepared TiO_2-SiO_2-C composite with anatase via the sol–gel method and indicated that plenty of OVs were prepared by simply annealing through the heterojunction exhibited unexpected catalytic activity in dye degradation in the presence of H_2O_2 [65]. Recently, Zhang et al. developed a CeO_2/ZnO heterojunction with controlled OVs via the sol–gel technique and studied its photocatalytic activity to degrade RhB under visible light irradiation [66]. Under visible light illumination, efficient separation and transfer of photogenerated electron–hole pairs were achieved while effectively suppressing recombination, due to the synergy between the interfacial electric field and OVs employing $CuFeO_2$@VO-CuO core-shell heterostructure as reported by Li et al. [67].

4.3.3 Co-Precipitation Method

The co-precipitation method is another chemical approach for synthesizing the heterojunction materials with controlled morphology of the materials. The controlled morphology can be altered with variation of pH and temperature.

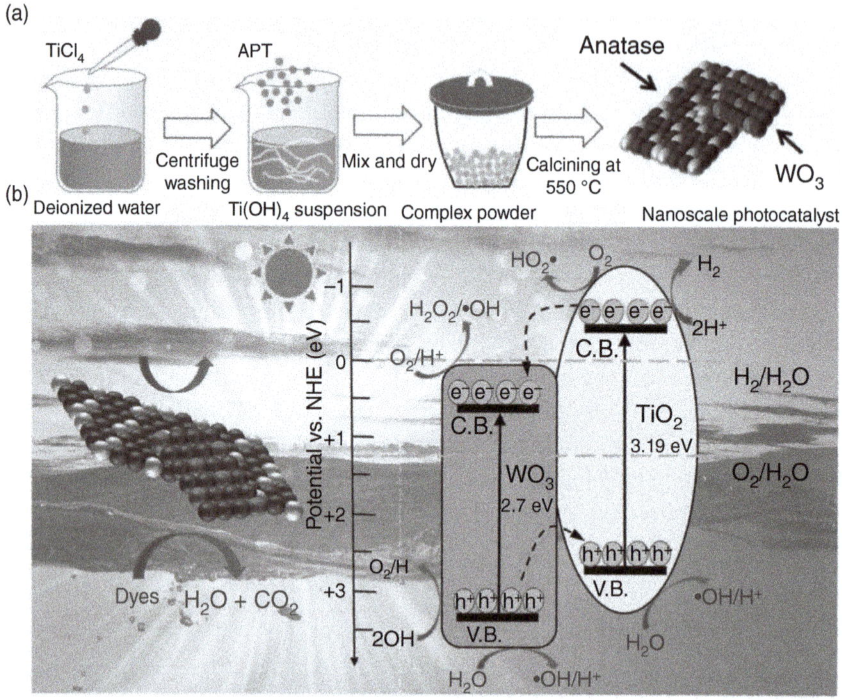

Figure 4.5 (a) Fabrication routes for TW-x photocatalyst. (b) The proposed possible photocatalytic mechanism of the prepared TW-2 photocatalyst. *Source:* Reproduced from Huang et al. [37] / with permission of Elsevier.

This method requires adding a precipitating agent to a solution containing the desired compounds, leading to the formation of solid phase materials. For example, Paramanik et al. have synthesized the oxygen-vacancy-rich BiOI/$PbTiO_3$ *p-n* heterojunction using the precipitation–deposition method [68]. They indicated that the effect of combining a *p-n* heterojunction and OVs effectively modulated the physicochemical, electronic, and catalytic properties of the photocatalyst simultaneously, resulting in superior degradation of the antibiotic tetracycline, i.e., 94% in two hours under solar harvest. In_2O_3 nanoparticle/$Bi_4O_5Br_2$ nanosheet S-scheme heterojunction with interface OVs was constructed by a hydrothermal-calcination approach, and a simple chemical precipitation process is for tetracycline hydrochloride degradation is reported by Zhang et al. [69]. Yu et al. adopted a co-precipitation method to synthesize CoAl-LDH/(110) facet-exposed BiOBr Z-type heterojunction rich-OVs [70]. In this investigation, the authors indicated that OVs contribute to the adsorption of dissolved oxygen in water, thereby facilitating the generation of $\bullet O_2^-$. And $Bi^{(3-x)+}$ Lewis acid sites promote the adsorption and degradation of ciprofloxacin molecules (Figure 4.6). In another study, surface oxygen vacancy defects on 2D/2D coupled Bi_2WO_6/BiOI nanosheets were reported by Kong et al. for photocatalytic CO_2 reduction into CH_4 [71]. The synthesized heterojunction exhibited a remarkably high production yield of CH_4, on exposure to visible light.

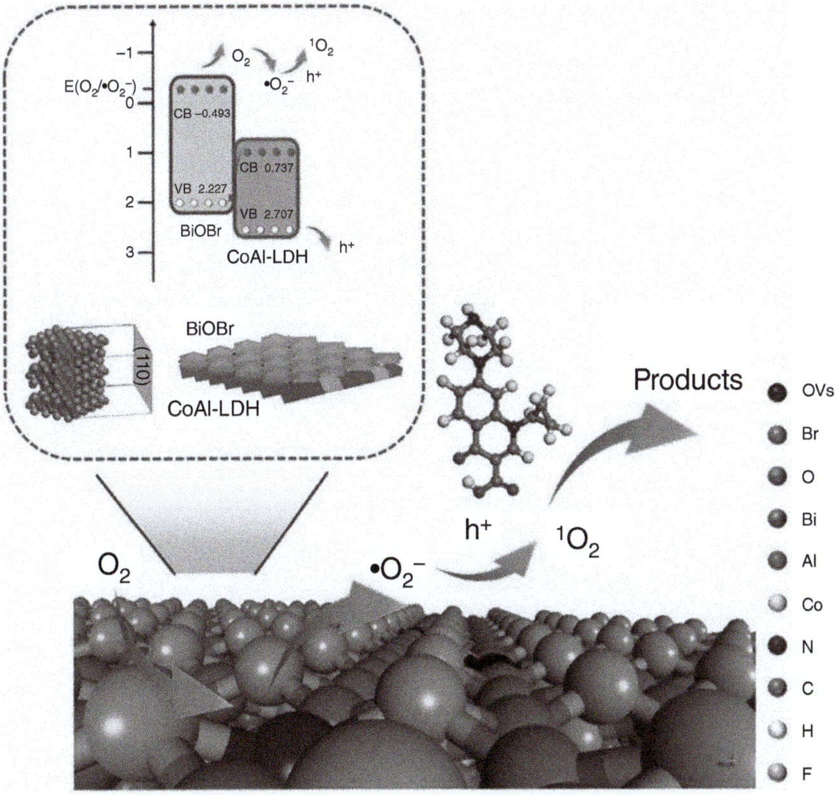

Figure 4.6 Fabrication route of CoAl-LDH/(110) facet-exposed BiOBr Z-type heterojunction rich-OVs for photocatalytic ciprofloxacin degradation. *Source:* Reproduced from Zhang et al. [70] / with permission of Elsevier.

4.4 Characterization of Defects

Characterization of defective heterojunctions is essential for understanding their properties and performance, especially in photocatalysis and energy storage applications. Techniques are discussed below to characterize the defective heterojunctions.

4.4.1 High Resolution-Transmission Electron Microscopy (HR-TEM)

High-resolution transmission electron microscopy (HR-TEM) is one of the most used characterization tools to confirm the formation of heterojunction photocatalysts by providing atomic-scale insights into their structural and interfacial properties. Additionally, HR-TEM plays a key role in the detection of structural defects such as grain boundaries, stacking faults, dislocations at the atomic level, lattice strain, and distortions at the heterojunction interface. These flaws have a significant impact on the charge carrier dynamics and in turn, the photocatalytic efficiency.

4.4.2 Kelvin Probe Force Microscopy (KPFM)

Kelvin probe force microscopy (KPFM) is a tool for characterizing heterojunction photocatalysts and gives a high-resolution mapping of surface potential and work function variations. It measures the contact potential difference and provides insights into band bending, built-in electric field, and charge carrier distribution at the heterojunction interface. It detects defect-induced surface potential changes, aiding defect engineering to enhance photocatalytic efficiency. KPFM also evaluates light-induced surface potential shifts, correlating photo response with interfacial properties. It provides charge carrier dynamics and interfacial charge transfer processes, which are critical for assessing photocatalytic efficiency. Moreover, its nanoscale resolution makes it ideal for studying localized charge transport phenomena in complex heterojunction systems, such as those involving nanostructures or multi-phase composites.

4.4.3 X-ray Photoelectron Spectroscopy

X-ray photoelectron spectroscopy (XPS) is a non-destructive technique for analysis of the oxidation state of the present elements, and it provides detailed information about the surface composition. The basic principle of XPS is based on the photoelectric effect. The XPS spectra reveal binding energy shifts resulting from electronic interactions at the heterojunction interface. Detecting defects such as OVs helps assess their impact on charge segregation and recombination, which are crucial for charge transfer and band alignment at the interface. Additionally, depth profiling examines compositional variations across the interface, and valence band spectra offer insights into the band alignment. These capabilities make the XPS an invaluable tool for optimizing the design and performance of the heterojunction photocatalyst.

4.4.4 Electron Paramagnetic Resonance (EPR)

Electron paramagnetic resonance (EPR) spectroscopy technique is used to study species with unpaired electrons in photocatalysis. It gives detailed insights into paramagnetic defect states, like OVs, metal ion centers, and trapped charge carriers, which play a crucial role in charge segregation, recombination dynamics, and interfacial charge transfer in heterojunction photocatalysts. This technique also provides the characteristics of reactive oxygen species (ROS), which are generated during photocatalytic reactions. These species are directly corroborated with the photocatalytic activity and efficiency of the heterojunction photocatalyst. The ability to quantify and analyze such species provides insights into reaction pathways and the role of interfacial charge dynamics.

4.5 Role of Defects

Based on the above discussed examples, defective heterojunction photocatalysts are a promising class of materials for various applications, including water splitting, environmental remediation, and organic synthesis. It offers several advantages over traditional photocatalysts, such as enhanced light absorption, improved charge segregation, and increased surface reaction sites. A defect in a heterojunction is created due to an inherent lattice mismatch between the two semiconductors, resulting in strain, disorder, interstitials, and other types of defects in the material's structure.

4.5.1 Energy Band Alignment

The band structure alignment is the key factor for developing an efficient heterojunction material. It refers to the relative positioning of the conduction and valence bands of both the semiconductors at the interface. This alignment critically influences the photocatalytic property of the heterojunction material. Considering the influence of oxygen vacancy on the band alignment of each semiconductor, a certain amount of oxygen vacancy is required to shift the potential of the energy levels. For example, Liu et al. have reported the band gap engineering by boron (B) doping strategy into A-TiO_2/R-TiO_2 heterojunction material [72]. This study shows that with varying the B dopant concentration, the band alignment is optimized, resulting in a positive shift of the band toward the potential. The incorporation of B into heterostructure material introduced the defect energy level, leading to a reduction in the photo-excitation energy and a red-shift in the absorption spectrum edge. Moreover, the defect-energy level served as a pathway to increase the charge density and segregation efficiency of charges. The surface OVs increase with B doping, which provides the reactive sites for rapid surface reactions. The highest photocatalytic performance is reported for 10% B doped into A-TiO_2/R-TiO_2 heterojunction due to the improved charge carrier segregation and oxidation-reduction ability. Uddin et al. determine the band energy alignment of NiO/anatase TiO_2 heterojunction photocatalysts at the interface by XPS and ultraviolet photoelectron spectroscopy (UPS) [73]. Under light illumination, the photogenerated electrons are transferred from the CB of NiO to the CB of TiO_2, and the holes are transferred from the VB of TiO_2 to the VB of NiO. The accumulation of the charge carriers in the VB of NiO and CB of TiO_2 reacts with the species present in the solution, resulting in the formation of H_2. The effect of energy band alignment on C-doped ZnO/TiO_2 heterojunction has been studied by Backer et al. [74]. Using the UPS spectra, a favourable band alignment which accounts for *n*-type C doping and type-II staggered heterojunction formation is reported. This study reveals the occurrence of redox reactions on the surface resulting in enhanced photodegradation and mineralization, which in turn is due to the newly formed band level within the redox potential of $\bullet O_2^-$ and •OH.

4.5.2 Wide Range of Optical Absorption

The photocatalytic reactions are significantly influenced by light absorption and the band structure of the material. However, the generation of photoinduced electron–holes depends on the band gap. For improving the photocatalytic performance, it is required to have a suitable band gap for absorption of the broad spectrum. The wide-band gap semiconductors typically absorb UV light (<400 nm), while narrow-band gap semiconductors absorb visible light (400–700 nm) and some near-infrared radiation (>700 nm). There is another well-known strategy, known as the defect engineering technique to improve the optical properties of materials. In this technique, the introduction of OVs plays an important role in creating defects in the structure. The incorporation of high-valency metal ions in place of low valency or the incorporation of a dopant is responsible for creating an oxygen vacancy. These defects can tune the properties such as electron distribution and localized sites for trapping the charge carriers, thereby improving the light absorption of the photocatalyst.

4.5.3 Segregation and Migration of Photogenerated Charge Carriers

In order to promote the oxidation and reduction reaction, the photogenerated charge carriers need to be segregated from the inner side of the photocatalyst and migrate to the surface (over a distance in the range of nanometers to micrometers). At the equilibrium, the fermi level energies of the two semiconductors align at the interface. This alignment causes electron migration and establishes the built-in electric field that modulates charge carriers through the interface. A built-in electric field at the interface can provide the driving force to facilitate the efficient segregation and migration of photogenerated charge carriers. Therefore, it improves the efficiency of photogenerated charge carriers segregation and migration. For example, Yang et al. have synthesized Er-$BiOIO_3$/g-C_3N_4 heterojunction material using a hydrothermal technique and studied the photocatalytic oxidation behavior of gaseous mercury [75]. The photocatalytic performance of Er^{+3} doped $BiOIO_3$/g-C_3N_4 found to be better than g-C_3N_4, $BiOIO_3$, and $BiOIO_3$/g-C_3N_4. The Er doping facilitated the creation of oxygen vacancy and improved the photogeneration of the charge carrier, resulting in improved photocatalytic efficiency.

4.5.4 Kinetics of Surface Redox Reactions

Defects have a significant impact on the catalyst surface and on photogenerated charge carriers, which participate in the redox reactions. In heterojunction photocatalysts, the chemical interactions at the interface of atoms have a crucial role in modifying the charge density around the active sites, which in turn influences the redox reactions occurring on the surface. Point defects in these heterojunctions serve to tune the interface characteristics at the atomic scale and function as an active site in the reaction. The interplay between point defects and heterojunctions enhances the transport of photogenerated electron–hole pairs at the surface. This facilitates their effective segregation, increases the number of surface-active sites, and regulates the kinetics of surface reactions, leading to more efficient photocatalytic redox processes.

4.6 Conclusions and Perspectives

Over the past decades, research has focused on developing efficient heterojunction photocatalyst materials. Defect-engineered heterojunction photocatalysts have exhibited significant advantages for enhanced catalytic activity by modulating light absorption capability, charge segregation efficiency, and surface reaction activity. This chapter summarizes the implementation of materials design, synthesis, and characterization, showcasing the importance and potential of photocatalysis. Despite the advancements, achieving practical application and commercialization of the heterojunction photocatalyst requires significant improvements in their efficiency and scalability. The development of high-quality heterojunction photocatalysts using an efficient and economical method is a major challenge on a large scale. Therefore, some prospects are proposed below for advancement in heterojunction photocatalysts.

The controlled regulation of defects is essential for their effective role in heterojunction photocatalysts. Advanced synthesis methods must be developed to control the type of defect, concentration, and distribution. Characterization techniques with high resolution and accuracy are required to quantify defect states and understand their impact on interfacial properties, band structure, and charge transfer. Ultimately, strategically utilizing point defects can optimize heterojunction geometry and electronic properties to enhance photocatalytic efficiency. Understanding the role of defects in heterojunctions is challenging due to the complexity of photocatalytic reactions. High-resolution in situ techniques can provide insights into defect evolution and reaction dynamics. Defect engineering at the atomic scale can introduce catalytically active sites, enhancing band structure modulation, charge separation, and transfer. A synergistic optimization of photocatalysis in terms of thermodynamics and kinetics can be achieved by leveraging these effects. Defect significantly enhances the absorption range of photocatalysts, making them highly suitable for near-infrared (NIR) light harvesting. Effective solar energy utilization improves the charge transfer pathway and redox capability, leading to better photocatalytic reactions. Strategies such as defect-induced narrow bandgap semiconductors or integrating localized surface plasmon resonance (LSPR)-active photocatalysts with visible light responsive materials can enhance NIR absorption and conversion, improving overall photocatalytic performance. Rapid charge recombination limits photocatalytic efficiency, and the role of defects in charge transfer remains under researched till date. Advanced spectroscopic techniques, such as in situ scanning probe microscopy and X-ray photoelectron spectroscopy, have provided insights into charge migration during the photocatalytic reaction. Future research should enhance detection resolution to deepen the understanding of defect-assisted charge dynamics in heterojunctions, bridging the gap between structure and photocatalytic activity. Hence, the development of innovative materials is crucial for designing better heterojunction photocatalysts. Many existing researched materials have limitations, such as high costs, wide band gaps, and low surface activity. Ideal materials should be cost-effective, absorb visible light, exhibit high solar energy conversion efficiency, have suitable band structures for redox reactions, maintain stability over a long period, and be scalable for commercial production.

Conflict of Interest

The authors declare that they have no conflict of interest.

References

1 Al-Nuaim, M.A., Alwasiti, A.A., and Shnain, Z.Y. (2023). The photocatalytic process in the treatment of polluted water. *Chem. Pap.* 77 (2): 677–701. https://doi.org/10.1007/s11696-022-02468-7.
2 Mohamadpour, F. and Amani, A.M. (2024). Photocatalytic systems: reactions, mechanism, and applications. *RSC Adv.* 14 (29): 20609–20645. https://doi.org/10.1039/D4RA03259D.
3 Asefa, G., Negussa, D., Lemessa, G., and Alemu, T. (2024). The study of photocatalytic degradation kinetics and mechanism of malachite Green dye on Ni-TiO_2 surface modified with polyaniline. *J. Nanomater.* https://doi.org/10.1155/2024/5259089.
4 Dharmana, G., Nadikatla, S.K., Gurugubelli, T.R., and Viswanadham, B. (2024). Enhanced photo-catalytic degradation of MB dye over hydrothermal grown-up of ZnO/SnO_2 catalyst. *J. Chem.* https://doi.org/10.1155/2024/9135923.
5 Teli, M.D. (2008). Textile coloration industry in India. *Color. Technol.* 124 (1): 1–13. https://doi.org/10.1111/j.1478-4408.2007.00114.x.
6 Robinson, T., Mcmullan, G., Marchant, R., and Nigam, P. Remediation of dyes in textile effluent: a critical review on current treatment technologies with a proposed alternative. *Bioresour. Technol.* 77 (3): 247–255. https://doi.org/10.1016/S0960-8524(00)00080-8.
7 Rafatullah, M., Sulaiman, O., Hashim, R., and Ahmad, A. (2010). Adsorption of methylene Blue on low-cost adsorbents: a review. *J. Hazard. Mater.* 177 (1–3): 70–80. https://doi.org/10.1016/j.jhazmat.2009.12.047.
8 Honda, K. and FUJISHIMA, A. (1972). Electrochemical photolysis of water at a semiconductor electrode. *Nature* 238 (5358): 37–38. https://doi.org/10.1038/238037a0.
9 Fei, J., Han, Z., Deng, Y. et al. (2021). Enhanced photocatalytic performance of iron phthalocyanine/TiO_2 heterostructure at joint fibrous interfaces. *Colloids Surf. A Physicochem. Eng. Asp.* 625. https://doi.org/10.1016/j.colsurfa.2021.126901.
10 Bharati, B., Sonkar, A.K., Singh, N. et al. (2017). Enhanced photocatalytic degradation of dyes under sunlight using biocompatible TiO_2 nanoparticles. *Mater Res Express* 4 (8). https://doi.org/10.1088/2053-1591/aa6a36.

11 Khan, M.M., Ansari, S.A., Pradhan, D. et al. (2014). Band gap engineered TiO_2 nanoparticles for visible light induced photoelectrochemical and photocatalytic studies. *J. Mater. Chem. A. Mater.* 2 (3): 637–644. https://doi.org/10.1039/C3TA14052K.

12 Chen, H., Nanayakkara, C.E., and Grassian, V.H. (2012). Titanium dioxide photocatalysis in atmospheric chemistry. *Chem. Rev.* 112 (11): 5919–5948. https://doi.org/10.1021/cr3002092.

13 Sathiyan, K., Bar-Ziv, R., Mendelson, O., and Zidki, T. (2020). Controllable synthesis of TiO_2 nanoparticles and their photocatalytic activity in dye degradation. *Mater. Res. Bull.* 126. https://doi.org/10.1016/j.materresbull.2020.110842.

14 Wen, J., Li, X., Liu, W. et al. (2015). Photocatalysis fundamentals and surface modification of TiO_2 nanomaterials. *Chin. J. Catal.* 36 (12): 2049–2070. https://doi.org/10.1016/S1872-2067(15)60999-8.

15 Yu, X.Y. and (David) Lou, X.W. (2018). Mixed metal sulfides for electrochemical energy storage and conversion. *Adv. Energy Mater.* 8: 1701592. https://doi.org/10.1002/aenm.201701592.

16 Zhang, P., Wang, S., Guan, B.Y., and Lou, X.W.D. (2019). Fabrication of CdS hierarchical multi-cavity hollow particles for efficient visible light CO_2 reduction. *Energy Environ. Sci.* 12 (1): 164–168. https://doi.org/10.1039/C8EE02538J.

17 He, J., Chen, L., Wang, F. et al. (2016). CdS nanowires decorated with ultrathin MoS_2 nanosheets as an efficient photocatalyst for hydrogen evolution. *ChemSusChem* 9 (6): 624–630. https://doi.org/10.1002/cssc.201501544.

18 Kumari, S., Sharma, K., Korpal, S. et al. (2024). A comprehensive study on photocatalysis: materials and applications. *CrystEngComm*. https://doi.org/10.1039/D4CE00630E.

19 Ren, J.T., Yuan, K., Wu, K. et al. (2019). A robust CdS/In_2O_3 hierarchical heterostructure derived from a metal-organic framework for efficient visible-light photocatalytic hydrogen production. *Inorg. Chem. Front.* 6 (2): 366–375. https://doi.org/10.1039/c8qi01202d.

20 Huang, H., Shen, K., Chen, F., and Li, Y. (2020). Metal-organic frameworks as a good platform for the fabrication of single-atom catalysts. *ACS Catal.* 10 (12): 6579–6586. https://doi.org/10.1021/acscatal.0c01459.

21 Zhang, W., Ou, J.Z., Tang, S.Y. et al. (2014). Liquid metal/metal oxide frameworks. *Adv. Funct. Mater.* 24 (24): 3799–3807. https://doi.org/10.1002/adfm.201304064.

22 Wang, L., Zeng, T., Liao, G. et al. (2019). Syntheses, structures and catalytic mechanisms of three new MOFs for aqueous Cr(VI) reduction and dye degradation under UV light. *Polyhedron* 157: 152–162. https://doi.org/10.1016/j.poly.2018.09.064.

23 Guesh, K., Caiuby, C.A.D., Mayoral, Á. et al. (2017). Sustainable preparation of MIL-100(Fe) and its photocatalytic behavior in the degradation of methyl Orange in water. *Cryst. Growth Des.* 17 (4): 1806–1813. https://doi.org/10.1021/acs.cgd.6b01776.

24 Ai, L., Zhang, C., Li, L., and Jiang, J. (2014). Iron terephthalate metal-organic framework: revealing the effective activation of hydrogen peroxide for the degradation of organic dye under visible light irradiation. *Appl. Catal. B: Environ.* 148–149: 191–200. https://doi.org/10.1016/j.apcatb.2013.10.056.

25 Falcaro, P., Ricco, R., Yazdi, A. et al. (2016). Application of metal and metal oxide nanoparticles at MOFs. *Coord. Chem. Rev.* 307: 237–254. https://doi.org/10.1016/j.ccr.2015.08.002.

26 Mahmoodi, N.M. and Abdi, J. (2019). Nanoporous metal-organic framework (MOF-199): synthesis, characterization and photocatalytic degradation of basic Blue 41. *Microchem. J.* 144: 436–442. https://doi.org/10.1016/j.microc.2018.09.033.

27 Wang, A., Zhou, Y., Wang, Z. et al. (2016). Titanium incorporated with UiO-66(Zr)-type metal-organic framework (MOF) for photocatalytic application. *RSC Adv.* 6 (5): 3671–3679. https://doi.org/10.1039/C5RA24135A.

28 Du, J.J., Yuan, Y.P., Sun, J.X. et al. (2011). New photocatalysts based on MIL-53 metal-organic frameworks for the decolorization of methylene Blue dye. *J. Hazard. Mater.* 190 (1–3): 945–951. https://doi.org/10.1016/j.jhazmat.2011.04.029.

29 Kimijima, T., Kanie, K., Nakaya, M., and Muramatsu, A. (2014). Hydrothermal synthesis of size- and shape-controlled $CaTiO_3$ fine particles and their photocatalytic activity. *CrystEngComm* 16 (25): 5591–5597. https://doi.org/10.1039/C4CE00376D.

30 Olagunju, M.O., Poole, X., Blackwelder, P. et al. (2020). Size-controlled $SrTiO_3$ nanoparticles photodecorated with Pd cocatalysts for photocatalytic organic dye degradation. *ACS Appl. Nano Mater.* 3 (5): 4904–4912. https://doi.org/10.1021/acsanm.0c01086.

31 Zhong, W., Jiang, T., Dang, Y. et al. (2018). Mechanism studies on methyl Orange dye degradation by perovskite-type $LaNiO_{3-\Delta}$ under dark ambient conditions. *Appl. Catal. A. Gen.* 549: 302–309. https://doi.org/10.1016/j.apcata.2017.10.013.

32 Jayapandi, S., Lakshmi, D., Premkumar, S. et al. (2018). Augmented photocatalytic and electrochemical activities of Ag tailored $LaCoO_3$ perovskite semiconductor. *Mater. Lett.* 218: 205–208. https://doi.org/10.1016/j.matlet.2018.02.015.

33 Gildo-Ortiz, L., Rodríguez-Betancourtt, V.M., Blanco-Alonso, O. et al. (2019). A simple route for the preparation of nanostructured $GdCoO_3$ via the solution method, as well as its characterization and its response to certain gases. *Results Phys.* 12: 475–483. https://doi.org/10.1016/j.rinp.2018.11.072.

34 Djurišić, A.B., Leung, Y.H., and Ng, A.M.C. (2014). Strategies for improving the efficiency of semiconductor metal oxide photocatalysis. *Mater. Horiz.* 1 (4): 400–410. https://doi.org/10.1039/C4MH00031E.

35 Zu, D., Wei, H., Lin, Z. et al. (2024). The role of point defects in heterojunction photocatalysts: perspectives and outlooks. *Adv. Funct. Mater.* 34 (48). https://doi.org/10.1002/adfm.202408213.

36 Balapure, A., Ray Dutta, J., and Ganesan, R. (2024). Recent advances in semiconductor heterojunctions: a detailed review of the fundamentals of photocatalysis, charge transfer mechanism and materials. *RSC Appl. Interfaces* 1 (1): 43–69. https://doi.org/10.1039/D3LF00126A.

37 Huang, S., Bao, R., Wang, J. et al. (2023). Synergistic effect of oxygen vacancy defects and TiO_2/WO_3 heterostructures in photocatalytic hydrogen production and dye degradation. *J. Alloys Compd.* 961. https://doi.org/10.1016/j.jallcom.2023.170945.

38 Su, M., Chen, Y., Wang, L. et al. (2024). A Z-scheme WO_3/Bi2MoO_6 heterostructure with improved photocatalytic activity: the synergistic effect of heterojunction and oxygen vacancy defects. *J. Phys. Chem. Solids* 188. https://doi.org/10.1016/j.jpcs.2024.111947.

39 Gnanaguru, M.V.L., Parida, V.K., Ghangrekar, M.M. et al. (2024). Insights into the performance of binary heterojunction photocatalysts for degradation of refractory pollutants. *Environ. Sci. Pollut. Res.* 31 (8): 11349–11370. https://doi.org/10.1007/s11356-023-31592-4.

40 Sasikala, R., Gaikwad, A.P., Jayakumar, O.D. et al. (2015). Nanohybrid MoS_2-PANI-CdS photocatalyst for hydrogen evolution from water. *Colloids Surf. A Physicochem. Eng. Asp.* 481: 485–492. https://doi.org/10.1016/J.COLSURFA.2015.06.027.

41 Xu, H., Yi, J., She, X. et al. (2018). 2D heterostructure comprised of metallic 1T-MoS_2/monolayer O-g-C_3N_4 towards efficient photocatalytic hydrogen evolution. *Appl. Catal. B: Environ.* 220: 379–385. https://doi.org/10.1016/j.apcatb.2017.08.035.

42 Liu, M., Li, P., Wang, S. et al. (2021). Hierarchically porous hydrangea-like In_2S_3/In_2O_3 heterostructures for enhanced photocatalytic hydrogen evolution. *J. Colloid Interface Sci.* 587: 876–882. https://doi.org/10.1016/j.jcis.2020.11.048.

43 Huang, Y., Zhu, D., Zhang, Q. et al. (2018). Synthesis of a $Bi_2O_2CO_3/ZnFe_2O_4$ heterojunction with enhanced photocatalytic activity for visible light irradiation-induced NO removal. *Appl. Catal. B: Environ.* 234: 70–78. https://doi.org/10.1016/j.apcatb.2018.04.039.

44 Assis, G.C., Silva, I.M.A., Dos Santos, T.V. et al. (2021). Photocatalytic properties of SnO_2/MoO_3 mixed oxides and their relation to the electronic properties and surface acidity. *J. Photochem. Photobiol. A Chem.* 407: 113035. https://doi.org/10.1016/j.jphotochem.2020.113035.

45 Li, J., Han, M., Guo, Y. et al. (2016). Fabrication of $FeVO_4/Fe_2TiO_5$ composite catalyst and photocatalytic removal of norfloxacin. *Chem. Eng. J.* 298: 300–308. https://doi.org/10.1016/j.cej.2016.03.107.

46 Zhao, J., Zhao, Z., Li, N. et al. (2018). Visible-light-driven photocatalytic degradation of ciprofloxacin by a ternary $Mn_2O_3/Mn_3O_4/MnO_2$ valence state heterojunction. *Chem. Eng. J.* 353: 805–813. https://doi.org/10.1016/j.cej.2018.07.163.

47 Jia, Y., Zhang, W., Yeon Do, J. et al. (2020). Z-scheme $SnFe_2O_4/\alpha\text{-}Fe_2O_3$ micro-octahedron with intimated interface for photocatalytic CO_2 reduction. *Chem. Eng. J.* 402. https://doi.org/10.1016/j.cej.2020.126193.

48 Inceesungvorn, B., Teeranunpong, T., Nunkaew, J. et al. (2014). Novel $NiTiO_3/Ag_3VO_4$ composite with enhanced photocatalytic performance under visible light. *Catal. Commun.* 54: 35–38. https://doi.org/10.1016/j.catcom.2014.05.015.

49 Rafiq, U., Mehraj, O., Lone, S. et al. (2020). Solvothermal synthesis of Ag_2WO_4/Sb_2WO_6 heterostructures for enhanced charge transfer properties and efficient visible-light-driven photocatalytic activity and stability. *J. Environ. Chem. Eng.* 8 (5): 104301. https://doi.org/10.1016/j.jece.2020.104301.

50 Hu, T., Dai, K., Zhang, J. et al. (2019). One-pot synthesis of step-scheme Bi_2S_3/porous g-C_3N_4 heterostructure for enhanced photocatalytic performance. *Mater. Lett.* 257: 126740. https://doi.org/10.1016/j.matlet.2019.126740.

51 Mei, F., Dai, K., Zhang, J. et al. (2019). Construction of Ag SPR-promoted step-scheme porous g-C_3N_4/Ag_3VO_4 heterojunction for improving photocatalytic activity. *Appl. Surf. Sci.* 488: 151–160. https://doi.org/10.1016/j.apsusc.2019.05.257.

52 Wang, R., Shen, J., Zhang, W. et al. (2020). Build-in electric field induced step-scheme $TiO_2/W_{18}O_{49}$ heterojunction for enhanced photocatalytic activity under visible-light irradiation. *Ceram. Int.* 46 (1): 23–30. https://doi.org/10.1016/j.ceramint.2019.08.226.

53 Jia, X., Han, Q., Zheng, M., and Bi, H. (2019). One pot milling route to fabricate step-scheme AgI/I-BiOAc photocatalyst: energy band structure optimized by the formation of solid solution. *Appl. Surf. Sci.* 489: 409–419. https://doi.org/10.1016/j.apsusc.2019.05.361.

54 Xu, Q., Ma, D., Yang, S. et al. (2019). Novel g-C_3N_4/g-C_3N_4 S-scheme isotype heterojunction for improved photocatalytic hydrogen generation. *Appl. Surf. Sci.* 495: 143555. https://doi.org/10.1016/j.apsusc.2019.143555.

55 Mu, J., Teng, F., Miao, H. et al. (2020). In-situ oxidation fabrication of 0D/2D SnO_2/SnS_2 novel Step-scheme heterojunctions with enhanced photoelectrochemical activity for water splitting. *Appl. Surf. Sci.* 501: 143974. https://doi.org/10.1016/j.apsusc.2019.143974.

56 Wang, J., Zhang, Q., Deng, F. et al. (2020). Rapid toxicity elimination of organic pollutants by the photocatalysis of environment-friendly and magnetically recoverable step-scheme $SnFe_2O_4/ZnFe_2O_4$ nano-heterojunctions. *Chem. Eng. J.* 379: 122264. https://doi.org/10.1016/j.cej.2019.122264.

57 He, F., Meng, A., Cheng, B. et al. (2020). Article (special issue on photocatalytic H_2-production and CO_2 reduction) enhanced photocatalytic H_2-production activity of WO_3/TiO_2 step-scheme heterojunction by graphene modification. *Chin. J. Catal.* 41: 9–20. https://doi.org/10.1016/S1872-2067(19)63382-6.

58 Li, J., Wang, X., Fang, H. et al. (2023). Unraveling the role of surface and interfacial defects in hydrogen production to construct an all-in-one broken-gap photocatalyst. *J. Mater. Chem. A. Mater.* 11 (46): 25639–25649. https://doi.org/10.1039/D3TA03079B.

59 Li, M., Wang, P., Ji, Z. et al. (2021). Efficient photocatalytic oxygen activation by oxygen-vacancy-rich CeO_2-based heterojunctions: synergistic effect of photoexcited electrons transfer and oxygen chemisorption. *Appl. Catal. B: Environ.* 120020. https://doi.org/10.1016/j.apcatb.2021.120020.

60 Shi, J., Chao, C., Gao, Y. et al. (2023). Oxygen vacancies assisted ferroelectric heterojunction for enhanced photocatalytic activity. *Appl. Surf. Sci.* 624: 157144. https://doi.org/10.1016/j.apsusc.2023.157144.

61 Zu, D., Ying, Y., Wei, Q. et al. (2024). Oxygen vacancies trigger rapid charge transport channels at the engineered interface of S-scheme heterojunction for boosting photocatalytic performance. *Angew. Chem. Int. Ed.* 63 (31): e202405756. https://doi.org/10.1002/anie.202405756.

62 Wu, S., Yu, X., Zhang, J. et al. (2021). Construction of BiOCl/$CuBi_2O_4$ S-scheme heterojunction with oxygen vacancy for enhanced photocatalytic diclofenac degradation and nitric oxide removal. *Chem. Eng. J.* 411: 128555. https://doi.org/10.1016/j.cej.2021.128555.

63 Liu, G., Hou, G., Mao, X. et al. (2022). Rational design of CeO_2/$Bi7O_9I_3$ flower-like nanosphere with Z-scheme heterojunction and oxygen vacancy for enhancing photocatalytic activity. *Chem. Eng. J.* 431: 133254. https://doi.org/10.1016/j.cej.2021.133254.

64 Nie, J., Zhu, G., Zhang, W. et al. (2021). Oxygen vacancy defects-boosted deep oxidation of NO by β-Bi_2O_3/$CeO_{2-\delta}$ *p-n* heterojunction photocatalyst in situ synthesized from Bi/Ce(CO_3)(OH) precursor. *Chem. Eng. J.* 424: 130327. https://doi.org/10.1016/j.cej.2021.130327.

65 Xie, Z., Yang, J., Wang, K. et al. (2022). Facile fabrication of TiO_2-SiO_2-C composite with anatase/rutile heterostructure via sol-gel process and its enhanced photocatalytic activity in the presence of H_2O_2. *Ceram. Int.* 48 (7): 9114–9123. https://doi.org/10.1016/j.ceramint.2021.12.096.

66 Zhang, Q., Zhao, X., Duan, L. et al. (2020). Controlling oxygen vacancies and enhanced visible light photocatalysis of CeO_2/ZnO nanocomposites. *J. Photochem. Photobiol, A* 392: 112156. https://doi.org/10.1016/j.jphotochem.2019.112156.

67 Li, R.P., Yu, S.M., and Zhao, Z.Y. (2023). Boosting photoelectrochemical performance of $CuFeO_2$/CuO photocathode by modulating heterojunction architecture and oxygen vacancies. *Appl. Surf. Sci.* 640: 158392. https://doi.org/10.1016/j.apsusc.2023.158392.

68 Paramanik, L., Reddy, K.H., and Parida, K.M. (2019). Stupendous photocatalytic activity of *p*-BiOI/*n*-$PbTiO_3$ heterojunction: the significant role of oxygen vacancies and interface coupling. *J. Phys. Chem. C* 123 (35): 21593–21606. https://doi.org/10.1021/acs.jpcc.9b05747.

69 Zhang, X., Zha, X., Luo, Y. et al. (2023). In_2O_3 nanoparticle/$Bi_4O_5Br_2$ nanosheet S-scheme heterojunctions with interfacial oxygen vacancies for photocatalytic degradation of tetracycline. *ACS Appl. Nano Mater.* 6 (13): 11877–11887. https://doi.org/10.1021/acsanm.3c01720.

70 Yu, S., Zhang, H., and Wu, D. (2025). Highly efficient CoAl-LDH/(110) facet-exposed BiOBr catalysts: promotional effect of Z-type heterojunction and oxygen vacancies in photocatalytic ciprofloxacin degradation. *Mater. Sci. Semicond. Process.* 186: 109105. https://doi.org/10.1016/j.mssp.2024.109105.

71 Kong, X.Y., Lee, W.Q., Mohamed, A.R., and Chai, S.P. (2019). Effective steering of charge flow through synergistic inducing oxygen vacancy defects and pn heterojunctions in 2D/2D surface-engineered Bi_2WO_6/BiOI cascade: towards superior photocatalytic CO_2 reduction activity. *Chem. Eng. J.* 372: 1183–1193. https://doi.org/10.1016/j.cej.2019.05.001.

72 Liu, C., Xu, C., Wang, W. et al. (2023). Oxygen vacancy mediated band-gap engineering via B-doping for enhancing Z-scheme A-TiO_2/R-TiO_2 heterojunction photocatalytic performance. *Nanomaterials* 13 (5): 794. https://doi.org/10.3390/nano13050794.

73 Uddin, M.T., Nicolas, Y., Olivier, C. et al. (2017). Band alignment investigations of heterostructure NiO/TiO_2 nanomaterials used as efficient heterojunction earth-abundant metal oxide photocatalysts for hydrogen production. *Phys. Chem. Chem. Phys.* 19 (29): 19279–19288. https://doi.org/10.1039/C7CP01300K.

74 Backer, S.N., Oussadou, S.E., Almanassra, I.W. et al. (2023). Effect of energy band alignments in carbon doped ZnO/TiO_2 hybrid heterojunction photocatalyst on the photodegradation of ofloxacin. *Results Eng.* 20: 101432. https://doi.org/10.1016/j.rineng.2023.101432.

75 Yang, X., Zhang, Z., Wu, J. et al. (2023). Fabrication of $BiOIO_3$/g-C_3N_4/Er^{3+} containing oxygen vacancies for photocatalytic oxidation of gaseous elemental mercury. *Mater. Res. Bull.* 167: 112376. https://doi.org/10.1016/j.materresbull.2023.112376.

5

Defect-Engineered Metal-Organic Frameworks for Photocatalytic Implementations

Zhiting Chen and Qiuhui Zhu

Wuyi University, College of Ecology and Resources Engineering, Fujian Provincial Key Laboratory of Eco-Industrial Green Technology, No. 358 Baihua Road, Wuyishan, 354300, PR China

5.1 Introduction

Metal-organic frameworks (MOFs) are a groundbreaking class of hybrid materials characterized by their crystalline porous architectures, composed of metal ions or clusters interconnected by organic ligands [1, 2]. The modular nature of MOFs allows for unparalleled control over their structural parameters [3, 4]. For instance, MOF-5 ($Zn_4O(BDC)_3$) exhibits a cubic structure with a Brunauer–Emmett–Teller (BET) surface area of ~3800 $m^2\,g^{-1}$, while NU-1101, a zirconium-based MOF, holds the record for the highest surface area at 7140 $m^2\,g^{-1}$, surpassing traditional porous materials like zeolites and activated carbon [5, 6]. These exceptional properties have enabled MOFs to excel in diverse applications, such as gas storage [7, 8], heterogeneous catalysis [9], and molecular separation [10] (Figure 5.1). However, conventional MOFs face inherent limitations: (i) their periodic, highly ordered structures often lack accessible active sites for catalytic reactions, particularly in large-molecule transformations; (ii) insulating organic linkers hinder efficient charge transport, limiting their utility in energy-intensive applications like photocatalysis and electrocatalysis [11]. To address these challenges, defect engineering has emerged as a transformative strategy, introducing controlled structural imperfections to enhance functionality while preserving structural integrity.

Defect engineering involves the deliberate introduction of structural imperfections – such as oxygen vacancies, missing ligands, metal node vacancies, or even extended defects like grain boundaries – into MOFs to tailor their physicochemical properties. These defects disrupt the ideal crystalline lattice, creating localized electronic states that enhance surface reactivity, charge separation efficiency, and mass transport [13]. Beyond enhancing existing properties, defects can impart novel functionalities. For instance, light-responsive defects in azobenzene-functionalized MOFs enable dynamic pore opening/closing under ultraviolet (UV) irradiation, offering potential for smart gas separation systems [15]. By strategically balancing defect density and structural stability, defect engineering bridges the gap between

Defect Engineering in Photocatalysis: From Fundamentals to Applications,
First Edition. Edited by Chuanyi Wang, Abdelkader Labidi, and Qiuhui Zhu.

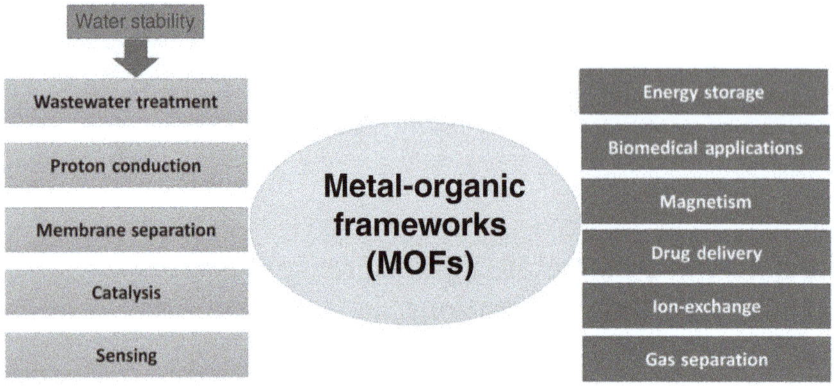

Figure 5.1 Various applications of pristine MOFs and water-stable MOFs. *Source:* Reproduced from Khan et al. [12] / with permission of Springer Nature.

idealized MOF models and real-world performance requirements, unlocking applications in energy conversion [16], environmental remediation [17], and biomedicine [18].

This chapter will offer knowledge about the process of designing oxygen vacancies in MOFs and depict their application in the photocatalysis process for various purposes, including energy conversion and environmental remediation processes, with a focus on their advantages and disadvantages.

5.2 Synthesis Strategies and Characterizations

5.2.1 Synthesis Strategies of Defective MOFs

Defect engineering in MOFs is achieved through two primary methodologies: "post-synthetic modification (PSM)" and "in situ defect introduction." Each approach offers unique advantages and challenges, as detailed below.

PSM of MOFs is a crucial technique. It modifies synthesized MOFs through covalent or non-covalent methods. Covalent modification adds functional groups to ligands, while non-covalent modification introduces guest molecules via weak interactions. PSM offers precise control over MOF structures and properties without disrupting their frameworks (Figure 5.2). This technique is applied in gas adsorption and separation, enhancing selectivity; in catalysis, improving catalytic performance; and in sensing, increasing sensitivity to target analytes.

Chemical etching: acidic or alkaline treatments selectively remove ligands or metal nodes. For example, treating MIL-125-NH_2(Ti) with propionic acid at 90 °C simultaneously introduces oxygen vacancies and hierarchical pores, enhancing the photocatalytic degradation of rhodamine B as the rate constant of H-MIL-125-NH_2-V_O is about eight times that of pristine MIL-125-NH_2 [14]. However, over-etching of Ti-MOF leads to the growth and coalescence of pre-existing defects and the creation of new small defects. This phenomenon significantly

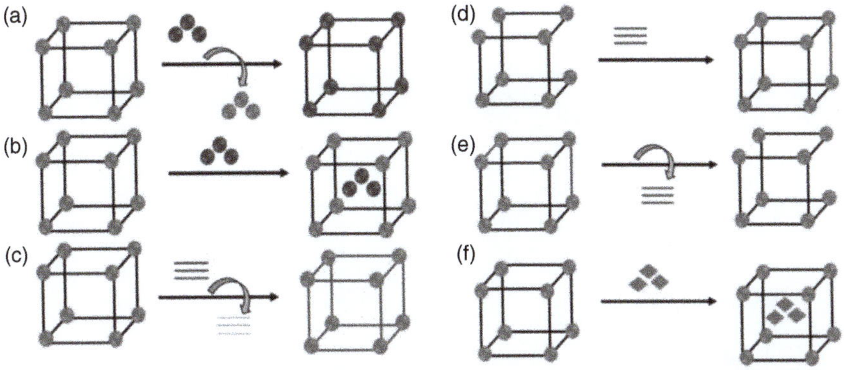

Figure 5.2 Possible pathways of PSM in MOFs: (a) metal exchange, (b) metal incorporation, (c) ligand exchange, (d) ligand installation, (e) ligand removal, and (f) guest incorporation inside the pores. *Source:* Reproduced from Mandal et al. [22] / with permission of John Wiley & Sons.

impairs the photocatalytic hydrogen evolution performance, as evidenced by the decreased hydrogen generation rate and AQY in over-etched samples. Moreover, it also alters the material's surface properties, such as increasing its hydrophobicity and roughness [19].

Thermal annealing: heating MOFs under inert or reducing atmospheres induces ligand decomposition. Three iron-based MOFs (MIL-53, MIL-88B, and MIL-101) annealed at 350 °C lose partial dicarboxylate ligands, exposing Fe–O clusters and forming oxygen vacancies. This structural change leads to an increase in the active sites of Fe–O clusters on the catalyst surface. The annealed derivatives show significantly improved thermal catalytic degradation performance of tetracycline (TC) than the raw materials, with the optimum sample achieving a 94.22% degradation efficiency for TC within 60 minutes at 70 °C [20].

Plasma treatment: non-thermal plasma selectively breaks metal–ligand bonds. For instance, the MO@Fe/Ce-MOFs obtained after dielectric barrier discharge (DBD) plasma treatment and calcination exhibited more efficient visible-light photocatalytic performance than Fe/Ce-MOFs. Specifically, MO@Fe/Ce-MOFs-b had a photocatalytic nitrogen-fixing efficiency of 229 $\mu mol\ L^{-1}\ h^{-1}$ under visible light and high cyclic stability, with the NH_4^+ generation rate remaining approximately 95% after three cycles [21].

In situ defect introduction in MOFs is a method of creating defects directly during the synthesis of MOFs. By adjusting reaction conditions like reactant ratios, temperature, time, or adding specific additives, various defects can be precisely introduced. This approach increases the number of active sites in MOFs, thus enhancing their performance in gas adsorption and catalysis. For instance, it can improve the catalytic activity and gas adsorption selectivity. Also, in situ-introduced defects are well-integrated with the MOF structure, ensuring the material's stability. It shows great potential in energy and environmental applications.

Solvent and pH modulation: adjusting solvent polarity or pH during synthesis destabilizes coordination equilibria. High-concentration oxygen vacancies were in

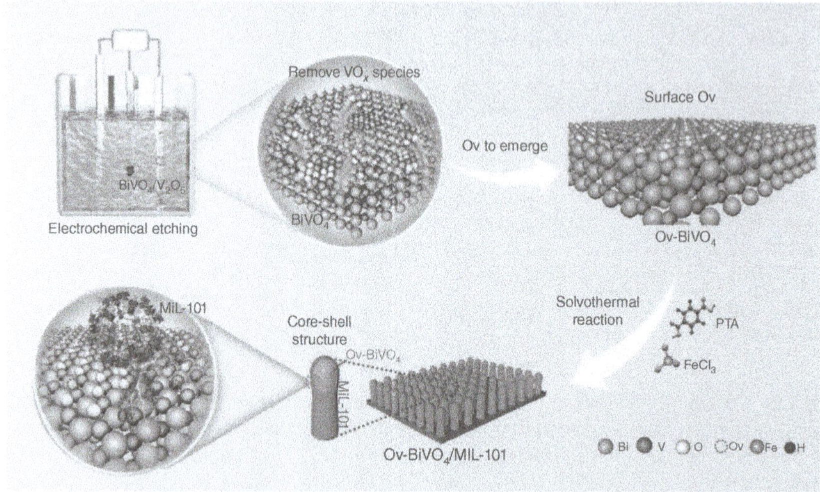

Figure 5.3 Schematic illustration of the synthesis process of Ov-BiVO4/MIL-101 photoanodes. *Source:* Reproduced from Xin et al. [23] / with permission of John Wiley & Sons.

situ-generated on the surface of $BiVO_4$ via electrochemical-assisted pH modulation, resulting in the formation of Ov-$BiVO_4$ (Figure 5.3). Subsequently, MIL-101 was grown on the surface of Ov-$BiVO_4$, successfully preparing the Ov-$BiVO_4$/MIL-101 composite. The as-prepared Ov-$BiVO_4$/MIL-101 composite demonstrates outstanding photocatalytic properties. It attains a photocurrent density of 5.91 mA cm^{-2} at 1.23 V_{RHE}, exhibits excellent stability, and reaches a solar-to-hydrogen conversion efficiency of 4.33% when integrated into an unbiased tandem device [23].

Ligand defect design: the Ni-MOF-74 material was modified through temperature and solvent modulation to prepare Ni-74-Am with a highly porous structure. This material contains more unsaturated Ni active sites generated by defects. Ni-74-Am exhibits excellent photocatalytic performance. Under visible light, it has a CO generation rate of 1380 μmol g^{-1} h^{-1} and a CO selectivity of 94%. Moreover, it has a higher photogenerated electron–hole separation efficiency and a faster carrier migration rate, significantly outperforming most of the reported MOF-based photocatalysts [24].

5.2.2 Characterization Techniques

Defects in MOFs directly influence porosity, stability, and reactivity, yet their uncontrolled presence can compromise performance. Precise characterization bridges the gap between defect engineering and functional outcomes, enabling tailored material design. Moreover, correlating defect characteristics with performance metrics guides synthesis strategies, minimizing undesired structural flaws. As defect engineering evolves, robust characterization becomes indispensable for unlocking MOFs' full potential, ensuring reproducibility, and driving innovations in next-generation functional materials.

X-ray diffraction (XRD) is mainly used in the characterization of oxygen vacancies to analyze lattice distortion, phase transition behavior, and doping effects and indirectly reflects the existence and distribution of oxygen vacancies. Chen et al. prepared a carbon-coated Ti^{3+} and oxygen vacancy co-doped anatase/rutile TiO_2 (A/R-TiO_2) composite via a one-step calcination method for photocatalytic degradation of antibiotics. XRD characterization confirmed the presence of Ti^{3+} in the material, indicating that the A/R mixed-phase structure might introduce abundant defect sites, providing indirect evidence for the existence of oxygen vacancies [25].

Electron paramagnetic resonance (EPR) is a simple and highly sensitive characterization technique for oxygen vacancies, based on paramagnetic samples with unpaired electrons [26]. Liu et al. analyzed by electron paramagnetic resonance (EPR) and indicated that during the ligand-selective pyrolysis of MIL-53, the calcination caused the Fe—O bond to break, generating oxygen vacancies and coordinatively unsaturated iron sites. This demonstrates that the pyrolysis treatment successfully introduced oxygen vacancies into the H-MIL-53 structure. The results of EPR analysis provide an important basis for understanding the formation mechanism of oxygen vacancies in the H-MIL-53 catalyst and their role in catalytic reactions [27].

X-ray absorption fine structure (XAFS) is an effective method for determining electronic structures and identifying oxygen vacancies, providing structural insights into defects, atomic coordination numbers, and additional structural details in materials [28]. Through XAFS analysis, after introducing oxygen vacancies on the surface of $BiVO_4$ (forming Ov-$BiVO_4$), the coordination environment around Bi atoms changes. Compared with the original $BiVO_4$, the Bi—O bond length in Ov-$BiVO_4$ increases. This indicates that the generation of oxygen vacancies disrupts the original crystal structure of $BiVO_4$ and alters the interaction between Bi atoms and the surrounding oxygen atoms, providing a structural basis for the subsequent molecular epitaxial growth with MIL-101 [23].

Density function theory (DFT) is a computational method for investigating the electronic structure of materials and is widely applied in physics, computational chemistry, and materials science [29]. Using DFT simulation analysis, Hussain et al. found that more electrons accumulate at the Pt-Ov site. This means that the oxygen vacancy changes the electron distribution of the system, making this site an electron-enriched region. This electron-accumulation phenomenon has a significant impact on the photocatalytic reaction and provides favorable conditions for subsequent charge transfer and chemical reactions [30].

5.3 Photocatalytic Applications

Defective MOFs exhibit exceptional potential in photocatalysis for energy and environmental applications. In energy conversion, defects enhance light absorption and charge separation, enabling efficient hydrogen evolution and CO_2-to-fuel conversion. In environmental remediation, defective MOFs degrade organic pollutants and adsorb heavy metals via tailored porosity and redox-active defect sites. Their mixed-metal or ligand-functionalized variants achieve visible-light degradation of persistent contaminants. Synergistic catalytic systems amplify performance: coupling defective MOFs with semiconductors forms Z-scheme heterojunctions, improving charge transfer, while embedding metal nanoparticles enhances interfacial kinetics for complex reactions. Challenges include precise defect control and stability optimization, yet advances in defect engineering and hybrid designs promise scalable solutions for sustainable energy and pollution mitigation.

5.3.1 Energy Conversion

Hydrogen production: a sustainable method utilizing light-absorbing catalysts to split water into hydrogen under solar energy, enabling clean, renewable fuel generation with zero carbon emissions. Pt-modified UiO-66-NH_2 materials with different oxygen vacancy contents were synthesized via the solvothermal method and high-temperature annealing method [30]. High-concentration oxygen vacancies not only make the Pt nanoparticles more stable on the UiO-66-NH_2 substrate but also enhance the interaction between the Pt nanoparticles and the substrate. As shown in Figure 5.4a, ESR of all photocatalysts shows a peak at $g = 2.0023$, with intensity increasing in the order UiO-66-NH_2 > Pt/UN-Ovl > Pt/UN-Ovh. PL

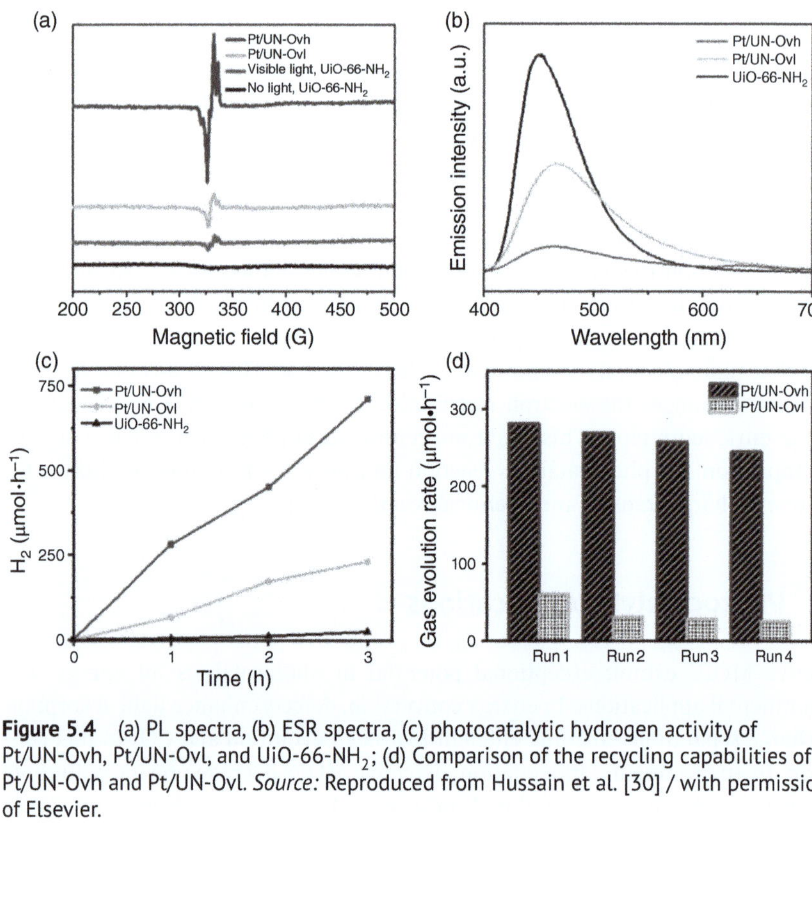

Figure 5.4 (a) PL spectra, (b) ESR spectra, (c) photocatalytic hydrogen activity of Pt/UN-Ovh, Pt/UN-Ovl, and UiO-66-NH_2; (d) Comparison of the recycling capabilities of Pt/UN-Ovh and Pt/UN-Ovl. *Source:* Reproduced from Hussain et al. [30] / with permission of Elsevier.

emission spectroscopy provided insights into photoexcited charge transfer and recombination. The PL spectrum of UiO-66-NH_2 is weaker than Pt/UN-Ovl and significantly suppressed in Pt/UN-Ovh, suggesting better suppression of electron–hole recombination in Pt/UN-Ovh (Figure 5.4b). Pt/UN-Ovl has a photocatalytic efficiency ~20 times higher than UiO-66-NH_2 (66.2 mmol g^{-1}), and Pt/UN-Ovh has an efficiency ~82 and 4.2 times higher than UiO-66-NH_2 and Pt/UN-Ovl, respectively (281.13 mmol g^{-1}) (Figure 5.4c). Recycling tests for Pt/UN-Ovh show consistent hydrogen production rates over four catalytic rounds (Figure 5.4d). In summary, the introduction of oxygen vacancies contributed to the separation of photocatalytic carriers, thereby improving the photocatalytic H_2 evolution.

CO_2 reduction: a catalytic process converting carbon dioxide into hydrocarbons or value-added chemicals using renewable energy-driven systems, offering a sustainable strategy to combat climate change and close the carbon cycle. Zhang et al. [31] synthesized a chemically stable La-MOF material modified by Ni-doped metal-oxygen clusters through a one-pot solvothermal reaction and applied it to photocatalytic CO_2 reduction. As shown in Figure 5.5a, EPR showed no signal in La-MOF and Ni-MOF, but La/Ni-MOF-X exhibited a significant EPR signal ($g = 2.004$) increasing with Ni content, indicating O vacancy formation due to Ni^{2+} substitution. Then, UV-Vis diffuse reflectance spectroscopy (DRS) revealed La-MOF and La/Ni-MOF-3 absorb below 400 nm, while Ni-MOF has bands at 400–500 and 600–700 nm (Figure 5.5b). As shown in Figure 5.5c, photocatalytic performance for CO_2 reduction was evaluated with acetonitrile, $[Ru(bpy)_3]Cl_2{\cdot}6H_2O$, and TEOA as sacrificial agents. Only gaseous hydrocarbons were observed, with CO yields of 252.9 and 136.3 μmol g^{-1} for La-MOF and Ni-MOF, respectively. La/Ni-MOF-3 achieved an optimal CO production rate (669.3 μmol g^{-1}), with 96.8% CO selectivity. Control experiments confirmed the catalyst's role, the necessity of CO_2, and the essential conditions for the reaction (Figure 5.5e). In addition, cycling tests showed La/Ni-MOF-3's stability and durability in CO_2 reduction (Figure 5.5f) [31].

5.3.1.1 Nitrogen Fixation

Nitrogen fixation: The pursuit of technologies for ammonia production under mild conditions represents a critically important scientific goal. In photocatalytic nitrogen fixation reactions, solar energy serves as the sole driving force, while water can function as both the reaction solvent and proton donor. The entire process eliminates the need for costly electron donors, organic solvents, strong acids, as well as thermal or electrical energy supplies, positioning it as one of the promising pathways for future green ammonia production. MOFs as photocatalysts, characterized by their porous structures and abundant surface-active sites, are widely utilized in photocatalytic nitrogen fixation reactions. Researchers have found that the introduction of oxygen vacancies helps to enhance the activity of MOFs in photocatalytic nitrogen fixation.

Pang and coworkers [32] employed a thermal calcination method to generate oxygen vacancies in proximity to the metal nodes of the material, ensuring the structural integrity remained undisturbed. The process was accompanied by an initial surge and subsequent decline in porosity, as illustrated in Figure 5.6a. And then, EPR

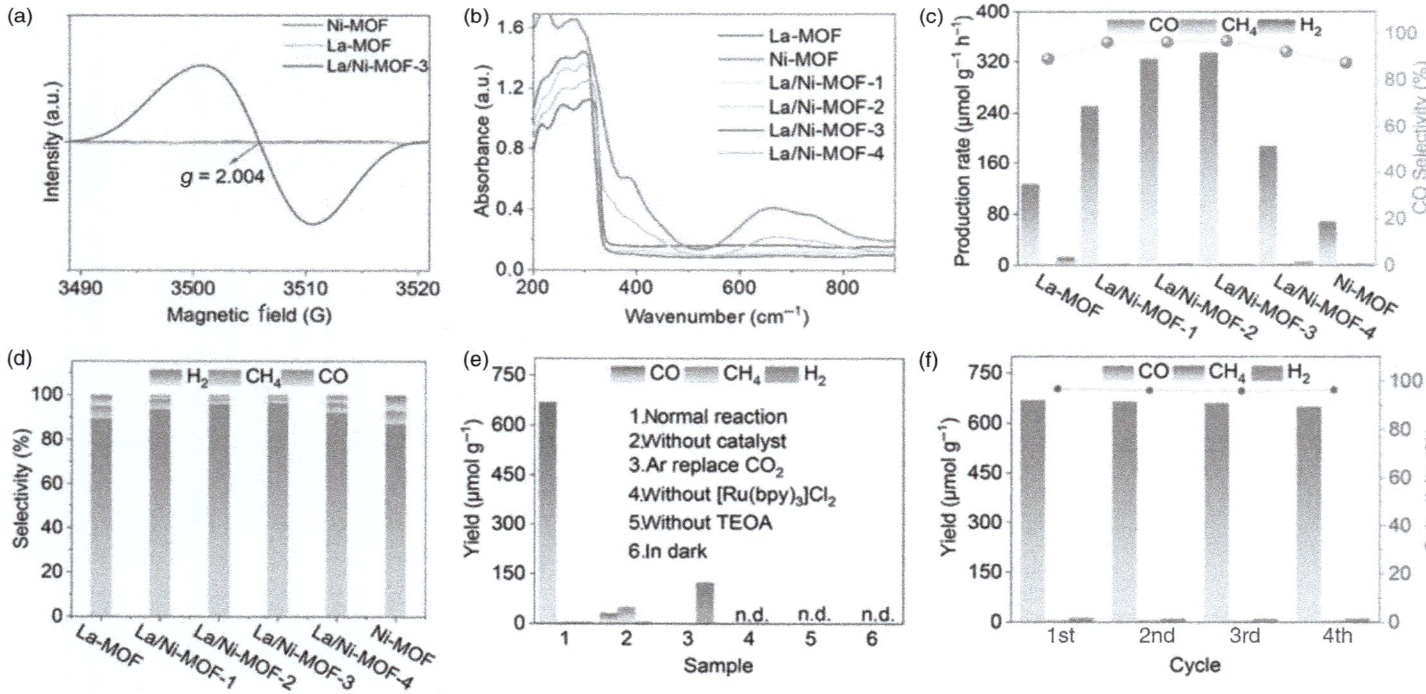

Figure 5.5 (a) EPR spectra and (b) DRS of as-prepared samples; (c) photocatalytic activity for CO_2 reduction of La-MOF, Ni-MOF, and La/Ni-MOF-X; (d) photocatalytic selectivity for CO_2 reduction of La-MOF, Ni-MOF, and La/Ni-MOF-X; (e) control experiments of the photocatalytic CO_2 reduction performance over La/Ni-MOF-3; (f) the photocatalytic yield and CO selectivity of the reduction products obtained on La/Ni-MOF-3 in four cycling tests. *Source:* Reproduced from Zhang et al. [31] / with permission of Royal Society of Chemistry.

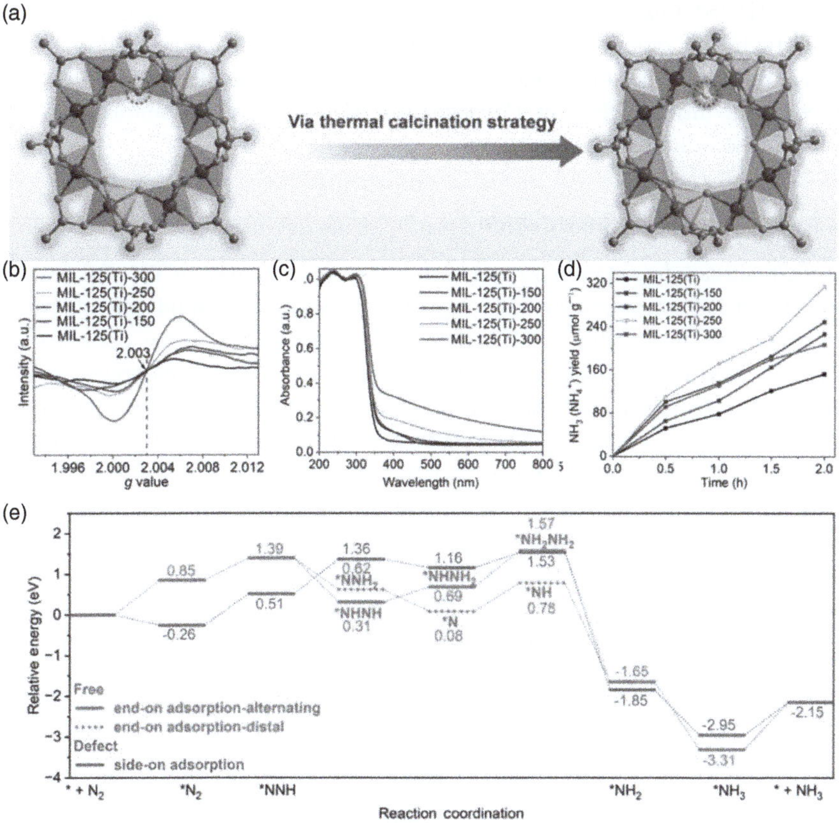

Figure 5.6 (a) The solvothermal preparation process of MIL-125(Ti); (b) UV-Vis DRS, (c) EPR spectra and (d) photocatalytic NH_3 evolution of MIL 125(Ti) and MIL125(Ti) T; (e) Computed free energy diagrams for intermediates in the conversion from N_2 to NH_3. *Source:* Reproduced from Sun et al. [32] / with permission of John Wiley & Sons.

was conducted to quantify the oxygen vacancies across all MIL-125(Ti)-T samples (Figure 5.6b). The results revealed a gradual intensification at 2.003, indicative of an escalating concentration of oxygen vacancies in MIL-125(Ti) under oxygen-deficient heating conditions. Furthermore, DRS was utilized to assess the optical properties. Figure 5.6c demonstrates that the absorption edge of MIL-125(Ti)-T shifts from 350 to 380 nm with an increase in heating temperature. Photocatalytic experiments were conducted, with the concentration of produced NH_4^+ ions determined using Nessler's reagent. The photocatalytic NH_3 production rate for pristine MIL-125(Ti) was found to be 76.2 μmol g^{-1}. Notably, MIL-125(Ti)-250 exhibited the highest ammonia generation rate of 156.9 μmol g^{-1} among all MIL-125(Ti)-T samples, approximately double that of pristine MIL-125(Ti) (Figure 5.6d). To gain insights into the role of oxygen vacancies in photocatalytic nitrogen fixation, theoretical calculations were performed. As shown in Figure 5.6e, the last step (*NH_3 to NH_3) was identified as the rate-determining step. For both the alternating and distal hydrogenation pathways, the energy barrier for MIL-125(Ti) was 1.16 eV. However, for

MIL-125(Ti)-250, this energy barrier was reduced to 0.80 eV. This underscores the synergistic impact of enhanced porosity and oxygen vacancy-induced Ti^{3+}, which significantly diminishes the challenge associated with the rate-determining step in photocatalytic nitrogen fixation. Consequently, MIL-125(Ti)-250 demonstrated an improved photocatalytic efficiency.

5.3.2 Environmental Remediation

Organic pollutant degradation: Zhang et al. synthesized a SnO_{2-x}/CD-MOF (SCM) heterostructured photocatalyst with oxygen vacancies via a mild solvothermal method for the degradation of Rhodamine B (RhB) and TC. The presence of oxygen vacancies can adjust the bandgap of SnO_2 and enhance its optical properties. Under visible-light irradiation, its removal efficiency for RhB is as high as 93.5% with a degradation rate constant of 0.0484 min^{-1}, and the removal efficiency for TC is 83.1% with a degradation rate constant of 0.0246 min^{-1} [33] (Figure 5.7). Qu et al. synthesized $Co_xFe_{3-x}O_4$ nanocrystals with oxygen vacancies via a one-step pyrolysis method, using cobalt-iron bimetallic organic framework (Co_xFe_{3-x} bi-MOF) as a template. These nanocrystals were then assembled onto

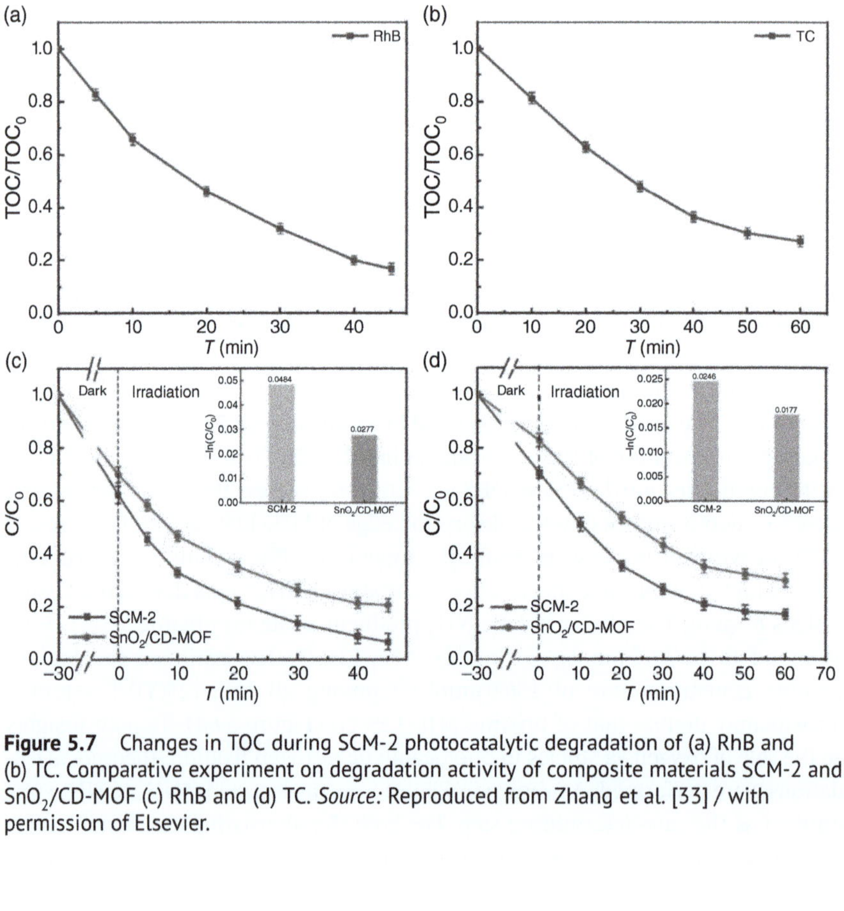

Figure 5.7 Changes in TOC during SCM-2 photocatalytic degradation of (a) RhB and (b) TC. Comparative experiment on degradation activity of composite materials SCM-2 and SnO_2/CD-MOF (c) RhB and (d) TC. *Source:* Reproduced from Zhang et al. [33] / with permission of Elsevier.

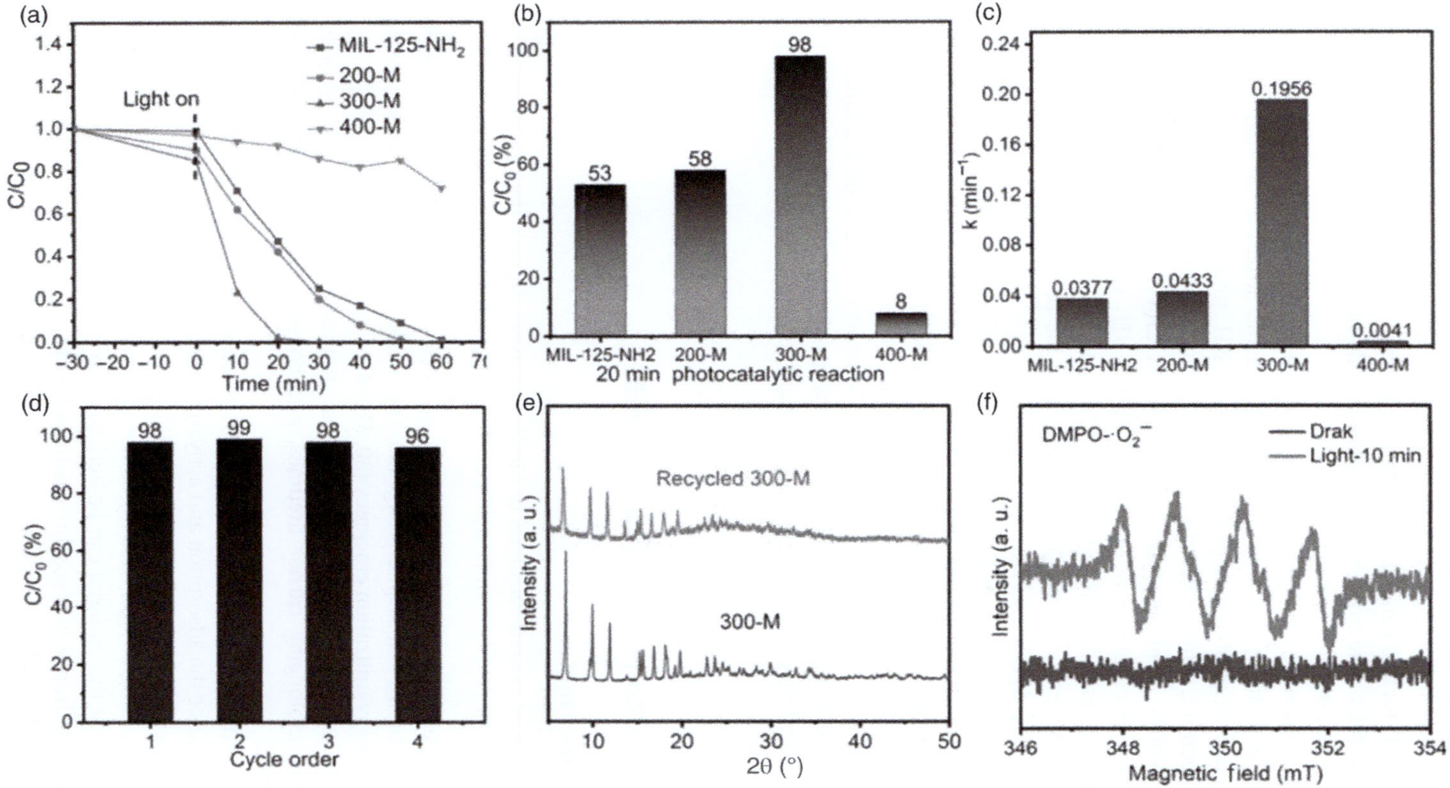

Figure 5.8 Photocatalytic efficiency (a, b) and corresponding *K* values (c) of MIL-125-NH2, 200-M, 300-M, and 400-M. (d) Recycle performance. (e) PXRD of 300-M and the recycled one. (f) ESR of 300-M. *Source:* Reproduced from Ma et al. [35] / with permission of Elsevier.

a polyvinylidene fluoride (PVDF) membrane to prepare the $Co_xFe_{3-x}O_4$/PVDF composite membrane. The composite membrane was used to activate peroxymonosulfate (PMS) for the catalytic degradation of emerging pollutants, and achieved favorable results. The prepared composite membrane has oxygen vacancies and exhibits good anti-interference ability [34].

Heavy metal removal: by controlling the heat treatment temperature, oxygen vacancies and mesopores were introduced into the structure of MIL-125-NH_2, resulting in a series of oxygen-deficient MIL-125-NH_2 photocatalysts with different defect structures, which were used for the photocatalytic reduction of Cr(VI) in sewage. Within 20 minutes, the removal rate of Cr(VI) was as high as 98%. In contrast, the removal rate of untreated MIL-125-NH_2 was only 53% (Figure 5.7). This indicates that the oxygen vacancies and mesopores introduced by controlled heat treatment play a crucial role in promoting the photoreduction of Cr(VI) by MIL-125-NH_2 [35] (Figure 5.8).

Volatile organic compound (VOC) capture: Yu et al. synthesized UiO-66 with cluster defects (reo-UiO-66) by adding a small amount of water during the synthesis process and applied it to the photocatalytic degradation of oxygenated volatile organic compounds (OVOCs). DFT calculations show that acetaldehyde thermodynamically tends to adsorb on the sites with reo-topology. Moreover, reo-UiO-66 is conducive to the separation of charge carriers and the generation of active free radicals. It also has an increased affinity for water and oxygen, thus improving the removal efficiency of OVOCs [36].

5.4 Advantages and Disadvantages

5.4.1 Advantages

The presence of oxygen vacancies in MOFs positively impacts their photocatalytic performance, primarily manifested in the following aspects:

i) **Enhanced light absorption.** Oxygen vacancies introduce defect energy levels and extend their absorption range into the visible spectrum, thereby improving solar energy utilization efficiency [30, 37, 38].
ii) Oxygen vacancies act as electron traps, suppressing electron–hole recombination, prolonging carrier lifetimes, and boosting photocatalytic efficiency [25, 39].
iii) **Active site provision.** Local electronic structure alterations near oxygen vacancies enhance the adsorption and activation of reactants (e.g., H_2O, CO_2), facilitating surface catalytic reactions [40, 41].
iv) By modulating oxygen vacancy concentrations, the conduction/valence band positions of the material can be optimized to better align with the thermodynamic requirements of specific reactions [31, 42].
v) **Improved stability.** Moderate oxygen vacancies in certain MOFs enhance lattice stability, mitigate photo-corrosion, and extend operational durability [43].

5.4.2 Disadvantages

The presence of oxygen vacancies in MOFs enhances photocatalytic performance, but several challenges limit their development:

i) **Complex synthesis control.** Introducing oxygen vacancies requires precise regulation of synthesis conditions (e.g., reducing atmosphere and thermal treatment temperatures); otherwise, structural disorder or non-uniform defect distribution may occur [44, 45].
ii) **Negative effects of excessive defects.** High oxygen vacancy densities may form recombination centers, accelerating electron–hole recombination and reducing catalytic activity [46, 47].
iii) **Insufficient long-term stability.** Under light exposure or oxidizing environments, oxygen vacancies may gradually refill or cause structural collapse, leading to performance degradation [48].
iv) **Unclear mechanisms.** The specific roles of oxygen vacancies in photocatalytic processes require further theoretical investigation [49].

5.5 Conclusion

Defect engineering, particularly oxygen vacancy incorporation, has revolutionized MOFs by tailoring their physicochemical properties for high-performance photocatalysis. Strategies like post-synthetic modification and in situ defect generation enable precise control over defect density and distribution, enhancing light absorption, charge transport, and catalytic activity. Applications in hydrogen production, CO_2 reduction, and pollutant degradation demonstrate the potential of defective MOFs to address energy and environmental challenges. However, challenges such as structural instability under prolonged irradiation, over-defect-induced recombination, and incomplete mechanistic understanding hinder practical implementation. Future advancements should focus on AI-guided synthesis, hybrid stabilization coatings, and eco-friendly defect engineering to improve reproducibility and scalability. By addressing these limitations, defective MOFs can transition from laboratory prototypes to sustainable solutions for global energy and environmental needs.

Conflict of Interest

The authors declare that they have no conflict of interest.

References

1 Juma, A.K., Merican, Z.M.A., and Haruna, A. (2024). Recent progress of MOF-based photocatalysts for environmental application and sustainability considerations. *Chem. Eng. Res. Des.* 208: 391–435.

2 Sepehrmansourie, H., Alamgholiloo, H., Noroozi Pesyan, N., and Zolfigol, M.A. (2023). A MOF-on-MOF strategy to construct double Z-scheme heterojunction for high-performance photocatalytic degradation. *Appl. Catal. B Environ.* 321: 122082.

3 Xia, J., Gao, Y., and Yu, G. (2021). Tetracycline removal from aqueous solution using zirconium-based metal-organic frameworks (Zr-MOFs) with different pore size and topology: Adsorption isotherm, kinetic and mechanism studies. *J. Colloid Interface Sci.* 590: 495–505.

4 Glasby, L.T., Cordiner, J.L., Cole, J.C., and Moghadam, P.Z. (2024). Topological characterization of metal–organic frameworks: a Perspective. *Chem. Mater.* 36: 9013–9030.

5 Hirscher, M. and Hirose, K. (2010). *Handbook of Hydrogen Storage: New Materials for Future Energy Storage.* Wiley.

6 Farha, O.K., Eryazici, I., Jeong, N.C. et al. (2012). Metal–organic framework materials with ultrahigh surface areas: is the sky the limit? *J. Am. Chem. Soc.* 134: 15016–15021.

7 Yusuf, M., Kumar, R., Ali Khan, M. et al. (2023). Metal-organic framework-based composites for biogas and natural gas uptake: An overview of adsorption and storage mechanisms of gaseous fuels. *Chem. Eng. J.* 478: 147302.

8 Jia, T., Gu, Y., and Li, F. (2022). Progress and potential of metal-organic frameworks (MOFs) for gas storage and separation: a review. *J. Environ. Chem. Eng.* 10: 108300.

9 Habib, N.R., Asedegbega-Nieto, E., Taddesse, A.M., and Diaz, I. (2021). Non-noble MNP@MOF materials: synthesis and applications in heterogeneous catalysis. *Dalton Trans.* 50: 10340–10353.

10 Fakhraei Ghazvini, M., Vahedi, M., Najafi Nobar, S., and Sabouri, F. (2021). Investigation of the MOF adsorbents and the gas adsorptive separation mechanisms. *J. Environ. Chem. Eng.* 9: 104790.

11 Liu, D., Zou, D., Zhu, H., and Zhang, J. (2018). Mesoporous metal–organic frameworks: synthetic strategies and emerging applications. *Small* 14: 1801454.

12 Khan, M.S., Li, Y., Li, D.-S. et al. (2023). A review of metal–organic framework (MOF) materials as an effective photocatalyst for degradation of organic pollutants. *Nanoscale Adv.* 5: 6318–6348.

13 Shan, Y., Zhang, G., Shi, Y., and Pang, H. (2023). Synthesis and catalytic application of defective MOF materials. *Cell Rep. Phys. Sci.* 4.

14 Chang, Y., Huang, H., Yang, T. et al. (2021). Simultaneous introduction of oxygen vacancies and hierarchical pores into titanium-based metal-organic framework for enhanced photocatalytic performance. *J. Colloid Interface Sci.* 599: 785–794.

15 Gafiullina, A., Ladewig, B.P., and Zhang, J. (2023). Photoresponsive polymer and polymer composite membranes for gas separation. *ACS Appl. Polym. Mater.* 5: 1–30.

16 Liu, F. and Fan, Z. (2023). Defect engineering of two-dimensional materials for advanced energy conversion and storage. *Chem. Soc. Rev.* 52: 1723–1772.

17 Daliran, S., Oveisi, A.R., Kung, C.-W. et al. (2024). Defect-enabling zirconium-based metal–organic frameworks for energy and environmental remediation applications. *Chem. Soc. Rev.* 53: 6244–6294.

18 Yuan, M., Kermanian, M., Agarwal, T. et al. (2023). Defect engineering in biomedical sciences. *Adv. Mater.* 35: 2304176.

19 Gonuguntla, S., Sk, S., Tripathi, A. et al. (2022). Anisotropic phenanthroline-based ruthenium polymers grafted on a titanium metal-organic framework for efficient photocatalytic hydrogen evolution. *Commun. Chem.* 5: 165.

20 Chen, H., Cai, T., Dong, W. et al. (2023). Catalytic thermal degradation of tetracycline based on iron-based MOFs and annealed derivative in dark condition. *J. Clean. Prod.* 406: 136976.

21 Huang, L., Kong, X., Chang, K. et al. (2023). DBD plasma assisted synthesis of MO@Fe/Ce-MOFs with rich oxygen defects as efficient photocatalysts for nitrogen fixation. *J. Environ. Chem. Eng.* 11: 109836.

22 Mandal, S., Natarajan, S., Mani, P., and Pankajakshan, A. (2021). Post-synthetic modification of metal–organic frameworks toward applications. *Adv. Funct. Mater.* 31: 2006291.

23 Xin, Y., Tian, J., Xiong, X. et al. (2025). Enhanced photocatalytic efficiency through oxygen vacancy-driven molecular epitaxial growth of metal–organic frameworks on $BiVO_4$. *Adv. Mater.* 37: 2417589.

24 Dong, Y.-L., Jiang, Y., Ni, S. et al. (2024). Ligand defect-induced active sites in Ni-MOF-74 for efficient photocatalytic CO_2 reduction to CO. *Small* 20: 2308005.

25 Chen, X., Peng, X., Jiang, L. et al. (2022). Photocatalytic removal of antibiotics by MOF-derived Ti^{3+}- and oxygen vacancy-doped anatase/rutile TiO_2 distributed in a carbon matrix. *Chem. Eng. J.* 427: 130945.

26 Sun, Y., Gao, S., Lei, F., and Xie, Y. (2015). Atomically-thin two-dimensional sheets for understanding active sites in catalysis. *Chem. Soc. Rev.* 44: 623–636.

27 Liu, W., Zhou, J., Liu, D. et al. (2021). Fe-MOF by ligand selective pyrolysis for Fenton-like process and photocatalysis: Accelerating effect of oxygen vacancy. *J. Taiwan Inst. Chem. Eng.* 127: 327–333.

28 Sun, Z., Liu, Q., Yao, T. et al. (2015). X-ray absorption fine structure spectroscopy in nanomaterials. *Science China Materials* 58: 313–341.

29 Liu, L., Liu, Q., Zheng, Y. et al. (2014). O_2 adsorption and dissociation on a hydrogenated anatase (101) surface. *J. Phys. Chem. C* 118: 3471–3482.

30 Hussain, M.B., Kang, B., Cheng, X. et al. (2023). Oxygen vacancy induced Pt-decorated MOF photocatalyst for hydrogen production. *Int. J. Hydrog. Energy* 48: 13780–13790.

31 Zhang, T., Huang, Z., Xie, G. et al. (2024). A non-equivalent Ni doped La-MOF for enhanced photocatalytic CO_2 reduction through oxygen vacancy regulation and electronic structure optimization. *Inorg. Chem. Front.* 11: 8890–8901.

32 Sun, Y., Ji, H., Sun, Y. et al. (2024). Synergistic effect of oxygen vacancy and high porosity of nano MIL-125(Ti) for enhanced photocatalytic nitrogen fixation. *Angew. Chem. Int. Ed.* 63: e202316973.

33 Zhang, L., Zhang, Y., Jiao, Y. et al. (2025). SnO_{2-x}/CD-MOF S-scheme heterostructure photocatalysts synergized by oxygen vacancies and highly porous for rhodamine B and tetracycline degradation: process, and mechanism analysis. *Colloids Surf. A Physicochem. Eng. Asp.* 705: 135625.

34 Qu, W., Wen, H., Qu, X. et al. (2022). Enhanced Fenton-like catalysis for pollutants removal via MOF-derived $Co_xFe_{3-x}O_4$ membrane: oxygen vacancy-mediated mechanism. *Chemosphere* 303: 135301.

35 Ma, J., Fan, L., Wang, X. et al. (2024). Defect engineering on metal-organic frameworks for enhanced photocatalytic reduction of Cr(VI). *Surf. Interf.* 54: 105174.

36 Yu, J., Wang, X., Lu, G. et al. (2024). Improving the removal efficiency of oxygenated volatile organic compounds by defective UiO-66 regulated with water. *J. Hazard. Mater.* 469: 134055.

37 Zhu, Y.-P., Yin, J., Abou-Hamad, E. et al. (2020). Highly stable phosphonate-based MOFs with engineered bandgaps for efficient photocatalytic hydrogen production. *Adv. Mater.* 32: 1906368.

38 Li, J., Li, X., Wu, G. et al. (2021). Construction of 2D Co-TCPP MOF decorated on B-TiO_{2-X} nanosheets: oxygen vacancy and 2D–2D heterojunctions for enhancing visible light-driven photocatalytic degradation of bisphenol A. *J. Environ. Chem. Eng.* 9: 106723.

39 Zhao, K., Zhang, Z., Feng, Y. et al. (2020). Surface oxygen vacancy modified Bi_2MoO_6/MIL-88B(Fe) heterostructure with enhanced spatial charge separation at the bulk & interface. *Appl. Catal. B Environ.* 268: 118740.

40 Han, W., Chen, Y., Jiao, Y. et al. (2023). Oxygen vacancy-rich S-scheme CeO_2@$Ni_{1-x}Co_xSe_2$ hollow spheres derived from NiCo-MOF for remarkable photocatalytic CO_2 conversion. *ACS Sustain. Chem. Eng.* 11: 7787–7797.

41 Rao, F., Zhu, G., Zhang, W. et al. (2021). Maximizing the formation of reactive oxygen species for deep oxidation of NO via manipulating the oxygen-vacancy defect position on $(BiO)_2CO_3$. *ACS Catal.* 11: 7735–7749.

42 Shen, M., Zhu, X., Lin, L. et al. (2025). MOFs-derived S-scheme ZnO/BiOBr heterojunction with rich oxygen vacancy for boosting photocatalytic CO_2 reduction. *Sep. Purif. Technol.* 353: 128620.

43 Wang, L., Wang, S., Li, M. et al. (2022). Constructing oxygen vacancies and linker defects in MIL-125 @TiO_2 for efficient photocatalytic nitrogen fixation. *J. Alloys Compd.* 909: 164751.

44 Dong, Q., Li, X., Sun, J. et al. (2024). Regulating concentration of surface oxygen vacancies in Bi_2MoO_6/Bi-MOF for boosting photocatalytic ammonia synthesis. *J. Catal.* 433: 115489.

45 More, G.S., Kar, A.K., and Srivastava, R. (2022). Cu–Ce bimetallic metal–organic framework-derived, oxygen vacancy-boosted visible light-active Cu_2O–CeO_2/C heterojunction: an efficient photocatalyst for the sonogashira coupling reaction. *Inorg. Chem.* 61: 19010–19021.

46 Hao, X., Cui, Z., Zhou, J. et al. (2018). Architecture of high efficient zinc vacancy mediated Z-scheme photocatalyst from metal-organic frameworks. *Nano Energy* 52: 105–116.

47 Guo, C., Wang, T., Zhang, L. et al. (2022). Precise regulation of defect concentration in MOF and its influence on photocatalytic overall water splitting. *Nanoscale* 14: 15316–15326.

48 Shi, J., Zhang, J., Cui, Z. et al. (2022). In situ growth of MOF-derived sulfur vacancy-rich CdS nanoparticles on 2D polymers for highly efficient photocatalytic hydrogen generation. *Dalton Trans.* 51: 5841–5858.

49 Zhan, F., Wen, G., Li, R. et al. (2024). A comprehensive review of oxygen vacancy modified photocatalysts: synthesis, characterization, and applications. *Phys. Chem. Chem. Phys.* 26: 11182–11207.

6

Defect Behavior of Carbon Nitride (g-C_3N_4) as Efficient Feedstock in Photocatalytic Reactions

Debika Gogoi[1], Rajeshvari Samatbhai Karmur[2], and Rakesh K. Sharma[1]

[1] *Indian Institute of Technology, Department of Chemistry, NH 62, Surpura Bypass Road, Karwar, Jodhpur 342030, India*
[2] *Nanomaterials Lab, BITS Pilani, K K Birla Goa Campus, Department of Chemistry, BITS Pilani, K K Birla Goa Campus, NH 17B, Bypass Road, Zuarinagar, Goa 403726, India*

6.1 Introduction

The exponential rise in global energy demand and environmental degradation occurring from the reliance on fossil fuel-based energy sources have revitalized the awareness as well as the quest for remediation in the past few decades [1]. The global population reached eight billion in 2022 and with this rapid population growth, rising standards of human civilization, and industrialization, it is anticipated that the global energy demand will rise by twice its current energy supply by 2050 [1, 2]. Currently, ~80% of the global energy requirement is primarily reliant on non-renewable fossil fuels, such as petroleum, coal, and natural gas, which are depleting expeditiously. Also, severe environmental deterioration arising due to the consumption of fossil fuel-based energies is also forcing mankind to opt for alternate green energy sources [2]. Consequently, groundbreaking advancements in materials science and engineering have been explored to address challenges in achieving efficient energy conversion while ensuring environmental conservation. Since the landmark discovery of the Honda–Fujishima effect in 1972, semiconductor-based photocatalysis, a renewable energy technology that harnesses clean and unlimited solar energy, has attracted significant interdisciplinary focus for its wide-ranging applications in energy and sustainability solutions [1, 3]. The direct transformation of solar energy into fuels and chemical energy has been recognized as a sustainable and green approach to addressing future energy and environmental crises [4]. Photocatalysis, driven by solar light, also requires a suitable semiconductor to facilitate various catalytic reactions, including the degradation of organic pollutants, water splitting to produce H_2 and O_2, bacterial disinfection, the selective synthesis of organic compounds, and the conversion of CO_2 into hydrocarbon fuels. After photocatalytic water splitting by TiO_2 using ultraviolet (UV) light was reported by Fujishima and Honda in 1972 [3], Carey et al. documented the first photocatalytic

Defect Engineering in Photocatalysis: From Fundamentals to Applications,
First Edition. Edited by Chuanyi Wang, Abdelkader Labidi, and Qiuhui Zhu.

degradation of organic pollutants using TiO_2 in aqueous suspensions, in 1976 [5]. Three years later, Inoue et al. explored the photocatalytic conversion of CO_2 into a range of organic compounds using different semiconductors like TiO_2, SiC, ZnO, CdS, and GaP in aqueous solutions [6]. Since then, substantial advancements have been made in the design and development of very effective semiconductor-based photocatalysts, as highlighted in various reviews. According to the literature, promising photocatalysts active under UV and visible light include TiO_2, Fe_2O_3, ZnO, Bi_2WO_6, CdS, Ta_2O_5, TaON, Ta_3N_5, and $BiVO_4$ [1]. The development of high-quality semiconductor photocatalysts for addressing energy shortages and environmental challenges has emerged as a prominent area of research.

Over the past decades, researchers have extensively pursued the design of visible-light-responsive photocatalysts to maximize the effective utilization of the solar spectrum, approximately 43% of which is visible light [1, 7]. However, the large band-gap energy of traditional semiconductor photocatalysts, such as TiO_2, remains a significant bottleneck, limiting their efficiency in visible-light applications due to low solar energy utilization. In the quest for robust, visible-light-active photocatalysts, the polymeric semiconductor graphitic carbon nitride (g-C_3N_4) has garnered considerable attention as a next-generation material. Its appeal lies in its simple synthesis, favorable electronic band structure, high physicochemical stability, and natural abundance [8]. Notably, g-C_3N_4 can be easily fabricated through the thermal polymerization of nitrogen-rich precursors such as melamine, urea, thiourea, ammonium thiocyanate, cyanamide, and dicyandiamide. Carbon nitride (C_3N_4) is not a recent discovery; it is regarded as one of the earliest artificial polymers documented in scientific literature. The origins of C_3N_4 polymers and their precursors trace back to 1834, when melon (precursor), a linear polymer composed of interconnected tri-s-triazine units linked via secondary nitrogen, was first synthesized by Berzelius and later named by Liebig [1]. Research on the use of g-C_3N_4 in heterogeneous catalysis began approximately 19 years ago, in 2006 [9]. In 2009, Wang et al. first reported the discovery of g-C_3N_4 as a metal-free conjugated semiconductor photocatalyst for H_2 evolution [10]. This discovery marked a shift in research focus from inorganic to polymeric conjugated semiconductor photocatalysts.

Seven distinct phases of C_3N_4 are known: g-h-heptazine, g-o-triazine, g-h-triazine, pseudo-cubic C_3N_4, cubic C_3N_4, β-C_3N_4, and α-C_3N_4, with corresponding band gaps of approximately 2.88, 0.93, 2.97, 4.13, 4.30, 4.85, and 5.49 eV, respectively [11]. It has been determined that the fundamental structural units forming the allotropes of g-C_3N_4 are triazine (C_3N_3) and heptazine, also termed as tri-s-triazine (C_6N_7) rings, which are shown in the Figure 6.1 [1]. In this context, the nitride pore size and the distinct electronic environments of nitrogen atoms play a crucial role in determining the energetic stability of different C_3N_4 phases. Among these, the tri-s-triazine-based g-C_3N_4 emerged as the most energetically favorable and stable phase under ambient conditions [12]. This finding is consistent with the theoretical calculations reported by Kroke et al. [13] Numerous studies have indicated that the polycondensation of precursors such as melamine, urea, dicyandiamide, or cyanamide leads to the formation of a melon polymer consisting of melem units, confirming this structure as the most stable configuration. As a result, tri-s-triazine is widely acknowledged as the primary building unit for the standard synthesis of g-C_3N_4 [1].

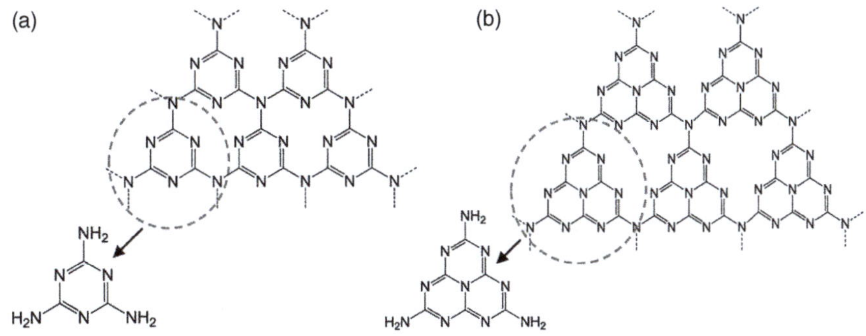

Figure 6.1 Structures of g-C_3N_4: (a) Triazine and (b) heptazine or tri-s-triazine. *Source:* Ong et al. [1] / American Chemical Society.

Despite its potential, the real-world applications of bare g-C_3N_4 are still constrained by many limitations, including a high-charge carrier recombination rate, poor electrical conductivity, and restricted light absorption beyond 460 nm. To overcome these challenges and optimize its photoactivity, extensive efforts have been made, focusing on modifications such as improved synthesis methods, electronic structure tuning, and nanostructure engineering [7]. In fact, copolymerization and doping are proficient strategies for incorporating impurities or defects into the g-C_3N_4 matrix, thereby altering its electronic structure and energy band alignment. In the field of nanostructure design, diverse g-C_3N_4 morphologies with distinct geometrical shapes have been developed using techniques such as supramolecular preorganization, hard templating, soft templating, and exfoliation strategies [14]. This chapter gives a comprehensive review on the recent development of defect-engineered g-C_3N_4 in photocatalytic reactions.

6.2 Synthesis of Defective g-C_3N_4

It is well established that the photocatalytic performance of g-C_3N_4 is closely related to its electronic properties, which play a vital role in light absorption and charge transfer processes. The reaction atmosphere significantly impacts the unique properties and chemical structures of g-C_3N_4 by inducing defects, structural disorders, and carbon or nitrogen vacancies. Defects, in particular, are vital for heterogeneous catalysis reactions, as they provide catalytic sites for reactant molecules and modify the electronic band configurations. This modification introduces extra energy bands between the valence band (VB) and the conduction band (CB), enabling greater absorption of visible light [1, 15]. Defects and lattice disorders in semiconductors generate mid-gap states, often referred to as band tail states, which contribute to electron-hole pair excitations and shape the optical response of the photocatalyst. Additionally, disordered defects offer the advantage of providing more trapping sites, effectively slowing the recombination of charge carriers [1, 16]. Although oxygen vacancies have been widely studied, the role of carbon and nitrogen vacancies in modulating optical and electronic properties is attaining significant interest in the development of g-C_3N_4-based photocatalysts. As an organic polymer, g-C_3N_4's polymerization is highly influenced by annealing conditions, allowing for substitution of atoms in the tri-s-triazine unit with foreign atoms such as metals (K, Na, Fe, Co, Cu, Ni, etc.) and nonmetals (P, S, O, C, etc.), also creating vacancies. Generally, the incorporation of heteroatoms from higher-periodic elements leads to smaller band gaps and lower highest occupied molecular orbital (HOMO) levels, thereby enhancing visible-light absorption and promoting faster electron mobility [17]. As a result, defect engineering, which introduces dopants and vacancies into the g-C_3N_4 matrix, has attracted considerable attention.

6.2.1 N and C Vacancies

Niu et al. successfully introduced N-vacancies into the g-C_3N_4 structure by controlling the polymerization temperature in static air [18]. Figure 6.2A depicts a schematic atomic model of an incompletely condensed graphitic carbon nitride sheet, formed from melon units and having NH/NH_2 groups. The XPS results revealed that nitrogen vacancies tend to form at the two-coordinated nitrogen sites (N_{2c}) due to their unsaturated coordination and higher abundance. This nitrogen deficiency generated extra electrons, leading to the formation of N-vacancy-related C_3^+ states within the band gap, consequently reducing the intrinsic band gap from 2.74 to 2.66 eV (Figure 6.2B) [18]. Additionally, they also formed N-vacancies by heating g-C_3N_4 in a hydrogen atmosphere [8c]. They emphasized the significance of homogeneous self-modification through vacancies in entirely reshaping the electronic structure of a layered g-C_3N_4 system. Hydrogenation, a common post-treatment approach, is known to create anion vacancies in the compound's surface layer with non-layered structures, a process referred to as surface self-modification through vacancies. In contrast, the layered structure of g-C_3N_4 allows H_2 to diffuse from the surface into the bulk via the periodic stacking of layers in the direction of the *c*-axis of melon, enabling homogeneous self-modification through vacancies throughout the material. This was evident when approximately 7% of N-atoms became vacant in a g-C_3N_4 sample produced at 540 °C under a hydrogen atmosphere with a C/N molar ratio of 3:3.87 (designated as R-melon, where "R" denotes "reduced"), compared to pure g-C_3N_4 having a C/N molar ratio of 3:4.14. However, at a higher synthesis temperature of 560 °C, N-loss increased dramatically to 41%, resulting in significant structural damage to g-C_3N_4, as confirmed by nuclear magnetic resonance (NMR) (Figure 6.2C) and Raman spectroscopic analyses (Figure 6.2D). The R-melon displayed a reduced band gap of 2.03 eV compared to pure g-C_3N_4 (2.78 eV), indicating a red-shift in the absorption edge caused by N-vacancies. This effect is thought to arise from H_2 post-treatment, which converts lattice nitrogen into NH_3, leading to the formation of nitrogen vacancies [8c]. In 2015, Tay and colleagues reported a study on the incorporation of N-vacancies in g-C_3N_4 via H_2 treatment at 550 °C during thermal polycondensation, which decreased the band gap from 2.7 to 2.0 eV [8b]. Elemental analysis revealed an excess of hydrogen atoms beyond theoretical predictions, indicating an increased concentration of protons. This was attributed to hydrogen atoms bonding with adjacent carbon atoms and occupying the two-coordinated

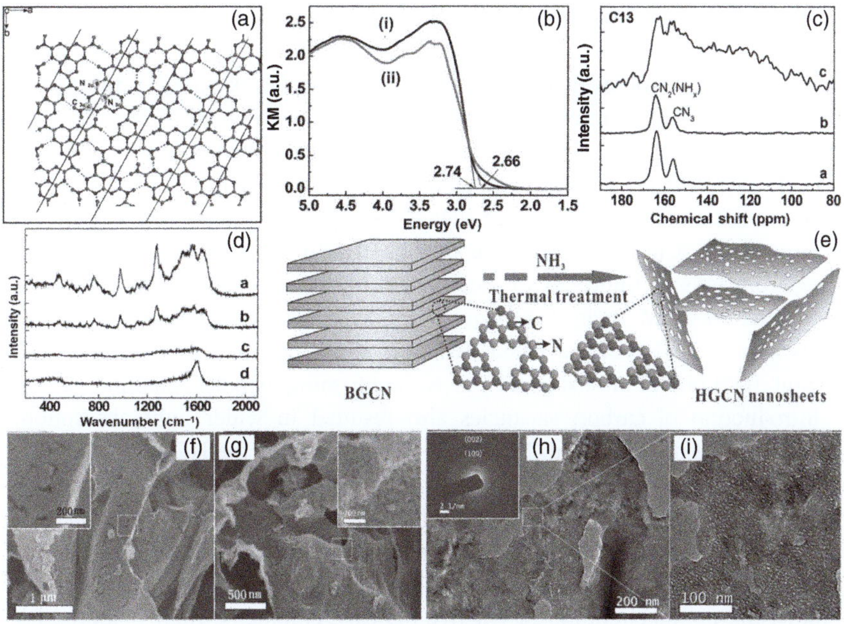

Figure 6.2 (a) A schematic atomic model of an incompletely condensed g-C_3N_4 sheet, formed from melon units and having NH/NH_2 groups. (N2c, N3c, and C3c indicate two-coordinated N, three-coordinated N, and three-coordinated C). (b) Kubelka–Munk (KM) function vs the energy of the absorbed light for (i) g-C_3N_4 and (ii) nitrogen-deficient g-C_3N_4. *Source:* Reproduced from Niu et al. [18] / with permission of American Chemical Society. (c) UV Raman spectra (excited at 244 nm) were recorded for: (a) pristine melon, (b) R-melon with a minor nitrogen atom loss (~7%), (c) R-melon with a significant nitrogen atom loss (~41%), and (d) reduced graphene oxide. (d) Solid-state ^{13}C NMR spectra were obtained for: (a) melon, (b) melon having minor nitrogen loss (~7%) prepared by heating at 540 °C in a H_2 atmosphere, and (c) R-melon with a substantial nitrogen loss (~41%) prepared by heating at 560 °C in a H_2 atmosphere. *Source:* Reproduced from Liang et al. [19] / with permission of John Wiley & Sons (e) Schematic demonstration of the synthesis route of HGCN nanosheets (f, g) Representative SEM images and (h, i) TEM images of HGCN nanosheets. The inset in panel (h) shows the corresponding SAED pattern. *Source:* Reproduced from Liang et al. [19] / with permission of John Wiley & Sons.

N-vacancy sites. The band gap narrowing was ascribed to the formation of extra electronic states below the conduction band, contributed by the carbon atoms surrounding the defects [8b]. In the same year, Li et al. also introduced hydrogenated defects in g-C_3N_4 by treating it at 550 °C. The combined effect of layer-separation and the induced defects resulted in a narrower band gap, which enhanced the photocatalytic H_2 evolution performance [20].

In addition to the nitrogen vacancies previously discussed, Liang et al. synthesized holey g-C_3N_4 nanosheets (HGCN) by treating bulk g-C_3N_4 thermally in an NH_3 atmosphere, as depicted in Figure 6.2E [19]. The in-plane holes not only introduced numerous boundaries, reducing van der Waals interactions and preventing excessive aggregation, but also revealed more cross-plane diffusion pathways and active edges, which significantly facilitated the transfer and movement of photoinduced electron-hole pairs. The presence of plentiful in-plane holes resulted in a larger specific surface area for HGCN (196 $m^2 g^{-1}$) in comparison to pristine g-C_3N_4 (6 $m^2 g^{-1}$). Moreover, the large pore volume of 1.38 $cm^3 g^{-1}$, due to the crumpled structure and presence of in-plane holes, further enhanced mass transfer kinetics and photo-activity. This was corroborated by scanning electron microscopy (SEM) (Figure 6.2F,G) and transmission electron microscopy (TEM) analyses (Figure 6.2H,I). SEM images showed that the HGCN nanosheets had a wrinkled structure with ~20 nm thickness and contained numerous in-plane nano-holes ranging from tens to hundreds of nanometers in diameter. These findings were corroborated by TEM images. Moreover, the selected area electron diffraction (SAED) pattern of HGCN revealed two diffraction rings corresponding to the (100) and (002) planes of hexagonal g-C_3N_4, confirming its polycrystalline nature. The introduction of carbon vacancies also resulted in wider light absorption, ranging up to the near-infrared (NIR) region. Furthermore, the HGCN nanosheets demonstrated an improved capacity to delocalize π–π conjugated electronic systems across interlayers. This was attributed to the lone pair of electrons on the carbon atoms of the tri-s-triazine rings within a p-bonded nano-sized cluster, as verified by electron paramagnetic resonance (EPR) analysis. This delocalization effectively facilitated the photogeneration of reactive radicals, enhancing their potential for driving photocatalytic reactions [19]. Yang et al. developed highly porous g-C_3N_4 with an enhanced specific surface area and carbon vacancies using a straightforward NH_3 etching thermal treatment [21]. Building on earlier findings, they identified that reactive radicals such as NH_2• and NH• are formed during NH_3 decomposition at elevated temperatures, facilitating carbon etching and subsequent gasification. At high temperatures, these radicals selectively etched the C–N=C and N–$(C)_3$ lattice sites of g-C_3N_4, resulting in the formation of C-vacancies and a porous structural framework [21]. Beyond the formation of C-vacancies and porous frameworks as discussed previously, Cui et al. introduced a one-step NH_3 thermal polymerization method to develop low-defect g-C_3N_4 with well-defined crystallinity from ammonium thiocyanate. The NH_3-treated g-C_3N_4 featured a higher surface area of 46 $m^2 g^{-1}$, more porosity, and fewer sp^2 nitrogen defects. Additionally, it exhibited an improved in-plane π–π conjugated system within the tri-s-triazine heterocycle, which played a key role in enhancing H_2 evolution efficiency [22].

Recent studies have investigated the structural distortion of g-C_3N_4 as an efficient method to improve its optical properties by facilitating more $n \rightarrow \pi^*$ electron transitions and extending visible-light absorption beyond 460 nm [23]. Chen et al. fabricated structure-distorted g-C_3N_4 nanosheets by carefully regulating the thermal polymerization of melamine, without relying on additional species [23b]. Achieving a broad visible-light response to initiate photoredox processes remains challenging, as $\pi \rightarrow \pi^*$ transitions exhibit inherently higher photoactivity compared to $n \rightarrow \pi^*$ transitions. Density functional theory (DFT) calculations were used to simulate the average optical absorption spectra through dielectric function calculations. The findings indicated that the distorted g-C_3N_4 monolayer demonstrated enhanced visible-light absorption, extending from roughly 410 to 600 nm, with a newly emerged absorption peak near 450 nm.

To align the theoretical calculations with experimental data, three variations of g-C_3N_4, monolayer, bilayer, and multilayer, were analyzed. These forms corresponded to the variations in the g-C_3N_4 thickness as annealing temperatures increased, establishing consistency between the theoretical calculations and the experimental outcomes [23b]. Ho et al. applied a H_2 post-calcination treatment to introduce structural distortion in g-C_3N_4 layers, resulting in adjustable optical properties [23c]. The H_2-treated g-C_3N_4 displayed enhanced optical absorption beyond 460 nm in the visible-light spectrum compared to its O_2-treated counterpart. This improvement is attributed to the smaller size and higher reactivity of the H atom, which allows it to more easily access the channels between layers and interact with the framework atoms in g-C_3N_4, specifically, H atoms bonded with nitrogen within the g-C_3N_4 framework, resulting in the generation of amino groups. These amino groups reinforced hydrogen bonding interactions between the g-C_3N_4 layers. The interlayer spacing of g-C_3N_4 measures 0.324 nm, considerably greater than the 0.1 nm N—H bond length of amino groups, implying the potential incorporation of amino groups between the layers. The enhanced hydrogen bonding interactions between these layers caused distortion in the planar structure of g-C_3N_4, leading to the formation of lone pairs on the nitrogen atoms at the edges of the tri-s-triazine rings embedded in the π-conjugated network. This structural distortion facilitated $n \rightarrow \pi^*$ electronic transitions, increasing the electron density in the π-conjugated system and promoting excitation to the CB, thereby enhancing photoactivity. These observations were confirmed through EPR analysis. Moreover, the separation of photoinduced charge carriers in the H_2-modified g-C_3N_4 was enhanced, with the electron-deficient H atoms serving as electron sinks. Together with the previous studies, these results highlight the promising potential of enhancing the intrinsic electronic properties of g-C_3N_4 through structural distortion, providing a valuable strategy to boost its photocatalytic performance [23c].

6.2.2 Metal Ion Doping

The g-C_3N_4 structure is marked by three nitrogen triangles, with each nitrogen atom containing a lone pair of electrons that contribute to its electronic properties. This structural characteristic makes g-C_3N_4 capable of enduring doping with metal ions [7]. Elemental doping plays a significant role in enhancing its photochemical performance by reducing the semiconductor's band gap. Specifically, doping lowers the rate of recombination of photoinduced charge carriers and extends the range of wavelengths within the visible-light spectrum that can drive photocatalytic reactions. Doping of alkali metals like Li^+, Na^+, and K^+ into g-C_3N_4 further boosts its photocatalytic activity, promoting redox reactions and improving its overall efficiency [7, 24]. For instance, Jasni et al. fabricated lithium-doped g-C_3N_4 by mixing a LiOH solution with a g-C_3N_4 dispersion, which was then stirred and dried to get the final product. This lithium-doped g-C_3N_4 exhibited improved performance compared to pure g-C_3N_4. This enhancement was ascribed to changes in the electronic configuration and a reduction in the recombination rate of photoinduced charge carriers [24a]. Similarly, Jiang et al. developed alkali metal

(Li, Na, or K)-doped g-C_3N_4 by mixing MOH (M = Li, Na, or K) solutions with a g-C_3N_4 dispersion, followed by stirring, drying, and heating the mixture at 600 °C for one hour in an N_2 atmosphere. It was observed that the Na-doped g-C_3N_4 displayed the highest efficiency in photocatalytic H_2 production. This was because of its smaller band-gap and larger surface area, both of which contributed to improved photocatalytic activity [24c]. Yan et al. synthesized K, Na, Mg, and Ca-doped g-C_3N_4 (Figure 6.3A) by first dispersing urea in XBr or YBr_2 aqueous solution (X = Na, K; Y = Ca, Mg) [25]. The mixture was then stirred vigorously overnight at 80 °C. Afterward, the resultant solid precursor was annealed at 550 °C for two hours in a capped alumina crucible. It was found that doping g-C_3N_4 with K and Na notably enhances its performance in H_2 evolution. This doping process modifies the g-C_3N_4 morphology by increasing the retention of O_2 within its structure. The interaction between the doped Na or K and the g-C_3N_4 layers resulted in the development of a stable lamellar-aggregated structure, making the material difficult to disassemble. In contrast, doping with Mg and Ca resulted in larger pores within the g-C_3N_4, which was because of partial polymerization. The reduced band gap in Na and K-doped g-C_3N_4 is attributed to the delocalization of electrons within the layers, while the presence of oxygen causes electron accumulation around the nitrogen atoms, which in turn downshifts the valence band maximum (VBM) [25].

Transition metal-doped g-C_3N_4 has gained considerable focus in recent years owing to its potential to enhance catalytic performance. For instance, Ma et al. prepared Ag-doped g-C_3N_4 by first placing melamine in a lidded crucible and annealing it at 550 °C for four hours, followed by a mixed-solvent exfoliation method to obtain g-C_3N_4 nanosheets [26]. To synthesize the Ag/g-C_3N_4 plasmonic photocatalyst, the g-C_3N_4 was dispersed in water, and $AgNO_3$ aqueous solution was gradually added. The suspension was then stirred under light irradiation for two hours, centrifuged, and rinsed with deionized water to yield the final product. Ag doping in g-C_3N_4 induced changes in its crystal structure and morphology, significantly enhancing the absorption of visible light and the separation of the charge carrier, thereby improving its photocatalytic activity. Overall, transition metal doping enhances g-C_3N_4 by improving charge carrier separation, reducing the band gap, and increasing the surface area [26]. Hussain et al. prepared Mn, Cu, and Zn-doped g-C_3N_4 using a high-temperature treatment approach [14]. The SEM micrographs highlighted the distinct morphologies of g-C_3N_4 and its doped variants. The synthesized g-C_3N_4 (Figure 6.3B) comprises round-shaped grains with non-uniform size distribution, likely due to grain agglomeration. The Cu-doped g-C_3N_4 (Figure 6.3C) shows larger, irregular grains with non-uniform size distribution and increased surface area. The Mn-doped g-C_3N_4 (Figure 6.3D) exhibits a fibrous morphology, enhancing photocatalytic activity and stability. Finally, the Zn-doped g-C_3N_4 (Figure 6.3E) presents an elongated flake-like structure with visible pores within the flakes, as revealed by high-resolution imaging. The doping of Cu, Mn, and Zn significantly enhanced the electronic properties of g-C_3N_4, extended its visible-light absorption range, and effectively reduced the band gap (Figure 6.3F), with Zn-doped g-C_3N_4 demonstrating the highest photocatalytic activity [14]. In another study, Ding et al. synthesized a series of metal-doped g-C_3N_4 catalysts (Co, Fe, Cu, V, and Ni-g-C_3N_4)

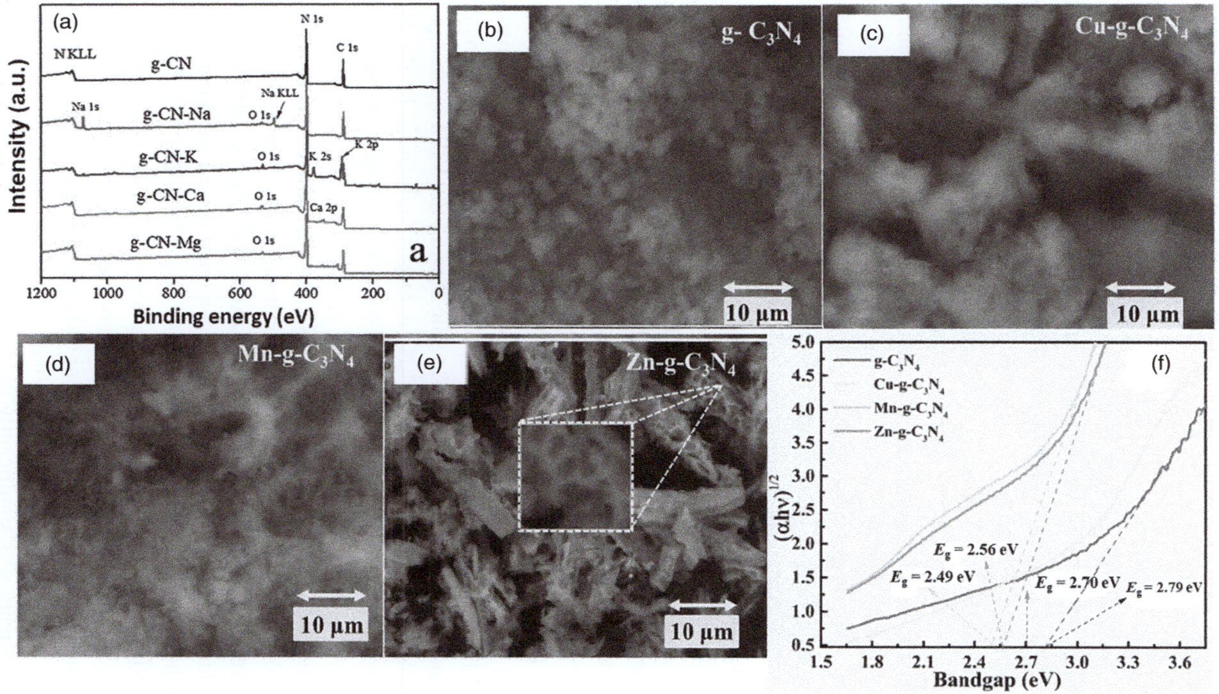

Figure 6.3 (a) XPS survey spectra of pure and alkali-doped g-C_3N_4. *Source:* Reproduced from Yan et al. [25] / with permission of Elsevier. (b–e). SEM Images (f) $(\alpha h\nu)^{1/2}$ vs. bandgap energy plot of pure g-C_3N_4, Cu-g-C_3N_4, Mn-g-C_3N_4, and Zn-g-C_3N_4. *Source:* Reproduced from Hussain et al. [14] / with permission of Elsevier.

using urea as the precursor via a simple annealing method in an N_2 atmosphere. Among them, V-g-C_3N_4 showed the highest efficiency for phenol synthesis from benzene using H_2O_2 as the oxidant [27].

6.2.3 Nonmetal Doping

The incorporation of metals into g-C_3N_4 can cause self-degradation, thermal instability, and corrosion. To mitigate these issues, researchers have investigated nonmetallic doping, which involves elements such as C, O, P, S, and halogens. These elements possess high ionization energy and strong electron affinity, creating robust covalent bonds in the lattice. Nonmetal doping modifies the surface, electronic, and optical properties of g-C_3N_4, significantly enhancing its photocatalytic performance [7, 28]. Li et al. reported that oxygen-doped g-C_3N_4 (O-doped g-C_3N_4), prepared via a simple H_2O_2 hydrothermal treatment, demonstrated an extended visible-light absorption, reaching up to 498 nm [28]. This doping improved the material's surface area and charge-separation efficiency for enhanced H_2 evolution. XPS analysis showed that oxygen atoms replaced some of the sp^2-hybridized N atoms in the g-C_3N_4 structure, leading to a reduced $N(sp^2)/N(sp^3)$ ratio and an enhanced C/N atomic ratio. The g-C_3N_4 exhibits an O 1s core level at 532.9 eV, which corresponds to adsorbed water. No extra signals associated with N—O or C—O bonds are detected in the pure catalyst. After H_2O_2 modification, two new core levels at 531.6 and 534.0 eV appear in the doped sample, credited to the formation of N—C—O moieties and adsorbed O_2, respectively. Additionally, after 60 seconds of Ar^+ etching, which removes approximately 2–4 nm of material, the N—C—O signal persists, indicating that the oxygen-containing species are not just adsorbed on the surface but are evenly distributed throughout the matrix (Figure 6.4A–C) [28]. Oxygen doping in O-doped g-C_3N_4 results in a downshift of the conduction band minimum by 0.21 eV, while leaving the VBM unchanged. This indicates that the VBM in O-doped g-C_3N_4 is mostly influenced by the N_{2p} orbitals. Since oxygen atoms have more electrons than nitrogen, substituting nitrogen with oxygen introduces extra electrons in the structure. The electrons become delocalized within the π-bonds of g-C_3N_4 and are transferred to neighboring carbon atoms, creating a defect state just beneath the CB minimum [28]. Huang et al. developed O-doped g-C_3N_4 with a extensively porous structure by first reacting melamine with H_2O_2 to create hydrogen bond-driven supramolecular aggregates, and then performing thermal calcination at 550 °C under nitrogen flow (Figure 6.4D) [29]. DFT calculations revealed two key insights: (i) O-doping preferentially occurred at the two-coordinated nitrogen sites, as demonstrated by the lowest creation energy, and (ii) O-doping enhanced light absorption and promoted charge separation. The band gap of the O-doped g-C_3N_4 decreased upon doping and the electron density was significantly reduced at the carbon atoms near the O dopant, while it increased at the neighboring nitrogen atoms. This created an internal electric field that facilitated electron-hole separation. This work emphasizes that pre-treatment of the starting material influences both structural morphology and the incorporation of foreign atoms into the g-C_3N_4 matrix, contributing to

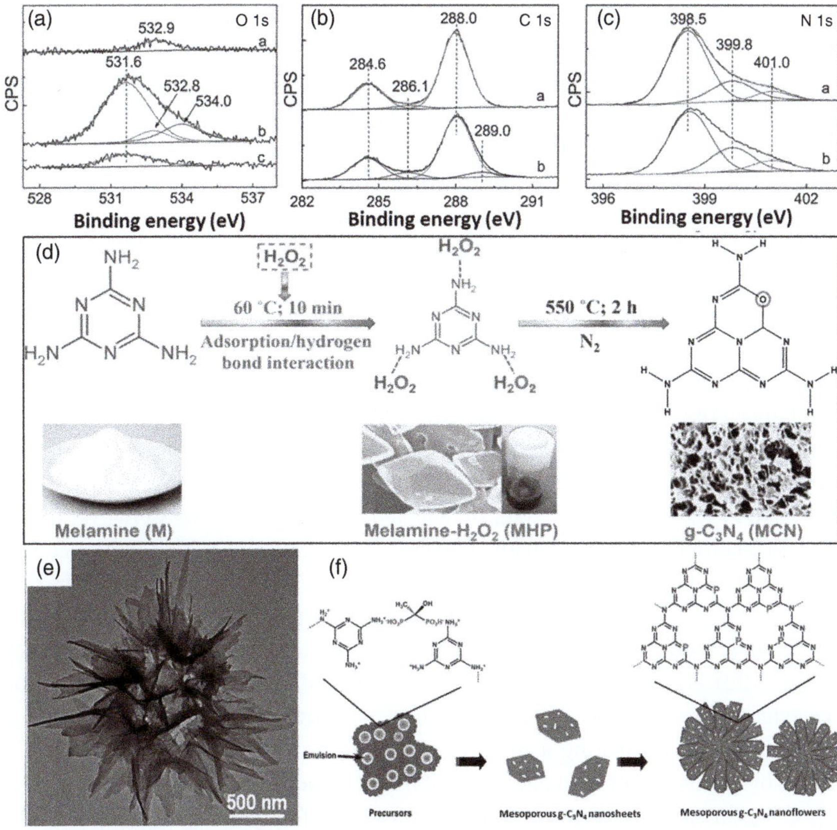

Figure 6.4 (a–c) XPS spectra of O 1s, C 1s, and N 1s for (a) pristine g-C_3N_4 and (b) O-doped g-C_3N_4. (c) The O 1s profile of O-doped g-C_3N_4 was obtained after 60 seconds of Ar^+ ion etching. *Source:* Reproduced from Li et al. [28] / with permission of Royal Society of Chemistry. (D) Schematic representation of the synthesis of porous and O-doped g-C_3N_4 via hydrogen bond-induced supramolecular assembly. *Source:* Reproduced from Huang et al. [29] / with permission of Elsevier. (e) TEM images of P-CN, (f) Schematic of the proposed formation mechanism of mesoporous P-doped g-C_3N_4 (P-CN) flowers. *Source:* Reproduced from Zhu et al. [30] / with permission of American Chemical Society.

the progress of next-generation photocatalysts [29]. Zhao et al. synthesized carbon self-doped g-C_3N_4 photocatalysts using polyporous carbon foam as a soft-template and observed the narrowing of band gaps [31]. In another approach, N-doped g-C_3N_4 was integrated into the g-C_3N_4 matrix through a straightforward one-pot thermal polymerization process, where urea and citric acid were used as the starting materials. During calcination, urea generated an amine-rich environment that reacted with citric acid, serving as a carbon source, to produce N-doped g-C_3N_4. This doping improved the delocalization of π-conjugated electrons, enhancing visible-light absorption and charge separation efficiency [32].

Hu et al. synthesized P-doped g-C_3N_4 using dicyandiamide as the starting material and diammonium hydrogen phosphate ($(NH_4)_2HPO_4$) as the P-source [33]. P atoms were incorporated into the g-C_3N_4 matrix, forming P—N bonds. This doping reduced the band gap and enhanced charge separation efficiency. Zhou et al. synthesized an alternative P-doped g-C_3N_4 using guanidinium hydrochloride and hexachlorocyclotriphosphazene ($Cl_6N_3P_3$) in a thermally induced copolymerization method [34]. The P atoms were integrated into the lattice, primarily at the corner and bay-carbon sites, influencing the electronic properties and preventing charge recombination. The phosphorus precursor affected the doping site (interstitial or substitutional), and the electron-rich P-doped g-C_3N_4 exhibited enhanced electron delocalization in the π-conjugated system [34]. Zhu et al. synthesized mesoporous P-doped g-C_3N_4 nano-flowers (Figure 6.4E) using a co-condensation method with HEDP ((hydroxyethylidene)diphosphonic acid) without the need for a template (Figure 6.4F) [30]. Phosphorus doping enhanced the material's properties, including a significant increase in specific surface area due to its high porosity. This modification enhanced the mass transfer of reactants and products, boosted light absorption, and facilitated efficient charge transfer and separation, ultimately leading to enhanced hydrogen production under visible light [30]. Liu et al. synthesized sulfur-doped g-C_3N_4 photocatalysts by exposing g-C_3N_4 to an H_2S environment, leading to an enhanced VB width and a lowered CB minimum, which improved H_2 generation [8a]. Additionally, using trithiocyanuric acid as a sulfur precursor, the –SH groups acted as leaving groups during thermal polymerization, resulting in sulfur doping through a sulfur-mediated method. The similarity in electronegativity between sulfur and nitrogen made the substitution of nitrogen atoms with sulfur at the edges of g-C_3N_4 favorable [35]. Yan et al. developed boron-doped g-C_3N_4 (B-doped g-C_3N_4) by heating a mixture of melamine and boron oxide at 500 °C, followed by further heating to 520 °C, which facilitated the reaction between B_2O_3 and g-C_3N_4, introducing boron into the N sites [36]. This process formed C—NB_2 and C—NB groups, confirmed by XPS analysis. The doping slightly reduced the band gap and increased the VBM.

6.2.4 Co-doping

Doping g-C_3N_4 with various elements improves charge carrier separation and light absorption, leading to enhanced photocatalytic performance. Co-doping with metals and nonmetals effectively reduces the band gap and further promotes charge carrier separation, boosting the material's photocatalytic efficiency [7]. Zhu et al. demonstrated that co-doping g-C_3N_4 with K and Fe induced structural changes, including (i) the formation of K junctions between interlayers, (ii) C—F bond creation, and (iii) C—N bond formation. The K junctions enhanced the electric field within the interlayers, significantly improving charge transport [37]. Additionally, halogen atoms were found to have a greater effect on reducing the band gap by creating mid-gaps compared to other dopants. Guo et al. developed K and I-doped g-C_3N_4, where iodine substituted nitrogen atoms and was positioned within the interlayers. The mesoporous structure promoted rapid charge transfer and reduced recombination [38]. In contrast, sulfur doping alone did not improve g-C_3N_4's performance, as it created spaces within the interlayers, restraining charge carrier mobility. This issue was resolved by co-doping with S and K, where the molten salt reduced interlayer distances and enhanced crystallinity, with sulfur reducing charge carrier recombination [39]. Metals were typically used to replace nitrides, while nonmetals were placed in positions enhancing electron movement. Some metal atoms substituted carbon or nitrogen atoms, increasing electron delocalization and altering the band edge. These metal atoms also inserted themselves within interlayers, acting as bridges to improve charge transport. While metal/nonmetal co-doping is more common, metal/metal co-doping is less frequent. Paul and group investigated metal/metal doping in g-C_3N_4 by incorporating Li and Mg into its matrix. The combination of these metals enhanced the absorption capacity, reduced the band gap, and suppressed charge carrier recombination in the co-doped g-C_3N_4 [40]. Meanwhile, Mo et al. prepared B and S co-doped g-C_3N_4 nanotubes for effective photocatalysis. The B dopant facilitated H_2O adsorption, while the S dopant inhibited charge carrier recombination. FT-IR spectra and XRD patterns confirmed the successful synthesis of the B, S co-doped g-C_3N_4, with band alignment showing a shift in the conduction band level after doping [41].

6.3 Photocatalytic Applications

This section highlights the latest advancements in photocatalytic reactions using defect-engineered g-C_3N_4, including H_2 and O_2 evolution via water splitting, reduction of CO_2 to hydrocarbon fuels, the degradation of organic pollutants, and bacterial disinfection. While these applications were briefly discussed in the previous section, this part provides a more concise and comprehensive summary of photocatalytic processes involving defect-induced g-C_3N_4-based catalysts.

6.3.1 Photocatalytic Water Splitting

H_2 generation through water splitting using photocatalysts and solar energy offers a sustainable and clean energy alternative, minimizing dependence on fossil fuels. With an energy density of 140 MJ kg^{-1}, hydrogen surpasses traditional hydrocarbon fuels like diesel and petrol, which typically range between 40–50 MJ kg^{-1} [1, 42]. This makes hydrogen an attractive candidate for future energy applications. Recent years have witnessed significant advancements in g-C_3N_4-based nanomaterials, which have demonstrated improved efficiency in photocatalytic water splitting. The process involves light irradiation that excites g-C_3N_4, generating photogenerated electron-hole pairs. These charge carriers migrate to the surface of the photocatalyst, where they participate in redox reactions to split water molecules into hydrogen and oxygen gases. As depicted in Figure 6.5A, the transfer of charge carriers to active sites without recombination is crucial for achieving efficient water splitting [1]. This section highlights the potential of defect-engineered g-C_3N_4 in driving clean energy solutions and advancing hydrogen-based fuel technologies.

Ran et al. synthesized innovative porous P-doped g-C_3N_4 through a combination of P-doping and thermal exfoliation, achieving a remarkable visible-light photocatalytic H_2-generation activity of 1596 μmol h^{-1} g^{-1} and an apparent quantum yield

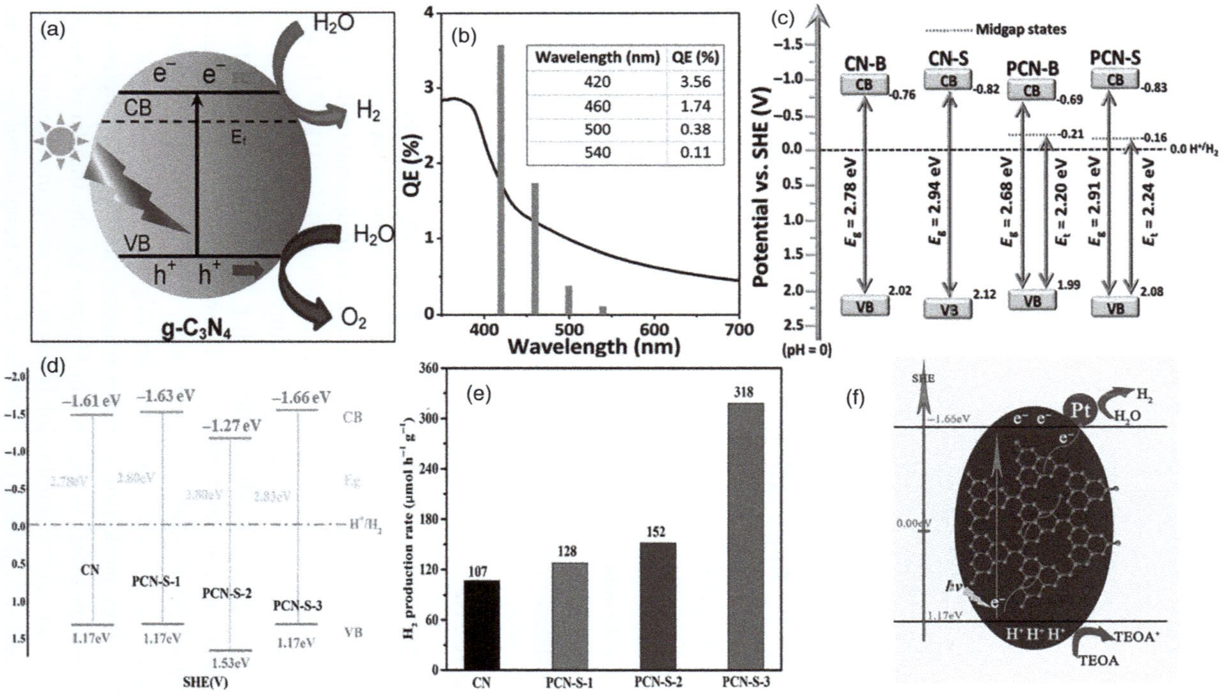

Figure 6.5 (a) Schematic illustration of photocatalytic water splitting for H_2 and O_2 evolution using pristine g-C_3N_4 as a reference. *Source:* Reproduced from Ong et al. [1] / with permission of American Chemical Society. (b) QE of PCN-S at wavelengths of 420, 460, 500, and 540 nm. (c) Electronic band structures of CN-B, CN-S, PCN-B, and PCN-S. *Source:* Reproduced from Ran et al. [43] / with permission of Royal Society of Chemistry. (d) Schematic representation of the band structures, (e) H_2 evolution rate of pure and P-doped g-C_3N_4 samples. (f) Plausible mechanism of charge separation and H_2 evolution. *Source:* Reproduced from Lin et al. [44] / with permission of Chinese Journal of Chemical Engineering.

(AQY) of 3.56% at 420 nm (Figure 6.5B) [43]. P doping introduced empty mid-gap states, extending the light-responsive range up to 557 nm, as confirmed by DFT and experimental studies. Additionally, the macroporous structure enhanced light harvesting and mass transfer. The inherent band gap of g-C_3N_4 reduced from 2.98 to 2.66 eV after P-doping, with mid-gap states positioned at −0.16 eV vs. SHE, enabling efficient H_2 evolution due to their higher energy than the reduction potential of H^+ to H_2 (Figure 6.5C). The findings indicated that photo-excited electrons transitioning from the VB to the mid-gap states possessed sufficient thermodynamic potential to effectively reduce protons, facilitating H_2 production [43]. Guo et al. prepared P-doped hexagonal tubular g-C_3N_4 with a stacked-layer configuration, reaching a hydrogen evolution rate of 67 μmol h^{-1} upon visible-light irradiation and an AQY of 5.68% at 420 nm [45]. The tubular architecture enhanced light scattering and offered more active sites, while P-doping reduced the band gap and improved electrical conductivity. Lin et al. demonstrated that incorporating phosphorus into the lattice of g-C_3N_4 using sodium hypophosphite in varying mass ratios (1:2, 1:10, and 1:20) effectively tuned its properties [44], significantly improving its photocatalytic activity toward H_2 evolution. P-doped g-C_3N_4 samples (PCN-S-1, PCN-S-2, and PCN-S-3) exhibited conduction band potentials more negative than the standard reduction potential of protons (0 V vs. NHE), ensuring favorable conditions for H_2 evolution (Figure 6.5D). The optimized PCN-S sample achieved a hydrogen production rate of 318 μmol h^{-1} g^{-1}, which is 2.98 times greater than that of pure g-C_3N_4 (Figure 6.5E). The proposed photocatalytic mechanism involves visible-light excitation (≥420 nm), where electrons transition from the VB to the CB, leaving holes behind (Figure 6.5F). TEOA, serving as a sacrificial agent, traps photo-generated holes and forms $TEOA^+$, facilitating charge carrier separation. Electrons in the CB subsequently reduce H^+ to generate H_2 [44]. Likewise, Lin and coworkers employed [Bmim][BF_4] as a multifunctional agent during the self-polymerization of urea to create B, F-doped porous g-C_3N_4 nanosheets, resulting in a photocatalytic H_2 production rate 3.9 times higher than pure g-C_3N_4 [46]. Furthermore, Zhang and group reported I-doped g-C_3N_4 solids with a H_2 production rate of approximately 38 μmol h^{-1} doubling the performance of pristine g-C_3N_4. Interestingly, I-doped g-C_3N_4 remained active up to 600 nm, unlike pristine g-C_3N_4, which became inactive beyond 500 nm [47]. Bi et al. prepared Ni-doped g-C_3N_4 via a solvothermal approach, utilizing melamine and acetylacetone nickel as starting materials [48]. Analysis of work function measurements and Mott-Schottky plots revealed changes in surface band bending upon interaction with Ni. The calculated work functions of Ni and pristine g-C_3N_4 were 4.86 and 4.34 eV, respectively. When the two materials were brought into contact, electron transfer occurred from g-C_3N_4 to Ni until their Fermi levels reached equilibrium. This interaction created an interfacial electric field directed from g-C_3N_4 to Ni, enhancing charge separation efficiency. The incorporation of Ni effectively intensified the surface band bending, which contributed to the improved proficiency of photogenerated electron-hole pair separation and boosted H_2 evolution activity [48]. Liu et al. developed S-doped g-C_3N_4 ($C_3N_{4-X}S_X$), which exhibits an exceptional electronic structure with an expanded VB width and a slightly reduced absorbance. This modification resulted in photoreactivity for H_2 evolution that was 7.2 and 8.0 times higher than C_3N_4 under $\lambda > 300$ and 420 nm, respectively. Notably, it also enabled the total oxidation of phenol under $\lambda > 400$ nm, which C_3N_4 could not achieve even under $\lambda > 300$ nm. The superior photoreactivity is attributed to sulfur substituting nitrogen in the lattice and the resulting quantum confinement effect [8a].

6.3.2 Photocatalytic CO_2 Reduction

The rising concentration of CO_2 (a major greenhouse gas) is a critical issue driving global warming and climate change. The primary source of anthropogenic CO_2 emissions is the burning of fossil fuels for energy, a result of industrial and societal development. Given the projected increase in global energy demand and fossil resource consumption over the coming decades, mitigating CO_2 emissions has become essential. Significant efforts have been made to reduce CO_2 emissions both regionally and globally. Inspired by the natural photosynthesis phenomenon in plants, photocatalytic CO_2 reduction into energy-rich products offers a promising solution to both mitigate the greenhouse effect and provide an alternative energy source. This approach effectively addresses environmental protection while simultaneously generating energy [1, 49]. In nature, plants convert CO_2 and H_2O into carbohydrates and O_2 under sunlight at room temperature. Similarly, photocatalytic CO_2 reduction with H_2O integrates both the reductive route of CO_2 fixation and the oxidative route of H_2O splitting, creating a carbon-neutral cycle. In recent years, g-C_3N_4 has emerged as a promising photocatalyst for facilitating CO_2 reduction, largely due to its high conduction band minimum, which supports the reduction half-reaction (Figure 6.6A) [1]. Studies have revealed that the CO_2 photoreduction process is a complex, proton-assisted, multistep reaction involving multiple electrons. Thermodynamically, CO_2 undergoes progressive reduction through the acquisition of 2, 4, 6, or 8 electrons and hydrogen radicals, producing a range of gaseous and liquid hydrocarbons, including HCOOH (liquid), CO (gas), HCHO (liquid), CH_3OH (liquid), and CH_4 (gas) [1, 7].

Dou and colleagues fabricated willow-shaped, Na-doped g-C_3N_4 to boost its photocatalytic activity [52]. The incorporation of Na ions extended the π-conjugated network in g-C_3N_4, enhancing electron transport and significantly improving photocatalytic activity. The CO_2 reduction rate for Na-doped g-C_3N_4 was 6.6 times greater than that of the undoped counterpart. Furthermore, Na doping increased the surface area and promoted the efficient transfer of photoinduced charge carriers [52]. Similarly, Singh et al. synthesized K-doped g-C_3N_4, which was employed for CO_2 conversion into formate [53]. The g-C_3N_4 sheets facilitated CO_2 adsorption and generated charge carriers that supported the electron transfer processes required for the conversion. Additionally, when combined with spirulina and Chlamydomonas reinhardtii, K-doped g-C_3N_4 enhanced formate production rates to 751 and 1192 L^{-1}, respectively, representing increases of 13% and 59% compared to the non-catalytic algal reactions [53]. Arumugam et al. synthesized B, O, P, and S-doped g-C_3N_4 for CO_2 reduction applications [50]. Among these, S-doped g-C_3N_4 exhibited superior performance, generating 55.10 nmol (mL $H_2O{\cdot}g_{cat.})^{-1}$ of CH_4 under UV

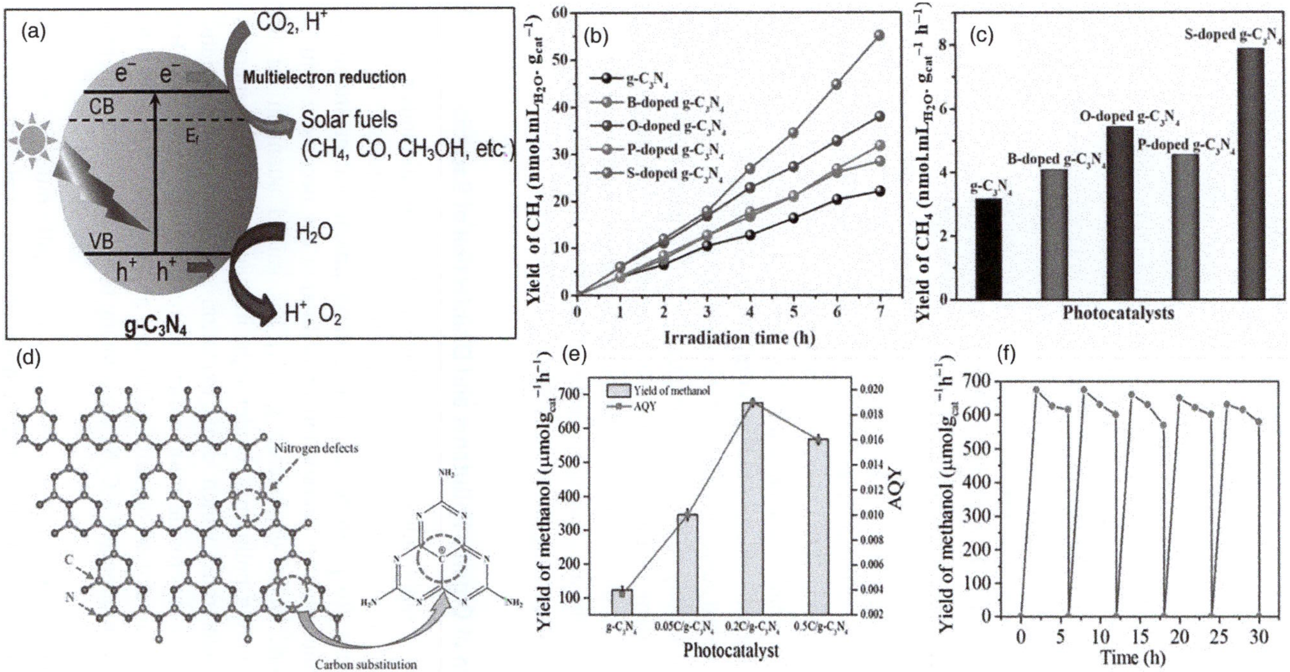

Figure 6.6 (a) Schematic representation of photocatalytic CO_2 reduction with H_2O into solar fuels using g-C_3N_4 as the catalyst under solar irradiation. *Source:* Reproduced from Ong et al. [1] / with permission of American Chemical Society. (b) Photocatalytic CO_2 reduction to CH_4 at different irradiation times. (c) Maximum CH_4 formation rates using pure, B, O, P, and S-doped g-C_3N_4 photocatalysts. *Source:* Reproduced from Arumugam et al. [50] / with permission of Elsevier. (d) Schematic representation of C atom substitution for N atom in 0.2C/g-C_3N_4. (e) Methanol yield of pristine g-C_3N_4 and C/g-C_3N_4 (0.05, 0.2, and 0.5 wt%) after two hours of irradiation. (f) Recycling performance of the 0.2C/g-C_3N_4 sample for photocatalytic CO_2 reduction over six hours. *Source:* Reproduced from Nor et al. [51] / with permission of Elsevier.

light for seven hours, which was 1.9, 1.4, 1.7, and 2.4 times higher than the B, O, P-doped, and pristine g-C_3N_4 photocatalysts, respectively (Figure 6.6B,C). The CH_4 production remained stable over three cycles, indicating excellent durability. Although S doping did not enhance the surface area or light absorption capabilities, the marked improvement in catalytic activity was attributed to enhanced charge carrier separation and migration [50]. Wang et al. synthesized sulfur-doped g-C_3N_4 photocatalysts using thiourea as the sulfur source for CO_2 reduction to CH_3OH [49]. DFT studies revealed that sulfur doping introduced impurity states, enabling easier electron excitation from the valence band to the impurity state or from the impurity state to the conduction band. This doping induced spin polarization and narrowed the band gap from 2.7 to 2.63 eV, extending light absorption. As a result, sulfur-doped g-C_3N_4 demonstrated improved photocatalytic performance, achieving a CH_3OH yield of 1.12 $\mu mol\,g^{-1}$ compared to 0.81 $\mu mol\,g^{-1}$ for pristine g-C_3N_4 [49]. Nor et al. investigated the influence of carbon doping on the optical and electronic properties of g-C_3N_4 using experimental techniques and DFT analysis [51]. By replacing a nitrogen atom with a carbon atom at a defect level of 12.7%, the photocatalyst exhibited enhanced charge carrier separation and superior activity compared to bare g-C_3N_4. The defect generated at the C_{N1} site extended light absorption (Figure 6.6D), sustained a narrow band gap, and provided a negative CB potential, enabling efficient reduction of CO_2 to CH_3OH. The optimized 0.2C/g-C_3N_4 photocatalyst achieved a CH_3OH yield of 651.7 $\mu mol\,g_{cat.}{}^{-1}\,h^{-1}$ with an AQY of 0.019, approximately 40% higher than pure g-C_3N_4 (Figure 6.6e). Stability tests confirmed consistent performance over five catalytic cycles (Figure 6.6f). Carbon doping enhanced the surface area, increased active sites, optimized energy levels, and extended light absorption, as revealed by DFT studies, making it a promising strategy for improving photocatalytic CO_2 reduction [51].

6.3.3 Degradation of Organic Pollutants and Disinfection of Bacteria

Population growth and rapid industrialization have led to the continuous release of hazardous pollutants into the environment, posing significant risks to human health and ecosystems. To promote sustainable development and environmental preservation, photocatalytic degradation of pollutants has become a critical focus for researchers. Defect-engineered g-C_3N_4 has gained attention for its effectiveness in degrading pollutants. Under light irradiation, g-C_3N_4 generates reactive holes and electrons, which drive oxidation and reduction reactions. These reactions produce superoxide anion radicals ($\bullet O_2{}^-$), hydroxyl radicals ($\bullet OH$), and protons, which mineralize organic pollutants and detoxify harmful substances, making defect-engineered g-C_3N_4 a promising material for environmental remediation [1, 7].

Zhang et al. synthesized C-doped g-C_3N_4 with a C/N molar ratio of 0.8 which showed ~100% degradation of 4-NP after three hours [54]. Yan et al. synthesized a B-doped g-C_3N_4 photocatalyst by heating a mixture of melamine and boron oxide [36]. EPR analysis and comparative experiments revealed that Rhodamin B (RhB) degradation primarily resulted from photogenerated hole oxidation, while methyl orange (MO) degradation involved an overall reaction, favoring photoreduction over photooxidation. Boron doping enhanced the photocatalytic activity for RhB degradation by improving both dye and light absorption properties of the catalyst [36]. Peng et al. developed an single atom (SA) Fe- g-C_3N_4 (600) catalyst for efficient and stable peroxymonosulfate-assisted degradation of tetracycline. The active nonradical iron-oxo species played a key role in the degradation process [55]. Similarly, Faryad et al. prepared a spinach-derived B-doped g-C_3N_4/TiO_2 nanocomposite using a hydrothermal method. This composite exhibited remarkable efficacy and stability in the degradation of methylene blue (MB) pollutants [56]. Liu et al. synthesized efficient S-doped g-C_3N_4 by heating dicyanamide in an H_2S atmosphere at 450 °C, enabling uniform S-atom incorporation into the interlayer galleries of g-C_3N_4 [8a]. This caused a blue shift in the absorption edge and a reduction in the band gap from 2.12 to 1.79 eV due to an upshift in VB maxima. The S-doping weakened hydrogen bonds, partially replaced N-atoms, and improved photocatalytic performance by enhancing light absorption, narrowing the band gap, shifting CB and VB potentials negatively, and increasing charge transfer. These changes enabled S-doped g-C_3N_4 to achieve superior photoreduction ability, as demonstrated by the 80% phenol degradation within 10 hours [8a]. Similarly, Ke et al. developed S-doped g-C_3N_4 from urea and benzyl disulfide under an Ar atmosphere, achieving red-shifted absorption edges which facilitate the photocatalytic reduction of UO_2^{2+} pollutants [57]. Numerous studies have highlighted defect-engineered g-C_3N_4 as highly effective catalysts for removing organic pollutants, particularly in dye degradation from wastewater.

Doped g-C_3N_4 photocatalysts have shown significant potential in bacterial disinfection by generating reactive oxidative species (ROS) under light exposure. These ROS (•OH and $•O_2^-$), effectively damage bacterial cell walls and structures, leading to their deactivation. Additionally, g-C_3N_4 facilitates the formation of H_2O_2, further enhancing its bactericidal properties [1, 7]. For example, Chen et al. synthesized sulfur-doped g-C_3N_4 nanosheets (SCNNSs) and improved their disinfection efficiency by combining them with Tetrastigma hemsleyanum polysaccharide-3 (THDG-3) through electrostatic interactions [58]. Likewise, Yang et al. developed oxygen-doped g-C_3N_4/carbon dots (O-CN/CDs) nanosheets using a hydrothermal process followed by electrostatic self-assembly. These modifications enhanced light utilization and charge separation, significantly improving bacterial disinfection in water. The process involves valence band holes oxidizing water to produce hydroxyl radicals (•OH), while oxygen reduction generates superoxide radicals ($•O_2^-$) and H_2O_2, which collectively disrupt and kill bacterial cells [59].

6.4 Conclusion and Perspectives

In conclusion, advancements in defect-engineered g-C_3N_4 photocatalysts mark a significant milestone in renewable energy and environmental remediation technologies. By strategically introducing various kinds of defects, researchers have successfully enhanced the light absorption, charge carrier separation, and surface reactivity of g-C_3N_4, enabling remarkable performance in applications like hydrogen generation, CO_2 conversion, pollutant degradation, and bacterial disinfection. These breakthroughs underscore the immense potential of g-C_3N_4-based materials in addressing the global challenges of energy sustainability and environmental health.

Moving forward, the focus must shift toward refining doping or defect-induced techniques, optimizing electronic and structural properties, and tailoring photocatalysts for targeted applications. Improving stability under harsh conditions and achieving scalability are critical steps toward industrial adoption. Combining g-C_3N_4 with other semiconductors or co-catalysts holds potential for further efficiency enhancement, while computational modeling can expedite material design by reducing reliance on trial-and-error experiments. These advancements will not only accelerate the adoption of g-C_3N_4-based systems but also pave the way for innovative solutions in energy conversion and environmental protection, driving progress toward a sustainable future.

Conflicts of Interest

There are no known conflicts of interest to declare.

References

1 Ong, W.-J., Tan, L.-L., Ng, Y.H. et al. (2016). Graphitic carbon nitride (g-C_3N_4)-based photocatalysts for artificial photosynthesis and environmental remediation: are we a step closer to achieving sustainability? *Chem. Rev.* 116: 7159–7329.

2 Gogoi, D., Karmur, R.S., Das, M.R., and Ghosh, N.N. (2023). Spent tea-waste-derived porous carbon-supported truncated octahedral Cu_2O for highly efficient energy storage devices. *Energy Fuel* 37: 14350–14364.

3 Fujishima, A. and Honda, K. (1972). Electrochemical photolysis of water at a semiconductor electrode. *Nature* 238: 37–38.

4 (a) Kim, D., Sakimoto, K.K., Hong, D., and Yang, P. (2015). Artificial photosynthesis for sustainable fuel and chemical production. *Angew. Chem. Int. Ed.* 54: 3259–3266. (b) Ashford, D.L., Gish, M.K., Vannucci, A.K. et al. (2015). Molecular chromophore–catalyst assemblies for solar fuel applications. *Chem. Rev.* 115: 13006–13049.

5 Carey, J.H., Lawrence, J., and Tosine, H.M. (1976). Photodechlorination of PCB's in the presence of titanium dioxide in aqueous suspensions. *Bull. Environ. Contam. Toxicol.* 16: 697–701.

6 Inoue, T., Fujishima, A., Konishi, S., and Honda, K. (1979). Photoelectrocatalytic reduction of carbon dioxide in aqueous suspensions of semiconductor powders. *Nature* 277: 637–638.

7 Khan, M.A., Mutahir, S., Shaheen, I. et al. (2025). Recent advances over the doped g-C_3N_4 in photocatalysis: A review. *Coord. Chem. Rev.* 522: 216227.

8 (a) Liu, G., Niu, P., Sun, C. et al. (2010). Unique electronic structure induced high photoreactivity of sulfur-doped graphitic C_3N_4. *J. Am. Chem. Soc.* 132: 11642–11648. (b) Tay, Q., Kanhere, P., Ng, C.F. et al. (2015). Defect engineered g-C_3N_4 for efficient visible light photocatalytic hydrogen production. *Chem. Mater.* 27: 4930–4933. (c) Niu, P., Yin, L.-C., Yang, Y.-Q. et al. (2014). Increasing the visible light absorption of graphitic carbon nitride (melon) photocatalysts by homogeneous self-modification with nitrogen vacancies. *Adv. Mater. (Deerfield Beach, Fla.)* 26: 8046–8052.

9 Goettmann, F., Fischer, A., Antonietti, M., and Thomas, A. (2006). Metal-free catalysis of sustainable Friedel–Crafts reactions: direct activation of benzene by carbon nitrides to avoid the use of metal chlorides and halogenated compounds. *Chem. Commun.* 4530-4532.

10 Wang, X., Maeda, K., Thomas, A. et al. (2009). A metal-free polymeric photocatalyst for hydrogen production from water under visible light. *Nat. Mater.* 8: 76–80.

11 Teter, D.M. and Hemley, R.J. (1996). Low-compressibility carbon nitrides. *Science* 271: 53–55.

12 Zheng, Y., Lin, L., Wang, B., and Wang, X. (2015). Graphitic carbon nitride polymers toward sustainable photoredox catalysis. *Angew. Chem. Int. Ed.* 54: 12868–12884.

13 Kroke, E., Schwarz, M., Horath-Bordon, E. et al. (2002). From trichloro-tri-s-triazine to graphitic C_3N_4 structures. *New J. Chem.* 26: 508–512.

14 Hussain, A., Maqsood, S., Ji, R. et al. (2023). Investigation of transition metal-doped graphitic carbon nitride for MO dye degradation. *Diam. Relat. Mater.* 132: 109648.

15 (a) Nowotny, M.K., Sheppard, L.R., Bak, T., and Nowotny, J. (2008). Defect chemistry of titanium dioxide. Application of defect engineering in processing of TiO_2 based photocatalysts. *J. Phys. Chem. C* 112: 5275–5300. (b) Hong, Z., Shen, B., Chen, Y. et al. (2013). Enhancement of photocatalytic H_2 evolution over nitrogen-deficient graphitic carbon nitride. *J Mater Chem A* 1: 11754–11761. (c) Li, S., Dong, G., Hailili, R. et al. (2016). Effective photocatalytic H_2O_2 production under visible light irradiation at g-C_3N_4 modulated by carbon vacancies. *Appl. Catal. B Environ.* 190: 26–35.

16 Thompson, T.L. and Yates, J.T. (2006). Surface science studies of the photoactivation of TiO_2 new photochemical processes. *Chem. Rev.* 106: 4428–4453.

17 Ma, X., Lv, Y., Xu, J. et al. (2012). A strategy of enhancing the photoactivity of g-C_3N_4 via doping of nonmetal elements: a first-principles study. *J. Phys. Chem. C* 116: 23485–23493.

18 Niu, P., Liu, G., and Cheng, H.-M. (2012). Nitrogen Vacancy-Promoted Photocatalytic Activity of Graphitic Carbon Nitride. *J. Phys. Chem. C* 116: 11013–11018.

19 Liang, Q., Li, Z., Huang, Z.H. et al. (2015). Holey graphitic carbon nitride nanosheets with carbon vacancies for highly improved photocatalytic hydrogen production. *Adv. Funct. Mater.* 25: 6885–6892.

20 Li, X., Hartley, G., Ward, A.J. et al. (2015). Hydrogenated defects in graphitic carbon nitride nanosheets for improved photocatalytic hydrogen evolution. *J. Phys. Chem. C* 119: 14938–14946.

21 Yang, P., Zhao, J., Qiao, W. et al. (2015). Ammonia-induced robust photocatalytic hydrogen evolution of graphitic carbon nitride. *Nanoscale* 7: 18887–18890.

22 Cui, Y., Zhang, G., Lin, Z., and Wang, X. (2016). Condensed and low-defected graphitic carbon nitride with enhanced photocatalytic hydrogen evolution under visible light irradiation. *Appl. Catal. B Environ.* 181: 413–419.

23 (a) Jorge, A.B., Martin, D.J., Dhanoa, M.T. et al. (2013). H_2 and O_2 evolution from water half-splitting reactions by graphitic carbon nitride materials. *J. Phys. Chem. C* 117: 7178–7185. (b) Chen, Y., Wang, B., Lin, S. et al. (2014). Activation of n→ π* transitions in two-dimensional conjugated polymers for visible light photocatalysis. *J. Phys. Chem. C* 118: 29981–29989. (c) Ho, W., Zhang, Z., Xu, M. et al. (2015). Enhanced visible-light-driven photocatalytic removal of NO: Effect on layer distortion on g-C_3N_4 by H_2 heating. *Appl. Catal. B Environ.* 179: 106–112.

24 (a) Jasni, N., Iqbal, A., Ahmad, M.N. et al. (2022). Synthesis and characterisation of Li-modified g-C_3N_4. *Mater Today Proc* 57: 1154–1161. (b) Hayat, A., Syed, J.A.S., Al-Sehemi, A.G. et al. (2022). State of the art advancement in rational design of g-C_3N_4 photocatalyst for efficient solar fuel transformation, environmental decontamination and future perspectives. *Int. J. Hydrog. Energy* 47: 10837–10867. (c) Jiang, J., Cao, S., Hu, C., and Chen, C. (2017). A comparison study of alkali metal-doped g-C_3N_4 for visible-light photocatalytic hydrogen evolution. *Chin. J. Catal.* 38: 1981–1989.

25 Yan, W., Yan, L., and Jing, C. (2019). Impact of doped metals on urea-derived g-C_3N_4 for photocatalytic degradation of antibiotics: structure, photoactivity and degradation mechanisms. *Appl. Catal. B Environ.* 244: 475–485.

26 Ma, S., Zhan, S., Jia, Y. et al. (2016). Enhanced disinfection application of Ag-modified g-C_3N_4 composite under visible light. *Appl. Catal. B Environ.* 186: 77–87.

27 Ding, G., Wang, W., Jiang, T. et al. (2013). Highly selective synthesis of phenol from benzene over a vanadium-doped graphitic carbon nitride catalyst. *ChemCatChem* 5: 192–200.

28 Li, J., Shen, B., Hong, Z. et al. (2012). A facile approach to synthesize novel oxygen-doped g-C_3N_4 with superior visible-light photoreactivity. *Chem. Commun.* 48: 12017–12019.

29 Huang, Z.-F., Song, J., Pan, L. et al. (2015). Carbon nitride with simultaneous porous network and O-doping for efficient solar-energy-driven hydrogen evolution. *Nano Energy* 12: 646–656.

30 Zhu, Y.-P., Ren, T.-Z., and Yuan, Z.-Y. (2015). Mesoporous phosphorus-doped g-C_3N_4 nanostructured flowers with superior photocatalytic hydrogen evolution performance. *ACS Appl. Mater. Interfaces* 7: 16850–16856.

31 Zhao, Z., Sun, Y., Dong, F. et al. (2015). Template synthesis of carbon self-doped gC_3N_4 with enhanced visible to near-infrared absorption and photocatalytic performance. *RSC Adv.* 5: 39549–39556.

32 Zhou, Y., Zhang, L., Huang, W. et al. (2016). N-doped graphitic carbon-incorporated g-C_3N_4 for remarkably enhanced photocatalytic H_2 evolution under visible light. *Carbon* 99: 111–117.

33 Hu, S., Ma, L., You, J. et al. (2014). A simple and efficient method to prepare a phosphorus modified gC_3N_4 visible light photocatalyst. *RSC Adv.* 4: 21657–21663.

34 Zhou, Y., Zhang, L., Liu, J. et al. (2015). Brand new P-doped gC_3N_4: enhanced photocatalytic activity for H_2 evolution and Rhodamine B degradation under visible light. *J Mater Chem A* 3: 3862–3867.

35 Zhang, J., Sun, J., Maeda, K. et al. (2011). Sulfur-mediated synthesis of carbon nitride: band-gap engineering and improved functions for photocatalysis. *Energy Environ. Sci.* 4: 675–678.

36 Yan, S., Li, Z., and Zou, Z. (2010). Photodegradation of rhodamine B and methyl orange over boron-doped g-C_3N_4 under visible light irradiation. *Langmuir* 26: 3894–3901.

37 Zhu, Q.-H., Chen, Z., Tang, L.-N. et al. (2019). K and halogen binary-doped graphitic carbon nitride (g-C_3N_4) toward enhanced visible light hydrogen evolution. *Int. J. Hydrog. Energy* 44: 27704–27712.

38 Guo, Y., Liu, Q., Li, Z. et al. (2018). Enhanced photocatalytic hydrogen evolution performance of mesoporous graphitic carbon nitride co-doped with potassium and iodine. *Appl. Catal. B Environ.* 221: 362–370.

39 Chen, L., Zhu, D., Li, J. et al. (2020). Sulfur and potassium co-doped graphitic carbon nitride for highly enhanced photocatalytic hydrogen evolution. *Appl. Catal. B Environ.* 273: 119050.

40 Paul, D.R., Sharma, R., Singh, S. et al. (2023). Mg/Li Co-doped g-C_3N_4: an excellent photocatalyst for wastewater remediation and hydrogen production applications towards sustainable development. *Int. J. Hydrog. Energy* 48: 37746–37761.

41 Mo, Z., Miao, Z., Yan, P. et al. (2023). Electronic and energy level structural engineering of graphitic carbon nitride nanotubes with B and S co-doping for photocatalytic hydrogen evolution. *J. Colloid Interface Sci.* 645: 525–532.

42 Edwards, P.P., Kuznetsov, V.L., and David, W.I. (2007). Hydrogen energy. *Philos. Trans. R. Soc. A Math. Phys. Eng. Sci.* 365: 1043–1056.

43 Ran, J., Ma, T.Y., Gao, G. et al. (2015). Porous P-doped graphitic carbon nitride nanosheets for synergistically enhanced visible-light photocatalytic H_2 production. *Energy Environ. Sci.* 8: 3708–3717.

44 Lin, Q., Li, Z., Lin, T. et al. (2020). Controlled preparation of P-doped g-C_3N_4 nanosheets for efficient photocatalytic hydrogen production. *Chin. J. Chem. Eng.* 28: 2677–2688.

45 Guo, S., Deng, Z., Li, M. et al. (2016). Phosphorus-doped carbon nitride tubes with a layered micro-nanostructure for enhanced visible-light photocatalytic hydrogen evolution. *Angew. Chem. Int. Ed.* 55: 1830–1834.

46 Lin, Z. and Wang, X. (2014). Ionic liquid promoted synthesis of conjugated carbon nitride photocatalysts from urea. *ChemSusChem* 7: 1547–1550.

47 Zhang, G., Zhang, M., Ye, X. et al. (2013). Iodine modified carbon nitride semiconductors as visible light photocatalysts for hydrogen evolution. *Adv. Mater. (Deerfield Beach, Fla.)* 26: 805–809.

48 Bi, L., Xu, D., Zhang, L. et al. (2015). Metal Ni-loaded gC_3N_4 for enhanced photocatalytic H_2 evolution activity: the change in surface band bending. *Phys. Chem. Chem. Phys.* 17: 29899–29905.
49 Wang, K., Li, Q., Liu, B. et al. (2015). Sulfur-doped g-C_3N_4 with enhanced photocatalytic CO_2-reduction performance. *Appl. Catal. B Environ.* 176-177: 44–52.
50 Arumugam, M., Tahir, M., and Praserthdam, P. (2022). Effect of nonmetals (B, O, P, and S) doped with porous g-C_3N_4 for improved electron transfer towards photocatalytic CO_2 reduction with water into CH_4. *Chemosphere* 286: 131765.
51 Nor, N.U.M., Mazalan, E., and Amin, N.A.S. (2021). Insights into enhancing photocatalytic reduction of CO_2: Substitutional defect strategy of modified g-C_3N_4 by experimental and theoretical calculation approaches. *J. Alloys Compd.* 871: 159464.
52 Dou, Q., Hou, J., Hussain, A. et al. (2022). One-pot synthesis of sodium-doped willow-shaped graphitic carbon nitride for improved photocatalytic activity under visible-light irradiation. *J. Colloid Interface Sci.* 624: 79–87.
53 Singh, A., Alam, U., Chakraborty, P. et al. (2023). A sustainable approach for the production of formate from CO_2 using microalgae as a clean biomass and improvement using potassium-doped g-C_3N_4. *Chem. Eng. J.* 454: 140303.
54 Zhang, P., Li, X., Shao, C., and Liu, Y. (2015). Hydrothermal synthesis of carbon-rich graphitic carbon nitride nanosheets for photoredox catalysis. *J Mater Chem A* 3: 3281–3284.
55 Peng, X., Wu, J., Zhao, Z. et al. (2022). Activation of peroxymonosulfate by single-atom Fe-g-C_3N_4 catalysts for high efficiency degradation of tetracycline via nonradical pathways: role of high-valent iron-oxo species and Fe–Nx sites. *Chem. Eng. J.* 427: 130803.
56 Faryad, S., Azhar, U., Tahir, M.B. et al. (2023). Spinach-derived boron-doped g-C_3N_4/TiO_2 composites for efficient photo-degradation of methylene blue dye. *Chemosphere* 320: 138002.
57 Ke, L., Li, P., Wu, X. et al. (2017). Graphene-like sulfur-doped g-C_3N_4 for photocatalytic reduction elimination of UO22+ under visible Light. *Appl. Catal. B Environ.* 205: 319–326.
58 Chen, X., Weng, M., Lan, M. et al. (2021). Superior antibacterial activity of sulfur-doped g-C_3N_4 nanosheets dispersed by Tetrastigma hemsleyanum Diels & Gilg's polysaccharides-3 solution. *Int. J. Biol. Macromol.* 168: 453–463.
59 Yang, X., Sun, J., Sheng, L. et al. (2023). Carbon dots cooperatively modulating photocatalytic performance and surface charge of O-doped g-C3N4 for efficient water disinfection. *J. Colloid Interface Sci.* 631: 25–34.

7

Structures, Properties, and Applications of Defective Graphene-Based Photocatalysts

Hanedi Elhleli[1,2] *and Younes Moussaoui*[2,3]

[1] *University of Gafsa, Faculty of Sciences of Gafsa, Laboratory for the Application of Materials to the Environment, Water and Energy (LR21ES15), Sidi Ahmed Zarroug, 2112, Gafsa, Tunisia*
[2] *University of Gafsa, Faculty of Sciences of Gafsa, Sidi Ahmed Zarroug, 2112, Gafsa, Tunisia*
[3] *University of Sfax, Faculty of Sciences of Sfax, Organic Chemistry Laboratory (LR17ES08), Soukra km 4, 3038, Sfax, Tunisia*

7.1 Introduction

Graphene, a two-dimensional material consisting of carbon atoms arranged in a honeycomb lattice, is renowned for its excellent electrical conductivity, thermal stability, and mechanical strength. However, the highly ordered structure of pristine graphene limits its efficiency in many applications, especially photocatalysis, which requires increased reactivity and tunable features. To overcome this, researchers and scientists have introduced defects, including vacancies, dislocations, and grain boundaries (Figure 7.1), or doping with heteroatoms like nitrogen, boron, sulfur, and phosphorus, which significantly modify their chemical, physical, and electronic characteristics [2–5]. Although concerns exist that defects could undermine its structural integrity, they actually offer significant benefits by boosting reactivity, thereby widening the range of potential applications [6]. In catalysis, defective graphene shows a lot of promise because it has catalytic active sites that make electrochemical reactions like the oxygen reduction reaction (ORR) and the hydrogen evolution reaction (HER) more efficient. These reactions are essential in sustainable energy systems, including fuel cells and water splitting technologies. Yang et al. [7], for example, reported that nitrogen-doped defective graphene had better ORR performance than platinum-based catalysts, especially in terms of durability and cost-effectiveness [7]. Similarly, Suragtkhuu et al. [8] demonstrated that boron-doped defective graphene enhances HER, highlighting its potential for large-scale hydrogen production [8]. Additional reported studies further support the role of defective graphene in improving catalytic efficiency for CO_2 reduction, illustrating its versatility in clean energy technologies [9, 10]. In the domain of energy storage, defective graphene displays remarkable improvements in the performance of lithium-ion batteries, sodium-ion batteries, and supercapacitors. Defects increase ion diffusion pathways and provide more active

Defect Engineering in Photocatalysis: From Fundamentals to Applications,
First Edition. Edited by Chuanyi Wang, Abdelkader Labidi, and Qiuhui Zhu.

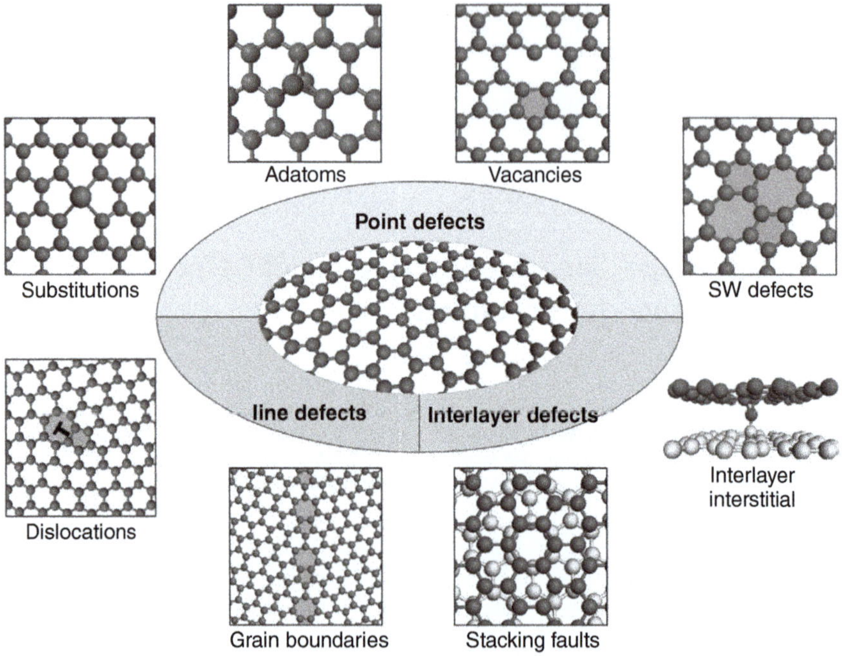

Figure 7.1 Schematic illustration of different defects in graphene-based materials. *Source:* Reproduced from Xu and Sun [1] / with permission of Elsevier.

sites for ion adsorption, enhancing both energy density and charge/discharge rates. For instance, Guo et al. [11] revealed that defective graphene-based anode materials greatly enhanced lithium storage capacity, while research by Zhang et al. [12] highlighted the improvements in sodium storage, broadening its potential for next-generation batteries. Additionally, defective graphene has improved the cycling stability of supercapacitors as indicated by Bokhari et al. [13], and offers faster charge–discharge cycles, which are critical for high-efficiency energy storage systems [14–16]. The enhanced chemical reactivity and increased surface area introduced by defects in graphene also significantly benefit sensor technologies. The high sensitivity of defective graphene to chemical and environmental variations makes it an ideal material for detecting gases like ammonia, nitrogen dioxide, and carbon monoxide. Souza et al. [17] assessed the gas sensing capabilities of defective graphene, showing how its reactivity with gas molecules enhances sensitivity, leading to rapid detection of gaseous compounds even at trace amounts. Other reported works indicated that defect-engineered graphene sensors outperform pristine graphene in terms of sensitivity, stability, and response time, particularly in hazardous gas detection and environmental remediation [18, 19]. In water purification, defective graphene has garnered significant attention due to its ability to adsorb a wide range of pollutants, including heavy metals, organic dyes, and pharmaceutical residues. The increased number of adsorption sites introduced by defects enhances its interaction with contaminants. For example, Ghenaatian et al. noted that defective graphene effectively removes heavy metals from wastewater [20], while other studies highlighted its role in removing dyes and organic pollutants [21, 22]. Graphene-based membranes with tailored defect structures are also being developed for advanced filtration and desalination technologies, offering potential solutions to global water scarcity [23, 24]. In electronics and optoelectronics, the introduction of defects into graphene allows researchers to tailor its electronic properties for specific applications [25]. By controlling defect density and type, the bandgap of graphene can be tuned, enabling its use in transistors, photodetectors, and flexible electronic devices [25]. In 2018, Zeng et al. [26] emphasized that defect-engineered graphene is particularly suited for transparent conductive films in solar cells and flexible displays, as it offers a combination of tunable conductivity and flexibility. Further studies by Lee et al. [27] showed that the electrical and optical properties of this material can be optimized for next-generation optoelectronic devices, such as light-emitting diodes and lasers. One rapidly growing application area for defective graphene is photocatalysis. Through introducing defects (i.e., line defects, point defects, and interlayer defects) (Figure 7.1) [1], particularly through doping with elements like nitrogen, boron, or sulfur, its ability to absorb light and separate charge carriers was greatly enhanced, making it an excellent material in photocatalytic reactions. This is crucial for environmental remediation, such as degrading organic pollutants, and for solar-driven hydrogen production.

For example, Van Dao et al. [28] indicated that nitrogen-doped defective graphene significantly improves photocatalytic hydrogen evolution due to the enhanced charge separation and increased reactive sites in graphene. Similarly, Tibodee et al. [29] reported that defective graphene-TiO_2 composites exhibited superior photocatalytic activity in organic pollutant degradation, demonstrating the potential of defective graphene in solar energy conversion and environmental remediation. In summary, defective graphene offers unique advantages across a wide range of applications due to its enhanced reactivity, tunable electronic properties, and high surface area. The ability to introduce and control defects allows researchers to optimize its performance for specific applications, from energy storage and catalysis to sensing and water purification. The development of defect-engineered graphene materials is crucial for advancing current technologies and driving innovation in emerging fields, particularly in energy and environmental applications. This chapter offers knowledge about the process of designing defective graphene-based photocatalysts and depicts their application in the photocatalysis process for various purposes, including energy conversion and environmental remediation, with a focus on their advantages and disadvantages.

7.2 Preparation Methods and Characterizations

7.2.1 Preparation Methods

Defect creation in graphene-based materials is essential in enhancing their photocatalytic performance due to their ability to generate active sites for catalytic reactions. Some reported methods for preparing defective graphene are stated below:

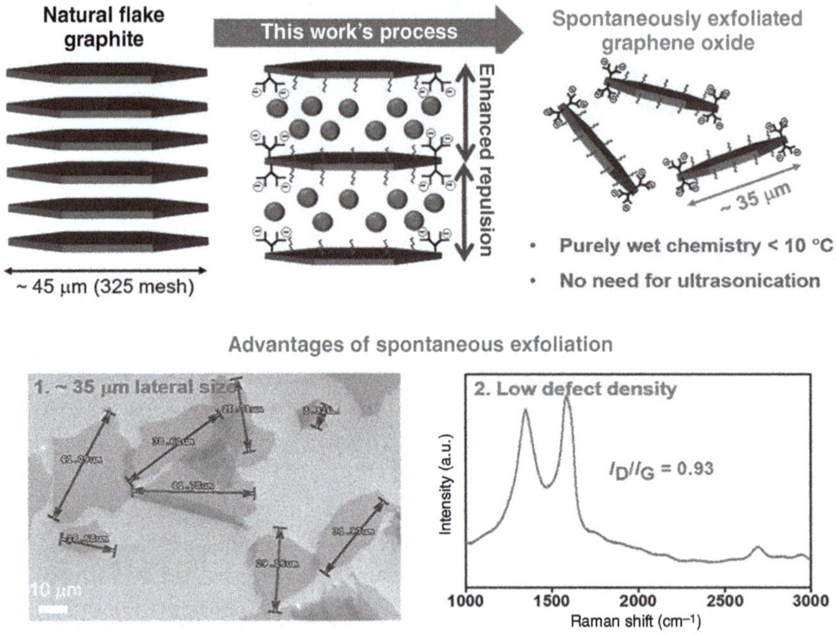

Figure 7.2 Synthesis scheme of high-quality graphene oxide with spontaneous exfoliation of graphene oxide without the need for ultrasonication. *Source:* Reproduced from Lee et al. [30] / with permission of Elsevier.

Chemical exfoliation and reduction have mostly been used to oxidize graphite to produce graphene oxide (GO), which is shown in Figure 7.2. GO is then reduced, usually by hydrazine or Hummers' method, or other chemical reducers. The oxidation stage introduces defects into the graphene structure, primarily in the form of oxygen-containing functional groups (carboxyl, hydroxyl, and epoxy groups). Depending on the chosen method, the reduction process either retains or partially removes these defects [30].

Ball milling is a mechanical method that generates defects in graphite through high-energy ball milling, which involves the repeated impact and friction between graphite particles and milling media (Figure 7.3). The intense forces applied during the milling process result in the exfoliation of graphite into graphene sheets while simultaneously introducing various defects such as vacancies, structural distortions, and functional groups. The severity and type of defects can be controlled by adjusting the milling time, speed, and other process parameters [31].

Plasma treatment is a process that involves exposing graphene to a reactive plasma (such as oxygen, hydrogen, or nitrogen), inducing etching and functionalization of the graphene surface. The high-energy ions and radicals in the plasma interact with the graphene, leading to the removal of carbon atoms and the introduction of vacancies and defects (Figure 7.4). Additionally, the process can functionalize the graphene by attaching various chemical groups (oxygen-containing or nitrogen-containing groups) to its surface, modifying its electronic, chemical, and

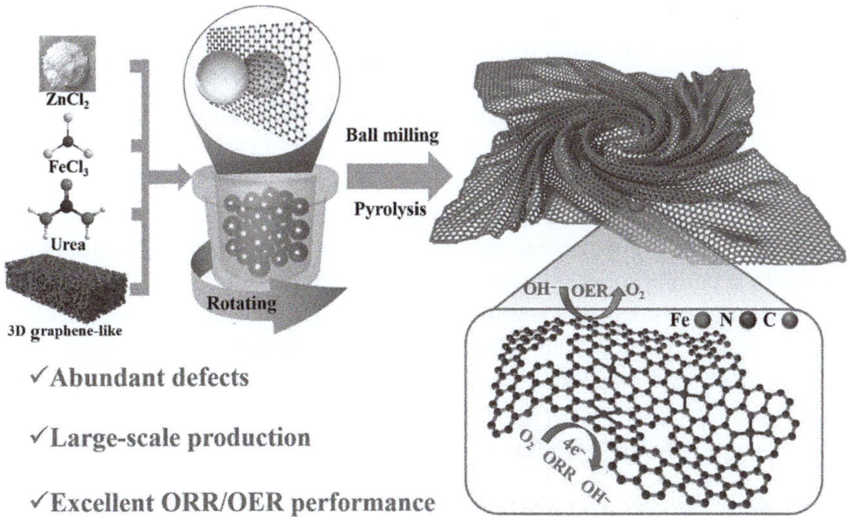

Figure 7.3 Preparation of defect-rich iron/nitrogen co-doped graphene-based materials employing ball-milling and $ZnCl_2$-induced defects strategy using $FeCl_3$ and urea as iron source and nitrogen source. *Source:* Reproduced from Liu et al. [31] / with permission of Elsevier.

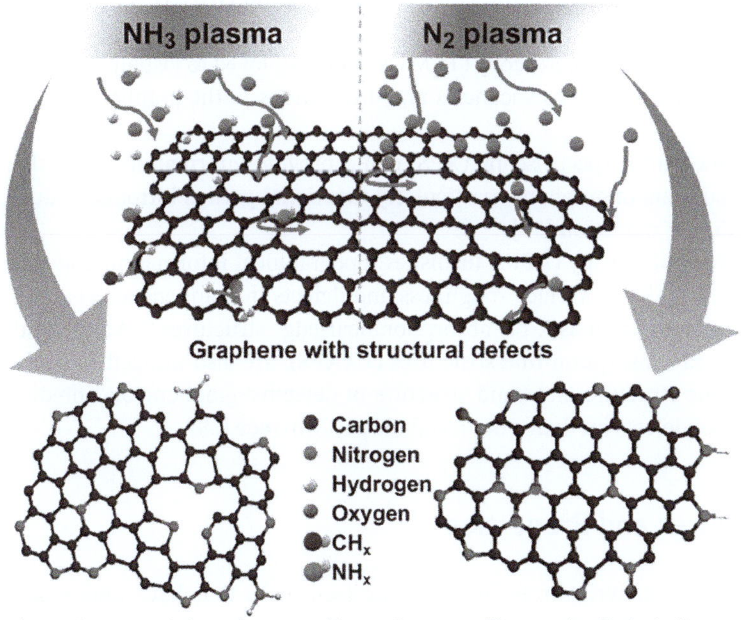

Figure 7.4 Cold gaseous plasma post-treatment for nitrogen incorporation in graphene nanowalls (CNWs). *Source:* Santhosh et al. [32] / Springer Nature / CC BY 4.0.

mechanical properties. Adjusting plasma composition, power, and exposure time can control the type and extent of defects, as well as the nature of the functional groups [32].

Thermal annealing involves subjecting graphene to controlled heating in an inert or reducing atmosphere, typically at high temperatures [33]. This process helps get rid of oxygen-containing functional groups (hydroxyl, epoxy, carboxyl) or even single carbon atoms from the graphene lattice. This creates flaws like vacancies or structural distortions. The thermal treatment can also enhance the sp^2 bonding in graphene by repairing some defects or causing further disorder, depending on the temperature, duration, and the specific atmosphere utilized. This method provides a versatile means of tuning the defect density and modifying the electronic and chemical properties of graphene.

7.2.2 Characterization Techniques

Characterization of defective graphene is crucial to understanding its properties and determining the nature and density of defects. The following are some examples of characterization techniques for defective graphene:

i) Raman spectroscopy is widely used to study defects in graphene. Generally speaking, in defective graphene, the D-band (~1350 cm^{-1}) indicates the presence of defects, while the G-band (~1580 cm^{-1}) represents the sp^2-hybridized carbon atoms. The intensity ratio of the D-band to the G-band (ID/IG) is a measure of defect density [34].
ii) Transmission electron microscopy (TEM) is often employed to visualize defects at the atomic level, such as vacancies and dislocations in the graphene lattice [35].
iii) X-ray photoelectron spectroscopy (XPS) could provide information about the chemical bonding of graphene, particularly functional groups introduced during oxidation or reduction [36].
iv) Atomic force microscopy (AFM) furnishes topographical information and can be employed to detect surface roughness and defects in graphene sheets [37]. Other techniques, such as X-ray absorption near-edge structures (XANES) and extended X-ray absorption fine structures (EXAFS), are also utilized to study the electronic structure and band structure of defective graphene and predict the impact of defects on their photocatalytic performance [38].

7.3 Structures and Properties

Defective graphene materials refer to graphene feedstocks that have undergone structural alterations, like vacancies, dopants, grain boundaries, and functional groups, which can significantly enhance their photocatalytic features compared to pristine graphene. These defects create localized electronic states that facilitate the separation and transport of photogenerated charge carriers, thereby improving their reactivity and efficiency in photocatalytic processes. Unlike pristine graphene, which suffers from limited surface reactivity and poor charge carrier separation, it was proven that defective graphene exhibits increased light absorption, more active sites, and enhanced electron–hole pair separation, making it an attractive research topic, especially in photocatalysis processes [39, 40].

7.3.1 Type of Defects

Vacancies – vacant sites where carbon atoms are missing from the graphene structure – create localized electronic states that can trap charge carriers generated during photocatalytic transformations. It has been reported that vacancy defects in graphene materials can be divided into single vacancy (SV), double vacancy (DV), and multiple vacancy (MV), referring to the number of atoms absent from the graphene lattice (Figure 7.5a–g). This helps improve charge separation and reduces recombination, boosting the overall efficiency. Additionally, these vacancies make the material more reactive by providing extra sites for efficient catalytic reactions [1].

Substitutional defects: substitutional defects in graphene materials are a type of impurity defect that can occur when one or more carbon atoms in the graphene lattice are replaced by different atoms [1]. It should be noted that introducing heteroatoms (i.e., nitrogen, boron, or sulfur, and metallic substitution) into the graphene lattice can significantly alter its electronic structure and increase its photocatalytic efficiency and enhance its chemical reactivity(Figure 7.6a–e).

Edge defects: edges in graphene material, especially those with broken bonds (zigzag or armchair edges), are naturally more reactive than the basal plane. These defects in graphene play an essential role in photocatalysis by providing additional active sites for reactant adsorption and by promoting electron transfer [41]. Grain boundaries: in polycrystalline graphene, the boundaries between graphene domains are often regions of disorder that can act as additional active sites for photocatalytic reactions. These grain boundaries can facilitate electron transport and enhance the catalytic efficiency of bare graphene [41].

Stone–Wales defects: Stone–Wales defects occur when a C—C bond in graphene rotates by 90°, creating a local curvature in the structure. This changes the electronic properties of graphene and forms new electronic states, which can enhance its photocatalytic performance [41].

Functional groups and oxygen-containing defects: graphene oxide and reduced graphene oxide, with their oxygen-containing groups (hydroxyl, carboxyl, and epoxy), exhibit high surface reactivity and hydrophilicity, making them highly suitable for photocatalytic applications. These functional groups enhance the material's interaction with water and organic molecules, increasing its efficiency in applications like pollutant degradation and hydrogen production [42].

7.3.2 Impact of Defects on Graphene Properties

One of the key advantages of introducing defects into graphene is the opening of a bandgap in its structure, which is essential for photocatalysis. Pristine graphene

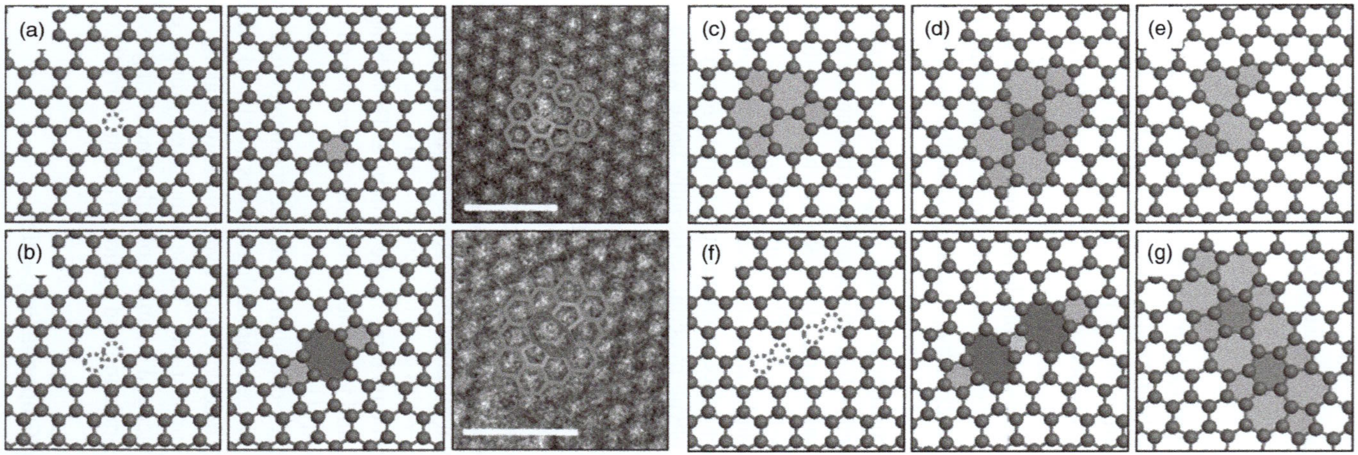

Figure 7.5 Vacancies in graphene, including single vacancy (SV), double vacancy (DV), and multiple vacancy (MV). (a) Atomic models of single vacancy before and after reconstruction and corresponding TEM image, scale bar51 nm. (b) Atomic models of 5-8-5divacancy before and after reconstruction and corresponding TEM micrograph, scale bar51 nm.5-8-5 defect can transform to some other configurations, including 555-777 (c), 5555-6-7777(D) and 5-7-7-5 (e) by C C bond rotations. Tetra vacancy generated by removal of two adjacent dimmers along the same armchair axis (f) or by bond rotation (G). *Source:* Reproduced from Xu and Sun [1] / with permission of Elsevier.

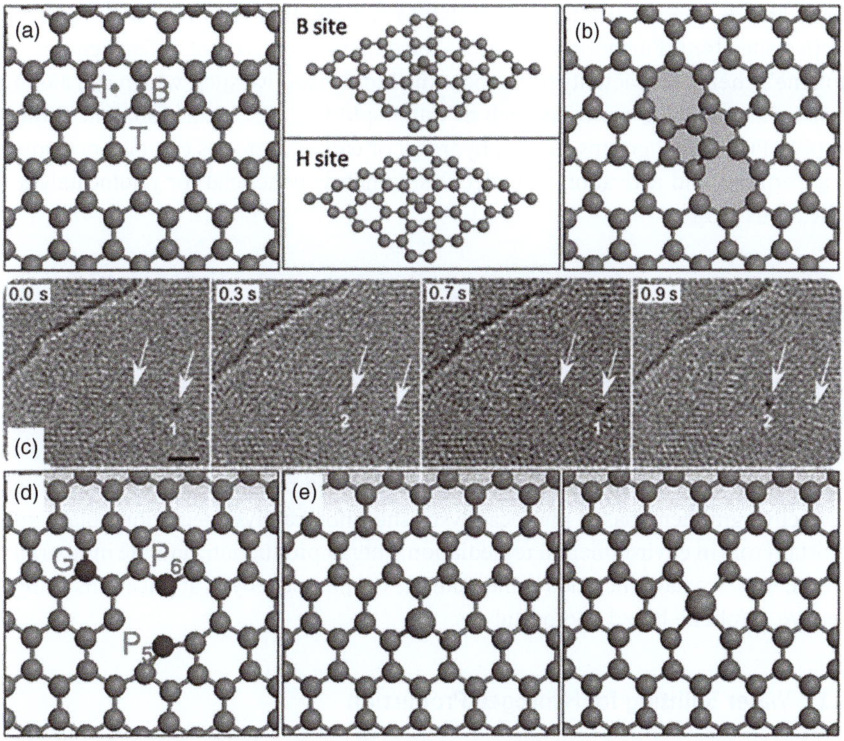

Figure 7.6 Carbon adatoms incorporation and substitutional graphitic, pyrrolic, and pyridinic-nitrogen into the lattice of graphene. (a) Configurations of a carbon adatom at different symmetry sites. (b) Inverse Stone Wales defects (7-5-5-7) generated when two carbon adatoms are incorporated into the lattice. (c) Oscillations of a W atom between two trapping sites 1 and 2 (arrowed), scale bar51 nm. (d) Types of N doping, sites G, P5, and P6 denote substitutional graphitic-, pyrrolic-, and pyridinic-nitrogen, respectively. (e) Typical atomic configurations of foreign atoms adsorbed on single and double vacancies. *Source:* Reproduced from Xu and Sun [1] / with permission of Elsevier.

is a semimetal with no bandgap, making it unsuitable for visible light absorption. But defects like vacancies and dopant atoms introduce localized states within the bandgap. These states make it easier for graphene to absorb light more efficiently in the UV–visible range, which improves its photocatalytic performance [43]. Besides, defects such as vacancies, dopants, and grain boundaries create localized electronic states that can trap charge carriers, preventing their recombination and extending their lifetimes. This is critical for photocatalytic reactions, where long-lived charge carriers are necessary for efficient photocatalytic applications [44, 45]. More importantly, the introduction of defects into graphene alters its electronic structure, allowing it to absorb light across a broader range of wavelengths, especially in the visible spectrum. This expanded absorption is crucial for photocatalytic reactions that rely on solar energy [46, 47]. Also, defects make graphene's surface area bigger by adding empty spaces, pores, and functional groups. These give reactant molecules more places to attach themselves. This increased surface area is beneficial for photocatalytic reactions that require a high density of active sites [48]. Furthermore, studies have demonstrated that defects enhance graphene's chemical reactivity by introducing highly reactive sites, which can aid in various photocatalytic reactions such as water splitting, CO_2 reduction, and pollutant [49]. Functional groups such as hydroxyl or carboxyl groups can also promote the adsorption and activation of molecules, which is beneficial for photocatalytic applications [50].

7.4 Photocatalytic Applications

The photocatalytic applications of defective graphene-based materials have been employed in order to facilitate chemical reactions under light irradiation, eventually leading to the degradation of various organic/inorganic pollutants and the conversion of renewable energy sources for high value-added product generation. By way of defective graphene-based photocatalysts, the photocatalysis technique plays an important role in environmental remediation, energy production, and the industrial sector. Below, we have included some examples of photocatalytic applications using defective graphene-based photocatalysts.

7.4.1 Water Splitting for Hydrogen Production

At present, photocatalytic water splitting is among the most active research areas for renewable energy production and addressing the global energy crisis. In this regard, researchers and industrialists have made significant efforts to improve the efficiency, cost-effectiveness, and scalability of defective graphene for clean hydrogen fuel production while contributing to environmental protection [51]. In 2015, Nguyen-Phan et al. [52] prepared a hydrogenated–graphene–TiO_2 prototype defective for photocatalytic water splittingwithout the assistance of a noble metal co-catalyst. Nguyen–Phan's group demonstrated that the defects in graphene improve electron mobility, reduce charge recombination, and increase the absorption of visible light, boosting hydrogen and oxygen evolution. They indicated that the obtained defective graphene exhibits a high photocatalytic H_2 evolution performance. They observed that the introduction of oxygen vacancies and Ti^{3+} states hindered the recombination of photogenerated charge carriers and then facilitated the charge transfer from TiO_{2-x} to the carbonaceous sheet of graphene. They mentioned that the defects in graphene improved the photocatalytic H_2 evolution by facilitating better charge separation and transportation over the utilized defective graphene [52].

7.4.2 Degradation of Organic Pollutants

In wastewater treatment processes, photodegradation of organic pollutants is one of the innovative and environmentally friendly approaches to mitigating water pollution [53]. In this case, the development of advanced photocatalytic materials, photoreactor engineering, and a deeper understanding of degradation mechanisms of organic pollutants have attracted the attention of researchers in the past few decades [54]. It can be noted that harnessing photodegradation processes can significantly contribute to cleaner water, thereby protecting the ecosystem and human health [55]. Even though there has been continuous progress in wastewater remediation employing the photocatalysis process, the fabrication of efficient photocatalysts is still much needed. For instance, Ahmed et al. developed graphene oxide (GO)-based hybrid photocatalyst $ZnO/(Fe_3O_4@SiO_2)/rGO$ (GZF) composites employing a microwave method for methylene blue degradation [56]. Ahmed and his colleagues discovered that the hybridization with the graphene material led to a significant decrease in the vacancy sites in the ZnO composite,the healing of oxygen vacancies, and the appearance of oxygen interstitial defect (Oi) levels as a new defect in the resulting photocatalyst (Figure 7.7). The defective graphene demonstrated 3.6 times higher efficiency than ZnO nanoparticles in breaking down methylene blue through oxidation when exposed to visible light. They also indicated that the XPS and photoluminescence (PL) spectra prove the existence of defects, and intimate interfacial contact leads to visible light photocatalytic activity of the defective graphene [56].

7.4.3 CO_2 Reduction to Fuels

The implementation of the photocatalytic reduction of CO_2 into value-added chemicals is a very promising strategy, but many unfavorable factors, including low spectral utilization and the short lifetime of photogenerated charges through the utilized photocatalytic feedstocks, have been restricting the industrial application of graphene photocatalysts [57]. Therefore, in order to meet the large-scale production requirements, it is necessary to develop a low-cost and simple defective graphene with a simple preparation method to fabricate highly efficient and stable graphene photocatalysts. In this regard, Kuang et al. [58] reported simulated sunlight (SS) and UV-irradiated GO (GOSS and GOUV) for efficient photocatalytic reduction of CO_2 to CO. They suggested that irradiation serves two crucial functions in the photocatalytic reaction process. In the first step, irradiating GO material to remove CO released under photolysis from photocatalytic reaction systems, and then in the second step, activating GO material to create defects and restore the large π-conjugated network, resulting in photolysis-saturated and photo-activated GO for photocatalytic reduction of CO_2 to CO (Figure 7.8). They observed that the order of CO generation was as follows: GOSS > GOUV > GO. Besides, the CO formation rate was found to be 1.23 μmol $g_{cat.}^{-1}$ h^{-1} for GOSS and 0.95 μmol $g_{cat.}^{-1}$ h^{-1} for GOUV at 4 hours of reaction time, which was 2.7-fold and 2.1-fold greater than that of pristine GO (0.46 μmol $g_{cat.}^{-1}$ h^{-1} at 4 hours of photocatalytic reaction time). These findings clearly show that defect formation/creation in graphene enhances the generation of reactive oxygen species through its surface, which plays a crucial role in the CO_2 conversion to CO [58].

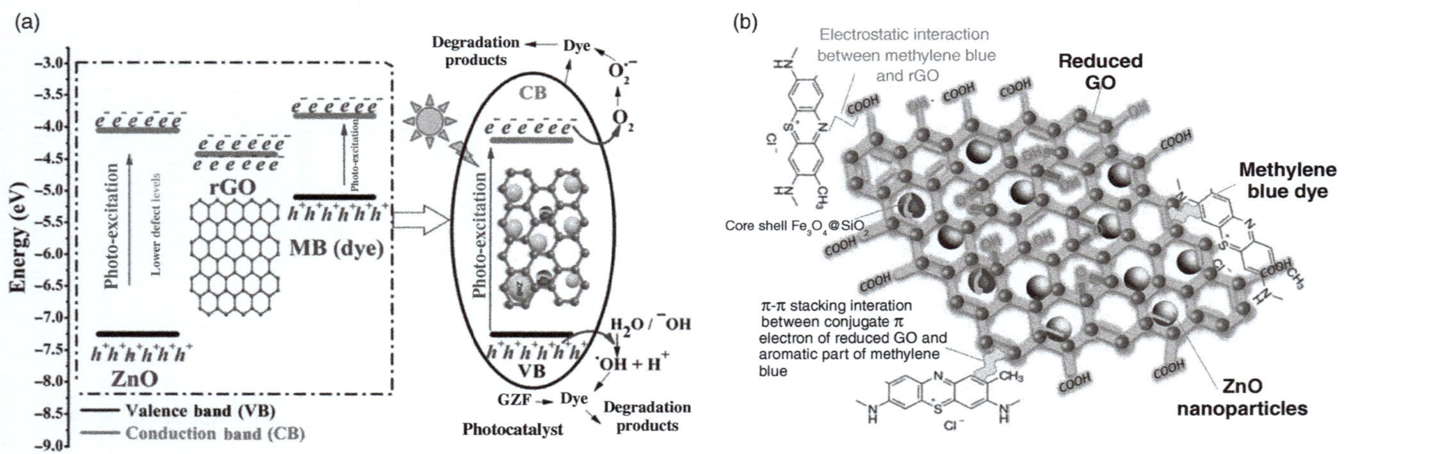

Figure 7.7 Mechanism for the MB photodegradation over GZF (a) defective photocatalyst; (b) interactions involved in the adsorption of methylene blue on GZF composites photocatalyst. *Source:* Reproduced from Ahmed et al. [56] / with permission of Elsevier.

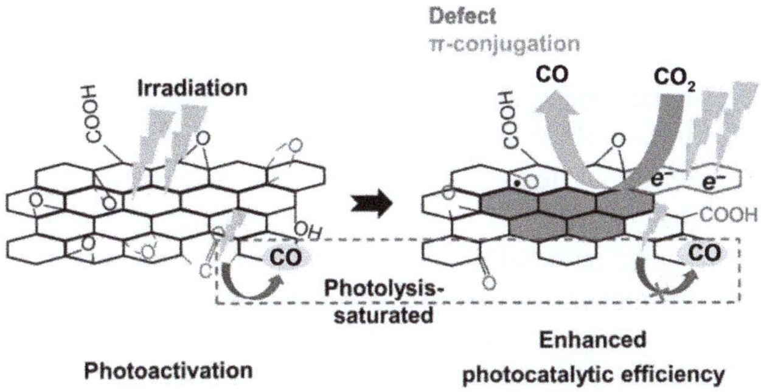

Figure 7.8 Photocatalytic efficiency of irradiated GO toward CO_2 conversion to CO. *Source:* Reproduced from Kuang et al. [58] / with permission of American Chemical Society.

7.4.4 Antibacterial and Self-cleaning Applications

Antibacterial disinfection is an important process of infection control for public health protection, with a wide range of photocatalytic disinfection methods and photocatalytic agents that are widely used to combat microbial contamination [59, 60]. As is well known, bacterial resistance presents a real issue at present; therefore, novel research and innovation in disinfection technologies are essential to enhance photocatalytic efficiency and mitigate the challenges related to antibacterial disinfection. Employing effective disinfection strategies, defective graphene-based photocatalysts could be an excellent alternative for the photocatalysis application in antibacterial disinfection, thereby ensuring environmental security. For example, in 2017, Chen et al. successfully fabricated graphene oxide/WS_2/Mg-doped ZnO nanocomposites (rGOMZ) as antimicrobial agents utilizing mechanical shear exfoliation [61]. As a result, they observed that GO-WS_2 and Mg/ZnO samples possess excellent bactericidal activity, suggesting that WS_2 nanoplate can further improve the bactericidal activity of the rGOMZ sample against the *S. aureus* colony or *E. coli* colony. Importantly, they demonstrated that the enhancement in photocatalytic and antibacterial efficiency is assigned to the effective synergistic effect between rGO and WS_2 materials and oxygen vacancies in ZnO, which are conducive to the formation of reactive oxygen species, further for environmental protection and self-cleaning using graphene material [61].

7.5 Advantages and Limitations

Nowadays, defective graphene-based photocatalysts present a promising alternative for advanced materials engineering. Thanks to their properties, they have offered significant opportunities for photocatalytic applications. Nevertheless, defect-engineered graphene still necessitates careful synthesis control to ensure effective usage in the photocatalysis processes. It is worth mentioning that balancing their advantages and limitations is important to unlocking the full potential of defective graphene-based photocatalysts in future innovations. It should be noted that structural imperfections such as vacancies, doping, and dislocations in graphene present both advantages and limitations, which are given in Table 7.1.

To sum up, despite the aforementioned limitations, defective graphene presents an excellent platform for photocatalytic implementations owing to its unique structural and electronic features influenced by different defects created on its surface. Even though the substantial efforts made in the photocatalysis process using several organic and inorganic materials [72–81], through defects engineering as photocatalytic materials transformations, researchers and scientists could tailor the efficiency of graphene-based photocatalysts for various applications, from environmental remediation to renewable energy conversion. In the future, advanced techniques and optimization of defective graphene are still needed for specific photocatalytic applications to attain sustainability goals in large-scale photocatalyst manufacture and industrial implementations of the photocatalysis.

7.6 Conclusion and Future Prospects

This chapter examines the atomic structure of various defects (i.e., Stone–Wales defects, vacancies, dislocations, and grain boundaries) in graphene material and their influence on its physicochemical properties and photocatalytic efficiency. In literature, several fabrication methods have been utilized for the synthesis of defective graphene. Techniques such as XANES, EXAFS, Raman spectra, and AFM have been widely utilized for the control and defect configurations in graphene material. It should be noted that various kinds of defects, such as vacancies, substitutional impurities, adatoms, etc., are inevitably formed during the growth of graphene because of the imperfection of the graphene formation strategies. Moreover, one can intentionally incorporate several defects into graphene through ion-beam irradiation or chemical treatments. Even though some defects in graphene can be helpful in some photocatalytic applications, including enhancing chemical reactivity and generating magnetism, defects in graphene could also affect the thermal conductivity and reduce its mechanical strength. In addition, edge modifications can improve the physicochemical features of graphene, which is a significant issue for specific photocatalytic applications. However, it is worth mentioning that the decrease in the transport properties of graphene can be attributed to defects. Nowadays, the large-scale production of materials with controlled defects is challenging for photocatalytic applications. It is important to note that density and the distribution of defects in graphene play a crucial role in its property modification, resulting in new functionalities, broadening its photocatalytic applications. Till now, the study of the influence of defects on graphene properties is still unsatisfactory; despite the existing theoretical data, experimental and control of defects in graphene is still needed. Therefore, this chapter only contains some examples of defects in graphene for photocatalytic applications without providing solutions for controlling defects in graphene. On the one hand, introducing intrinsic defects into graphene is crucial for achieving sustainability goals in photocatalysis applications across various fields. Even with the existing

Table 7.1 Advantages and limitations of defective graphene in photocatalysis.

Advantages	Limitations
Defective graphene introduces active sites for photocatalysis, making it highly effective in several photocatalytic reactions, such as the oxygen reduction reaction (ORR) and the hydrogen evolution reaction (HER) [62].	Defects muss up the lattice of graphene, making it less able to conduct electricity, which can be bad for some photocatalytic uses [63].
The presence of defects enhances ion adsorption and diffusion, generating reactive oxygen species and significantly improving the performance of graphene [64].	Some defects in graphene can make it less stable chemically and mechanically, particularly in harsh photocatalytic conditions like UV light and strong oxidizing environments. This makes it less useful for mechanical uses [65].
The enhanced adsorption capacity of defective graphene, due to defects like vacancies and functional groups, makes it valuable for environmental applications such as water treatment. These defects create more active sites for adsorbing pollutants, including organic/inorganic pollutants, improving the graphene's effectiveness in removing contaminants from wastewater [66].	In some cases, the methods used to introduce defects or functionalize graphene may involve toxic chemicals or harsh conditions, which could pose environmental risks. For example, the use of certain metal dopants or aggressive oxidation processes can lead to the production of toxic by-products, raising concerns about the sustainability and environmental impact of these materials for large-scale applications [67].
Defects in graphene enable the tuning of its electronic properties, making it suitable for applications such as photocatalysis and photoreactor engineering. Introducing defects like vacancies or dopants adjusts its electronic behavior, enhancing conductivity or light absorption. This flexibility makes defective graphene ideal for photocatalytic implementations [68].	A major challenge in using defective graphene for photocatalysis is controlling the defect density. Too many defects can weaken the material and reduce conductivity, while too few defects may not improve photocatalytic activity. Achieving the right balance of defects is key to optimizing performance, but this remains difficult to control nowadays [69].
Defects in graphene increase its reactivity by creating sites for functionalization, which are useful in photocatalytic applications. These defects provide active sites for attaching functional groups, allowing for surface modifications that enhance interaction with molecules. This makes graphene more suitable for targeted pollutant photodegradation and energy conversion [70].	The process of creating defect-engineered graphene is challenging and difficult to scale for large-scale photocatalytic applications. Methods like chemical vapor deposition (CVD) and plasma treatment are challenging to adapt for large production, and maintaining consistent defects across batches is problematic. These issues limit the scalability and practical applications of defective graphene in photocatalysis processes [71].

challenge of controllable synthesis of well-defined defect structures in graphene, we urgently need more experimental and theoretical studies in the future on defects and their impact on the physicochemical properties. Conversely, it is crucial to acknowledge that extrinsic defects, such as foreign atoms, significantly impact the physicochemical properties of graphene and its potential for photocatalysis. Overall, controlling the location distribution and total amount of dopants on graphene is crucial for the design of graphene-based photocatalysts. Also, graphene that is broken tends to be more chemically reactive than graphene that is perfect because it has more surface area and bonds that are not attached properly. While this improved reactivity can be conducive in certain photocatalytic applications, it may also lead to undesirable side reactions, degradation, and instability. This instability can limit the long-term efficiency of defective graphene. Also, the type and number of defects in defective graphene can be very different depending on the method used to make it, which makes its photocatalytic properties less consistent. Graphene defects can also block the pathways and channels that electrons use to move through it, which makes charge transfer less efficient, especially in photocatalytic applications. Finally, graphene is known for its good thermal conductivity; the presence of some defects can significantly impact its thermal transport features. In some cases, defective graphene may display reduced phonon mobility, resulting in increased thermal resistance, limiting its applications in photocatalysis. Overall, this chapter is expected to provide insights into the benefits and limitations of defects in graphene and thereby guide researchers and scientists to investigate defective graphene for potential photocatalytic applications.

Acknowledgments

The authors are thankful for the financial support of the University of Gafsa to carry out this work.

Conflict of Interest

The authors declare that they have no conflict of interest.

References

1 Xu, T. and Sun, L. (2018). Structural defects in graphene. In: *Defects in Advanced Electronic Materials and Novel Low Dimensional Structures* (ed. J. Stehr, I. Buyanova, and W. Chen), 137–160. Elsevier.

2 Geim, A.K. and Novoselov, K.S. (2007). The rise of graphene. *Nat. Mater.* 6 (3): 183–191.

3 Boukhvalov, D. and Katsnelson, M. (2009). Chemical functionalization of graphene. *J. Phys.: Condens. Matter* 21: 344205.

4 Huang, H. and Wang, X. (2012). Pd nanoparticles supported on low-defect graphene sheets: for use as high-performance electrocatalysts for formic acid and methanol oxidation. *J. Mater. Chem.* 22: 22533–22541.

5 Sturala, J., Luxa, J., Pumera, M., and Sofer, Z. (2018). Chemistry of graphene derivatives: synthesis, applications, and perspectives. *Chem. Eur. J.* 24: 5992–6006.

6 Hu, H., Xin, J.H., Hu, H. et al. (2015). Metal-free graphene-based catalyst—insight into the catalytic activity: a short review. *Appl. Catal., A* 492: 1–9.

7 Yang, H.B., Miao, J., Hung, S.-F. et al. (2016). Identification of catalytic sites for oxygen reduction and oxygen evolution in N-doped graphene materials: development of highly efficient metal-free bifunctional electrocatalyst. *Sci. Adv.* 2: e1501122.

8 Suragtkhuu, S., Bat-Erdene, M., Bati, A.S. et al. (2020). Few-layer black phosphorus and boron-doped graphene based heteroelectrocatalyst for enhanced hydrogen evolution. *J. Mater. Chem. A* 8: 20446–20452.

9 Ali, S., Razzaq, A., and In, S.-I. (2019). Development of graphene based photocatalysts for CO_2 reduction to C1 chemicals: a brief overview. *Catal. Today* 335: 39–54.

10 Low, J., Yu, J., and Ho, W. (2015). Graphene-based photocatalysts for CO_2 reduction to solar fuel. *J. Phys. Chem. Lett.* 6: 4244–4251.

11 Guo, H., Long, D., Zheng, Z. et al. (2017). Defect-enhanced performance of a 3D graphene anode in a lithium-ion battery. *Nanotechnology* 28: 505402.

12 Zhang, Y., Xia, X., Liu, B. et al. (2019). Multiscale graphene-based materials for applications in sodium ion batteries. *Adv. Energy Mater.* 9: 1803342.

13 Bokhari, S.W., Siddique, A.H., Sherrell, P.C. et al. (2020). Advances in graphene-based supercapacitor electrodes. *Energy Rep.* 6: 2768–2784.

14 Zhu, J., Yang, D., Yin, Z. et al. (2014). Graphene and graphene-based materials for energy storage applications. *Small* 10: 3480–3498.

15 Yadav, S., Zhu, Z., and Singh, C.V. (2014). Defect engineering of graphene for effective hydrogen storage. *Int. J. Hydrogen Energy* 39: 4981–4995.

16 Zhou, Y., Chu, W., Jing, F. et al. (2017). Enhanced hydrogen storage on Li-doped defective graphene with B substitution: a DFT study. *Appl. Surf. Sci.* 410: 166–176.

17 de Souza, F.A., Amorim, R.G., Prasongkit, J. et al. (2018). Topological line defects in graphene for applications in gas sensing. *Carbon* 129: 803–808.

18 Haridas, V., Sukhananazerin, A., Sneha, J.M. et al. (2020). α-Fe_2O_3 loaded less-defective graphene sheets as chemiresistive gas sensor for selective sensing of NH_3. *Appl. Surf. Sci.* 517: 146158.

19 Ricciardella, F., Vollebregt, S., Polichetti, T. et al. (2017). Effects of graphene defects on gas sensing properties towards NO_2 detection. *Nanoscale* 9: 6085–6093.

20 Ghenaatian, H.R., Shakourian-Fard, M., Moghadam, M.R. et al. (2019). Tailoring of graphene quantum dots for toxic heavy metals detection. *Appl. Phys. A.* 125: 754.

21 Meng, X., Yu, C., Song, X. et al. (2018). Scrutinizing defects and defect density of selenium-doped graphene for high-efficiency triiodide reduction in dye-sensitized solar cells. *Angew. Chem.* 130: 4772–4776.

22 Khan, M.E., Khan, M.M., and Cho, M.H. (2017). Ce^{3+}-ion, surface oxygen vacancy, and visible light-induced photocatalytic dye degradation and photocapacitive performance of CeO_2-graphene nanostructures. *Sci. Rep.* 7: 5928.

23 Han, Z.-Y., Huang, L.-J., Qu, H.-J. et al. (2021). A review of performance improvement strategies for graphene oxide-based and graphene-based membranes in water treatment. *J. Mater. Sci.* 56: 9545–9574.

24 Liu, G., Jin, W., and Xu, N. (2015). Graphene-based membranes. *Chem. Soc. Rev.* 44: 5016–5030.

25 Lee, S.-M., Kim, J.-H., and Ahn, J.-H. (2015). Graphene as a flexible electronic material: mechanical limitations by defect formation and efforts to overcome. *Mater. Today* 18: 336–344.

26 Zeng, W., Tao, X.-M., Lin, S. et al. (2018). Defect-engineered reduced graphene oxide sheets with high electric conductivity and controlled thermal conductivity for soft and flexible wearable thermoelectric generators. *Nano Energy* 54: 163–174.

27 Chang, H. and Wu, H. (2013). Graphene-based nanomaterials: synthesis, properties, and optical and optoelectronic applications. *Adv. Funct. Mater.* 23 (15): 1984–1997.

28 Van Dao, D., Jung, H.D., Nguyen, T.T. et al. (2021). Defect-rich N-doped CeO_2 supported by N-doped graphene as a metal-free plasmonic hydrogen evolution photocatalyst. *J. Mater. Chem.* 9: 10217–10230.

29 Tibodee, A., Pannak, P., Akkarachaneeyakorn, K. et al. (2019). Use of the graphite intercalation compound to produce low-defect graphene sheets for the photocatalytic enhancement of graphene/TiO_2 composites. *Mater. Chem. Phys.* 235: 121755.

30 Lee, W.-J., Kim, C.-S., Yang, S.-Y. et al. (2021). Spontaneous exfoliation of large-sized graphene oxide with low defect concentration by simple wet chemistry. *Carbon* 182: 214–222.

31 Liu, Y., Bao, J., Li, Z. et al. (2022). Large-scale defect-rich iron/nitrogen co-doped graphene-based materials as the excellent bifunctional electrocatalyst for liquid and flexible all-solid-state zinc-air batteries. *J. Colloid Interface Sci.* 607: 1201–1214.

32 Santhosh, N.M., Filipič, G., Kovacevic, E. et al. (2020). N-graphene nanowalls via plasma nitrogen incorporation and substitution: the experimental evidence. *Nano Micro Lett.* 12: 1–17.

33 Yeom, D.-Y., Jeon, W., Tu, N.D.K. et al. (2015). High-concentration boron doping of graphene nanoplatelets by simple thermal annealing and their supercapacitive properties. *Sci. Rep.* 5: 9817.

34 Ferrari, A.C. and Basko, D.M. (2013). Raman spectroscopy as a versatile tool for studying the properties of graphene. *Nat. Nanotechnol.* 8: 235–246.

35 Meyer, J.C., Geim, A.K., Katsnelson, M.I. et al. (2007). The structure of suspended graphene sheets. *Nature* 446: 60–63.

36 Ganesan, K., Ghosh, S., Krishna, N.G. et al. (2016). A comparative study on defect estimation using XPS and Raman spectroscopy in few layer nanographitic structures. *Phys. Chem. Chem. Phys.* 18: 22160–22167.

37 Gómez-Navarro, C., Meyer, J.C., Sundaram, R.S. et al. (2010). Atomic structure of reduced graphene oxide. *Nano Lett.* 10: 1144–1148.

38 Caretti, I. and Jiménez, I. (2011). Point defects in hexagonal BN, BC_3, and BC*x*N compounds studied by X-ray absorption near-edge structure. *J. Appl. Phys.* 110: 121301.

39 Zhang, D., Hu, B., Guan, D., and Luo, Z. (2016). Essential roles of defects in pure graphene/Cu_2O photocatalyst. *Catal. Commun.* 76: 7–12.

40 Mateo, D., Garcia-Mulero, A., Albero, J., and Garcia, H. (2019). N-doped defective graphene decorated by strontium titanate as efficient photocatalyst for overall water splitting. *Appl. Catal., B* 252: 111–119.

41 Bhatt, M.D., Kim, H., and Kim, G. (2022). Various defects in graphene: a review. *RSC Adv.* 12: 21520–21547.

42 Xiang, J., Xu, J., Li, H. et al. (2024). Distribution of oxygen-containing functional groups on defective graphene: properties engineering and Li adsorption. *Phys. Chem. Chem. Phys.* 26: 12764–12777.

43 Tang, B., Chen, H., He, Y. et al. (2017). Influence from defects of three-dimensional graphene network on photocatalytic performance of composite photocatalyst. *Compos. Sci. Technol.* 150: 54–64.

44 Banhart, F., Kotakoski, J., and Krasheninnikov, A.V. (2011). Structural defects in graphene. *ACS Nano* 5: 26–41.

45 Nakaharai, S., Iijima, T., Ogawa, S. et al. (2013). Conduction tuning of graphene based on defect-induced localization. *ACS Nano* 7: 5694–5700.

46 Mehta, M., Chandrabose, G., Krishnamurthy, S. et al. (2022). Improved photoelectrochemical properties of TiO_2-graphene nanocomposites: effect of defect induced visible light absorption and graphene conducting channel for carrier transport. *Appl. Surf. Sci. Adv.* 11: 100274.

47 Kaliannan, T., Balasubramaniam, S., and Sang-Jae, K. (2015). Gate-tunable photoresponse of defective graphene: from ultraviolet to visible. *Appl. Phys. Lett.* 107: 111107.

48 Prusty, K., Barik, S., and Swain, S.K. (2019). A correlation between the graphene surface area, functional groups, defects, and porosity on the performance of the nanocomposites. In: *Functionalized Graphene Nanocomposites and Their Derivatives* (ed. K. Prusty, S. Barik, and S.K. Swain), 265–283. Elsevier.

49 Jayaprakash, G.K., Casillas, N., Astudillo-Sánchez, P.D., and Flores-Moreno, R. (2016). Role of defects on regioselectivity of nano pristine graphene. *J. Phys. Chem. A.* 120: 9101–9108.

50 Li, X., Yu, J., Wageh, S. et al. (2016). Graphene in photocatalysis: a review. *Small* 12: 6640–6696.

51 Datteo, M., Liu, H., and Di Valentin, C. (2018). Water on graphene-coated TiO_2: role of atomic vacancies. *ACS Appl. Mater. Interfaces* 10: 5793–5804.

52 Nguyen-Phan, T.-D., Luo, S., Liu, Z. et al. (2015). Striving toward noble-metal-free photocatalytic water splitting: the hydrogenated-graphene–TiO_2 prototype. *Chem. Mater.* 27: 6282–6296.

53 Ramalingam, G., Perumal, N., Priya, A., and Rajendran, S. (2022). A review of graphene-based semiconductors for photocatalytic degradation of pollutants in wastewater. *Chemosphere* 300: 134391.

54 Kausor, M.A. and Chakrabortty, D. (2021). Graphene oxide-based semiconductor photocatalysts for degradation of organic dye in wastewater: a review on fabrication, performance enhancement and challenges. *Inorg. Chem. Commun.* 129: 108630.

55 Thakre, K.G., Barai, D.P., and Bhanvase, B.A. (2021). A review of graphene-TiO_2 and graphene-ZnO nanocomposite photocatalysts for wastewater treatment. *Water Environ. Res.* 93: 2414–2460.

56 Ahmed, G., Hanif, M., Zhao, L. et al. (2016). Defect engineering of ZnO nanoparticles by graphene oxide leading to enhanced visible light photocatalysis. *J. Mol. Catal. A: Chem.* 425: 310–321.

57 Xiang, Q., Yu, J., and Jaroniec, M. (2012). Graphene-based semiconductor photocatalysts. *Chem. Soc. Rev.* 41: 782–796.

58 Kuang, Y., Shang, J., and Zhu, T. (2019). Photoactivated graphene oxide to enhance photocatalytic reduction of CO_2. *ACS Appl. Mater. Interfaces* 12: 3580–3591.

59 He, J., Kumar, A., Khan, M., and Lo, I.M. (2021). Critical review of photocatalytic disinfection of bacteria: from noble metals-and carbon nanomaterials-TiO_2 composites to challenges of water characteristics and strategic solutions. *Sci. Total Environ.* 758: 143953.

60 He, J., Zheng, Z., and Lo, I.M. (2021). Different responses of gram-negative and gram-positive bacteria to photocatalytic disinfection using solar-light-driven magnetic TiO_2-based material, and disinfection of real sewage. *Water Res.* 207: 117816.

61 Chen, C., Yu, W., Liu, T. et al. (2017). Graphene oxide/WS_2/Mg-doped ZnO nanocomposites for solar-light catalytic and anti-bacterial applications. *Sol. Energy Mater. Sol. Cells.* 160: 43–53.

62 Albero, J., Mateo, D., and García, H. (2019). Graphene-based materials as efficient photocatalysts for water splitting. *Molecules* 24: 906.

63 Bie, C., Yu, H., Cheng, B. et al. (2021). Design, fabrication, and mechanism of nitrogen-doped graphene-based photocatalyst. *Adv. Mater.* 33: e2003521.

64 Liu, L., Qing, M., Wang, Y., and Chen, S. (2015). Defects in graphene: generation, healing, and their effects on the properties of graphene: a review. *J. Mater. Sci. Technol.* 31: 599–606.

65 Putri, L.K., Ong, W.-J., Chang, W.S., and Chai, S.-P. (2015). Heteroatom doped graphene in photocatalysis: a review. *Appl. Surf. Sci.* 358: 2–14.

66 Jilani, A., Othman, M.H.D., Ansari, M.O. et al. (2018). Graphene and its derivatives: synthesis, modifications, and applications in wastewater treatment. *Environ. Chem. Lett.* 16: 1301–1323.

67 Tian, W., Li, W., Yu, W., and Liu, X. (2017). A review on lattice defects in graphene: types, generation, effects, and regulation. *Micromachines* 8: 163.

68 Bai, X., Wang, L., Zong, R. et al. (2013). Performance enhancement of ZnO photocatalyst via synergic effect of surface oxygen defect and graphene hybridization. *Langmuir* 29: 3097–3105.

69 Melo, M.A. Jr., and Osterloh, F.E. (2018). Defect states control effective band gap and photochemistry of graphene quantum dots. *ACS Appl. Mater. Interfaces* 10: 27195–27204.

70 Meng, J., Chen, Q., Lu, J., and Liu, H. (2018). Z-scheme photocatalytic CO_2 reduction on a heterostructure of oxygen-defective ZnO/reduced graphene oxide/UiO-66-NH_2 under visible light. *ACS Appl. Mater. Interfaces* 11 (2018): 550–562.

71 Kim, S.-M., Cho, Y., Kim, J.-H. et al. (2014). Healing defective CVD-graphene through vapor phase treatment. *Nanoscale* 6: 5639–5644.

72 Labidi, A., Ren, H., Zhu, Q. et al. (2024). Coal fly ash and bottom ash low-cost feedstocks for CO_2 reduction using the adsorption and catalysis processes. *Sci. Total Environ.* 912: 169179.

73 Ren, H., Qi, F., Labidi, A. et al. (2023). Chemically bonded carbon quantum dots/Bi_2WO_6 S-scheme heterojunction for boosted photocatalytic antibiotic degradation: interfacial engineering and mechanism insight. *Appl. Catal. B: Environ.* 330: 122587.

74 Liang, X., Gao, T., Cui, Y. et al. (2024). Photoreforming of poly(ethylene-terephthalate) plastic into valuable chemicals and hydrogen over $BiVO_4$/MoO*x*: synergistic promotion of oxidation and reduction processes. *Appl. Catal. B: Environ.* 357: 124326.

75 Labidi, A., Ren, H., Sial, A. et al. (2023). Coal ash for removing toxic metals and phenolic contaminants from wastewater: a brief review. *Crit. Rev. Environ. Sci. Technol.* 53: 2006–2029.

76 Labidi, A., Ren, H., Liang, X. et al. (2024). Visible-light-driven photocatalytic degradation of ciprofloxacin antibiotic by novel heterostructured coal fly ash waste: mechanism insight, toxicity pathway and DFT calculation. *Mater. Today Chem.* 42: 102388.

77 Li, X., Dong, Q., Li, F. et al. (2024). Defective Bi@BiOBr/C microrods derived from Bi-MOF for efficient photocatalytic NO abatement: directional regulation of interfacial charge transfer via carbon-loading. *Appl. Catal. B: Environ.* 340: 123238.

78 Labidi, A., Ren, H., Sial, A. et al. (2024). Photocatalyst materials formed using organic radicals for water splitting. In: *Organic Radicals* (ed. C. Wang, A. Labidi, and L. Eric), 211–230. Elsevier.

79 Zhu, Q., Wang, Y., Wang, J. et al. (2024). Synergistic polarization and oxygen vacancies engineering for enhancing photocatalytic NO removal over $Bi_4Ti_3O_{12}$ nanowires. *Appl. Catal. B: Environ.* 346: 123734.

80 Labidi, A., Ren, H., Sial, A. et al. (2024). Organic radicals used to modify materials surface and their application to remove pharmaceutical residues from wastewater. In: *Organic Radicals* (ed. C. Wang, A. Labidi, and L. Eric), 321–339. Elsevier.

81 Cui, Y., Labidi, A., Liang, X. et al. (2024). Pivotal impact factors in photocatalytic reduction of CO_2 to value-added C_1 and C_2 products. *ChemSusChem* 17: e202400551.

8

Defective Photocatalysts in Organic Synthesis

Sudirman

University of Mataram, Department of Chemistry, Majapahit Street 62, Mataram, Indonesia

8.1 Introduction

Photocatalysis has emerged as a key approach in organic synthesis, harnessing light to drive chemical transformations under mild, environmentally friendly conditions. This process reduces energy consumption and minimizes the production of waste compared to traditional catalytic methods [1, 2]. Photocatalysis aligns with green chemistry principles by using solar or artificial light as a primary energy source, making it suitable for large-scale industrial applications such as pharmaceutical manufacturing and environmental remediation [3, 4]. Recent advances, particularly in visible-light photoredox catalysis and hydrogen atom transfer mechanisms, have broadened the scope of photocatalysis by allowing selective activation of chemical bonds and providing chemists with precise control over reaction pathway [5–8]. These innovations position photocatalysis as a sustainable tool for synthesizing complex organic molecules, particularly under mild conditions. In photochemistry, light directly interacts with the reactant molecules, exciting them to an energized state, where they undergo transformations or generate reactive intermediates (i.e., radicals and singlet oxygen). This process does not necessarily involve a catalyst. Photochemistry is often associated with specific, highly selective reactions that proceed via direct excitation of reactants, leading to photochemical rearrangements, cycloadditions, or cleavage reactions. However, these reactions require high-intensity light sources and can produce multiple by-products due to the high energy imparted directly to the reactants [3]. In photocatalysis, light activates a photocatalyst, which then transfers energy to the reactants through electron or energy transfer, rather than direct excitation of the reactants. The photocatalyst, which absorbs light to generate electron–hole pairs, facilitates the transformation of substrates via redox reactions and remains chemically unchanged at the end of the reaction cycle. This indirect method of excitation enables greater control over reaction conditions, lowers the energy requirements, and often leads to improved selectivity and yield compared to direct photochemistry. Photocatalysts, such as

Defect Engineering in Photocatalysis: From Fundamentals to Applications,
First Edition. Edited by Chuanyi Wang, Abdelkader Labidi, and Qiuhui Zhu.

TiO_2, g-C_3N_4, and metal–organic frameworks (MOFs), are commonly used to drive oxidative or reductive processes in organic synthesis with higher efficiency and reduced side reactions [9]. The foundations of photocatalysis trace back to the early 20th century, with Giacomo Ciamician's visionary work on using sunlight for chemical transformations. Although primarily theoretical, his ideas laid the groundwork for future advancements in light-driven chemistry [10–12]. Practical applications of photocatalysis began in the 1970s with metal oxides like TiO_2, which demonstrated the capacity of light to induce redox reactions. Despite their promise, early photocatalysts like TiO_2 had limitations due to their high bandgaps, restricting their use to UV light. A breakthrough occurred in the early 2000s with the development of visible-light photoredox catalysis. In 2008, Nicewicz and MacMillan pioneered the use of photoredox catalysis combined with organocatalyzed, achieving complex organic transformations, such as asymmetric alkylation and [2 + 2] cycloaddition, under mild conditions [8, 13]. Since then, advances in photocatalyst materials, such as organic chromophores, hybrid photocatalysts, and thermally activated delayed fluorescence materials, have enabled broader light absorption, greater efficiency, and improved stability. Materials such as g-C_3N_4, CMPs (Conjugated Microporous Polymers), and COF (Covalent Organic Frameworks) have emerged as stable and tuneable photocatalysts, suitable for diverse reactions across the visible spectrum [9]. More recently, the focus has shifted toward sustainable photocatalytic techniques, such as using aqueous media and eliminating photosensitizers, to make photocatalysis more compatible with industrial-scale applications.

In the recent few years, it was proven that defect-engineered photocatalysts enable a wide range of organic transformations by strategically utilizing both the valence band (VB) and conduction band (CB) for specific redox processes. Each transformation type – oxidative, reductive, and coupling reactions – relies on distinct reaction sites and mechanisms within the photocatalyst's electronic structure, which are influenced by the presence of defects [14]. On the one hand, oxidative reactions primarily occur in the VB, where photogenerated holes act as oxidizing agents. Upon light absorption, the catalyst generates electron–hole pairs, with holes residing in the VB. These holes participate in oxidative transformations by extracting electrons from substrates, facilitating the formation of oxidized products such as aldehydes or ketones from alcohols. In photocatalysts with oxygen vacancies (e.g., TiO_2), the VB holes interact more efficiently with adsorbed oxygen or reactant molecules, enhancing the oxidation process [15]. On the other hand, reductive reactions typically occur in the CB, where photogenerated electrons serve as reducing agents. When the photocatalyst absorbs light, electrons are excited into the CB, allowing them to reduce electron-accepting species, such as nitro groups or carbonyl compounds, transforming them into amines or alcohols, respectively. Defects like metal dopants (e.g., Cu in TiO_2) and oxygen vacancies help trap these electrons, extending their lifetime and increasing their likelihood of participating in reductive processes. It is worth mentioning that coupling reactions, which involve the formation of C–C and C–N bonds, can utilize both the VB and CB, depending on the specific reaction mechanism. For example, in Suzuki and Heck coupling reactions, electron transfer from the CB can activate substrates, while holes in the VB can promote bond formation steps. The use of heterojunctions or multi-defect sites in the photocatalyst helps balance electron and hole availability, providing an optimal environment for selective coupling reactions. Nitrobenzene was effectively and selectively reduced to aniline in an acidic aqueous suspension of titanium (IV) oxide photocatalyst in the presence of oxalic acid as a hole scavenger, and the aniline yield was improved in the presence of a small amount of dioxygen [16]. Pd-doped TiO_2, for instance, stabilizes electron–hole pairs through its dopant sites, making it effective for C–C and C–N bond formation under mild conditions. By strategically utilizing the VB or CB for different transformations, defect-engineered photocatalysts optimize redox processes, maximizing reaction efficiency and selectivity. This chapter illustrates how these transformations are achieved in practice, with each section detailing the role of defect modifications in enhancing the catalytic performance for oxidative, reductive, and coupling reactions in organic synthesis. Various strategies for defect creation in photocatalytic materials are presented. Then, photocatalytic transformations employing defective photocatalysts are outlined. Lastly, challenges and limitations in the application of defective photocatalysis for organic synthesis are also summarized.

8.2 Defects Creation in Photocatalytic Materials

In the context of photocatalysis, defect engineering has proven to be an effective strategy for enhancing the photocatalytic performance of various materials (Figure 8.1). Defects can modify electronic structures, increase light absorption, and improve charge separation, all of which are critical for efficient catalysis in organic synthesis [17, 18]. This section explores various strategies for creating and controlling defects in photocatalytic materials.

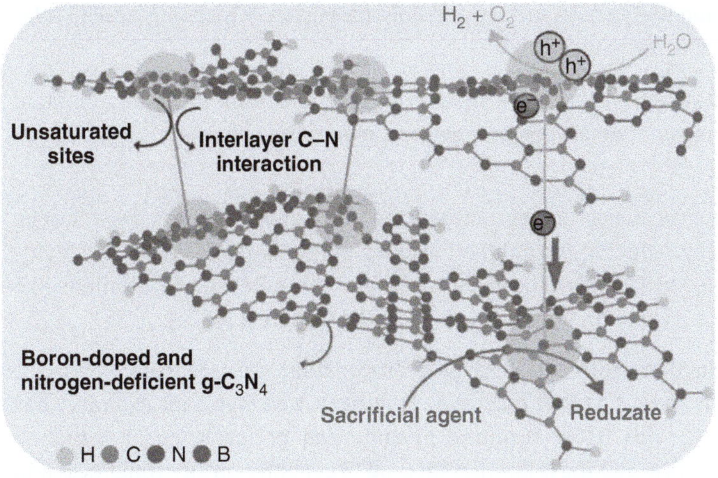

Figure 8.1 Defects creation and chemical bond formation in photocatalytic materials for efficient photocatalytic applications. *Source:* Reproduced from Khan et al. [17] / with permission of Elsevier.

8.2.1 Doping and Substitution

Doping is one of the most widely used strategies for creating defects in photocatalytic materials [19]. It involves introducing foreign atoms into the material's lattice, which disrupts the electronic structure and forms mid-gap states that allow visible-light absorption.

- **Metal doping.** Introducing transition metals such as Ag, Al, Mn, Ni, and Zn into materials such as TiO_2 and ZnO creates localized energy levels within the bandgap, which enhances visible-light absorption and facilitates charge separation [20]. Metal doping has been particularly effective in TiO_2 for applications requiring redox activity under visible light, enabling selective oxidation and reduction reactions in organic synthesis [19].
- **Non-metal doping.** Non-metal elements, such as nitrogen, sulfur, and carbon, have been used to shift the absorption spectrum further into the visible range and improve photocatalytic activity. For instance, nitrogen-doped TiO_2 [21] and sulfur-doped g-C_3N_4 have shown increased visible-light response due to the modified band structures and reduced recombination rates. These doped materials are especially valuable in oxidation reactions, where controlled redox behavior is essential for achieving high selectivity [22]. Doping, whether with metals or non-metals, also helps in stabilizing charge carriers. By creating electron or hole traps within the material, doping enhances the lifespan of photogenerated charge carriers, improving the efficiency in multistep organic transformations.

8.2.2 Thermal and Chemical Treatments

Thermal and chemical treatments are commonly used to introduce defects such as vacancies and dislocations without incorporating foreign atoms. These treatments are effective in modifying both surface and bulk properties, which play crucial roles in photocatalytic performance.

- **Thermal treatments.** Processes such as high-temperature annealing or rapid quenching can generate oxygen vacancies in materials such as TiO_2 (Figure 8.2) [23].

These oxygen vacancies act as active sites for redox reactions by trapping electrons, thereby reducing recombination rates and enhancing catalytic activity. Oxygen vacancies in TiO_2 have been shown to improve efficiency in reactions like alcohol oxidation, where high reactivity is essential (Figure 8.3) [24].

- Chemical reduction treatments, such as hydrogenation, are used to create surface defects and increase material reactivity. Hydrogen treatment, for example, has been used to modify TiO_2, resulting in enhanced performance in reduction reactions. Chemical etching is another method to generate surface roughness and defects, improving adsorption and reactivity in organic transformations. Both thermal and chemical treatments are effective for creating stable defect structures that enhance photocatalyst performance. These treatments are often used

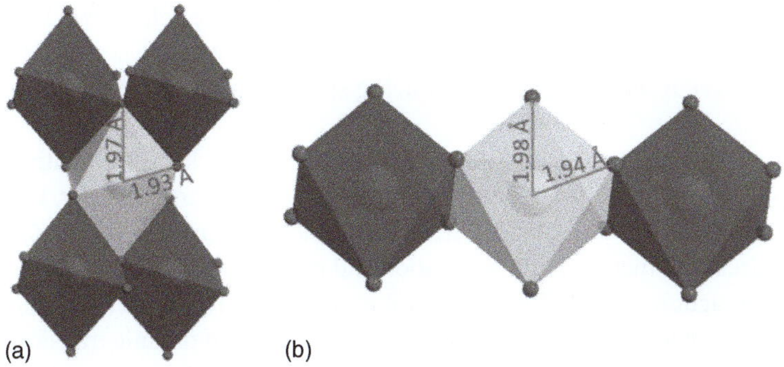

Figure 8.2 (a, b) Schematic TiO_6 octahedral units for anatase and rutile structures, respectively. Full chemical environment of a Ti octahedron is in both structures. *Source:* Allieta et al. [23] / MDPI / CC BY 4.0.

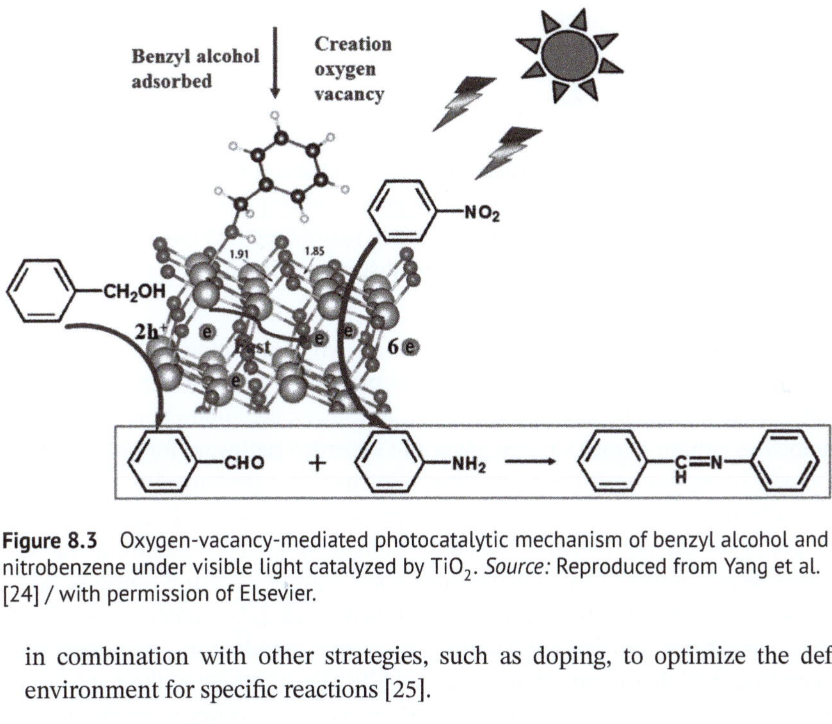

Figure 8.3 Oxygen-vacancy-mediated photocatalytic mechanism of benzyl alcohol and nitrobenzene under visible light catalyzed by TiO_2. *Source:* Reproduced from Yang et al. [24] / with permission of Elsevier.

in combination with other strategies, such as doping, to optimize the defect environment for specific reactions [25].

8.2.3 Surface Functionalization

Surface functionalization and control over morphology are crucial strategies for tailoring the photocatalytic properties of materials, particularly in reactions that depend on surface interactions.

- Functionalizing the surface with groups such as hydroxyl (–OH) or amine (–NH_2) can increase the material's affinity for reactants, improving reaction rates. Surface functionalization is especially effective in photocatalysts like g-C_3N_4, where functional groups enhance visible-light absorption and facilitate selective oxidation reactions. In organic synthesis, surface-functionalized photocatalysts provide greater control over reaction pathways, reducing side reactions and enhancing yield.

- Modifying the physical shape and structure of photocatalysts, such as creating nanorods, spheres, or sheets, influences light absorption and charge transfer. For example, 2D materials like MoS_2 benefit from morphology control that allows for layered structures, enhancing charge separation and providing high surface areas for reactant interaction. Morphological control is critical for reactions requiring large surface areas, such as selective oxidation and reduction in organic transformations. By controlling the surface functionalization and morphology, researchers can optimize photocatalysts for specific reactions, especially those that require high interaction between the catalyst and reactants.

8.2.4 Multi-defect Engineering

Multi-defect engineering and precise control over the morphology of photocatalytic materials, such as creating nanostructures like nanorods, nanowires, and nanosheets, play a crucial role in enhancing photocatalytic efficiency. The specific shape and size of these nanostructures influence light absorption, charge separation, and reactivity, which are essential for high-performance photocatalysis in organic synthesis [26, 27]. Nanostructures like nanorods, nanowires, and nanosheets exhibit unique optical and electronic properties that distinguish them from bulk materials. For example:

- **Nanorods and nanowires.** Studies have shown that nanowire structures of TiO_2 outperform nanoparticle (NP) counterparts in organic oxidation reactions due to their enhanced charge separation and lower recombination rates. TiO_2 nanowires exhibit higher surface areas and extended pathways for charge carriers, which are less prone to recombine, allowing for a prolonged reaction time with adsorbed molecules. The unique elongated morphology of nanowires also facilitates directional electron transport, enhancing charge mobility along the nanowire axis, which is beneficial for reactions requiring electron transfer [28].
- **Nanosheets and 2D materials.** Two-dimensional (2D) materials, such as MoS_2 nanosheets, provide large surface areas and abundant reactive edges that contribute to higher photocatalytic activity [29]. The thinness of the nanosheets shortens the distance that photogenerated charges must travel to reach the surface, reducing recombination and enabling efficient redox processes. This feature is particularly useful in photocatalytic organic transformations that require high reactivity and rapid electron transfer.

It is worth mentioning that combining multi-defect engineering with specific nanostructured morphologies creates synergistic effects that further enhance photocatalytic efficiency:

- Introducing dopants into nanorods and nanowires can create additional mid-gap states that facilitate visible-light absorption and extend the lifetime of photogenerated charge carriers. For instance, nitrogen-doped TiO_2 nanorods exhibit improved light absorption and enhanced photocatalytic activity in organic transformations by supporting redox processes under visible light.

- Forming heterojunctions between different 2D materials, such as MoS_2 and g-C_3N_4, provides a built-in electric field at the interface, which drives charge separation and enhances reactivity. These heterostructures have shown remarkable activity in photocatalytic CO_2 reduction and hydrocarbon transformations, which are highly relevant for organic synthesis applications. By integrating multi-defect engineering with morphology control, researchers can tailor photocatalytic materials for specific applications in organic synthesis, optimizing light absorption, charge separation, and surface reactivity to achieve higher efficiency and selectivity.

8.3 Photocatalytic Transformations

Oxidative transformations are essential in organic synthesis for converting alcohols, hydrocarbons, and other compounds into valuable intermediates and products. Defect-engineered photocatalysts enhance these reactions by facilitating electron transfer and stabilizing reactive intermediates. Table 8.1 provides examples of photocatalyst efficiencies for oxidative transformations. Among them, BiOCl, WO_3, $Bi_2WO_6/CdWO_4$, and TiO_2 with oxygen vacancies are widely used in the photocatalytic oxidation of alcohols to aldehydes and ketones (Figure 8.4) [38].

The oxygen vacancies in TiO_2 serve as active sites that enhance the adsorption of oxygen molecules, increasing their interaction with the photogenerated holes. This leads to the efficient oxidation of alcohols under visible light, with high selectivity and minimal by product formation. For example, benzyl alcohol can be selectively oxidized to benzaldehyde using oxygen-deficient TiO_2, which enhances reactivity while operating under ambient conditions. In addition to metal-oxide photocatalysts, metal-free semiconductors such as g-C_3N_4 has emerged as an effective photocatalyst for hydrocarbon oxidation under visible light. The nitrogen dopants in g-C_3N_4 extend its light absorption into the visible spectrum and improve charge separation, resulting in the efficient oxidation of hydrocarbons to the corresponding alcohols or ketones. This approach is especially relevant to pharmaceutical synthesis, where selective oxidation is crucial for generating bioactive molecules. It should be noted that defect-engineered MOFs, particularly those with missing metal sites, have demonstrated high efficiency in the oxidation of aromatic compounds. These defects act as catalytic sites that promote electron transfer, facilitating the oxidation process. MOFs such as Zr-based UiO-66 with ligand defects have been used to oxidize toluene to benzyl alcohol and benzaldehyde, providing high selectivity and yield under visible light [38].

Reductive transformations are commonly used in organic synthesis to convert functional groups into more reduced forms, such as reducing nitro groups to amines or carbonyl compounds to alcohols. Defect engineering enhances the reductive capabilities of photocatalysts by stabilizing photogenerated electrons and

Table 8.1 Examples of photocatalysts for oxidative transformations.

Photocatalyst materials	Reactant → Products	Selectivity (%)	Yield (%)	Mechanism	References
TiO_2/maghemite–silica composite	Benzyl alcohol → Benzaldehyde	90	45	Fe_3O_4 supports enhance adsorption; acetonitrile solvent prevents undesired side reactions (shield effect)	[30]
CdS/TiO_2	Benzyl alcohol → Benzaldehyde	99	98	CdS quantum dots improve visible-light absorption; heterojunction prevents charge recombination	[31]
VO@g-C_3N_4	Benzyl alcohol → Benzaldehyde	98	98	Oxygen vacancies enhance O_2 activation; visible light initiates $\bullet O_2^-$ species	[32]
g-C_3N_4	Benzyl alcohol → Benzaldehyde	99	57	Mesoporous structure improves reactant access; no •OH generation avoids over-oxidation	[33]
$BiVO_4$	Benzyl alcohol → Benzaldehyde	100	73	Visible-light activation with enhanced selectivity due to surface oxygen vacancies	[34]
Au-modified BiOCl	Benzyl alcohol → Benzaldehyde	99	75.6	Oxygen vacancies and plasmonic hot electrons facilitate high selectivity	[34]
Fe-g-C_3N_4	Benzene → Phenol	100	4.8	Fe^{3+}/Fe^{2+} cycle generates •OH radicals, enhancing phenol production with H_2O_2	[35]
Pt/WO_3	Benzene → Phenol	73.7	50.8	•OH radicals generated from water selectively oxidize benzene to phenol	[36]
$Bi_2WO_6/CdWO_4$	Benzene → Phenol	99.9	5.8	Reactive species formed on Bi_2WO_6 facilitate selective hydroxylation of benzene	[37]

facilitating charge transfer. Table 8.2 provides examples of photocatalyst efficiencies for reductive transformations.

- Copper-doped TiO_2 has shown notable efficiency in the reduction of nitro compounds to amines, a critical transformation in the production of pharmaceutical intermediates. The Cu dopants in TiO_2 introduce mid-gap states that allow visible-light absorption and act as electron traps, stabilizing photogenerated electrons. This stabilizing effect reduces nitrobenzene to aniline with high selectivity and under mild conditions, making it suitable for environmentally friendly synthetic applications.
- Defective ZnO nanostructures, such as ZnO nanowires with oxygen vacancies, exhibit strong reductive capabilities due to their high surface area and efficient

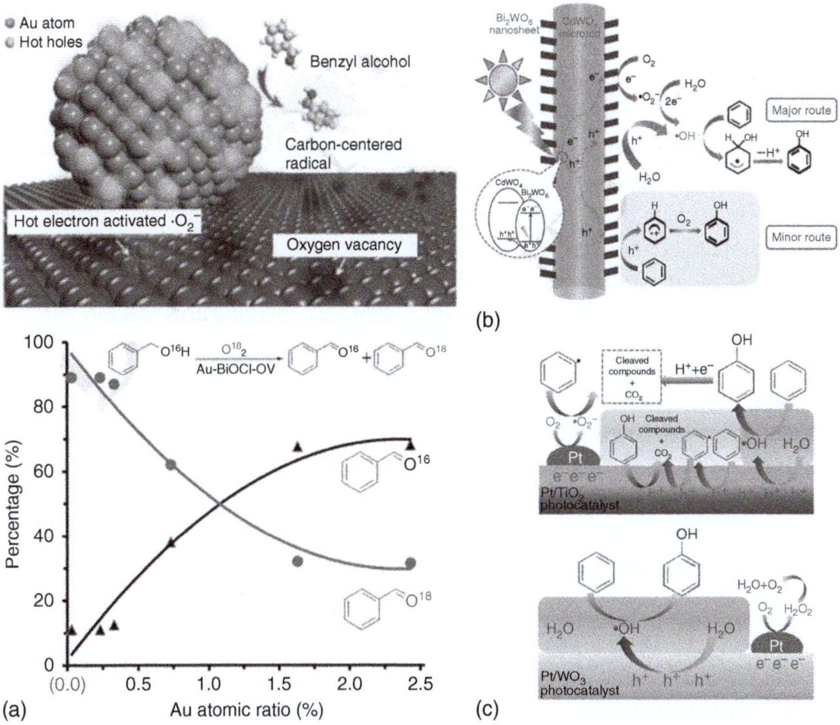

Figure 8.4 Oxidative transformations over (a) BiOCl, (b) $Bi_2WO_6/CdWO_4$, (c) TiO_2 and WO_3 photocatalysts for various organic compounds generation. *Source:* Reproduced from Chen et al. [38] / with permission of Elsevier.

Table 8.2 Examples of photocatalysts for reductive transformations.

Photocatalyst materials	Reactant → Product	Selectivity (%)	Yield (%)	Mechanism	References
TiO_2 NPs	Nitrobenzene → Aniline	~100	90	Photogenerated electrons in the CB reduce nitrobenzene; holes in the VB oxidize water	[16]
Ag-TiO_2 composite	CO_2 → Methanol	~90	50	Plasmonic Ag enhances visible-light absorption and charge separation	[39]
ZnO nanorods	Benzaldehyde → Benzyl alcohol	~95	60	Oxygen vacancies promote selective adsorption and electron transfer	[40]

charge separation. Oxygen vacancies trap photogenerated electrons, allowing them to participate in the reduction of carbonyl compounds to alcohols. This method is highly effective for converting benzaldehyde to benzyl alcohol, showcasing its utility in organic transformations that require selective reduction. Pd-doped molybdenum disulfide (MoS_2) is effective for hydrogenation reactions, leveraging the synergy between Pd and MoS_2 to facilitate electron transfer and stabilize hydrogen intermediates. This doped catalyst has shown success in reducing alkenes and alkynes, providing a selective and efficient pathway for hydrogenation in organic synthesis. The Pd dopants act as active sites for H_2 activation, while MoS_2 facilitates charge transport, making this system highly effective for hydrogenation under visible light.

Coupling reactions, such as C–C and C–N bond formation, are fundamental in constructing complex organic molecules. Defect-engineered photocatalysts enhance these reactions by providing active sites and improving charge separation, leading to higher yields and selectivity. Table 8.3 provides the examples of photocatalyst

Table 8.3 Examples of photocatalysts for coupling reactions.

Photocatalyst materials	Reactant → Product	Selectivity (%)	Yield (%)	Mechanism	References
Pd-NPs@TTT-COF	Suzuki coupling: Methyl-iodobenzene → Phenylboronic acid	~99	~99	Electron transfer facilitated by Pd-NPs immobilized in the COF structure.	[41]
Ir,Ni@Phen-COF	Aryl bromides → Potassium benzyltrifluoroborate salts	~90	86–90	Synergistic dual catalysis with Ir and Ni active sites for redox reactions	[42]
Tp-Acr COF	4-Bromobenzotri-fluoride → Pyrrolidine	~94	66–94	Broad light absorption and efficient charge separation via β-ketoenamine tautomerism	[43]
sp^2-COFdpy–Ni	4-Bromobenzonitrile → 4-Methoxybenzonitrile	~99	~99	Efficient C–O coupling via visible-light-induced Ni activation	[44]
Ace-COF–Ni	Thiophenol → Iodobenzene	~96	~96	Radical-mediated visible light-driven Ni catalysis for S–C coupling	[45]
TPT-COF–Cu	Terminal alkyne → H-phosphonate compounds	~95	85–95	Cu catalysis with single-electron transfer for P–C coupling	[46]

Source: Reproduced from Ohmatsu et al. [7] / with permission of American Chemical Society.

efficiencies for coupling transformations. For example, Pd-doped TiO_2 has demonstrated remarkable efficiency in facilitating Suzuki and Heck coupling reactions under mild conditions. The Pd dopants create active sites that enhance electron transfer, while the TiO_2 support stabilizes reaction intermediates. This combination allows for high selectivity in forming C–C bonds with minimal by-product formation, making it suitable for synthesizing pharmaceutical compounds and fine chemicals. Recently, heterojunctions between g-C_3N_4 and other semiconductor materials, such as TiO_2 or MoS_2, improve the photocatalytic performance in C–N coupling reactions. The heterojunctions provide a built-in electric field that promotes charge separation, facilitating the formation of amines from aryl halides and amines with high efficiency. This C–N coupling is particularly useful in the synthesis of nitrogen-containing pharmaceuticals and agrochemicals, where high selectivity is critical. In the recent few decades, defective Cu_2O nanostructures, especially those with surface oxygen vacancies, are effective photocatalysts for Sonogashira coupling reactions, enabling the formation of C–C bonds between alkynes and aryl halides. The oxygen vacancies on Cu_2O facilitate charge separation and stabilize reactive intermediates, leading to high catalytic activity and selectivity in forming carbon–carbon bonds.

8.4 Photocatalysis in Organic Synthesis

In photocatalytic organic synthesis, monitoring the reaction process, validating the products, and assessing the efficiency of the photocatalyst are crucial steps to ensure the effectiveness and reproducibility of the catalytic system. This chapter explores the methods used to track the progress of photocatalytic reactions, verify product formation, and evaluate the performance of photocatalysts in catalyzing organic reactions under light irradiation [47].

8.4.1 Monitoring Photocatalytic Reactions

Real-time monitoring of photocatalytic reactions is essential for understanding reaction kinetics, optimizing conditions, and determining reaction intermediates and pathways. Several analytical techniques provide insights into the progress of photocatalytic processes which are stated below:

- UV-Visible spectroscopy allows continuous monitoring of reactant conversion and product formation by measuring absorbance changes over time. For example, in the oxidation of alcohols, the formation of aldehydes or ketones can be tracked by observing the characteristic absorption peaks. This technique is useful for understanding reaction kinetics and determining the rate of reactant consumption, which is indicative of the efficiency of the photocatalyst under specific light conditions [33, 34, 48–50].
- Gas chromatography (GC) and high-performance liquid chromatography (HPLC) are widely used to analyze and quantify the concentration of reactants and products during the reaction. These methods are essential in reactions where multiple

products or intermediates may be formed, such as coupling or selective oxidation reactions. GC is particularly suitable for analyzing volatile products, while HPLC is ideal for non-volatile or polar compounds. By using these techniques, researchers can determine the selectivity and yield of the photocatalytic process in organic synthesis.

- In situ mass spectrometry (MS) provides valuable information on reaction intermediates and products formed during the photocatalytic process. For example, in photocatalytic CO_2 reduction or hydrogen evolution, MS can be used to detect gas-phase products in real-time, providing insights into reaction mechanisms. This method is also useful in organic transformations where unstable intermediates are generated, as it allows for real-time analysis without interrupting the reaction [51, 52].

8.4.2 Identification of By-products

To confirm the success and efficiency of photocatalytic reactions, it is essential to validate the products and intermediates formed during the process using some techniques. This includes ensuring the chemical identity, purity, and structural integrity of the final products. Examples of them are given as follows:

- Nuclear magnetic resonance (NMR) spectroscopy is used to confirm the structure and purity of the final products obtained from photocatalytic reactions. It provides detailed information about the molecular structure, functional groups, and purity of organic compounds. In complex organic transformations, NMR is crucial for validating the selective formation of desired products over side products, which is particularly important for pharmaceutical applications where product purity is critical [53, 54].
- Gas chromatography-mass spectrometry (GC-MS) and liquid chromatography-mass spectrometry (LC-MS) offer both separation and molecular identification capabilities, allowing researchers to validate product composition and detect trace impurities. GC-MS and LC-MS are commonly used in photocatalytic organic synthesis to verify the presence of specific products and intermediates, confirm reaction completion, and assess product selectivity. These techniques are especially useful in analyzing complex mixtures resulting from coupling or oxidation reactions [55].
- X-ray diffraction (XRD) for solid products is used to analyze crystalline products obtained from solid-state photocatalytic reactions. This technique identifies crystalline phases and helps determine the purity and phase composition of the products. XRD is also useful in validating the crystallographic structure of any solid photocatalytic by-products or intermediates, ensuring that the photocatalyst itself remains unchanged after the reaction [56].

8.4.3 Performance Evaluation

Evaluating the performance and efficiency of photocatalysts is essential to determine their effectiveness in light-driven organic synthesis. Various techniques measure key parameters like quantum efficiency, reaction rate, and stability, providing a comprehensive assessment of the photocatalyst's capabilities.

- Measuring the reaction rate or turnover frequency is critical in assessing a photocatalyst's activity. These metrics quantify how fast a photocatalyst converts reactants into products under specific light conditions. For instance, in the photocatalytic oxidation of alcohols, the rate at which benzyl alcohol is converted to benzaldehyde can be used to evaluate catalyst efficiency [57].
- Quantum yield and apparent quantum efficiency can quantify the number of photons utilized in the photocatalytic reaction relative to the total photons absorbed. These parameters are particularly important for evaluating photocatalysts designed to operate under visible light, as they indicate the photocatalyst's efficiency in converting light energy into chemical energy. By comparing quantum yield values under different wavelengths, researchers can determine which modifications (e.g., doping, defect engineering) have effectively enhanced the catalyst's light-harvesting capability [58].
- Stability tests are performed by recycling the photocatalyst through multiple reaction cycles and analyzing its performance retention. High durability and reusability are essential for practical applications, especially in industrial organic synthesis. Photocatalyst stability is evaluated by comparing activity levels after each cycle and assessing changes in structure or composition through techniques like XPS or TEM. A stable photocatalyst will retain its structural integrity and activity over repeated use, which is crucial for cost-effective and sustainable applications [59].
- UV–Vis diffuse reflectance spectroscopy (UV–Vis–DRS) is used to assess the photocatalyst's light absorption capacity across different wavelengths. By analyzing the spectral response, researchers can confirm the effectiveness of bandgap engineering modifications, such as doping or heterojunction formation, which allow the photocatalyst to utilize visible light. Improved light absorption capabilities generally correlate with higher photocatalytic activity, especially for the reactions performed under solar or artificial visible light sources [60].

8.4.4 Advanced Techniques

Advanced characterization methods provide deeper insights into the mechanism of photocatalytic reactions, linking defect properties with photocatalytic performance.

- Electron paramagnetic resonance (EPR) can detect unpaired electrons associated with defect sites, such as oxygen vacancies or dopant states, providing information about the electronic structure of the photocatalyst. EPR data can correlate defect concentration with photocatalytic performance, revealing how specific defects influence charge separation, recombination rates, and catalytic activity [61].
- Time-resolved photoluminescence (TRPL) measures the lifetimes of photogenerated electrons and holes, giving insights into charge recombination dynamics. By comparing TRPL lifetimes in pristine and defect-engineered photocatalysts, researchers can determine how effectively defects extend carrier lifetimes, which directly affects photocatalytic efficiency [62].

- Electrochemical impedance spectroscopy assesses charge transfer resistance and overall conductivity, providing information on the photocatalyst's ability to transport charges to the reaction interface. This technique helps evaluate how modifications, such as doping or the formation of heterojunctions, influence charge mobility, enabling efficient catalytic cycles under light irradiation [63].

8.5 Challenges and Limitations

While photocatalysis offers significant advantages for sustainable organic synthesis, several challenges and limitations need to be addressed to optimize its practical applications. These challenges include catalyst stability, scalability, efficiency under visible light, selectivity, and environmental impact. This section outlines these issues and discusses potential strategies to overcome them.

8.5.1 Stability and Durability

One of the main limitations in photocatalytic organic synthesis is the instability of photocatalysts under reaction conditions, particularly under prolonged light exposure. Photocatalysts, especially metal sulfides like CdS, are prone to photo-corrosion, where reactive species generated during the photocatalytic process degrade the material itself. For instance, CdS can decompose into Cd^{2+} ions under visible light, limiting its long-term usability in synthetic applications.

To enhance stability, researchers are exploring protective coating methods, such as adding a thin oxide layer or using heterojunctions that reduce surface reactivity. These coatings help protect the catalyst surface from reactive intermediates while allowing efficient charge transfer. Additionally, incorporating dopants like carbon or nitrogen has been shown to improve the stability of materials like g-C_3N_4 by increasing structural robustness.

8.5.2 Scalability and Economic Viability

Scaling up photocatalytic systems from laboratory settings to industrial applications presents significant challenges. The synthesis of high-performance photocatalysts often involves complex procedures, such as doping, annealing, and multi-step chemical treatments, which can be costly and time-consuming. Additionally, high-quality materials such as MOFs and COFs are generally expensive to produce at an industrial scale, which limits their economic feasibility for large-scale synthesis processes.

For practical implementation, cost-effective synthesis methods, such as green synthesis routes and the use of abundant, low-cost materials such as TiO_2, are being developed. Additionally, efforts are underway to simplify doping and defect-engineering processes, aiming to create high-performance photocatalysts through scalable and environmentally friendly methods.

8.5.3 Efficiency Under Visible Light

Many photocatalysts, particularly early materials such as TiO_2, have a wide bandgap that restricts them to UV light absorption, limiting their efficiency in natural sunlight, where UV constitutes only about 4% of the solar spectrum. Efficient visible-light absorption is critical for making photocatalysis viable under solar irradiation. However, visible-light-active photocatalysts often suffer from reduced charge separation efficiency and faster recombination of photogenerated electron–hole pairs.

Bandgap engineering techniques, such as doping and heterojunction formation, have been widely adopted to shift light absorption into the visible range. For instance, nitrogen-doped TiO_2 and heterostructures like g-C_3N_4/TiO_2 effectively utilize visible light, improving the efficiency of photocatalytic transformations. Further advancements in bandgap tuning and light-harvesting capabilities are essential for enhancing visible-light-driven photocatalysis.

8.5.4 Selectivity and Product Control

Achieving high selectivity in photocatalytic reactions is challenging, especially in complex organic transformations where multiple reaction pathways are possible. Selectivity is crucial in synthetic chemistry, as uncontrolled side reactions can lead to undesired by-products, reducing the overall efficiency and purity of the final product.

- Introducing specific types of defects, such as oxygen vacancies, has been shown to influence reaction pathways by creating preferential active sites that favor certain reactions over others. For example, oxygen vacancies in TiO_2 promote selective alcohol oxidation by stabilizing key intermediates.
- Modifying the photocatalyst surface with functional groups or organic linkers can provide steric and electronic control, guiding reactants toward specific reaction sites. Surface functionalization is especially beneficial in reactions such as selective oxidation and coupling reactions, where product purity is paramount.

These approaches help control product formation by modifying the reaction environment, enabling more selective transformations in complex organic synthesis.

8.6 Future Research Directions

As photocatalysis continues to evolve, several promising directions are emerging that could enhance the capabilities and broaden the applications of defect-engineered photocatalysts in organic synthesis. This chapter highlights key areas of focus for future research, including advanced defect design through computational methods, integration with renewable energy sources, and expanding the scope of organic transformations.

Computational methods, including machine learning (ML) and density functional theory (DFT) calculations, are increasingly used to predict and optimize defect structures in photocatalysts. These approaches allow researchers to model how different defects, such as dopant types, vacancies, and heterojunction configurations, affect photocatalytic activity before performing costly experimental procedures.

Machine learning for predictive modelling can analyze large datasets from previous photocatalytic studies to identify patterns in defect types, band structures, and reaction outcomes. This predictive modelling aids in designing photocatalysts with tailored defect properties for specific reactions, such as selective oxidation or coupling processes.

DFT calculations for defect engineering provide insights into the electronic structure and charge distribution in defect-engineered photocatalysts, enabling precise control over properties such as bandgap energy and charge mobility. By simulating potential defect configurations, DFT helps researchers optimize photocatalyst materials for enhanced light absorption and charge separation.

Using computational tools, researchers can streamline the development of high-performance photocatalysts, reducing experimental costs and accelerating the discovery of materials suited for specific organic transformations.

Integrating photocatalysis with renewable energy sources, particularly solar energy, represents a sustainable approach to power organic synthesis. Photocatalysts that operate efficiently under sunlight can significantly reduce the energy demands of synthetic processes, making them eco-friendlier and more scalable.

Solar-Driven Photocatalysis: Developing photocatalysts with optimized visible-light absorption is critical for solar-driven applications. Recent advancements in bandgap engineering, such as doping and heterojunction formation, enable materials such as TiO_2, g-C_3N_4, and perovskites to harness sunlight effectively. Implementing solar concentrators or reflective systems in reactor designs can further enhance the utilization of sunlight in large-scale photocatalysis.

Hybrid Systems with Photovoltaics: Combining photocatalysis with photovoltaic cells can create hybrid systems that enhance energy efficiency. For example, photocatalysts powered by photovoltaics can maintain consistent light intensity, improving reaction rates even in low-light conditions. Such hybrid systems can be valuable in industrial settings where stable and controlled reaction conditions are necessary.

Photocatalysis in organic synthesis is traditionally focused on reactions such as oxidation, reduction, and coupling. However, there is potential to expand its application to more complex transformations and multi-step synthesis pathways.

Cascade Reactions: Defect-engineered photocatalysts can be tailored for cascade reactions, where multiple transformations occur sequentially in a single reaction system. For instance, a catalyst with multi-functional defect sites could enable both oxidation and reduction steps in a complex reaction sequence. This approach is highly efficient for synthesizing complex molecules, reducing the need for intermediate purification steps.

Photocatalytic Asymmetric Synthesis: Asymmetric synthesis remains challenging in photocatalysis due to the need for precise control over stereoselectivity. By modifying defect structures or combining photocatalysts with chiral co-catalysts, researchers can potentially create systems capable of enantioselective transformations. This would be transformative for applications in pharmaceutical synthesis, where chiral purity is essential.

Sustainability is a growing focus in the development of photocatalysts. There is an increasing interest in creating photocatalysts from abundant, non-toxic materials that have minimal environmental impact.

Metal-free photocatalysts: Materials such as graphitic carbon nitride (g-C_3N_4) and carbon-based compounds are promising as metal-free photocatalysts. These materials can be engineered with heteroatoms or surface modifications to enhance their photocatalytic properties without introducing toxic elements. Metal-free catalysts are particularly attractive for applications in biomedicine and green chemistry.

Biocompatible Photocatalysts for Green Chemistry: Researchers are exploring biocompatible photocatalysts, such as polymeric photocatalysts or MOFs with organic linkers. These materials are safer for applications involving bioactive compounds and can be used in environmentally sensitive processes. Additionally, using renewable sources to synthesize these catalysts aligns with sustainable practices.

8.7 Conclusion

The advancements in defect-engineered photocatalysts discussed in this chapter underscore their transformative impact on the field of organic synthesis. By intentionally manipulating material properties through defect creation, doping, and structural modifications, researchers have unlocked new potentials for photocatalytic transformations under mild and sustainable conditions. These engineered defects, such as vacancies, dopants, and heterojunctions, have been shown to enhance light absorption, charge separation, and catalytic activity, leading to improved efficiency, selectivity, and yield across a wide array of chemical reactions. Practical applications of these photocatalysts span diverse fields, from selective oxidation and reduction reactions to complex C–C and C–N bond formations, demonstrating their utility in producing high-value organic compounds. Moreover, the ability to tailor catalytic properties for specific reactions through defect engineering has provided researchers with powerful tools to design systems optimized for green and sustainable chemistry. In conclusion, the field of defect-engineered photocatalysis holds immense promise for reshaping organic synthesis. By addressing existing challenges and exploring innovative pathways, researchers can further bridge the gap between laboratory breakthroughs and real-world industrial applications. This ongoing evolution not only contributes to the advancement of synthetic chemistry but also aligns with global efforts to adopt sustainable, eco-friendly technologies in chemical manufacturing.

Acknowledgments

The author expresses sincere gratitude to the Head of the Department of Chemistry, Faculty of Mathematics and Natural Sciences, University of Mataram, for their unwavering support and guidance throughout the preparation of this chapter. Their leadership and encouragement have been instrumental in fostering an environment conducive to academic and research excellence.

Conflict of Interest

The author declares that there are no conflicts of interest.

References

1 Hao, H., Zhang, L., Wang, W., and Zeng, S. (2018). Modification of heterogeneous photocatalysts for selective organic synthesis. *Catal. Sci. Technol.* 8 (5): 1229–1250. https://doi.org/10.1039/C7CY01853C.

2 Akita, M., Ceroni, P., Stephenson, C.R.J., and Masson, G. (2023). Progress in photocatalysis for organic chemistry. *J. Org. Chem.* 88 (10): 6281–6283. https://doi.org/10.1021/acs.joc.3c00812.

3 Reischauer, S. and Pieber, B. (2021). Emerging concepts in photocatalytic organic synthesis. *iScience* 24 (3): 102209. https://doi.org/10.1016/j.isci.2021.102209.

4 Gisbertz, S. and Pieber, B. (2020). Heterogeneous photocatalysis in organic synthesis. *ChemPhotoChem* 4 (7): 456–475. https://doi.org/10.1002/cptc.202000014.

5 Bryden, M.A., Pinosa, E., Suzuki, R., and Ooi, T. (2023). Assessments and advances in photocatalysis with pyrimidyl sulfone. *J. Org. Chem.* 88 (10): 6364–6373. https://doi.org/10.1021/acs.joc.2c01137.

6 Pinosa, E., Bassan, E., Cetin, S. et al. (2023). Light-induced access to carbazole-1,3-dicarbonitrile as a photocatalyst. *J. Org. Chem.* 88 (10): 6390–6400. https://doi.org/10.1021/acs.joc.2c01825.

7 Ohmatsu, K., Suzuki, R., Fujita, H., and Ooi, T. (2023). Zwitterionic diphenylphosphinyl amidate as a powerful catalyst. *J. Org. Chem.* 88 (10): 6553–6556. https://doi.org/10.1021/acs.joc.2c02362.

8 Nicewicz, D.A. and MacMillan, D.W.C. (2008). Merging photoredox catalysis with organocatalysis. *Science* 322 (5898): 77–80. https://doi.org/10.1126/science.1161976.

9 Xiao, J., Liu, X., Pan, L. et al. (2020). Heterogeneous photocatalytic organic transformation using conjugated polymers-based materials. *ACS Catal.* 10 (20): 12256–12283. https://doi.org/10.1021/acscatal.0c03480.

10 Albini, A. and Fagnoni, M. (2008). Giacomo Ciamician and the concept of green chemistry. *ChemSusChem* 1 (1–2): 63–66. https://doi.org/10.1002/cssc.200700015.

11 Furusawa, A. (2015). *Quantum States of Light*, SpringerBriefs in Mathematical Physics, vol. 10. Tokyo: Springer Japan https://doi.org/10.1007/978-4-431-55960-3.

12 Tucker, J.W. and Stephenson, C.R.J. (2012). Shining light on photoredox catalysis: theory and synthetic applications. *J. Org. Chem.* 77 (4): 1617–1622. https://doi.org/10.1021/jo202538x.

13 Prier, C.K., Rankic, D.A., and MacMillan, D.W.C. (2013). Visible-light photoredox catalysis with transition metal complexes. *Chem. Rev.* 113 (7): 5322–5363. https://doi.org/10.1021/cr300503r.

14 Rahman, A., Jennings, J.R., Tan, A.L., and Khan, M.M. (2022). Molybdenum disulfide-based nanomaterials for visible-light-induced photocatalysis. *ACS Omega* 7 (26): 22089–22110. https://doi.org/10.1021/acsomega.2c01314.

15 Markushyna, Y., Smith, C.A., and Savateev, A. (2020). Organic photocatalysis: carbon nitride semiconductors vs. molecular catalysts. *Eur. J. Org. Chem.* 2020 (10): 1294–1309. https://doi.org/10.1002/ejoc.201901112.

16 Kominami, H., Iwasaki, S., Maeda, T. et al. (2009). Photocatalytic reduction of nitrobenzene to aniline in an aqueous suspension of titanium (IV) oxide particles in the presence of oxalic acid as a hole scavenger and promotive effect of dioxygen in the system. *Chem. Lett.* 38 (5): 410–411. https://doi.org/10.1246/cl.2009.410.

17 Khan, I., Khan, S., Shayan, M. et al. (2024). Simultaneous dopants and defects modulating band structures in Z-scheme photocatalysts. *Carbon* 229: 119451. https://doi.org/10.1016/j.carbon.2024.119451.

18 Jamma, A., Jaksani, B., Vennapoosa, C.S. et al. (2024). Defect-rich UiO-66@g-C_3N_4/Ni frameworks as efficient water splitting photocatalysts. *Mater. Adv.* 5 (7): 2785–2796. https://doi.org/10.1039/D3MA00880K.

19 Eldoma, M.A., Alaswad, S.O., Mahmoud, M.A. et al. (2024)). Enhancing photocatalytic performance through defect engineering and doping. *J. Photochem. Photobiol. Chem.* 446: 115164. https://doi.org/10.1016/j.jphotochem.2023.115164.

20 Gao, X., Zhou, B., and Yuan, R. (2015). Doping metals on TiO_2 nanotubes and its effect on photocatalytic oxidation. *Environ. Eng. Res.* 20 (4): 329–335. https://doi.org/10.4491/eer.2015.062.

21 Du, S., Lian, J., and Zhang, F. (2022). Visible-light-responsive N-doped TiO_2 photocatalysis. *Trans. Tianjin Univ.* 28 (1): 33–52. https://doi.org/10.1007/s12209-021-00303-w.

22 Wang, K. et al. (2015). Sulfur-doped g-C_3N_4 with enhanced photocatalytic CO_2 reduction performance. *Appl. Catal. B: Environ.* 176–177: 44–52. https://doi.org/10.1016/j.apcatb.2015.03.045.

23 Allieta, M., Coduri, M., and Naldoni, A. (2024). Black TiO_2 and oxygen vacancies: unraveling the role in anatase-to-rutile transformation. *Appl. Nano.* 5 (2): 72–83. https://doi.org/10.3390/applnano5020007.

24 Yang, X., Tao, H., Leow, W.R. et al. (2019). Oxygen-vacancies-engaged efficient carrier utilization for coupling reactions. *J. Catal.* 373: 116–125. https://doi.org/10.1016/j.jcat.2019.03.022.

25 Yan, Y., Han, M., Konkin, A. et al. (2014). Slightly hydrogenated TiO_2 with enhanced photocatalytic performance. *J. Mater. Chem. A* 2 (32): 12708–12716. https://doi.org/10.1039/C4TA02192D.

26 Hassaan, M.A., El-Nemr, M.A., Elkatory, M.R. et al. (2023). Principles of photocatalysts and their different applications: a review. *Top. Curr. Chem.* 381 (6): 31. https://doi.org/10.1007/s41061-023-00444-7.

27 Mohamadpour, F. and Amani, A.M. (2024). Photocatalytic systems: reactions, mechanism, and applications. *RSC Adv.* 14 (29): 20609–20645. https://doi.org/10.1039/D4RA03259D.

28 Sudirman, S., Atmaja, L.D.P., Pratama, A.J. et al. (2023). Highly photocatalytic performance of TiO_2 nanowires in the conversion of benzaldehydes to benzoic acid. *Acta Chim. Asiana* 6 (2): 315–321. https://doi.org/10.29303/aca.v6i2.160.

29 Chava, R.K., Do, J.Y., and Kang, M. (2018). Hydrothermal growth of two-dimensional hierarchical MoS_2 nanospheres on one-dimensional CdS nanorods for high performance and stable visible photocatalytic H_2 evolution. *Appl. Surf. Sci.* 433: 240–248. https://doi.org/10.1016/j.apsusc.2017.09.260.

30 Colmenares, J.C., Ouyang, W., Ojeda, M. et al. (2016). Mild ultrasound-assisted synthesis of TiO_2 supported on magnetic nanocomposites for selective photo-oxidation of benzyl alcohol. *Appl. Catal. B: Environ.* 183: 107–112. https://doi.org/10.1016/j.apcatb.2015.10.034.

31 Li, X., Wang, J., Men, Y., and Bian, Z. (2016). TiO_2 mesocrystal with exposed (001) facets and CdS quantum dots as an active visible photocatalyst for selective oxidation reactions. *Appl. Catal. B: Environ.* 187: 115–121. https://doi.org/10.1016/j.apcatb.2016.01.034.

32 Verma, S., Baig, R.B.N., Nadagouda, M.N., and Varma, R.S. (2016). Selective oxidation of alcohols using photoactive VO@g-C_3N_4. *ACS Sustainable Chem. Eng.* 4 (3): 1094–1098. https://doi.org/10.1021/acssuschemeng.5b01163.

33 Su, F., Mathew, S.C., Lipner, G. et al. (2010). Mpg-C_3N_4-catalyzed selective oxidation of alcohols using O_2 and visible light. *JACS* 132 (46): 16299–16301. https://doi.org/10.1021/ja102866p.

34 Li, H., Qin, F., Yang, Z. et al. (2017). New reaction pathway induced by plasmon for selective benzyl alcohol oxidation on BiOCl possessing oxygen vacancies. *JACS* 139 (9): 3513–3521. https://doi.org/10.1021/jacs.6b12850.

35 Chen, X., Zhang, J., Fu, X. et al. (2009). Fe-g-C_3N_4-catalyzed oxidation of benzene to phenol using hydrogen peroxide and visible light. *JACS* 131 (33): 11658–11659. https://doi.org/10.1021/ja903923s.

36 Tomita, O., Ohtani, B., and Abe, R. (2014). Highly selective phenol production from benzene on a platinum-loaded tungsten oxide photocatalyst with water and molecular oxygen: selective oxidation of water by holes for generating hydroxyl radical as the predominant source of the hydroxyl group. *Catal. Sci. Technol.* 4 (11): 3850–3860. https://doi.org/10.1039/C4CY00445K.

37 Chen, P., Chen, L., Zeng, Y. et al. (2018). Three-dimension hierarchical heterostructure of $CdWO_4$ microrods decorated with Bi_2WO_6 nanoplates for high-selectivity photocatalytic benzene hydroxylation to phenol. *Appl. Catal. B: Environ.* 234: 311–317. https://doi.org/10.1016/j.apcatb.2018.04.028.

38 Chen, L., Tang, J., Song, L.-N. et al. (2019). Heterogeneous photocatalysis for selective oxidation of alcohols and hydrocarbons. *Appl. Catal. B: Environ.* 242: 379–388. https://doi.org/10.1016/j.apcatb.2018.10.025.

39 Liu, E., Kang, L., Wu, F. et al. (2014). Photocatalytic reduction of CO_2 into methanol over Ag/TiO_2 nanocomposites enhanced by surface plasmon resonance. *Plasmonics* 9 (1): 61–70. https://doi.org/10.1007/s11468-013-9598-7.

40 Nair, V., Colmenares, J.C., and Lisovytskiy, D. (2019). Ultrasound-assisted ZnO coating in a microflow-based photoreactor for selective oxidation of benzyl alcohol to benzaldehyde. *Green Chem.* 21 (6): 1241–1246. https://doi.org/10.1039/C8GC03131B.

41 Yang, Y., Niu, H., Zhao, W. et al. (2020). Ultrafine Pd nanoparticles loaded benzothiazole-linked covalent organic framework for efficient photocatalytic C–C cross-coupling reactions. *RSC Adv.* 10 (49): 29402–29407. https://doi.org/10.1039/D0RA03739G.

42 López-Magano, A., Ortín-Rubio, B., Imaz, I. et al. (2021). Photoredox heterobimetallic dual catalysis using engineered covalent organic frameworks. *ACS Catal.* 11 (19): 12344–12354. https://doi.org/10.1021/acscatal.1c03634.

43 Traxler, M., Gisbertz, S., Pachfule, P. et al. (2022). Acridine-functionalized covalent organic frameworks (COFs) as photocatalysts for metallaphotocatalytic C–N cross-coupling. *Angew. Chem. Int. Ed.* 61 (21): e202117738. https://doi.org/10.1002/anie.202117738.

44 Dong, W., Yang, Y., Xiang, Y. et al. (2021). A highly stable all-in-one photocatalyst for aryl etherification: the Ni II embedded covalent organic framework. *Green Chem.* 23 (16): 5797–5805. https://doi.org/10.1039/D1GC01902C.

45 Chen, H., Liu, W., Laemont, A. et al. (2021). A visible-light-harvesting covalent organic framework bearing single nickel sites as a highly efficient sulfur–carbon cross-coupling dual catalyst. *Angew. Chem. Int. Ed.* 60 (19): 10820–10827. https://doi.org/10.1002/anie.202101036.

46 Chen, Y.-X., Zhang, M., Zhang, S.-Z. et al. (2022). Copper-decorated covalent organic framework as a heterogeneous photocatalyst for phosphorylation of terminal alkynes. *Green Chem.* 24 (10): 4071–4081. https://doi.org/10.1039/D2GC00754A.

47 Madani, A. and Pieber, B. (2023). In situ reaction monitoring in photocatalytic organic synthesis. *ChemCatChem* 15 (7): e202201583. https://doi.org/10.1002/cctc.202201583.

48 Lu, H., Schmidt, M.A., and Jensen, K.F. (2001). Photochemical reactions and on-line UV detection in microfabricated reactors. *Lab Chip* 1 (1): 22–28. https://doi.org/10.1039/b104037p.

49 Yue, J., Falke, F.H., Schouten, J.C., and Nijhuis, T.A. (2013). Microreactors with integrated UV/Vis spectroscopic detection for online process analysis under segmented flow. *Lab Chip* 13 (24): 4855–4863. https://doi.org/10.1039/c3lc50876e.

50 Telegeiev, I., Thili, O., Lanel, A. et al. (2018). In situ FTIR reactor for monitoring gas-phase products during a (photo)catalytic reaction in the liquid phase. *Anal. Chem.* 90 (24): 14586–14592. https://doi.org/10.1021/acs.analchem.8b04754.

51 Han, Z., Gu, X.H., Wang, S.-W. et al. (2020). Time-resolved in situ monitoring of photocatalytic reactions by probe electrospray ionization mass spectrometry. *Analyst* 145 (9): 3313–3319. https://doi.org/10.1039/D0AN00305K.

52 Buzzetti, L., Crisenza, G.E.M., and Melchiorre, P. (2019). Mechanistic studies in photocatalysis. *Angew. Chem. Int. Ed.* 58 (12): 3730–3747. https://doi.org/10.1002/anie.201809984.

53 Feldmeier, C., Bartling, H., Magerl, K., and Gschwind, R.M. (2015). LED-illuminated NMR studies of flavin-catalyzed photooxidations reveal solvent control of the electron-transfer mechanism. *Angew. Chem. Int. Ed.* 54 (4): 1347–1351. https://doi.org/10.1002/anie.201409146.

54 Goez, M. (1995). Pulse techniques for CIDNP. *Concepts Magn. Reson.* 7 (4): 263–279. https://doi.org/10.1002/cmr.1820070403.

55 Zelenka, J., Cibulka, R., and Roithová, J. (2019). Flavinium catalysed photooxidation: detection and characterization of elusive peroxyflavinium intermediates. *Angew. Chem. Int. Ed.* 58 (43): 15412–15420. https://doi.org/10.1002/anie.201906293.

56 Li, C., Zhang, H., Wang, X. et al. (2022). A thiadiazolopyridine-functionalized Zr(iv)-based metal–organic framework for enhanced photocatalytic synthesis of tetrahydroquinolines under visible light. *RSC Adv.* 12 (3): 1638–1644. https://doi.org/10.1039/D1RA07363J.

57 Serpone, N., Salinaro, A., Emeline, A., and Ryabchuk, V. (2024). Turnovers and photocatalysis: a mathematical description. *J. Photochem. Photobiol. Chem.* 446: 115164. https://doi.org/10.1016/j.jphotochem.2023.115164.

58 Wang, Z., Hisatomi, T., Li, R. et al. (2021). Efficiency accreditation and testing protocols for particulate photocatalysts toward solar fuel production. *Joule* 5 (2): 344–359. https://doi.org/10.1016/j.joule.2021.01.001.

59 Akter, N., Ahmed, T., Haque, I. et al. (2024). XPS valence band observable light-responsive system for photocatalytic acid Red114 dye decomposition using a ZnO–Cu_2O heterojunction. *Heliyon* 10 (10): e30802. https://doi.org/10.1016/j.heliyon.2024.e30802.

60 Kalisamy, P., Lallimathi, M., Suryamathi, M. et al. (2020). ZnO-embedded S-doped g-C_3N_4 heterojunction: mediator-free Z-scheme mechanism for enhanced charge separation and photocatalytic degradation. *RSC Adv.* 10 (47): 28365–28375. https://doi.org/10.1039/D0RA04642F.

61 Zhang, L., Li, R.-H., Li, X.-X. et al. (2024). Photocatalytic aerobic oxidation of C(sp^3)-H bonds. *Nat. Commun.* 15 (1): 537. https://doi.org/10.1038/s41467-024-44833-y.

62 Ma, J., Miao, T.J., and Tang, J. (2022). Charge carrier dynamics and reaction intermediates in heterogeneous photocatalysis by time-resolved spectroscopies. *Chem. Soc. Rev.* 51 (14): 5777–5794. https://doi.org/10.1039/D1CS01164B.

63 Wang, Z., Jiao, X., Chen, D. et al. (2020). Porous copper/zinc bimetallic oxides derived from MOFs for efficient photocatalytic reduction of CO_2 to methanol. *Catalysts* 10 (10): 1127. https://doi.org/10.3390/catal10101127.

9

Photo-Therapy Based on Defective Photocatalysts

Jianmin Luo[1,2], *Jiahong Ye*[1,2], *and Bin Liu*[1]

[1] *Shaoguan University, School of Chemistry and Civil Engineering, No. 288, University Road, Zhenjiang District, Shaoguan City 512005, Guangdong Province, PR China*
[2] *Yili Normal University, College of Chemistry and Chemical Engineering, No. 448, Jiefang West Road (Jiefang Road), Yining City 835000, PR China*

9.1 Introduction

In recent years, phototherapy, the use of light to treat a variety of conditions, has received widespread attention due to its non-invasive and targeted therapeutic potential [1–3]. Conventional photocatalysts are mainly used for decomposition of water [4], carbon dioxide conversion [5], and degradation [6]; however, emerging research has demonstrated the potential of photocatalysts in the field of phototherapy through defect engineering [7]. Deliberate introduction of defects in photocatalysts can alter the optical properties and electronic structure of the photocatalysts for efficient and controllable light-activated therapeutic purposes [8]. Defective photocatalysts can facilitate the separation of charge carriers during the photocatalytic process, thereby enhancing the generation of ROSs for targeted and localized treatment of diseased tissues [9]. These ROSs play a crucial role in various phototherapy applications such as photodynamic therapy (PDT), photothermal therapy (PTT), and photochemical internalization (PCI) [10, 11]. By precisely controlling the type and concentration of introduced defects, researchers can tailor photocatalytic activity to effectively treat diseases such as cancer and bacterial infections and promote wound healing [12, 13]. In addition, defective photocatalysts are multifunctional, allowing the incorporation of other therapeutic agents such as drugs or nanoparticles into the photocatalyst. The combination of multimodal therapy further improves the therapeutic effect [14]. In conclusion, modification of photocatalysts through defect engineering is an important way to achieve efficient and precise phototherapy purposes [15]. Continued advances in this field will hold great potential for improving patient outcomes and expanding the use of phototherapy in healthcare.

Defect Engineering in Photocatalysis: From Fundamentals to Applications,
First Edition. Edited by Chuanyi Wang, Abdelkader Labidi, and Qiuhui Zhu.

9.2 Reactive Oxygen Species

Defect-based photocatalysis for phototherapy is an emerging research direction in medicine and biomedicine in recent years [16]. This technology combines the principle of photocatalysis and the needs of biomedicine and uses defective photocatalytic materials to generate ROSs under specific light conditions for the treatment of cancer [17], bacterial infections [18], or wound repair [19]. ROSs are redox products generated by the reaction of photogenerated electrons (e^-) and holes (h^+) on the surface of photocatalytic materials with high-energy carriers and oxygen-containing molecules in the phototherapy. ROSs have high reactivity, which plays an important role in maintaining the normal life activities of organisms [20]. For example, by precisely modulating the production of ROSs, scientists are able to exert their beneficial anti-tumor effects in therapy [21]. The ROSs with appropriate concentrations can act as signaling molecules involved in regulating a variety of physiological processes in cells, such as cell proliferation, immune response, cell repair, and anti-tumor [22]. The main types of ROSs include superoxide anion radical ($\bullet O_2^-$), hydrogen peroxide (H_2O_2), hydroxyl radical (•OH), and singlet oxygen ($^1O_2$) [23]. Figure 9.1 represents the transformational relationship between them.

9.2.1 Superoxide Anion Radical ($\bullet O_2^-$)

$\bullet O_2^-$ is a negative ion produced during electron transfer, usually from the reaction of a light-activated semiconductor surface with O_2 in a photocatalytic reaction. The $\bullet O_2^-$ is a relatively weak dioxide but has strong reducing properties. It is commonly used for its bactericidal and antiviral properties and for enhancing certain biological reactions [25].

9.2.2 Hydrogen Peroxide (H_2O_2)

H_2O_2 is a weaker oxidizing agent and is usually generated by the conversion of $\bullet O_2^-$ or $^1O_2$ in photocatalytic reactions [26]. It is an intermediate product in redox

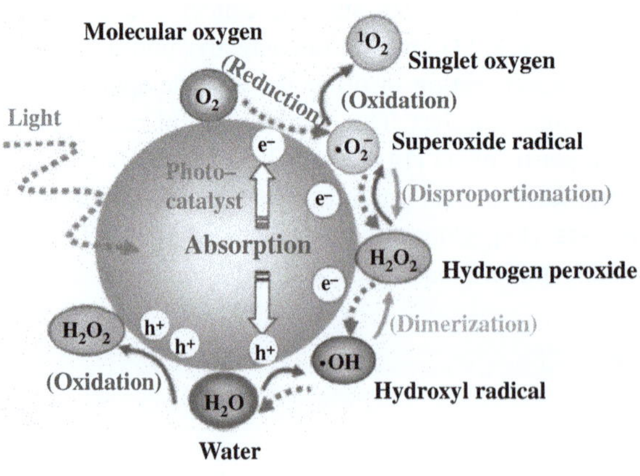

Figure 9.1 The generation of ROSs in the photocatalytic reduction and oxidation processes of O_2 and H_2O. *Source:* Reproduced from Nosaka and Nosaka [24] / with permission of American Chemical Society.

reactions, but it can be further converted to •OH under certain conditions. As a reaction intermediate, H_2O_2 serves to transfer oxidizing capability in photocatalytic reactions.

9.2.3 Hydroxyl Radical (•OH)

•OH is generated by further conversion of H_2O_2 or $•O_2^-$ during photocatalysis. The •OH, as a strong oxidant, is very effective in anti-bacterial and anti-viral applications [27].

9.2.4 Singlet Oxygen ($^1O_2$)

$^1O_2$ is an electronically excited state oxygen molecule, which is usually produced by the reaction of a semiconductor in the excited state with O_2 in the photocatalytic process, and has a high oxidizing capacity. The selectivity of $^1O_2$ is mainly reflected in its high activity and precise targeting, making it widely used in phototherapy for bactericidal and anti-tumor treatments, as well as for the removal of organic pollutants [28].

In conclusion, the roles of ROSs in phototherapy are diverse, producing oxidative effects on cancer cells, bacteria, or viruses through different reaction pathways and mechanisms [29]. However, when ROSs are produced in excess, they can cause damage to cells and tissues and even trigger a range of diseases [30, 31]. Therefore, maintaining a balance of ROSs is essential for good health.

9.3 Phototherapy Reactions

Phototherapy reactions are chemical processes that occur when light irradiation is used to produce ROSs on photocatalysts to treat various conditions [32]. These processes are commonly used in medical treatments such as PDT [33], PTT [34], and PCI [35]. In recent years, defect-engineered photocatalytic materials have achieved remarkable results in the application of phototherapy. Defects in the structure of photocatalysts can facilitate the separation of charge carriers during photocatalysis, thereby enhancing the generation of ROSs for targeted and localized treatment of diseased tissues, which plays an important role in applications such as the treatment of cancer [36], bacterial infections [37], or wound healing [38]. The mechanism and applications of PDT, PTT, and PCI will be introduced below in defective phototherapy.

9.3.1 Photodynamic Therapy

PDT is a method of treating disease by using the combined action of a photocatalyst and a specific wavelength of light to produce a biological effect [39]. At the heart of PDT is the fact that when light of a specific wavelength irradiates a photocatalyst, it absorbs light energy and excites to a high-energy state, subsequently releasing energy and generating ROSs, which are capable of directly damaging biomolecules (lipids, proteins, and DNA) within the cell, triggering cellular damage and death. It is worth noting that the photocatalyst has the feature of not causing damage to cells and normal body tissues in the absence of light itself. Xu et al. [40] constructed a synergistic system of Ag_3PO_4 nanoparticles, decorated with black urchin-like defective TiO_2 (BU-TiO_{2-X}/Ag_3PO_4), which can inactivate the bacteria for a short period of time under visible light. In addition, the nanospikes on the surface of BU-TiO_{2-X} can, from all directions, physically puncture bacterial cells, thus assisting the hybrid's photodynamic therapeutic effects, alongside the small amount of Ag^+ released from Ag_3PO_4 to inhibit the growth of bacteria for a long time. Zhang et al. [41] successfully synthesized rhombic dodecahedral Ni-ZnO@C by transition metal nickel doping and high-temperature carbonation of ZIF-8 precursor, which significantly improved the photocatalytic performance of Ni-ZnO@C and enhanced the ROSs production, aggravating oxidation-induced damage to bacterial cell membranes. The electrons escaping from the surface of Ni-ZnO@C could inhibit adenosine triphosphate (ATP) synthesis and further accelerate bacterial lysis through proton trapping and photogeneration. ROSs interfered with key enzymes in bacterial ATP synthesis, synergistically inhibiting ATP synthesis and further accelerating bacterial lysis and death, as shown in Figure 9.2. Meanwhile, the accumulation of Ni-ZnO@C and photogenerated ROSs in local tissues could effectively activate the anti-inflammatory response and initiate tissue regeneration, realizing highly efficient PDT antibacterial in vivo and ex vivo, and significantly promoting the healing of infected wounds.

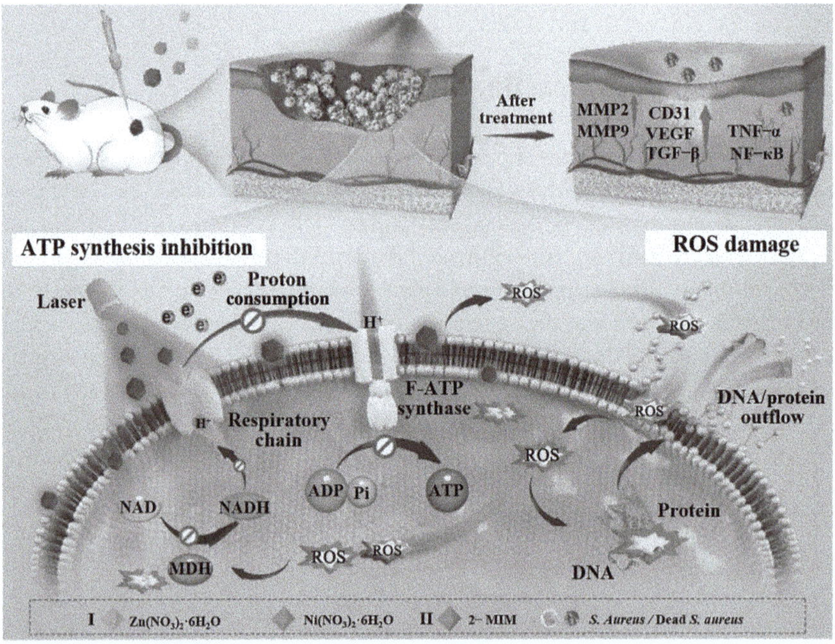

Figure 9.2 Schematic of the PDT process of Ni-ZnO@C for enhancing ROS damage and ATP synthesis inhibition in bacterial infections. *Source:* Reproduced from Zhang et al. [41] / with permission of Elsevier.

It is important to note that PDT is highly dependent on the presence of oxygen. The reaction of oxygen with the excited state photocatalyst to generate ROS is a key factor in triggering cell damage and killing. In the absence of oxygen, the reaction of the excited state photocatalyst with oxygen will be limited, thus greatly reducing the therapeutic effect. PDT can precisely overcome the oxygen limitation by modifying the photocatalyst through a defect-engineering strategy. Yang et al. [22] reported a metal doping strategy for the construction of a series of defect-rich metal-doped amorphous CoMo-layered double hydroxides (a-M-CoMo-LDH, M = Mn, Cu, Al, Ni, Mg, Zn) for NIR-II PDT. Under 1270 nm NIR-II laser irradiation, the defect-enriched a-Zn-CoMo-LDH nanosheets exhibited the best PDT performance, which is the highest among all the reported photosensitizers. After modification by polyethylene glycol (PEG), a-Zn-CoMo-LDH-PEG nanosheets could effectively induce apoptosis in vitro and successfully clear tumors in vivo.

9.3.2 Photothermal Therapy

PTT typically uses light in the near infrared region (NIR) region to induce a thermal effect in tissues. The mechanism of PTT is that nanoparticles are used to absorb the NIR light and convert it into heat, increasing the local temperature [42]. The localized heat generated by these nanoparticles selectively destroys diseased cells and kills bacteria, thus achieving a therapeutic effect [43]. One of the characteristics of photothermal therapy lies in the use of an NIR light, which can effectively penetrate deeper tissues and is able to work at deeper levels in the body. In recent years, defect-engineering PTT has emerged as a prominent research area [44]. By introducing defects, photocatalysts are able to generate higher efficacy and enhance therapeutic effects during photothermal therapy, mainly in terms of enhanced light absorption and photothermal conversion efficiencies, improved therapeutic targeting and selectivity, and multifunctional therapeutic effects [45]. Sun et al. [19] prepared a multifunctional hydrogel composite (γ-PGA/PDA/GOx/WO_x [PPGW]) enriched with oxygen vacancies by doping polydopamine (PDA), glucose oxidase (GOx), and tungsten oxide (WO_x) nanowires into a poly(γ-glutamic acid) (γ-PGA) framework for synergistic photocatalytic O_2 production and PTT to promote wound healing in bacterial-infected diabetic wounds, as shown in Figure 9.3.

9.3.3 Photochemical Internalization

PCI is a technology that uses light and photocatalysts to enhance the entry of therapeutic agents into cells [46]. The mechanism of PCI is as follows: the photocatalyst is introduced into the cell and activated by a specific wavelength of light, usually visible or near infrared light [47]. The ROSs generated during the activation process can cause localized damage to the cell membrane, altering its permeability, which can allow the therapeutic agent or bioactive molecule to enter the cell to achieve a therapeutic effect [48]. By combining defectively engineered photocatalysts with PCI, the therapeutic efficacy can be greatly enhanced by enhancing the photocatalytic reaction, improving targeting and local reactivity, improving light

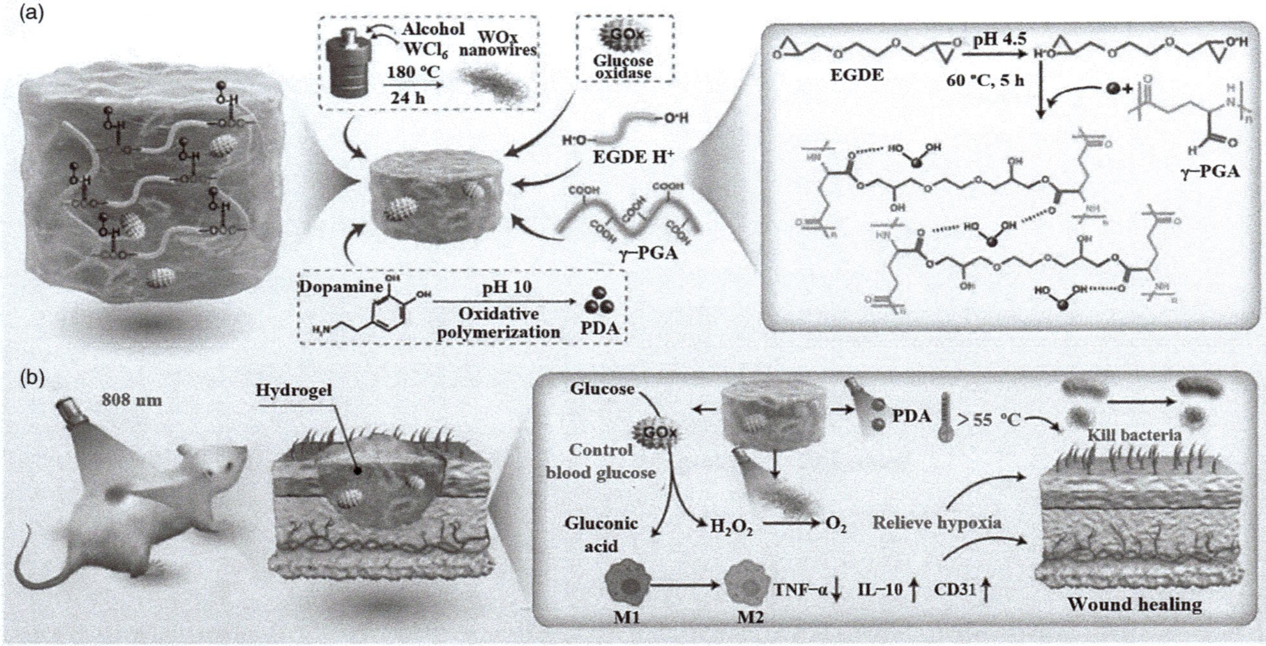

Figure 9.3 (a) Schematic illustration of the preparation of PPGW; (b) Schematic showing the diabetic wound healing mechanism of PPGW. *Source:* Reproduced from Sun et al. [19] / with permission of John Wiley & Sons.

absorption, reducing side effects, and enhancing the internalization efficiency of drugs or genes [49]. Wang et al. [50] reported a new strategy to promote lysosomal escape and anti-tumor efficacy by modifying cylindrical polymeric brush-based prodrugs via phenylboronic acid (PBA) and demonstrated that PBA modification is an effective method to improve the overall properties of nanomedicines.

9.4 Types of Defects

In the field of phototherapy research based on defect photocatalysis, photocatalysts with defects intentionally introduced into their structures have shown promising applications in enhancing phototherapy effects [51]. These defects not only affect the surface properties, electronic structure, and light absorption properties of the catalysts, but also promote the generation of reaction intermediates or the separation of electron–hole pairs for the purpose of efficient and controllable photoactivated therapy. The classification of defects in photocatalytic materials is beneficial to help researchers rationally design and regulate the defects in photocatalysts and to promote the development of photocatalytic technology in the field of phototherapy.

It is important to mention that in mono-component photocatalytic systems (Figure 9.4), all three processes, including charge generation, migration, and consumption, take place within the same semiconductor material [52]. The defects present in these systems can be found either on the surface or inside, which is why they are categorized as surface defects and body defects, respectively. In multi-component photocatalytic systems, multiple elements are integrated to create a shared interface where photocatalytic charge generation and consumption will occur across various components. Therefore, these systems also feature interfacial defects, primarily arising from the surface defects of the interacting components. These defects have varying impacts on the photocatalytic reaction.

9.4.1 Anionic Vacancies and Metal Vacancies

Anionic vacancies and metal vacancies are two prominent types of photocatalytic defects [53, 54], with anionic vacancies being one of the most widely studied defects

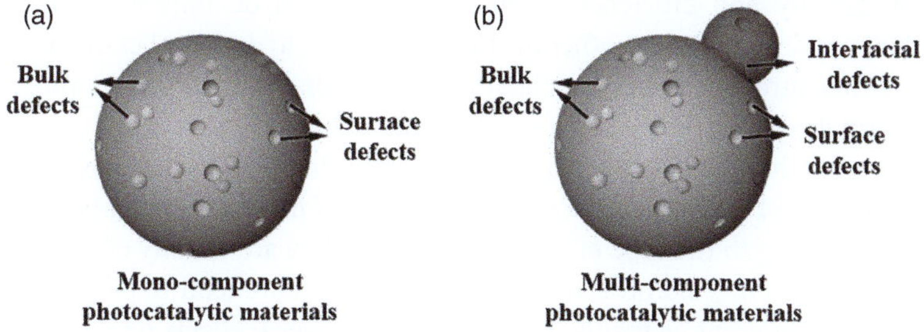

Figure 9.4 Schematic illustrating the defects at the various locations in (a) Mono-component and (b) Multi-component photocatalytic materials. *Source:* Reproduced from Bai et al. [52] / with permission of Elsevier.

in metal oxides. Typically, oxygen, nitrogen, and sulfur vacancies are the most prevalent anionic vacancies found in photocatalysts, and among these [55], oxygen vacancies (OVs) are the most extensively investigated for photocatalytic purposes [56]. OVs, as defect energy levels, contribute to the separation of photogenerated carriers, reduce the complexation of electron–hole pairs, and improve the photocatalytic reaction efficiency [57]. In addition, OVs promote the reaction of oxygen molecules with surfaces to generate •OH and $\bullet O_2^-$, which achieve phototherapy effects by destroying the structures of pathogens and viruses. Liu et al. [58] created OV-enriched bimetallic silicate nanozymes featuring Fe–Ca dual active sites for targeted aggregation-enhanced therapy by modifying oxidized sodium alginate and gallic acid (GA) loading (OFeCaSA-V@GA). Density functional theory analysis indicates that the Ca sites enhance the adsorption of H_2O_2, while the Fe sites facilitate the dissociation of H_2O_2, resulting in a synergistic catalytic effect. Notably, the targeted delivery of metal ions can trigger a morphological change at tumor sites, leading to a high retention rate (up to 36.3%) of theranostic components within tumor cells.

Creating metal vacancies in photocatalysis is a modification strategy that leverages metal-vacancy structures to improve photocatalytic efficiency, showing significant promise for phototherapy in photocatalysis [12]. Metal vacancy engineering has been shown to be an effective modulation strategy in promoting photogenerated carrier separation and small molecule activation [59]. Ma et al. [12] designed catalysts based on copper oxide (Cu_2O) by employing defect engineering and a heterogeneous composition of perylene-3,4,9,10-tetracarboximide (PDINH). Introducing transition metals like Ga into the Cu_2O lattice creates Cu vacancies to accelerate the kinetics of ROSs and induces a small size effect in Cu_2O nanoparticles to boost photocatalytic performance. The results indicated that the optimized $Cu_{2-x}Ga_xO$/PDINH ($x = 0.2$) nanomaterials exhibited outstanding photocatalytic antimicrobial capabilities, achieving an inhibition rate of more than 91% against *Staphylococcus aureus* and *Pseudomonas aeruginosa*.

9.4.2 Elemental Doping

Elemental doping plays an important role in phototherapy, and by doping different elements, the activity and efficiency of photocatalysts can be improved, thus enhancing their application in phototherapy [60]. According to the number of types of dopant elements introduced, they are classified into single doping and co-doping. Among them, single doping is usually applied to modulate one aspect of the material's performance, while co-doping focuses more on optimizing multiple aspects of the material's properties through the synergistic effect of multiple elements [61]. Furthermore, according to the type of doping elements, they are usually classified as transition metal doping, non-transition metal doping, non-metal doping, and lanthanide doping. Different elements doping will have diverse effects on the performance of photocatalysis. Li et al. [62] designed lanthanide-doped nanoparticles

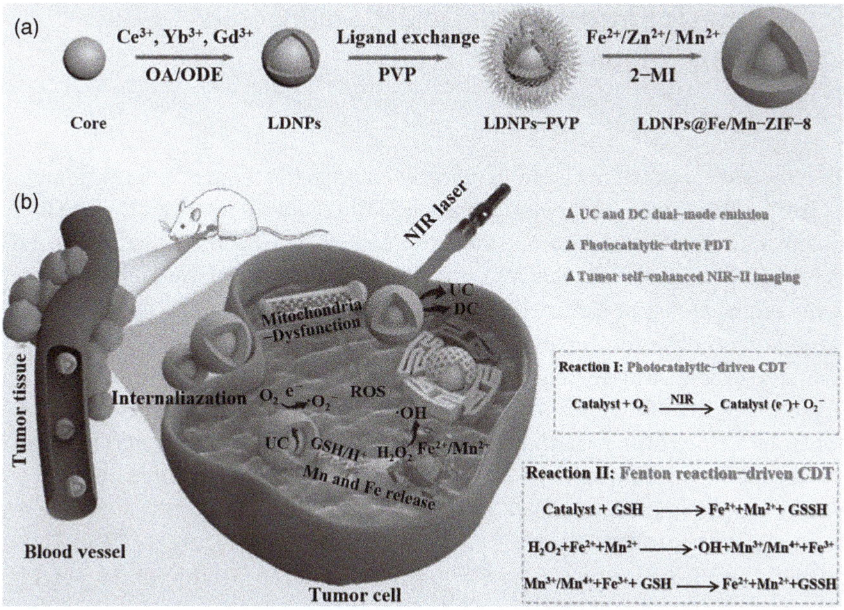

Figure 9.5 Schematic illustration for synthesis of LDNPs@Fe/Mn-ZIF-8 (a) and the mechanism of phototherapy and NIR-II imaging of LDNPs@Fe/Mn-ZIF-8 in the TME when exposed to NIR laser irradiation (b). *Source:* Reproduced from Li et al. [62] / with permission of American Chemical Society.

(LDNPs) coated with Fe/Mn bimetal-doped ZIF-8 (LDNPs@Fe/Mn-ZIF-8) for second near-infrared imaging-guided synergistic photodynamic/chemodynamic therapy (PDT/CDT), as shown in Figure 9.5. The dual doping of Fe^{2+}/Mn^{2+} not only significantly reduced the band gap of the ZIF-8, allowing it to be effectively excited by dual-mode red upconversion photons for photo-driven PDT, but also facilitated the release of these metals in the tumor microenvironment (TME) through the degradation of Fe/Mn-ZIF-8, leading to a Fenton/Fenton-like reaction that generates •OH for CDT.

9.4.3 Lattice Defects

Lattice defects refer to irregular or missing atomic arrangements of the crystal structure caused by stress, heat treatment, or other factors during the synthesis process, and usually include point defects, line defects (dislocations), and surface defects (grain boundaries) [63]. Lattice defects can change the electronic band structure of photocatalytic materials and produce localized defect energy levels, which are conducive to the separation of photogenerated electrons and holes, and thus can improve the photocatalytic activity by altering the lattice constants and energy band structure of the materials [64, 65].

9.5 Synthesis Strategies for Defect-Engineered Photocatalytic Materials

9.5.1 High-Temperature Annealing

High-temperature annealing is one of the most common methods of defect introduction. During the annealing process, the material undergoes prolonged heating in a high-temperature environment, and the movement of atoms in the lattice makes the structure of the material change, resulting in the creation of defects such as vacancies, interstitial atoms, or dislocations [66, 67]. Huang et al. [68] used a hydrothermal method to synthesize ultrasmall TiO_2 nanoparticles, which were annealed at high temperatures in a hydrogen-xenon environment to form OVs.

In addition, different annealing atmospheres may modulate the type of defects. OVs may be formed in an oxidizing atmosphere, whereas metallic sites may be introduced in a reducing atmosphere [70]. Cui and coworkers [69] prepared an H-TiO_2 with Ti–H bonds in the lattice. The Ti–H bonds are annealed in an Ar atmosphere and transformed into OVs to obtain OV-TiO_2, as shown in Figure 9.6. The Ti–H defects, on the other hand, are repaired when annealed in an O_2 atmosphere, obtaining TiO_2 that does not contain OVs.

9.5.2 Chemical Reduction Synthesis

Chemical reduction synthesis is a technique for the synthesis of materials by reduction reactions, often used to prepare nanomaterials, metal particles, or low-dimensional structures. In defect synthesis, the use of strong reducing agents (hydrogen, sodium borohydride) can induce the reduction of a portion of the oxygen

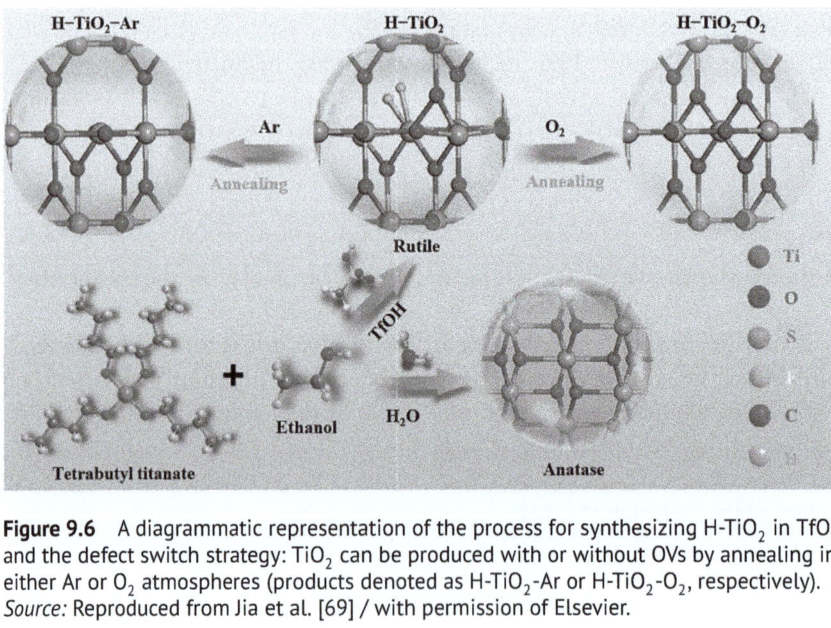

Figure 9.6 A diagrammatic representation of the process for synthesizing H-TiO_2 in TfOH and the defect switch strategy: TiO_2 can be produced with or without OVs by annealing in either Ar or O_2 atmospheres (products denoted as H-TiO_2-Ar or H-TiO_2-O_2, respectively). *Source:* Reproduced from Jia et al. [69] / with permission of Elsevier.

atoms or other components of the oxide material, resulting in the formation of oxygen vacancies, doping defects, or lattice dislocations [71, 72]. Zhang et al. [73] prepared a series of Cu/defective-support (Cu/ZrO-D and Cu/ZnO-D) catalysts with different defect concentrations by magnesium thermal reduction and high-pressure hydrogen reduction. In this work, strong metal-carrier interaction (SMSI) and electronic metal-carrier interaction (EMSI) enhancements were demonstrated after the introduction of oxygen vacancies in the Cu/ZrO_2 and Cu/ZnO catalyst carriers. The results showed that the Cu/ZrO_2-D and Cu/ZnO-D catalysts phase-assisted improved photocatalytic performance. Additionally, self-carbon-thermal reduction is reported to be a defect introduction strategy for engineering carbon defects at individual Fe-N_4 sites in ZnO-carbon nanoreactors [74].

9.5.3 Solvothermal and Hydrothermal Methods

In solvothermal and hydrothermal methods, which are conducted at high temperatures and pressures, the solvent molecules are able to promote the dissolution and recrystallization process of the material, and defects in the lattice (vacancies, distortions) are generated during the reaction [75]. Cai et al. [76] synthesized ultrathin Bi_4OsBr, nanosheets with controllable oxygen vacancy concentration by adjusting the ratio of water to ethylene glycol using a one-step solvothermal method. Wang et al. [77] synthesized Zn-doped MoS_2 nanosheets (Zn-MoS_2) by a simple hydrothermal process and ultrasonic solvent stripping in $Zn(NO_3)_2{\cdot}_6H_2O$ solution. The combination of Zn-MoS_2 and ZIF-8 (ZnDMZ) was carried out by a simple physical mixing process of a solution of Zn-MoS_2 with a solution of 2-methylimidazole in methanol. In this work, the ZnDMZ exhibited efficient bactericidal and wound healing effects due to the synergistic effect of photocatalytic and photothermal effects and released zinc ions, as shown in Figure 9.7.

9.5.4 Photo-Induced Synthesis

Light-induced methods use a specific wavelength of light, such as ultraviolet light, to stimulate electron leaps and thus induce defects in materials and are commonly used

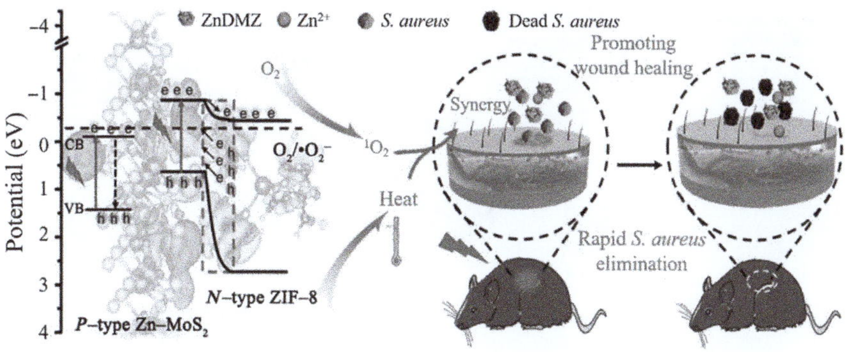

Figure 9.7 The antibacterial and wound healing mechanism of ZnDMZ P-N heterojunction under 660 nm light irradiation. *Source:* Reproduced from Wang et al. [77] / with permission of Elsevier.

in photocatalytic and photovoltaic materials [78]. Controlling the intensity of the light source and the duration of irradiation regulates the generation of defects. Chen et al. [79] synthesized $BiSbO_4$ and homogeneously dispersed it in a 12-cm-diameter glass petri dish using deionized water. After drying, $BiSbO_4$ with OVs was obtained by treating it under light irradiation at 280 nm (UV280, 6 W) for 3, 5, 10, 15, and 20 h. The experimental results showed that the introduction of OVs into $BiSbO_4$ could improve the light adsorption capacity and limit the recombination of photogenerated carriers.

9.5.5 Ball Milling Method

Ball milling synthesis crushes or mixes raw materials primarily through mechanical forces (impact, friction, and shear), which can deform the structure of the material, significantly reduce its volume, and expose defects, providing an economical, effective, and viable method to create void edge defects [80]. Dong et al. [81] reported the construction of Van der Waals heterojunctions of carbonized polymer dots and potassium-doped carbon nitride nanosheets by ball milling to promote photocatalytic performance. However, the external forces generated in the ball milling method of synthesis are prone to creating volume defects, which not only reduce carrier separation but also become carrier recombination centers, leading to low photocatalytic efficiency [72].

9.5.6 Molten Salt Synthesis

The molten salt synthesis is able to introduce different types of defects in the synthesis of photocatalytic materials through the chemical environment of a high-temperature molten salt solution, and its preparation process and synthesis are shown in Figure 9.8. In high-temperature molten salt, the redox reaction of the material is more intense, which can promote the generation of defects in photocatalytic materials [83]. In addition, the molten salt solution can dissolve

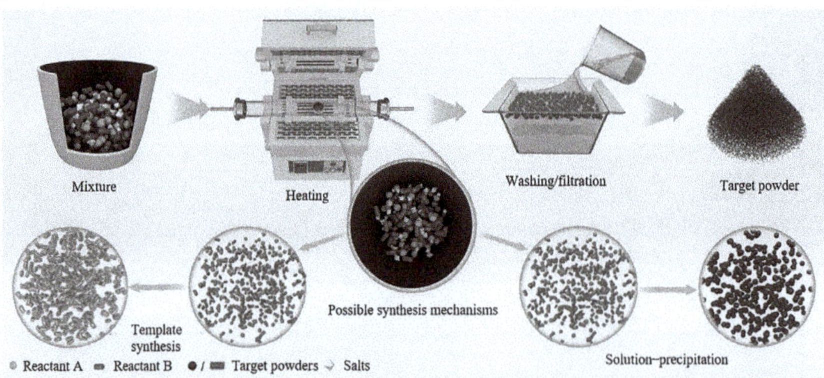

Figure 9.8 Process for preparing molten salt synthesis and schematic illustration of synthesis methods. *Source:* Reproduced from Li et al. [82] / with permission of John Wiley & Sons.

some components or impurities of the material, which can lead to lattice defects, especially in a high-temperature environment, where the lattice structure of the material is prone to distortion or fracture, leading to defects. Wang et al. [84] employed the molten salt technique to synthesize defect-rich ruthenium oxide/iridium oxide heterojunction nanosheets (RuO_2/IrO_2), utilizing Ru as the catalytic site to improve antibacterial effectiveness. The synthesis was carried out by adding $IrCl_3{\cdot}xH_2O$ and $RuCl_3{\cdot}xH_2O$ to molten $NaNO_3$, and then the mixture was introduced into a tube furnace and heated up to 350 °C at 10 °C min^{-1} (maintained for 10 min) and then heated up to 490 °C at 5 °C min^{-1} (maintained for 30 min) to obtain RuO_2/IrO_2. Benefiting from the typically high catalytic activity of Ru in various reactions and the distinctive Ir-mediated defect-rich heterostructure, RuO_2/IrO_2 demonstrates significantly boosted biocatalytic activities in generating reactive oxygen species essential for eliminating bacteria.

9.6 Conclusion and Future Prospects

Defective phototherapy has emerged in recent years as a new direction in the application of photocatalytic technology in the fields of medicine and environmental treatment. By introducing or modulating structural defects (anionic vacancies, metal vacancies, elemental doping, and lattice defects) in photocatalytic materials, the catalytic activity and reaction selectivity of the materials can be enhanced, thus boosting their potential for cancer therapy, antimicrobial, and wound healing. Defective photocatalytic materials hold significant promise for various applications, yet they still face challenges in terms of photocatalyst stability, defect controllability, the effect of defects on the photoelectric properties of the materials, large-scale production, and practical application. In the future, with the continuous development of materials science, nanotechnology, optoelectronics, and catalysis theory, defective phototherapy is expected to usher in more mature technological breakthroughs, which will promote the in-depth research of this technology in the medical field.

References

1 Li, X., Ding, B., Li, J. et al. (2024). Valence-tailored copper-based nanoparticles for enhanced chemodynamic therapy through prolonged ROS generation and potentiated GSH depletion. *Nano Res.* 17: 6342–6352.

2 Zhou, K., Du, L., Ding, R. et al. (2024). Photocatalytic therapy via photoinduced redox imbalance in biological system. *Nat. Commun.* 15: 10551.

3 Yang, C., Wang, M., Chang, M. et al. (2023). Heterostructural nanoadjuvant $CuSe/CoSe_2$ for potentiating ferroptosis and photoimmunotherapy through intratumoral blocked lactate efflux. *JACS* 145: 7205–7217.

4 Li, X., Liu, Q., Deng, F. et al. (2022). Double-defect-induced polarization enhanced OV-$BiOBr/Cu_{2-x}S$ high-low junction for boosted photoelectrochemical hydrogen evolution. *Appl. Catal. B: Environ.* 314: 121502.

5 Shi, X., Dai, W., Dong, X.A. et al. (2023). Dual Cu and S vacancies boost CO_2 photomethanation on Cu1.95S1-x: vacancy-regulated selective photocatalysis. *Appl. Catal. B Environ.* 339: 123147.

6 Luo, J., Han, H., Wang, X. et al. (2023). Constructing oxygen absorption and activation sites in Ce-doped g-C_3N_4 photocatalyst for effective removal of amoxicillin: performance, mechanism and degradation pathways. *Appl. Surf. Sci.* 611: 155808.

7 Shen, W., Hu, T., Liu, X. et al. (2022). Defect engineering of layered double hydroxide nanosheets as inorganic photosensitizers for NIR-III photodynamic cancer therapy. *Nat. Commun.* 13: 3384.

8 Shu, A., Qin, C., Li, M. et al. (2024). Electric effects reinforce charge carrier behaviour for photocatalysis. *Energy Environ. Sci.* 17: 4907–4928.

9 Liu, H.-X., Gao, Z., Yan, H. et al. (2023). Ensemble of single-atom catalysis and defect engineering in Cu_1/CeO_2 nanozymes for tumor therapy. *Sci. China Chem.* 66: 2590–2599.

10 Ran, B., Wang, Z., Cai, W. et al. (2021). Organic photo-antimicrobials: principles, molecule design, and applications. *JACS* 143: 17891–17909.

11 Tian, J., Li, B., Wu, C. et al. (2024). Programmable singlet oxygen battery for automated photodynamic therapy enabled by pyridone-pyridine tautomer engineering. *JACS* 146: 16458–16468.

12 Ma, C., Chen, S., Wang, C. et al. (2024). Defect engineering of $Cu_{2-x}Ga_xO$/PDINH nanomaterial for significantly improved photocatalytic antibacterial activities. *Nano Today* 56: 102230.

13 Zhang, Z., Wang, L., Liu, W. et al. (2021). Photogenerated-hole-induced rapid elimination of solid tumors by the supramolecular porphyrin photocatalyst. *Natl. Sci. Rev.* 8: nwaa155.

14 Fan, G., Hou, S., Zhang, W. et al. (2024). Polymer-DNA carriers Co-deliver photosensitizer and siRNA for light-promoted gene transfection and hypoxia-relieved photodynamic therapy. *Angew. Chem. Int. Ed. Engl.* 63: e202405600.

15 Zhang, T., Cheng, Q., Lei, J.H. et al. (2023). Constructing oxygen-related defects in carbon nanodots with Janus optical properties: noninvasive NIR fluorescent imaging and effective photocatalytic therapy. *Adv. Mater.* 35: e2302705.

16 Hu, D., Li, Y., Li, R. et al. (2024). Recent advances in reactive oxygen species (ROS)-responsive drug delivery systems for photodynamic therapy of cancer. *Acta Pharm. Sin. B.* 14: 5106–5131.

17 Zhao, B., Ye, J., Chen, Z. et al. (2025). Mitochondria-targeted photoredox catalysis activates pyroptosis for effective tumor therapy. *Adv. Funct. Mater.* 2417681.

18 Yang, Y., Cai, Q., Wang, L. et al. (2024). Oxygen-deficient Bi_2MoO_6@sRuO_2@HA heterojunction for photocatalytic treatment of drug-resistant bacterial infections. *Chem. Eng. J.* 502: 157828.

19 Sun, X., Wang, P., Tang, L. et al. (2024). Multifunctional hydrogel containing oxygen vacancy-rich WO_x for synergistic photocatalytic O_2 production and photothermal therapy promoting bacteria-infected diabetic wound healing. *Adv. Funct. Mater.* 34: 2411117.

20 Li, X., Gao, J., Wu, C. et al. (2024). Precise modulation and use of reactive oxygen species for immunotherapy. *Sci. Adv.* 10: eadl0479.

21 Sies, H., Belousov, V.V., Chandel, N.S. et al. (2022). Defining roles of specific reactive oxygen species (ROS) in cell biology and physiology. *Nat. Rev. Mol. Cell Biol.* 23: 499–515.

22 Yang, Y., Hu, T., Zhao, K. et al. (2025). Metal doping enabling defective CoMo-layered double hydroxide nanosheets as highly efficient photosensitizers for NIR-II photodynamic cancer therapy. *Adv. Mater.* 37: 2405847.

23 Wang, P., Gong, Q., Hu, J. et al. (2020). Reactive oxygen species (ROS)-responsive prodrugs, probes, and theranostic prodrugs: applications in the ROS-related diseases. *J. Med. Chem.* 64: 298–325.

24 Nosaka, Y. and Nosaka, A.Y. (2017). Generation and detection of reactive oxygen species in photocatalysis. *Chem. Rev.* 117: 11302–11336.

25 Fan, X., Fu, Q., Liu, G. et al. (2024). Applying molecular oxygen for organic pollutant degradation: strategies, mechanisms, and perspectives. *Environ. Sci. Ecotechnol.* 22: 100469.

26 Deng, Y., Liu, W., Xu, R. et al. (2024). Reduction of superoxide radical intermediate by polydopamine for efficient hydrogen peroxide photosynthesis. *Angew. Chem. Int. Ed.* 63: e202319216.

27 Chen, X., Sheng, X., Zhou, H. et al. (2024). Hydrophobicity promoted efficient hydroxyl radical generation in visible-light-driven photocatalytic oxidation. *Small* 20: e2310128.

28 Demyanenko, A.V., Bogomolov, A.S., Dozmorov, N.V. et al. (2019). Singlet oxygen $^1O_2$ in photocatalysis on TiO_2. where does it come from? *J. Phys. Chem. C.* 123: 2175–2181.

29 Yu, C., Xu, J., Wang, Y. et al. (2024). Generation of reactive oxygen species on MIL-100(Fe)/TiO_2 driven by mild natural light. *ChemCatChem* 16: e202400817.

30 Wu, K., El Zowalaty, A.E., Sayin, V.I. et al. (2024). The pleiotropic functions of reactive oxygen species in cancer. *Nat. Cancer* 5: 384–399.

31 Zhang, J., Wang, Z., Lin, X. et al. (2024). Mn-ce symbiosis: nanozymes with multiple active sites facilitate scavenging of reactive oxygen species (ROS) based on electron transfer and confinement anchoring. *Angew. Chem. Int. Ed. Engl.* 64: e202416686.

32 Wang, Y., Ma, K., Kang, M. et al. (2024). A new era of cancer phototherapy: mechanisms and applications. *Chem. Soc. Rev.* 53: 12014–12042.

33 Jia, Y., Yan, X., and Li, J. (2022). Schiff base mediated dipeptide assembly toward nanoarchitectonics. *Angew. Chem. Int. Ed.* 61: e202207752.

34 Zhou, Z., Li, X., Hu, T. et al. (2022). Molybdenum-based nanomaterials for photothermal cancer therapy. *Adv. NanoBiomed Res.* 2: 2200065.

35 Yu, G., Zhu, B., Shao, L. et al. (2019). Host-guest complexation-mediated codelivery of anticancer drug and photosensitizer for cancer photochemotherapy. *Proc. Natl. Acad. Sci. USA* 116: 6618–6623.

36 Li, Z., Bian, J., Xu, Z. et al. (2023). Cascaded antitumor therapy excited by dual nanozymes based on energy restriction and photocatalysis. *ACS Appl. Mater. Interfaces* 15: 56869–56880.

37 Yu, Q., Wang, C., Zhang, X. et al. (2024). Photochemical strategies toward precision targeting against multidrug-resistant bacterial infections. *ACS Nano* 18: 14085–14122.

38 Wang, Y., Shi, X., Hou, S. et al. (2025). A light-responsive multilayered 3D porous Ga_2O_3 hydrogel for photocatalytic antibacterial therapy promoting healing of MDR *S. aureus*-infected wounds. *J. Mater. Sci. Technol.* 225: 188–202.

39 Dougherty, T.J., Gomer, C.J., Henderson, B.W. et al. (1998). Photodynamic therapy. *J. Natl. Cancer Inst.* 90: 889–905.

40 Xu, Y., Liu, X., Zheng, Y. et al. (2021). Ag_3PO_4 decorated black urchin-like defective TiO_2 for rapid and long-term bacteria-killing under visible light. *Bioact. Mater.* 6: 1575–1587.

41 Zhang, R., Chen, Z., Li, Y. et al. (2024). Enhanced photodynamic therapy efficacy of Ni-doped/oxygen vacancy double-defect Ni-ZnO@C photosensitizer in bacteria-infected wounds based on ROS damage and ATP synthesis inhibition. *J. Mater. Sci. Technol.* 192: 173–189.

42 He, L., Di, D., Chu, X. et al. (2023). Photothermal antibacterial materials to promote wound healing. *J. Controlled Release* 363: 180–200.

43 Liu, Z., Liu, S., Liu, B. et al. (2024). Facile synthesis of Fe-based metal-quinone networks for mutually enhanced mild photothermal therapy and ferroptosis. *Angew. Chem. Int. Ed.* 64: e202414879.

44 Dong, S., Dong, Y., Zhao, Z. et al. (2023). "Electron transport chain interference" strategy of amplified mild-photothermal therapy and defect-engineered multi-enzymatic activities for synergistic tumor-personalized suppression. *JACS* 145: 9488–9507.

45 Jana, D., He, B., Chen, Y. et al. (2022). A defect-engineered nanozyme for targeted NIR-II photothermal immunotherapy of cancer. *Adv. Mater.* 36: 2206401.

46 Barnett, C., Joubert, F., Iliopoulou, A. et al. (2023). Photochemical internalization using natural anticancer drugs, antimetabolites, and nanoformulations: a systematic study against breast and pancreatic cancer cell lines. *Mol. Pharmaceutics* 20: 1818–1841.

47 Jeong, H., Kim, H., Kim, Y.A. et al. (2023). Intranasal mRNA delivery via customized RNA-polyplex nanoparticles enhancing gene expression through photochemical mechanisms. *ACS Appl. Mater. Interfaces* 15: 56749–56759.

48 Yi, X., Hu, J.J., Dai, J. et al. (2021). Self-guiding polymeric prodrug micelles with two aggregation-induced emission photosensitizers for enhanced chemo-photodynamic therapy. *ACS Nano* 15: 3026–3037.

49 Zeng, X., Wang, Y., Huang, Y.S. et al. (2022). Amphiphilic metallodrug assemblies with red-light-enhanced cellular internalization and tumor penetration for anticancer phototherapy. *Small* 18: 2205461.

50 Wang, R., Yin, C., Liu, C. et al. (2021). Phenylboronic acid modification augments the lysosome escape and antitumor efficacy of a cylindrical polymer brush-based prodrug. *JACS* 143: 20927–20938.

51 Tian, J., Li, J., Guo, Y. et al. (2024). Oxygen vacancy mediated bismuth-based photocatalysts. *Advanced Powder Materials* 3: 100201.

52 Bai, S., Zhang, N., Gao, C., and Xiong, Y. (2018). Defect engineering in photocatalytic materials. *Nano Energy* 53: 296–336.

53 He, M., Wang, X., Yu, H. et al. (2024). Nitrogen vacancy-rich carbon nitride anchored with iron atoms for efficient redox dyshomeostasis under ultrasound actuation. *Biomaterials* 305: 122446.

54 Han, D., Han, Y., Li, J. et al. (2020). Enhanced photocatalytic activity and photothermal effects of Cu-doped metal-organic frameworks for rapid treatment of bacteria-infected wounds. *Appl. Catal. B: Environ.* 261: 118248.

55 Zhang, Y.C., Afzal, N., Pan, L. et al. (2019). Structure-activity relationship of defective metal-based photocatalysts for water splitting: experimental and theoretical perspectives. *Adv. Sci.* 6: 1900053.

56 Chen, L., He, F., Huang, Y. et al. (2016). Hydrogenated nanoporous TiO_2 film on Ti-25Nb-3Mo-2Sn-3Zr alloy with enhanced photocatalytic and sterilization activities driven by visible light. *J. Alloys Compd.* 678: 5–11.

57 Wang, L.-W., Liu, L., You, Z. et al. (2023). Surface amorphization oxygen vacancy-rich porous Sn_3O_x nanosheets for boosted photoelectrocatalytic bacterial inactivation. *Rare Met.* 42: 1508–1515.

58 Liu, S., Sun, Y., Ye, J. et al. Targeted delivery of active sites by oxygen vacancy-engineered bimetal silicate nanozymes for intratumoral aggregation-potentiated catalytic therapy. *ACS Nano* 18 (2024): 1516–1530.

59 Chen, H., Xu, R., Chen, D. et al. (2024). Subsurface Mo vacancy in bismuth molybdate promotes photocatalytic oxidation of lactate to pyruvate. *ACS Catal.* 14: 1977–1986.

60 Jiang, G., Fan, J., Wan, Y. et al. (2024). Enhanced photocatalytic and antimicrobial activities of carbon-dots nanozymes modulated with P-doping. *Chem. Eng. J.* 480: 148216.

61 Zhang, S., Yi, X., Hu, G. et al. (2023). Configuration regulation of active sites by accurate doping inducing self-adapting defect for enhanced photocatalytic applications: a review. *Coord. Chem. Rev.* 478: 214970.

62 Li, C., Ye, J., Yang, X. et al. (2022). Fe/Mn bimetal-doped ZIF-8-coated luminescent nanoparticles with up/downconversion dual-mode emission for tumor self-enhanced NIR-II imaging and catalytic therapy. *ACS Nano* 16: 18143–18156.

63 Zafar, Z., Yi, S., Li, J. et al. (2021). Recent development in defects engineered photocatalysts: an overview of the experimental and theoretical strategies. *Energy Environ. Mater.* 5: 68–114.

64 Wang, F., Zhao, Q., Li, H. et al. (2024). Crystal defect engineering to construct oxygen vacancies in MXene-derived TiO_2 nanocomposites for boosting photocatalytic degradation of 2,4,6-trichlorophenol. *Chem. Eng. J.* 481: 148855.

65 Xie, S., Jiang, J., Zhou, G. et al. (2025). Optimizing electronic structure through lattice engineering for enhanced photocatalytic reduction of Cr(VI). *Inorg. Chem. Commun.* 173: 113845.

66 Han, X., Zhou, Y., Tai, X. et al. (2024). In-situ atomic tracking of intermetallic compound formation during thermal annealing. *Nat. Commun.* 15: 7200.

67 Xu, J., Shao, G., Tang, X. et al. (2022). Frenkel-defected monolayer MoS_2 catalysts for efficient hydrogen evolution. *Nat. Commun.* 13: 2193.

68 Huang, Z.-X., Wang, Z.-Y., Yuan, Y.-J. et al. (2024). Defect engineering of ultra-small TiO_2 nanoparticles enables highly efficient photocatalysts for Solar H_2 production from woody biomass. *Nano Lett.* 24: 11968–11975.
69 Jia, G., Wang, Y., Cui, X. et al. (2022). Wet-chemistry hydrogen doped TiO_2 with switchable defects control for photocatalytic hydrogen evolution. *Matter* 5: 206–218.
70 Wei, Z., Wang, W., Li, W. et al. (2021). Steering electron–hole migration pathways using oxygen vacancies in tungsten oxides to enhance their photocatalytic oxygen evolution performance. *Angew. Chem. Int. Ed.* 60: 8236–8242.
71 Horlyck, J., Nashira, A., Lovell, E. et al. (2019). Plasma treating mixed metal oxides to improve oxidative performance via defect generation. *Materials* 12: 2756.
72 He, Y., Lei, Q., Li, C. et al. (2021). Defect engineering of photocatalysts for solar-driven conversion of CO_2 into valuable fuels. *Mater. Today* 50: 358–384.
73 Zhang, C., Wang, L., Chen, Y. et al. (2023). Shifting CO_2 hydrogenation from producing CO to CH_3OH by engineering defect structures of Cu/ZrO_2 and Cu/ZnO catalysts. *Chem. Eng. J.* 475: 146102.
74 Wei, S., Sun, Y., Qiu, Y.-Z. et al. (2023). Self-carbon-thermal-reduction strategy for boosting the Fenton-like activity of single $Fe\text{-}N_4$ sites by carbon-defect engineering. *Nat. Commun.* 14: 7549.
75 Qi, Z., Li, L., and Xu, Z.P. (2022). Engineering lattice defects in 2D nanomaterials for enhancing biomedical performances. *Particuology* 64: 121–133.
76 Cai, H., Chen, F., Hu, C. et al. (2024). Oxygen vacancies mediated ultrathin $Bi_4O_5Br_2$ nanosheets for efficient piezocatalytic peroxide hydrogen generation in pure water. *Chin. J. Catal.* 57: 123–132.
77 Wang, C., Luo, Y., Liu, X. et al. (2022). The enhanced photocatalytic sterilization of MOF-Based nanohybrid for rapid and portable therapy of bacteria-infected open wounds. *Bioact. Mater.* 13: 200–211.
78 Lee, C.W., Lee, B.-H., Park, S. et al. (2024). Photochemical tuning of dynamic defects for high-performance atomically dispersed catalysts. *Nat. Mater.* 23: 552–559.
79 Chen, P., Wang, Y., Liao, J. et al. (2023). Enhanced adsorption and dissociation of H_2O facilitates photocatalytic CO_2 reduction on defective $BiSbO_4$. *ACS ES&T Eng.* 3: 1248–1257.
80 Shen, H., Yang, M., Hao, L. et al. (2021). Photocatalytic nitrogen reduction to ammonia: insights into the role of defect engineering in photocatalysts. *Nano Res.* 15: 2773–2809.
81 Dong, J., Zhao, J., Yan, X. et al. (2024). Construction of carbonized polymer dots/potassium doped carbon nitride nanosheets Van der Waals heterojunction by ball milling method for facilitating photocatalytic CO_2 reduction performance in pure water. *Appl. Catal. B: Environ.* 351: 123993.
82 Li, S., Song, J., Che, Y. et al. (2022). Advances in molten salt synthesis of non-oxide materials. *Energy Environ. Mater.* 6: e12339.

83 Quan, Y., Li, J., Li, X. et al. Molten salt-assisted synthesis of carbon nitride with defective sites as visible-light photocatalyst for highly efficient hydrogen evolution. *Appl. Catal. B Environ. Energy* 362 (2025): 124711.

84 Wang, D., Zhao, S., Li, J. et al. (2024). Defect-rich RuO_2/IrO_2 heterojunction with dual enzyme-mimic activities for boosting biocatalytic disinfection. *Chem. Eng. J.* 481: 148888.

10

Defected-Engineered Materials for Photocatalytic Sterilization

Haitao Ren[1], Abdelkader Labidi[2], Yuzhen Zhao[1], Atif Sial[2], and Chuanyi Wang[2,3]

[1]*Xijing University, Technological Institute of Materials & Energy Science (TIMES), Changan District, Xi'an 710123, PR China*
[2]*Shaanxi University of Science and Technology, School of Environmental Science and Engineering, Weiyang District, Xi'an, 710021, PR China*
[3]*Department of Environmental and Sustainable Engineering, Faculty of Engineering, Chulalongkorn University, 254 Phayathai Road, Pathumwan, Bangkok, 10330, Thailand*

10.1 Introduction

With the progress of human society and the development of technology, bacterial infectious diseases pose a huge burden on human health and economy and are still a serious public health issue around the world [1]. For instance, the outbreak of the coronavirus disease 2019 (COVID-19) caused serious public health risks [2]. At the same time, clinical research shows that many patients infected with COVID-19 die of secondary bacterial infection, especially multidrug-resistant bacterial infection [3]. Therefore, for these infectious diseases, antibiotics have greatly reduced mortality rates. During antibiotic treatment, antibiotics target the processes required for microbial growth or metabolism, including the preparation of cell walls and membranes, as well as the fabrication of proteins, ribonucleic acid (RNA), or deoxyribonucleic acid (DNA), ultimately leading to microbial cell death [4, 5]. However, traditional systemic antibiotic therapy often involves large doses of antibiotics or prolonged administration, which also leads to antibiotic resistance issues [6]. Increasing numbers of bacteria are developing resistance to traditional antibiotics, making infections more difficult to treat and posing a major challenge to public health [7]. Therefore, many researchers have been actively screening innovative antibacterial drugs and exploring new strategies to address the escalating problem of antibiotic resistance. Recently, photocatalytic technology has gained widespread attention due to its environmentally friendly nature, simple operation, and strong adaptability, and has been extensively employed as an effective strategy for inactivating pathogenic microorganisms [8, 9].

After Fujishima and Honda discovered, for the first time, the photocatalytic water splitting in 1972, photocatalysis has gained significant interest from both

Defect Engineering in Photocatalysis: From Fundamentals to Applications,
First Edition. Edited by Chuanyi Wang, Abdelkader Labidi, and Qiuhui Zhu.

researchers and industrialists [10]. Since then, photocatalysis has attracted widespread attention and achieved numerous benefits in various application fields, including organic synthesis, environmental remediation, and solar energy conversion [11]. In recent years, sterilization therapy based on photocatalytic technology has been widely utilized and has become a green and effective method for treating antibiotic-resistant pathogens [12]. Compared with antibiotics degradation methods, photocatalytic sterilization offers several benefits: (i) it possess effective antibacterial effects on numerous microbial species including bacteria, viruses, fungi, and parasites; (ii) due to the abundance of solar energy, photocatalytic sterilization is a low-cost and environmentally friendly sterilization method; (iii) more interestingly, the photogenerated reactive oxygen species (ROS) can efficiently inactivate microorganisms, which does not easily lead to pathogen resistance [13, 14].

In the basic process of photocatalytic sterilization, as shown in Figure 10.1, photocatalytic materials are activated by light of distinct wavelengths, such as ultraviolet (UV)/visible light/near-infrared (NIR), based on their bandgap, and then generate pairs of electrons (e^-) and holes (h^+) at the same time [15]. Afterward, water (H_2O) molecules can be oxidized by holes to generate hydroxyl radicals (•OH), while oxygen (O_2) molecules are reduced by e^- to form superoxide anion radicals (•O_2^-) [16]. At the same time, some •O_2^- radicals can combine with h^+ to produce singlet oxygen ($^1O_2$). These reactive species are highly active compounds that can kill bacterial cell membranes and organelles, potentially leading to microbial cell destruction [17, 18]. In photocatalytic sterilization, oxygen molecules are not required, while water molecules are abundantly present endogenously. Therefore, it overcomes the limitation of traditional phototherapy processes, which is the insufficient production of ROS in bacterial-infected tissues under hypoxic conditions.

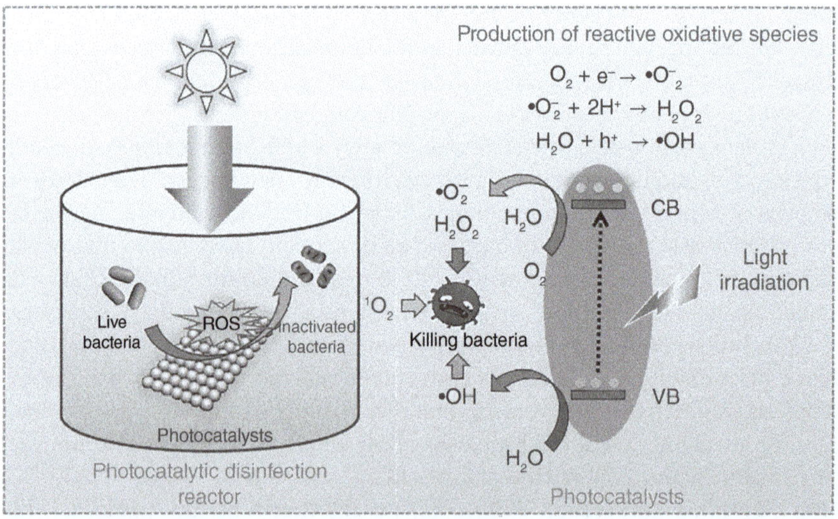

Figure 10.1 Schematic diagram of the photocatalytic sterilization process. *Source:* Reproduced from Ding et al. [15] / with permission of Elsevier.

In order to achieve efficient photocatalytic disinfection, it is necessary to explore photocatalysts with satisfactory light energy conversion efficiency and high photocatalytic activity [19]. So far, many researchers and industrials have devoted a lot of efforts to designing state-of-the-art photocatalytic feedstocks and applying them to various practical applications, including metal oxides (such as TiO_2, ZnO, SiO_2, Fe_2O_3, and CuO), metal sulfides (i.e., MoS_2, Bi_2S_3, and CuS), metal nanoparticles (such as Cu, Ag, and Au), carbon-based materials (such as graphene, carbon quantum dots, carbon nanotubes, and fullerenes), polymer materials (i.e., carbon nitrides, covalent organic frameworks [COFs], and metal organic frameworks [MOFS]), and other different semiconductor materials [20–22]. Figure 10.2 shows numerous photocatalytic semiconductor systems that have been utilized for photocatalytic bacterial destruction [23]. However, most of the reported photocatalysts for photocatalytic sterilization still face serious problems such as a narrow photo-response range, a great recombination rate of photogenerated charge carriers, and low efficiency of ROS generation, which limit their practical application in photocatalytic sterilization [24, 25].

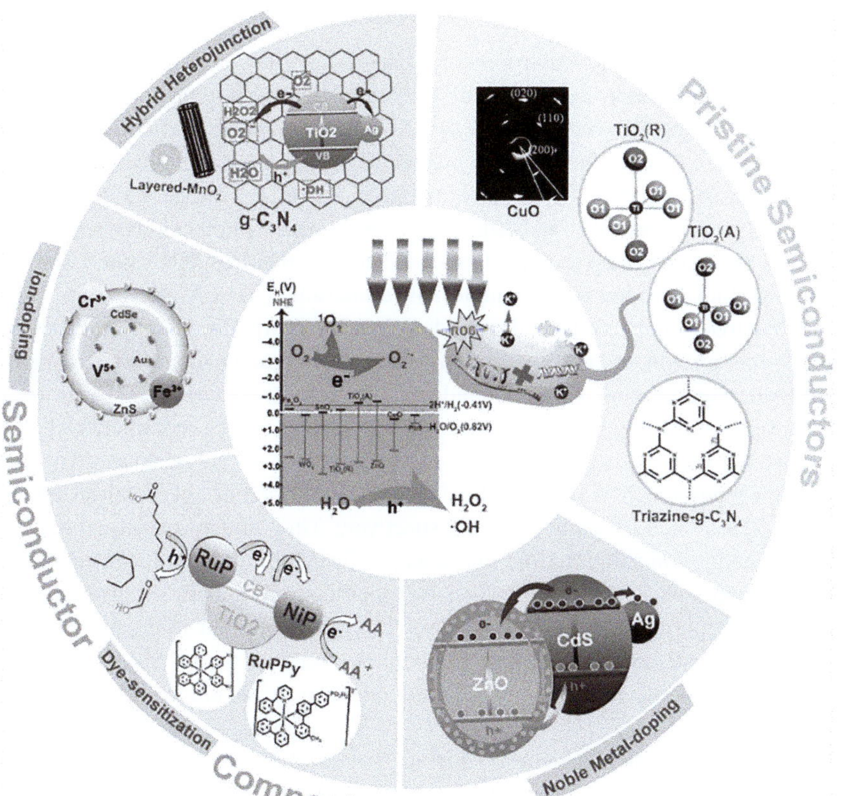

Figure 10.2 Different photocatalytic semiconductor materials for photocatalytic bacterial destruction. *Source:* Reproduced from Zeng et al. [23] / with permission of Royal Society of Chemistry.

Defects are known to have a significant impact on the physicochemical features as well as the functionality of materials. It has been proven that defect engineering can effectively enhance the physicochemical features of photocatalytic materials, thereby enhancing their efficiency in photocatalytic generation of ROS [26]. Defect engineering will alter the electronic and geometric structure of materials, providing potential opportunities for optimizing their photocatalytic performance [27]. Therefore, defect engineering has been widely used to enhance the photocatalytic performance of photocatalysts by adjusting the light absorption range, hindering recombination of photogenerated charge carriers, and surface reactions [28]. The correlation between the defects of catalysts and their catalytic sterilization activity has always been considered a good research topic in the field of photocatalysis processes. In this chapter, we provide a detailed introduction to the key role of defect engineering in enhancing the sterilization performance of semiconductor photocatalytic materials.

10.2 Sterilization Mechanism

Due to its high simplicity, environmental friendliness, low cost, and sustainability, solar-driven photocatalytic bacterial inactivation technology has attracted widespread interest in recent years [29]. ROS-based chemical oxidation is considered the most plausible working mechanism for photocatalytic sterilization, such as hydroxyl radicals (•OH), superoxide anions ($O_2{\bullet}^-$), singlet oxygen ($^1O_2$), and hydrogen peroxide (H_2O_2) [30]. Usually, ROS-based photocatalytic chemical oxidation for photocatalytic degradation of bacteria requires three main stages: (i) photoinduced photocatalytic materials excitation, (ii) ROS formation, and (iii) ROS stress on bacterial cells, leading to cell death. In the photocatalytic process, the photon energy absorbed by the photocatalytic material is exactly equal to or greater than its own bandgap. Electrons with negative charge (e^-) in the valence band (VB) of semiconductors jump to the conduction band (CB), leaving gaps with positive charge (h^+) in VB of semiconductors [31, 32]. The photogenerated h^+ and e^- diffusing on the surface of the catalyst act as strong oxidants and reducing agents, respectively. They undergo electrochemical oxidation–reduction reactions and macroscopically exhibit photoinduced chemical synthesis processes. These photo-responsive materials and catalytic semiconductors are named as photocatalysts. Nevertheless, the recombination of charge carriers is often unsatisfactory, impacting their photocatalytic performance. Relaxed electrons can directly return to the conduction band or recombine with holes in the trap.

The mechanism of ROS generation during photocatalysis is shown in Figure 10.3. When irradiating photocatalysts, charge carriers are generated, and $\bullet O_2^-$ appears through the interaction between oxygen atoms and electrons [33]. Also, •OH can be generated via the oxidation of H_2O or OH– [34]. Then, $\bullet O_2^-$ radicals are reduced to generate H_2O_2. Finally, •OH radicals are produced from H_2O_2. The $^1O_2$ can be generated directly from oxygen (O_2) by employing the energy from recombination [35].

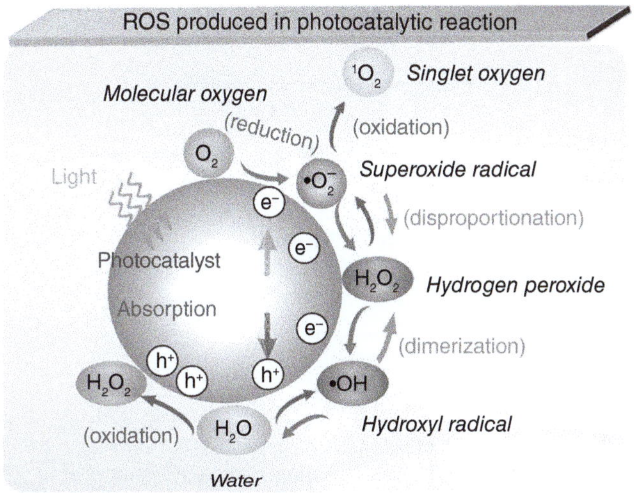

Figure 10.3 ROS production in the photocatalytic reduction and oxidation reactions. *Source:* Reproduced from Ran et al. [33] / with permission of American Chemical Society.

In addition, part of $\bullet O_2^-$ further interreact with h^+ to generate $^1O_2$. These processes can be presented by formulas (10.1)–(10.7) as follows:

$$O_2 + e^- \rightarrow O_2^{\bullet-} \tag{10.1}$$

$$O_2^{\bullet-} + 2H^+ + e^- \rightarrow H_2O_2 \tag{10.2}$$

$$H_2O_2 + e^- \rightarrow OH^- + \bullet OH \tag{10.3}$$

$$H_2O + h^+ \rightarrow H^+ + \bullet OH \tag{10.4}$$

$$OH^- + h^+ \rightarrow \bullet OH \tag{10.5}$$

$$O_2^{\bullet-} + h^+ \rightarrow 1O_2 \tag{10.6}$$

$$O_2^- \rightarrow 1O_2 \tag{10.7}$$

In summary, all of these ROS are excellent active oxidants that affect cell membranes, organelles, and redox substances associated with bacterial cell activity, eventually leading to microbial death in organisms, such as bacteria, fungi, viruses, and protozoa [36]. As is well known, the oxidation–reduction effect of photocatalytic reactions and the band position of photocatalysts are crucial to the possibility of photocatalytic oxidation–reduction reactions [37]. The occurrence of redox reactions also involves certain thermodynamic conditions. For example, the conduction band position of the photocatalyst should be lower than the potential of the O_2 reduction reaction, while the valence band position should be higher than the potential of the H_2O oxidation reaction [38]. Meanwhile, the bandgap energy, CB position, and VB position of common photocatalysts are shown in Figure 10.4. Therefore, the rational design and regulation of the band structure of photocatalytic materials are crucial for promoting the generation of ROS.

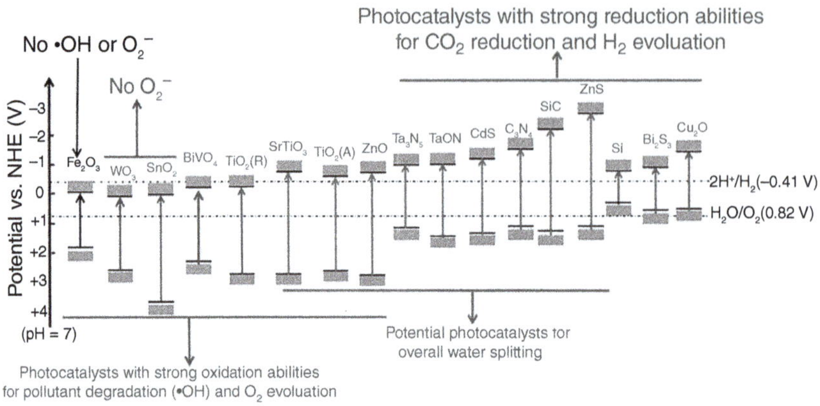

Figure 10.4 Summary of CB position, VB potential, and band gap energy of some photocatalytic feedstocks at physiological pH 7.0. *Source:* Reproduced from Zhang et al. [38] / with permission of Elsevier.

10.3 Types and Characterization of Defects

At present, various photocatalysts are used for sterilization, but the initial photoexcitation and photocarrier transfer processes are very complex. Interestingly, we have summarized the most challenging obstacles to the application of semiconductor photocatalytic materials in sterilization. (a) Insufficient light absorption is the primary challenge for most semiconductor materials in photocatalytic sterilization applications, where if more excited photocarriers are generated, more effective photocarriers can participate in the final oxidation–reduction reaction [39, 40]. So far, creating a richer specific surface area through defect engineering to improve multiple scattering, diffraction, and absorption of solar light has been proven to effectively improve the photocatalytic performance of semiconductor materials [41]. However, the bandgap of most semiconductors is about 2.5 eV, indicating that hole/electron pairs can be only excited under light with wavelengths shorter than 500 nm, which accounts for only about 20.0% of the solar spectrum. The low absorption of visible light and even near-infrared light longer than 500 nm can lead to a limited number of photoexcited electrons and holes, which greatly reduces the photocatalytic activity of semiconductor materials [42]. (b) In terms of charge transfer, the challenges are more complex. The slow transfer kinetics of photogenerated charge carriers and severe photocarrier recombination in the bulk phase are another important challenge that limits the bactericidal performance of photocatalytic materials [43]. The excited active electrons in the conduction band are not in thermodynamic equilibrium, so they can be returned to the ground state and release energy through non-radiative transitions. This is especially true for intrinsic semiconductor materials, as severe electronic localization greatly suppresses the transfer rate of photogenerated carriers, leaving a long time for photogenerated carrier bleaching [44]. In addition, there are no additional energy levels between the conduction band and the valence band, and there is a lack of temporary photogenerated carrier "reservoirs" to buffer the photoexcited electrons or holes in the conduction band and valence band, respectively. Consequently, the photogenerated electrons and holes cannot be effectively separated, leading to insufficient photocarriers for redox reactions, which urgently need improvement.

Currently, a large amount of research indicates that defect engineering is the main strategy to address the challenges faced by the aforementioned semiconductor materials. Defect engineering refers to introducing impurities into the substrate or adjusting the atomic periodicity of semiconductors, which has been considered as an effective strategy for customizing the electronic band structure, optical characteristics, and conductivity of photocatalytic materials [45, 46]. Interestingly, the external morphology of semiconductor materials can also be optimized in terms of precursor type, reaction conditions (reaction temperature, heating rate, reaction time, and pressure) [47]. Therefore, defective semiconductor samples can significantly improve the extended solar energy collection capability, efficient photogenerated carrier transfer process, and higher surface area with richer active sites, thereby bringing comprehensive activity enhancement to various photocatalytic applications. The defect engineering of photocatalysts is a thriving field. The creation/formation of defects regulates the electronic and surface properties of photocatalysts to improve their inherent activity. Generally speaking, defects in semiconductors can be divided into four types based on the size of the lattice (Figure 10.5): 0D point defects (such as vacancies and doping), 1D line defects (such as screw dislocations and edge dislocations), 2D plane defects (such as grain boundaries and twin boundaries), and 3D volume defects (i.e., lattice disorder and pits) [48]. Especially, 0D point defects are the easiest form and are simple to prepare compared to other types of defects. So far, a large amount of research has been conducted on the role of point defects, especially on the study of vacancies. Meanwhile, with the further development of defect creation/formation strategies and a deeper understanding of photocatalytic mechanisms, the positive role of defects in promoting photocatalytic activity is gradually being awakened. There are several methods for forming defects in photocatalytic materials, including chemical

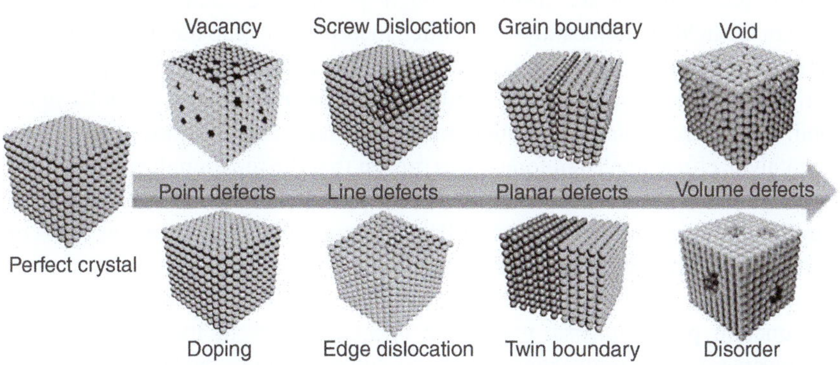

Figure 10.5 Types of defects in photocatalytic feedstocks. *Source:* Reproduced from Lei et al. [48] / with permission of Elsevier.

and physical methods, which can be divided into chemical/thermal reduction, peeling, chemical etching, plasma technology, and ball milling.

Generally speaking, defect engineering on semiconductor materials should follow three important principles, namely, creating abundant active sites, enhancing solar energy collection capabilities, and efficiently separating and transporting photo-generated charge carriers. For the rich active sites in the former, the synthesis methods mainly adopt precursor modification and thermal etching under the required gas atmosphere. Scanning electron microscopy (SEM), transmission electron microscopy (TEM), atomic force microscopy (AFM), and Brunauer–Emmett–Teller (BET) techniques have been widely utilized to characterize the porous structures of defective photocatalysts. For improved solar energy collection requirements, the bandgap calculated by UV/Vis-DRS should be applied through experimental feedback [49]. More interestingly, theoretical studies such as density functional theory (DFT) calculations are a good guiding tool for learning the advantages of defects. For the efficient transport of photo carriers, time-resolved fluorescence spectroscopy (TRPL), photocurrent, and electrochemical impedance spectroscopy (EIS) are excellent tools for evaluating the efficiency of photo carrier separation by obtaining lifetime, photocurrent, and capture resistance results.

In order to achieve defect photocatalytic materials with optimal bactericidal performance, more research work is needed in the near future, especially the precise control of defects, crystallinity, shallow defect states, and even optimization of surface states should be addressed. In order to attain defect customization, more advanced in situ detection techniques are needed, including in situ diffuse reflectance infrared Fourier transform spectroscopy (DRIFTS) and in situ femtosecond transient spectroscopy [50].

10.4 Defective Photocatalysts for Sterilization

Defect engineering has been applied to enhance the sterilization performance of photocatalytic materials. Next, we will analyze the specific application of defect engineering in photocatalytic sterilization through specific case studies. TiO_2 is widely utilized in the field of photocatalysis and often incorporated onto the surface of other organic/inorganic feedstocks. However, there are two main problems with pure TiO_2 material: first, the bandgap is large and can only be utilized by ultraviolet light; second, the fast recombination of photogenerated charge carriers leads to a decrease in its catalytic efficiency. Therefore, it was proven that the functionalization of TiO_2 by heterojunctions promotes the separation of photogenerated charge carriers on its surface, and expanding its range of light absorption has been a major focus of research in photocatalysis processes [51].

In 2024, Cong et al. [52] developed defective CDs@TiO_{2-x} direct Z-scheme heterojunction photocatalysts, which can effectively prevent the recombination of photogenerated charge carriers while exhibiting stronger oxidation–reduction ability, resulting in excellent photocatalytic antibacterial ability. In the preparation process, as shown in Figure 10.6a, TiO_2 was first reduced to defective TiO_2

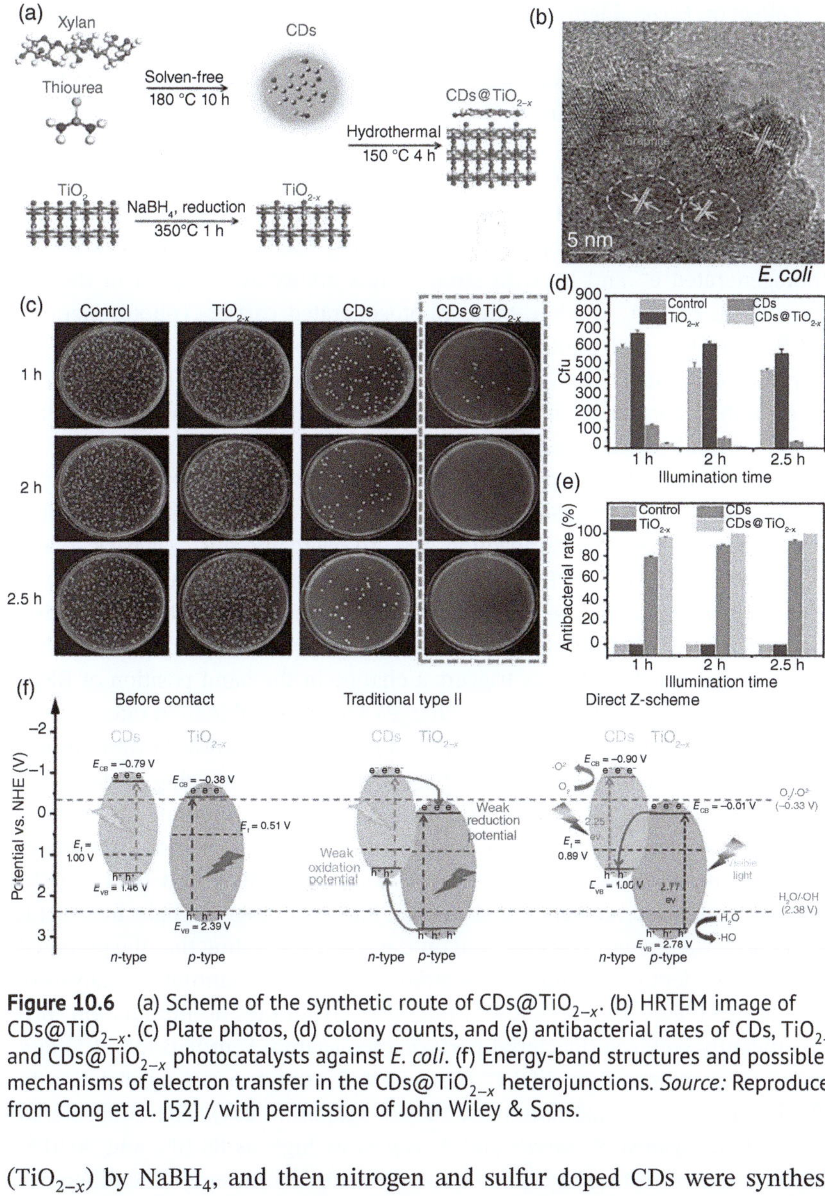

Figure 10.6 (a) Scheme of the synthetic route of CDs@TiO_{2-x}. (b) HRTEM image of CDs@TiO_{2-x}. (c) Plate photos, (d) colony counts, and (e) antibacterial rates of CDs, TiO_{2-x}, and CDs@TiO_{2-x} photocatalysts against *E. coli*. (f) Energy-band structures and possible mechanisms of electron transfer in the CDs@TiO_{2-x} heterojunctions. *Source:* Reproduced from Cong et al. [52] / with permission of John Wiley & Sons.

(TiO_{2-x}) by $NaBH_4$, and then nitrogen and sulfur doped CDs were synthesized by a solvent-free method. Then, CDs were combined with TiO_{2-x} using the hydrothermal method to prepare CDs@TiO_{2-x} direct Z-scheme heterojunction nanocomposites. As presented in Figure 10.6b, for CDs@TiO_{2-x}, the lattice stripe spacing shown in high-resolution TEM images belongs to 0.21 nm for CDs and 0.35 nm for TiO_{2-x}, respectively, proving the successful recombination of CDs and TiO_{2-x}. Figure 10.6c–e exhibits the antibacterial activities of CDs, TiO_{2-x}, and CDs@TiO_{2-x} against *Escherichia coli* (*E. coli*) under visible light irradiation. It is evident from the colony count on nutrition agar plates that both CDs and CDs@TiO_{2-x} have an antibacterial effect on *E. coli*, while CDs@TiO_{2-x} exhibits a stronger effect on *E. coli*. Figure 10.6f shows the band structures of CDs, TiO_{2-x}, and two kinds of heterojunction that might be formed based on their band structures: traditional type-II heterojunction and direct Z-scheme heterojunction, which have high charge separation performance. In the type-II heterojunction, photogenerated e^- gathered in the CB of TiO_{2-x} with a weak reduction potential, and photogenerated h^+ gathered in the VB of CDs with weak oxidation potential; these potentials could not reach −0.33 V ($O_2/{\bullet}O_2^-$) and 2.38 V for ($H_2O/{\bullet}OH$), respectively. While in the direct Z-scheme heterojunction, after being excited by light irradiation, the photogenerated e^- and h^+ with strong redox ability are retained in the CB of CDs and VB of TiO_{2-x}, while other photogenerated carriers (photogenerated e^- in CB of TiO_{2-x} and photogenerated h^+ in VB of CDs) can participate in carrier recombination. Therefore, the direct Z-scheme could solve the problem of carrier recombination while maintaining strong redox ability, achieving a better photocatalytic antibacterial effect.

The inherent narrow bandgap of bismuth oxyiodide (BiOI) always results in rapid recombination of photogenerated charge carriers, severely limiting its practical application in photocatalytic sterilization [53]. To overcome this drawback, Wang et al. designed a Bi/$BiOI_{1-x}$ (BB) nanosheet with edge heterostructure, which was modified by introducing iodine defects and in situ bismuth deposition employing high-temperature calcination and solvothermal method. It is worth noting that the introduction of iodine defects triggers a change in the band position of BiOI, resulting in an effective reduction of the charge carrier recombination rate.

The photocatalytic activity of the synthesized photocatalyst toward *Gram-positive Staphylococcus aureus* and *Gram-negative E. coli* was conducted using the standard plate counting method [54]. Figure 10.7a shows that in the dark, all groups have little bactericidal effect on *S. aureus and E. coli*. Under visible light irradiation, no significant bactericidal effect was shown in the BiOI (BIB) and $BiOI_{1-x}$ (BB) entities, but some bacteria were killed by BIB. However, after BB light treatment, almost no bacterial growth was observed on the plate, suggesting that the prepared BB has more efficient photocatalytic bacterial activity. The quantitative analysis of plate counting (Figure 10.7b,c) showed that after 60 minutes of illumination, BIB had significantly better antibacterial performance against *S. aureus* (30.61%) and *E. coli* (31.69%) compared to BiOI (7.5% against *S. aureus*, 6.21% against *E. coli*) and $BiOI_{1-x}$ (13.01% against *S. aureus*, 16.91% against *E. coli*). The antibacterial efficiency of BB against *S. aureus* and *E. coli* is as high as 88.10% and 90.01%, respectively, demonstrating high photocatalytic bacterial efficiency. The difference in antibacterial performance of the fabricated photocatalyst was further studied by confocal laser microscopy, with live and destructed bacteria indicated by DAPI (blue) and PI (red) diffuse light, respectively. The findings showed that only a low amount of blue fluorescence (live bacteria) was shown in the BB group, indicating that BB has the best antibacterial ability, which is in accordance with the plate counting results. In addition, the morphology of *S. aureus* and *E. coli* after interaction with photocatalytic material was qualitatively examined by SEM analysis (Figure 10.7d). In comparison with the control group, the bacteria treated with BiOI and $BiOI_{1-x}$ still maintained their original morphology, indicating that

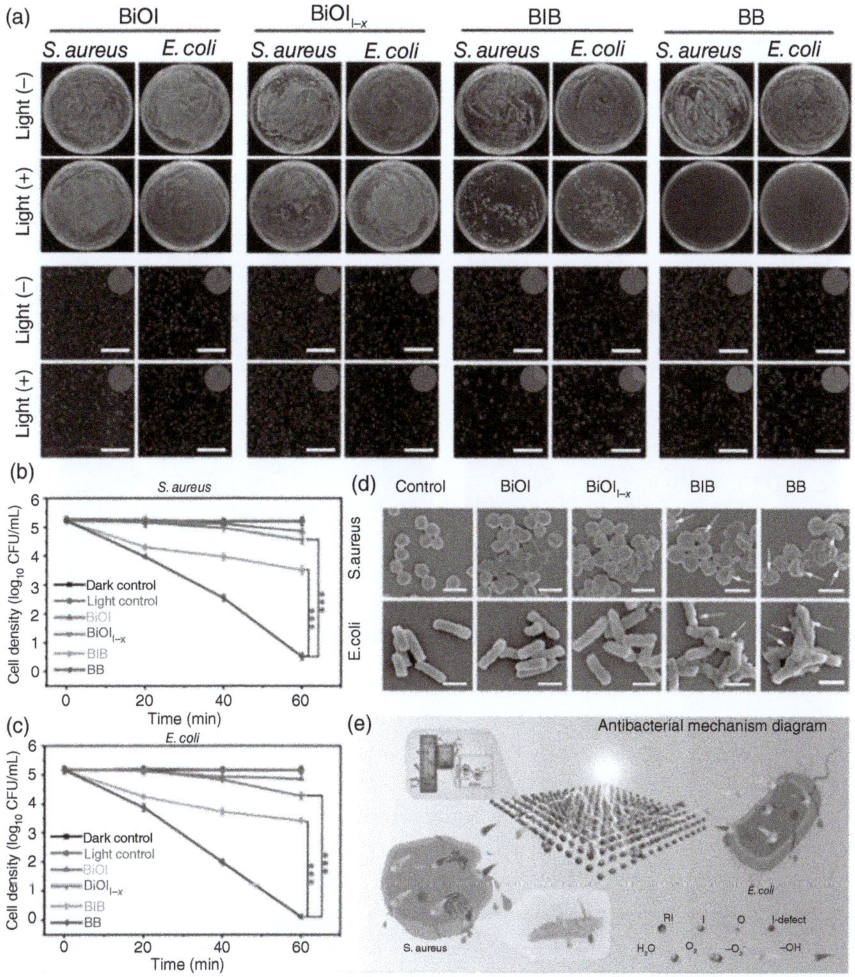

Figure 10.7 (a) Photographs display the bacterial culture plates of *S. aureus* and *E. coli* dosed by BiOI, $BiOI_{1-x}$, BIB, and BB materials under dark or visible-light illumination for 60 minutes of contact time. (b, c) Bacterial density of distinct samples toward *S. aureus* and *E. coli* under visible-light irradiation. (d) SEM micrographs of *S. aureus* and *E. coli* after BiOI, $BiOI_{1-x}$, BIB, and BB treatments under visible-light illumination. (e) Antibacterial schematic diagram of BB. *Source:* Reproduced from Fu et al. [54] / with permission of Elsevier.

the photocatalytic killing effect of BiOI and $BiOI_{1-x}$ on bacteria can be ignored. However, significant cell membrane indentation or rupture can be seen in the BIB and BB groups. The bacteria treated with BB showed higher morphological modifications, with some content leakage between the two groups. The photocatalytic destruction of these bacteria is mainly assigned to the high amount of ROS produced by BB photocatalysts under visible light irradiation (Figure 10.7e).

Tungsten trioxide (WO_3) is known as a nontoxic and photo-stable *n*-type semiconductor, which can generate a large amount of ROS under photoexcitation and is considered an attractive photocatalytic sterilizing agent [55]. Nevertheless, the low

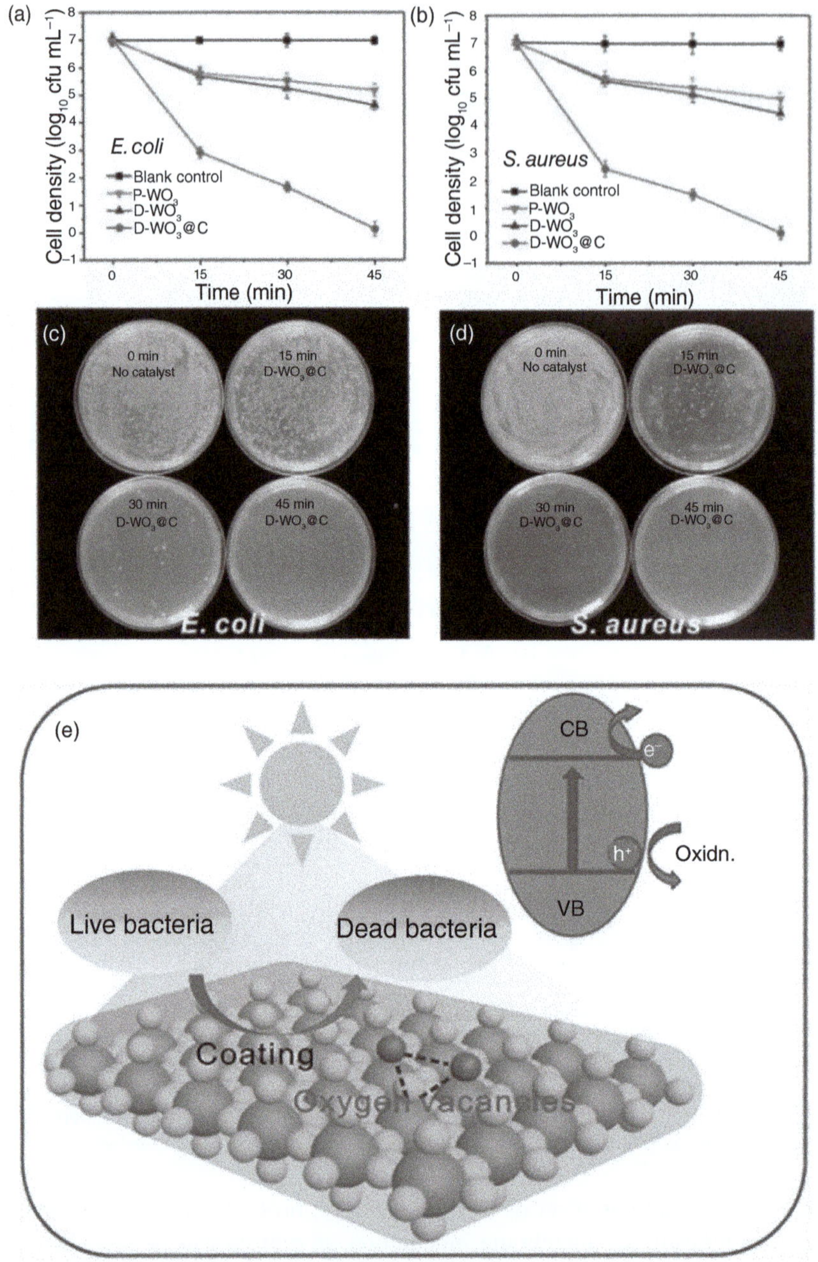

Figure 10.8 Cell density of *E. coli* (a) and *S. aureus* (b) with distinct photocatalysts under visible-light illumination. Bacterial colony growth in the existence of D-WO_3@C photocatalyst with *E. coli* (c) and *S. aureus* (d) at distinct irradiation times. (e) Schematic diagram of photocatalytic sterilization by D-WO_3@C. *Source:* Reproduced from Wang et al. [56] / with permission of Elsevier.

charge transfer and rapid recombination of charge carriers limit the photocatalytic sterilization performance of the original WO_3. According to reports, oxygen vacancies on two-dimensional WO_3 are an effective means to address the aforementioned limitations. Introducing high-density oxygen vacancies can efficiently increase the donor density of WO_3, thereby promoting carrier separation. Therefore, defect WO_3 achieves enhanced photocatalytic sterilization performance by effectively utilizing carrier separation.

Based on this, Wang et al. [56] constructed tungsten oxide carbon-based hybrid nanosheets with rich-oxygen vacancies (labeled as D-WO_3@C). As a robust photocatalyst for visible light collection. As expected, this material exhibits effective photocatalytic bactericidal activity toward *E. coli* and *S. aureus* under visible light illumination from a xenon lamp ($\lambda > 420$ nm). As two typical harmful microorganisms, *Gram-negative E. coli and Gram-positive S. aureus* were utilized to study the antibacterial efficiency of the prepared photocatalyst. As shown in Figure 10.8a–d, the activity of killing *E. coli* by the D-WO_3@C was compared with defect WO_3 (D-WO_3) and pure WO_3 (P-WO_3). Neglected antibacterial activity was detected in the blank control, suggesting that irradiation sterilization cannot be performed in the absence of photocatalysts.

Because of the low photocatalytic efficiency of P-WO_3 under visible light illumination, the cell density of *E. coli* over the P-WO_3 photocatalyst was approximately 5.5 $\log_{10}$ cfu mL^{-1} after 30 minutes. However, the cell density over D-WO_3@C photocatalyst was only 1.7 $\log_{10}$ cfu mL^{-1} after 30 minutes of irradiation time, which is significantly higher than that of P-WO_3. As shown in Figure 10.8e, the interaction between defective WO_3 and the carbon layer can promote the transfer and separation of photo-induced charge carriers from WO_3 to carbon. Electrons and holes participate in surface oxygen reduction reactions to generate a large amount of ROS, ultimately leading to bacterial death.

10.5 Advantages and Limitations

Research has shown that defects can provide a wide range of beneficial effects in photocatalytic materials and further enhance their performance in photocatalytic sterilization applications [57, 58], with the following benefits:

a) Despite the efforts made in the photocatalysis processes for several applications [59–68]. It is worth noting that defects can promote the separation of photogenerated charge carriers by providing pathways for charge transfer and transmission, thereby improving the photocatalytic efficiency of photocatalytic materials.
b) Introducing various defects can also alter the electronic band structure of photocatalytic materials, thereby designing new band gaps, enhancing light absorption, and further improving the efficiency of photocatalytic materials under visible light illumination.
c) The presence of defects can alter the surface energy of photocatalytic materials, thereby increasing their reactivity. This is particularly beneficial in specific

applications where surface interactions play an important role, such as photocatalytic sterilization processes.

d) Surface defects can promote the production of reactive oxygen species under light irradiation, which is crucial for the use of photocatalytic processes to oxidize and degrade organic pollutants and achieve sterilization effects.

Although defect engineering in photocatalytic materials provides a powerful option for optimizing their photocatalytic sterilization performance, it also has some drawbacks, as described below:

a) Unfortunately, in some cases, defects can become recombination centers for photogenerated electrons and holes, leading to a decrease in carrier concentration and thus reducing the efficiency of photocatalytic sterilization.
b) Defects may cause instability of photocatalysts, especially under prolonged exposure or harsh environmental conditions. Therefore, photocatalytic materials with excessive or uncontrolled defects may lose their photocatalytic activity for a period of time.
c) Introducing defects may generate new active sites, thereby improving the occurrence of side reactions and reducing the selectivity of photocatalysts for the desired products.
d) Surface defects may lead to the accumulation of reaction by-products, especially the blockage of active sites, thereby hindering photocatalytic activity and causing photocatalyst deactivation. In addition, body defects can also have a negative impact on the electronic properties of photocatalytic feedstocks.
e) Defect engineering can make photocatalytic materials more susceptible to chemical corrosion in complex reaction environments, affecting their lifespan and operational feasibility. Therefore, defects are a double-edged sword, and they need to be introduced reasonably according to specific reaction conditions in practical use.

10.6 Conclusion

In recent years, research on defect engineering has made certain progress in enhancing photocatalyst sterilization. Photocatalytic sterilization mainly generates photogenerated electron–hole pairs by absorbing photons. These electron holes react with surrounding water and oxygen to generate reactive oxygen species, including hydroxyl radicals and superoxide anions. These active substances have strong oxidizing properties and can destroy the cell membrane and protein structure of microorganisms, thereby inactivating them. Photocatalytic sterilization technology has significant characteristics of being green, low cost, adaptable to a wide range of reactions, and having minimal secondary pollution. Among them, defect engineering is an efficient strategy to improve the sterilization performance of photocatalysts. Defects can increase the utilization range of semiconductor materials for sunlight, thereby generating a large number of photogenerated charge carriers. Additionally, defects can capture electrons, promote the separation of photogenerated charge carriers, and generate a large amount of ROS in surface reactions, ultimately achieving efficient photocatalytic sterilization performance. Therefore, in this chapter, we focus on the mechanism of photocatalytic sterilization and point out the mechanism and advantages of defect engineering in enhancing the sterilization performance of photocatalytic feedstocks. At the same time, we emphasized the challenges still faced in this field and proposed strategic recommendations for the development of defective photocatalytic materials. This chapter is expected to make important contributions to promoting the development of defect-type photocatalytic feedstocks in the field of photocatalytic sterilization.

Acknowledgements

This work was financially supported by the National Natural Science Foundation of China (Nos. DL2023041004L and G2023041021L) and Xijing University Special Fund for High-Level Talents (No. XJ24B20).

Conflict of Interest

The authors declare that they have no conflict of interest.

References

1 Fisher, R.A., Gollan, B., and Helaine, S. (2017). Persistent bacterial infections and persister cells. *Nat. Rev. Microbiol.* 15: 453–464.
2 Hendaus, M.A. and Jomha, F.A. (2021). Covid-19 induced superimposed bacterial infection. *J. Biomol. Struct. Dyn.* 39: 4185–4191.
3 Zhang, H., Tang, W., Chen, Y., and Yin, W. (2020). Disinfection threatens aquatic ecosystems. *Science* 368: 146–147.
4 Kohanski, M.A., Dwyer, D.J., and Collins, J.J. (2010). How antibiotics kill bacteria: from targets to networks. *Nat. Rev. Microbiol.* 8: 423–435.
5 Li, P., Sun, L., Xue, S. et al. (2022). Recent advances of carbon dots as new antimicrobial agents. *SmartMat* 3: 226–248.
6 Ellabaan, M., Munck, C., Porse, A. et al. (2021). Forecasting the dissemination of antibiotic resistance genes across bacterial genomes. *Nat. Commun.* 12: 2435.
7 Zhou, Z., Zhu, H., Wu, S. et al. (2023). The combination of S-doped ZIF-8 with graphene oxide for enhanced near-infrared light photocatalytic and photothermal sterilization. *Chem. Eng. J.* 455: 140857.
8 Yang, R., Song, G., Wang, L. et al. (2021). Full solar-spectrum-driven antibacterial therapy over hierarchical Sn_3O_4/PDINH with enhanced photocatalytic activity. *Small* 17: 2102744.
9 Ren, H., Qi, F., Labidi, A. et al. (2023). Chemically bonded carbon quantum dots/Bi_2WO_6 S-scheme heterojunction for boosted photocatalytic antibiotic

degradation: interfacial engineering and mechanism insight. *Appl. Catal., B* 330: 122587.

10 Fujishima, A. and Honda, K. (1972). Electrochemical photolysis of water at a semiconductor electrode. *Nature* 238: 37–38.

11 He, D., Zhang, Z., Xing, Y. et al. (2020). Black phosphorus/graphitic carbon nitride: a metal-free photocatalyst for "green" photocatalytic bacterial inactivation under visible light. *Chem. Eng. J.* 384: 123258.

12 Wang, L., Zhang, X., Yu, X. et al. (2019). An all-organic semiconductor C_3N_4/PDINH heterostructure with advanced antibacterial photocatalytic therapy activity. *Adv. Mater.* 31: 1901965.

13 Kumari, N., Kumar, S., Karmacharya, M. et al. (2020). Surface-textured mixed-metal-oxide nanocrystals as efficient catalysts for ROS production and biofilm eradication. *Nano Lett.* 21: 279–287.

14 Ran, B., Wang, Z., Cai, W. et al. (2021). Organic photo-antimicrobials: principles, molecule design, and applications. *JACS* 143: 17891–17909.

15 Ding, Y., Yang, G., Zheng, S. et al. (2024). Advanced photocatalytic disinfection mechanisms and their challenges. *J. Environ. Manage.* 366: 121875.

16 Wang, L., Wu, J., Hu, Y. et al. (2018). Using porous magnetic iron oxide nanomaterials as a facile photoporation nanoplatform for macromolecular delivery. *J. Mater. Chem. B* 6: 4427–4436.

17 Nosaka, Y. and Nosaka, A.Y. (2017). Generation and detection of reactive oxygen species in photocatalysis. *Chem. Rev.* 117: 11302–11336.

18 Fu, J., Yu, J., Jiang, C., and Cheng, B. (2018). g-C_3N_4-based heterostructured photocatalysts. *Adv. Energy Mater.* 8: 1701503.

19 Liu, Y., Zeng, X., Hu, X. et al. (2019). Two-dimensional nanomaterials for photocatalytic water disinfection: recent progress and future challenges. *J. Chem. Technol. Biotechnol.* 94: 22–37.

20 Wang, W., Huang, G., Jimmy, C.Y., and Wong, P.K. (2015). Advances in photocatalytic disinfection of bacteria: development of photocatalysts and mechanisms. *J. Environ. Sci.* 34: 232–247.

21 Zhou, Z., Li, B., Liu, X. et al. (2021). Recent progress in photocatalytic antibacterial. *ACS Appl. Bio Mater.* 4: 3909–3936.

22 Levchuk, I., Kralova, M., Rueda-Márquez, J.J. et al. (2018). Antimicrobial activity of printed composite TiO_2/SiO_2 and TiO_2/SiO_2/Au thin films under UVA-LED and natural solar radiation. *Appl. Catal., B* 239: 609–618.

23 Zeng, J., Li, Z., Jiang, H., and Wang, X. (2021). Progress on photocatalytic semiconductor hybrids for bacterial inactivation. *Mater. Horiz.* 8: 2964–3008.

24 Zhang, Y.C., Afzal, N., Pan, L. et al. (2019). Structure-activity relationship of defective metal-based photocatalysts for water splitting: experimental and theoretical perspectives. *Adv. Sci.* 6: 1900053.

25 Mao, C., Xiang, Y., Liu, X. et al. (2018). Repeatable photodynamic therapy with triggered signaling pathways of fibroblast cell proliferation and differentiation to promote bacteria-accompanied wound healing. *ACS Nano* 12: 1747–1759.

26 Mao, C., Wang, J., Zou, Y. et al. (2019). Anion (O, N, C, and S) vacancies promoted photocatalytic nitrogen fixation. *Green Chem.* 21: 2852–2867.

27 Gao, D., Li, B., Yang, Y. et al. (2021). Defect-induced double-stranded DNA unwinding on graphene. *J. Phys. Chem. B* 125: 2833–2840.

28 Kayani, Z.N., Intizar, T., Riaz, S., and Naseem, S. (2021). Antibacterial, magnetic and dielectric properties of nano-structured V doped TiO_2 thin films deposited by dip coating technique. *Mater. Chem. Phys.* 267: 124659.

29 Qiao, Y., Liu, X., Li, B. et al. (2020). Treatment of MRSA-infected osteomyelitis using bacterial capturing, magnetically targeted composites with microwave-assisted bacterial killing. *Nat. Commun.* 11: 4446.

30 Qiu, P., Zhu, J., Zhang, C. et al. (2023). Visible light-supported efficient photocatalytic disinfection using a robust silver-doped boron photocatalyst. *J. Environ. Chem. Eng.* 11: 111058.

31 Kim, E.-J., Choi, M., Park, H.Y. et al. (2019). Thorn-like TiO_2 nanoarrays with broad spectrum antimicrobial activity through physical puncture and photocatalytic action. *Sci. Rep.* 9: 13697.

32 Lu, H.-T., Huang, G.-Y., Chang, W.-J. et al. (2022). Modification of chitosan nanofibers with CuS and fucoidan for antibacterial and bone tissue engineering applications. *Carbohydr. Polym.* 281: 119035.

33 Ran, B., Ran, L., Wang, Z. et al. (2023). Photocatalytic antimicrobials: principles, design strategies, and applications. *Chem. Rev.* 123: 12371–12430.

34 Xia, P., Cao, S., Zhu, B. et al. (2020). Designing a 0D/2D S-scheme heterojunction over polymeric carbon nitride for visible-light photocatalytic inactivation of bacteria. *Angew. Chem. Int. Ed.* 59: 5218–5225.

35 Lipovsky, A., Levitski, L., Tzitrinovich, Z. et al. (2012). The different behavior of rutile and anatase nanoparticles in forming oxy radicals upon illumination with visible light: an EPR study. *Photochem. Photobiol.* 88: 14–20.

36 Chen, S., Sun, Z., Zhang, L., and Xie, H. (2020). Photodegradation of gas phase benzene by SnO_2 nanoparticles by direct hole oxidation mechanism. *Catalysts* 10: 117.

37 Lee, K., Shin, S., Lee, W.J. et al. (2021). Sunlight-activatable ROS generator for cell death using TiO_2/c-Si microwires. *Nano Lett.* 21: 6998–7004.

38 Zhang, X., Zhang, S., Mathivanan, K. et al. (2025). Research progress and prospects in antifouling performance of photocatalytic sterilization: a review. *J. Mater. Sci. Technol.* 208: 189–201.

39 Hou, X., Shi, T., Wei, C. et al. (2020). A 2D-2D heterojunction Bi_2WO_6/WS_{2-x} as a broad-spectrum bactericide: sulfur vacancies mediate the interface interactions between biology and nanomaterials. *Biomaterials* 243: 119937.

40 Wang, H., Yong, D., Chen, S. et al. (2018). Oxygen-vacancy-mediated exciton dissociation in BiOBr for boosting charge-carrier-involved molecular oxygen activation. *JACS* 140: 1760–1766.

41 Jun, S., Choi, S., Choi, S. et al. (2021). Direct synthesis of molybdenum phosphide nanorods on silicon using graphene at the heterointerface for efficient photoelectrochemical water reduction. *Nano-Micro Lett.* 13: 81.

42 Li, Z., Xiao, C., Fan, S. et al. (2015). Dual vacancies: an effective strategy realizing synergistic optimization of thermoelectric property in BiCuSeO. *JACS* 137: 6587–6593.

43 Wu, G., Chen, X., Zhang, Z. et al. (2020). Vacancy-induced toxicity of $CoSe_2$ nanomaterials in rat lung macrophages. *Nanotoxicology* 14: 968–984.

44 Liu, M., Zhu, H., Zhu, N., and Yu, Q. (2021). Vacancy engineering of BiOCl microspheres for efficient removal of multidrug-resistant bacteria and antibiotic-resistant genes in wastewater. *Chem. Eng. J.* 426: 130710.

45 Liu, Y., Xiao, C., Li, Z., and Xie, Y. (2016). Vacancy engineering for tuning electron and phonon structures of two-dimensional materials. *Adv. Energy Mater.* 6: 1600436.

46 Liu, M., Spanos, P.D., and Yu, S.-H. (2019). Synthesis of ultrathin Bi_2Se_3 nanosheets/graphene nanocomposite with defects/vacancies-dependent transient photocurrent performance. *Nano Energy* 64: 103877.

47 Hou, S., Gao, X., Lv, X. et al. (2024). Decade milestone advancement of defect-engineered g-C_3N_4 for solar catalytic applications. *Nano Micro Lett.* 16: 70.

48 Lei, W., Yu, Y., Zhang, H. et al. (2022). Defect engineering of nanostructures: insights into photoelectrochemical water splitting. *Mater. Today* 52: 133–160.

49 Jiao, X., Chen, Z., Li, X. et al. (2017). Defect-mediated electron–hole separation in one-unit-cell $ZnIn_2S_4$ layers for boosted solar-driven CO_2 reduction. *JACS* 139: 7586–7594.

50 Chen, M., Sun, M., Cao, X. et al. (2024). Progress in preparation, identification and photocatalytic application of defective g-C_3N_4. *Coord. Chem. Rev.* 510: 215849.

51 Ren, Y., Han, Y., Li, Z. et al. (2020). Ce and Er Co-doped TiO_2 for rapid bacteria-killing using visible light. *Bioact. Mater.* 5: 201–209.

52 Cong, S., Cai, J., Li, X. et al. (2024). Direct Z-scheme xylan-based carbon dots@TiO_{2-x} nanocomposites for visible light driven photocatalytic of dye degradation and antibacterial. *Adv. Funct. Mater.* 34: 2401540.

53 Vahabirad, S. and Nezamzadeh-Ejhieh, A. (2022). Co-precipitation synthesis of $BiOI/(BiO)_2CO_3$: brief characterization and the kinetic study in the photodegradation and mineralization of sulfasalazine. *J. Solid State Chem.* 310: 123018.

54 Fu, B., Pan, Y., Zhao, P. et al. (2024). "Edge in-situ heterogeneous" BiOI based on defect engineering and non-noble metal deposition: boosting visible-light photocatalytic sterilization. *Chem. Eng. J.* 491: 152071.

55 Hameed, A., Ismail, I.M., Aslam, M., and Gondal, M. (2014). Photocatalytic conversion of methane into methanol: performance of silver impregnated WO_3. *Appl. Catal., A* 470: 327–335.

56 Wang, R., Zhang, W., Zhu, W. et al. (2018). Enhanced visible-light-driven photocatalytic sterilization of tungsten trioxide by surface-engineering oxygen vacancy and carbon matrix. *Chem. Eng. J.* 348: 292–300.

57 Wang, B., Liu, J., Yao, S. et al. (2021). Vacancy engineering in nanostructured semiconductors for enhancing photocatalysis. *J. Mater. Chem. A* 9: 17143–17172.

58 Wu, G., Sun, J., Zhang, Z. et al. (2022). Recent advances in biological applications of nanomaterials through defect engineering. *Sci. Total Environ.* 816: 151647.

59 Wang, H., Labidi, A., Ren, M. et al. (2024). Recent progress of microstructure-regulated g-C_3N_4 in photocatalytic NO conversion: the pivotal roles of adsorption/activation sites. *Acta Phys. Chim. Sin.* 41: 100039.

60 Labidi, A., Ren, H., Liang, X. et al. (2024). Visible–light–driven photocatalytic degradation of ciprofloxacin antibiotic by novel heterostructured coal fly ash waste: mechanism insight, toxicity pathway and DFT calculation. *Mater. Today Chem.* 42: 102388.

61 Ghasemi, S., Parastesh, A., Padervand, M. et al. (2025). Recent progress on Z-and S-scheme photocatalysis: mechanistic understanding toward green applications. *Curr. Opin. Chem. Eng.* 47: 101059.

62 Sial, A., Gao, T., Li, F. et al. (2024). Multifaceted synergism of dual active sites in oxygen vacancies enriched plasmonic Ag-BiOI nanosheets for enhanced piezo-photocatalytic degradation of trimethoprim. *ACS ES&T Eng.* 4: 2923–2937.

63 Liu, J., Du, S., Zhu, Q. et al. (2024). Recent progress on bismuth-based materials for full-spectrum-driven photocatalysis. *J. Environ. Chem. Eng.* 114790.

64 Ren, H., Labidi, A., Sial, A. et al. (2024). Carbon quantum dots modified Z and S-scheme heterojunctions for pharmaceutical contaminants photodegradation: state-of-the-art, benefits, and limitations. *Sep. Purif. Technol.* 346: 127508.

65 Padervand, M., Bargahi, A., Eftekhari-Sis, B. et al. (2024). In-situ synthesis of $SnO/CuSnO_3$ nanostructures to catalyze azo dye degradation, CO_2 reduction, and amines direct alkylation reactions under visible light. *Results Eng.* 23: 102515.

66 Liang, X., Gao, T., Cui, Y. et al. (2024). Photoreforming of poly(ethylene-terephthalate) plastic into valuable chemicals and hydrogen over $BiVO_4$/MoOx: synergistic promotion of oxidation and reduction processes. *Appl. Catal., B* 357: 124326.

67 Dong, Q., Li, X., Sun, J. et al. (2024). Regulating concentration of surface oxygen vacancies in Bi_2MoO_6/Bi-MOF for boosting photocatalytic ammonia synthesis. *J. Catal.* 433: 115489.

68 Labidi, A., Ren, H., Sial, A. et al. (2024). Photocatalyst materials formed using organic radicals for water splitting. In: *Organic Radicals* (ed. C. Wang, A. Labidi, and E. Lichtfouse), 211–230. Elsevier.

11

Nano- and Microplastics Degradation Using Defect-engineered Photocatalytic Feedstocks

Elham Jalilnejad[1], Mehran Alizadeh[1], and Sahar Rezaei[2]

[1] *Urmia University of Technology, Faculty of Chemical Engineering, Urmia, 57155419, Iran*
[2] *Sahand University of Technology, Chemical Engineering Faculty, Sahand New Town, Tabriz, 51335-1996, Iran*

11.1 Introduction

The global water crisis has escalated into one of the most critical environmental and societal challenges of the 21st century [1]. This issue is not limited to water scarcity but is compounded by the widespread contamination of water systems with persistent pollutants, particularly nanoplastics (NPs) and microplastics (MPs) [2]. These plastic fragments, generated through the degradation of larger plastic waste or directly introduced from industrial and domestic sources, have been identified in freshwater, marine environments, and even drinking water supplies [3]. Their small size allows them to evade conventional water treatment processes, while their chemical properties enable them to adsorb toxic substances such as pesticides, heavy metals, and hydrophobic organic pollutants [4]. These contaminants not only disrupt aquatic ecosystems by harming organisms at multiple trophic levels but also pose significant risks to human health through the food chain [2]. The persistence and ubiquity of microplastics and nanoplastics underscore the urgent need for innovative solutions to mitigate their environmental impact.

Among the emerging technologies to address plastic contamination, photocatalysis stands out as a sustainable and efficient method [5]. Photocatalytic processes employ semiconductor materials activated by light to generate reactive oxygen species capable of degrading complex organic pollutants [6]. This technique has demonstrated its ability to break down the robust polymer chains of micro- and nanoplastics into simpler, nontoxic molecules like carbon dioxide and water [7]. Unlike conventional methods, photocatalysis is environmentally benign, as it does not produce secondary pollutants and utilizes solar energy, a renewable resource [8]. Recent studies have shown that photocatalysis is not only effective in removing plastic contaminants but also excels in treating associated pollutants, such as organic dyes and pharmaceutical residues, making it a multifaceted solution for

Defect Engineering in Photocatalysis: From Fundamentals to Applications,
First Edition. Edited by Chuanyi Wang, Abdelkader Labidi, and Qiuhui Zhu.

water purification [9]. Its versatility and scalability position make it a promising approach for real-world applications in the fight against plastic pollution.

The application of defect engineering has revolutionized the field of photocatalysis by significantly enhancing the efficiency of photocatalytic materials [10]. Defects such as vacancies, dopants, and dislocations within the crystal lattice of semiconductors modify their electronic and surface properties, addressing critical limitations like insufficient light absorption and rapid electron–hole recombination [11]. For example, oxygen vacancies in TiO_2-based photocatalysts extend their light absorption range into the visible spectrum, while dopants introduce energy states that facilitate charge carrier separation [12]. Surface defects, meanwhile, create active sites that promote the adsorption and decomposition of plastic molecules [13]. These advancements enable defect-engineered photocatalysts to operate effectively under ambient conditions and with lower energy input, making them highly suitable for the degradation of micro- and nanoplastics in aquatic environments [14]. Ongoing research continues to optimize defect engineering strategies, pushing the boundaries of photocatalytic performance.

This chapter investigates the critical issues of microplastics and nanoplastics pollution, emphasizing the transformative role of photocatalysis and defect engineering in mitigating these challenges. Moreover, by thoroughly exploring defect classifications, synthesis methods, characterization techniques, and application of defect-engineered photocatalysts in the degradation of micro- and nanoplastics, it aims to provide a clear understanding of how these technologies can contribute to a cleaner and more sustainable water future.

11.2 Sources and Accumulation

11.2.1 Nano- and Microplastics in the Environment

The discovery of plastic in 1907 was a groundbreaking moment that transformed life after the Industrial Revolution. Plastics quickly became essential in many areas, including construction, household products, medical tools, agriculture, personal care, and cosmetics. Over the past 50 years, plastic production has grown rapidly, increasing by about 8.7% annually and reaching a total of 9.1 billion tons globally [15]. This rapid growth, especially in single-use plastics, has led to a serious environmental challenge. Many plastic products are not properly recycled, with recycling rates ranging from less than 10% to around 33% depending on the region. Improper disposal allows large amounts of plastic waste to end up in the environment [9]. Over time, this waste breaks down through physical forces, biological activity, and sunlight, forming smaller particles known as macro-, micro-, and nanoplastics. Microplastics are defined as plastic particles between 1 μm and 5 mm in size, while nanoplastics are even smaller, ranging from 1 to 1000 nm. Nanoplastics can form from the breakdown of larger plastics or during industrial production [16]. These tiny particles are now a growing concern due to their widespread presence and potential harm to ecosystems [17].

Table 11.1 Classification based on particle size.

Kind	Size range	Source	References
Macroplastics	≥5 mm	Complete items or pieces of packaging along with different objects made from polymer, primarily resulting from breakage processes or intentional release.	[18]
Primary microplastics	1–5 mm	Scrubbing agents of cosmetics, pharmaceuticals, and textile fibers, as well as the emission of fragments and fibers during the manufacturing and upkeep of polymer-based materials	[19]
Secondary microplastics	1 μm–1 mm	Originating from macro- and basic microplastics during various environmental decomposition proceeding, including biological degradation, photo-oxidation, thermo-oxidation, UV irradiation, and mechanical actions such as tire wear	[20]
Nanoplastics	1–1000 nm	Primarily similar to secondary microplastics, arise from activities like washing and drying of textiles, wear and tear of tires, and the breakdown of other materials, including polystyrene.	[21]

Microplastics and nanoplastics are introduced into the environment through a variety of pathways, originating from both primary and secondary sources (as outlined in Table 11.1). Primary sources refer to materials specifically manufactured as microplastics, such as those found in electronics, paints, adhesives, coatings, textiles, medical devices, and agricultural products [20]. Secondary sources, on the other hand, result from the breakdown of larger plastic items into smaller particles over time [22].

Natural events, including tsunamis, floods, and hurricanes, can also contribute significantly to the release of secondary microplastics into aquatic ecosystems, adding to the growing environmental burden of plastic pollution [21].

The per capita emissions of micro- and nanoplastics differ significantly across regions due to factors such as population size, economic status, and the effectiveness of waste management systems. It is estimated that about 75–90% of secondary plastic contaminants in aquatic environments originate from land-based sources, with the remaining 10–25% coming from ocean-based activities. Key land-based contributors include littering, improper disposal of plastic waste, and leakage from inadequately managed landfill sites. Ocean-based sources are linked to lost materials, aquaculture activities, waste from recreational and commercial fishing vessels, merchant ships, oil and gas platforms, and cargo spills from ships [23].

The variety and complexity of micro- and nanoplastic sources make it difficult to accurately determine the contribution of each to overall pollution. While larger plastics are widely regarded as the primary origin of secondary microplastics, establishing the precise pathways remains challenging. Current estimates are often based on the production or consumption of plastic products, but reliable quantitative data on specific sources remain limited [24].

11.2.2 Impacts on the Environment

Plastic degradation into its simplestmineralized form is an extremely slow process, often taking hundreds of years. This long lifespan makes microplastics and nanoplastics persistent pollutants that are difficult to eliminate and pose serious environmental challenges. Improperly managed plastic waste can block drains and waterways, leading to flooding and stagnant water, which often result in outbreaks of waterborne diseases [25]. In landfills, plastics release greenhouse gases like methane and carbon dioxide, exacerbating climate change, while harmful chemicals seep into the soil and groundwater, threatening ecosystems. Incineration introduces additional risks by releasing carbon dioxide, which accelerates global warming, and producing toxic by-products such as volatile organic compounds, dioxins, furans, heavy metals, and polycyclic aromatic hydrocarbons. These toxins pollute the air, soil, and water, causing significant harm to plants, aquatic life, soil organisms, and humans [26].

Compared to aquatic ecosystems, soil is a more complex medium, resulting in distinct differences in the transport, bioavailability, and impact of microplastics and nanoplastics on organisms. In soil ecosystems, the effects of plastic contamination are evident at both the organism level and within the soil microbiome [19]. A significant consequence of micro- and nanoplastics in soil is the increased mortality of earthworms, which negatively impacts soil porosity and overall soil health. The specific effects of micro- and nanoplastics on soil ecosystems depend on the type of plastic present. For example, polyester and polyacrylic particles reduce microbial diversity and activity, while fibrous polypropylene and membranous polyethylene appear to have the opposite effect. These interactions underline the complexity of plastic pollution in soil and its varied ecological consequences [21].

Plastic particles, including micro- and nanoplastics, are widespread in aquatic environments, causing harm to organisms when ingested. Smaller particles are more easily consumed, leach harmful chemicals, and can disrupt food webs by impairing the growth and reproduction of species like microalgae and zooplankton. High concentrations negatively affect the behavior and health of higher trophic-level species, while also creating habitats for microbial communities that can carry antibiotic-resistant genes, pathogens, and harmful metabolic pathways. This phenomenon is known as the plastisphere [27].

The effects of microplastics and nanoplastics on plants depend on factors like particle size, shape, material type, and plant species. Large microplastics may damage plant growth by blocking sunlight or harming roots, while smaller microplastics and nanoplastics can penetrate plant cells, causing oxidative stress and altering cellular structures [21]. Additionally, these particles can leach toxins due to weak bonds between plastics and additives, further harming plants. In animals, micro- and nanoplastics enter through air, water, and soil, accumulating in food chains and water supplies. These particles can damage digestive systems, disrupt metabolism, and accumulate in tissues, posing long-term risks, particularly for aquatic animals like fish, where they impair nutrient absorption, immune function, reproduction, and organ health [19].

Humans are exposed to micro- and nanoplastics through various pathways, including diet, drinking water, dermal contact, and inhalation [28]. The detection of micro- and nanoplastics in food crops, fruits, vegetables, animal products, and drinking water highlights dietary intake as a major route of exposure. Inhalation of airborne micro- and nanoplastics affects the respiratory system, with adverse effects on the lungs, potentially contributing to respiratory diseases, including lung cancer [18]. The toxicity of nanoplastics is influenced by their ability to accumulate within organisms and increase in concentration along the food chain. These processes determine the extent of nanoplastic transport within organisms and ecosystems [15]. Micro- and nanoplastics have been identified in human feces, blood, colon, placenta, and lungs, but their direct link to systemic diseases remains uncertain. Research using in vivo and in vitro models has demonstrated that NPs and MPs induce intestinal toxicity, interfere with metabolic processes, and cause reproductive, neurological, and immune-related toxicity [29]. These effects are driven by mechanisms such as oxidative stress, apoptosis, and the activation of specific molecular pathways [17].

11.2.3 Remediation Technologies

The widespread occurrence of nano- and microplastics in the environment necessitates effective remediation strategies, with various approaches, as outlined in Figure 11.1. Incineration is insufficient for eliminating synthetic plastic waste, leaving residual ash as a potential source of these pollutants. Sustainable upstream approaches, including enhanced wastewater treatment, bioplastics, and better waste management, are being investigated alongside downstream remediation methods, which encompass physical, chemical, and biological techniques across primary, secondary, and tertiary treatment stages [20]. Detailed comparisons of remediation methods for various plastics are summarized in Table 11.2.

A widely adopted physical method for the removal of nano- and microplastics involves coagulation, flocculation, and sedimentation, collectively known as CFS [20]. This process leverages the density differences between solids and liquids to facilitate phase separation, resulting in the sedimentation of particles. Initially, coagulant agents are added to the liquid phase to destabilize the plastic fragments. Following this, the solution undergoes flocculation, which is a slow mixing process that promotes the aggregation of particles to form clusters known as flocs, made up of plastic particles [47]. The formation of these flocs depends on the liquid's properties, including pH and ionic strength. In the sedimentation stage, the gravitational force causes the flocs to settle, with the efficiency of this step influenced by the density and shape of NPs and MPs. Irregularly shaped particles are particularly effective in forming dense flocs, enhancing their sedimentation. The effectiveness of the CFS

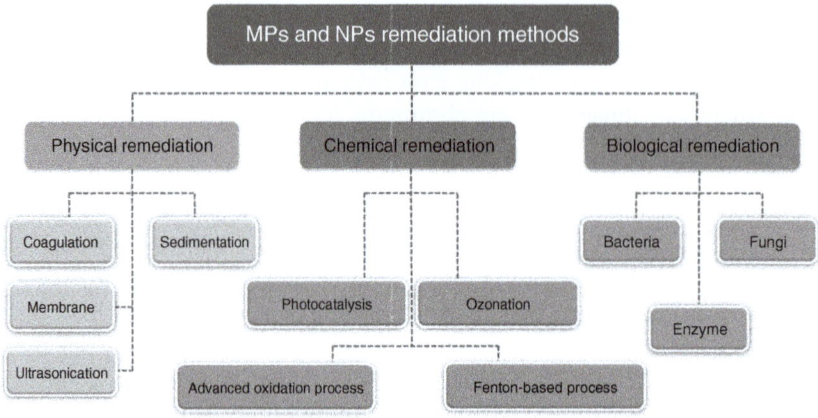

Figure 11.1 Schematic diagram of nano- and microplastics remediation methods.

Table 11.2 Types of plastics and utilized remediation methods.

Remediation type	Utilized technique or organism	Plastic material	References
Physical remediation	Sedimentation	PAM, PET, PS	[30, 31]
	Coagulation	PE, PAM, PS	[30, 32, 33]
	Flocculation	MPs	[34]
	Membrane	PE, MPs, fiber MPs	[32, 35, 36]
	Ultrasonication	PE	[37]
Chemical remediation	Photolysis	PET	[38]
	Photocatalysts	PE, PS	[39]
	Fenton-oxidation	PVC	[40]
	Ozonation	PS, PE, PP, PET	[41]
Bioremediation	Bacillus	Marine plastics, thermoset polymers, PU	[42, 43]
	Penicillium	PE, LDPE, HDPE, PET	[44, 45]
	Aspergillus	PE, LDPE, HDPE	[46]
	Ideonella sakaiensis	PET	[45]

process is primarily dictated by particle characteristics, such as shape, size, and surface features. To maximize removal efficiency, this method is often combined with other treatment techniques [31].

Membrane filtration methods play a critical role in the secondary and tertiary treatment of wastewater, effectively removing nano- and microplastics. Techniques such as ultrafiltration, nanofiltration, reverse osmosis, membrane bioreactors, and dynamic membranes vary in efficiency based on particle characteristics [32]. Reverse osmosis utilizes high pressure to filter out impurities, while ultrafiltration targets NPs and MPs at lower pressures. Nanofiltration serves as an intermediary, removing larger plastics through size and charge interactions. Membrane bioreactors demonstrate superior efficiency, achieving up to 99.4% removal, compared to 98.3% in traditional systems, and possess strong antifouling properties [34]. Dynamic membranes present a cost-effective, energy-efficient alternative by utilizing contaminants to create a filtration layer. Despite their advantages, challenges like inconsistent performance, fouling, and potential bypass of microplastics can hinder effectiveness, particularly with larger pore-sized materials employed in low-resistance, gravity-driven operations [36].

In wastewater treatment, chemical methods can be combined with other techniques to achieve complete removal of MPs from the environment [48]. Advanced oxidation processes (AOPs), initially proposed in the 1980s, are extensively studied for the chemical treatment of contaminants. These processes utilize light, heat, plasma, sonication, and catalysts to generate reactive oxygen species (ROS), commonly referred to as radicals, which play a key role in breaking down plastic particles. These radicals are typically formed through the reaction of catalysts with oxidizing agents like ozone (O_3) or hydrogen peroxide (H_2O_2) [49]. In this process, reactive radicals adhere to the plastic particles, breaking specific chemical bonds and creating intermediates. Further oxidation by the radicals then fragments the NPs and MPs into smaller particles or other by-products. The efficiency of chemical degradation depends on factors such as the type of plastic, operational conditions, and the specific techniques applied [50]. AOPs provide several benefits, including the removal of organic matter, enhancement of biodegradability, and the complete conversion of hazardous pollutants. While AOPs can be costly, they can serve as an effective pretreatment for biodegradation, focusing on surface degradation of NPs and MPs in an environmentally friendly manner. Different types of AOPs, such as UV/photocatalytic reactions, ozone-based processes, and Fenton-based methods, offer diverse approaches to tackling plastic pollution [9, 51].

Photocatalysis, an advanced oxidation process, has garnered significant attention for its capability to degrade micro- and nanoplastics effectively [52]. This method employs semiconductor materials, primarily metal oxides such as titanium dioxide (TiO_2), as photocatalysts to produce reactive oxygen species like hydroxyl and superoxide radicals under UV or visible light irradiation. These ROS break down NPs and MPs by rupturing polymer chains, branching, crosslinking, or even mineralizing them into CO_2 and H_2O [53]. As a mature green technology that utilizes abundant and renewable solar energy, photocatalysis presents a promising, eco-friendly, and cost-effective solution for plastic degradation. Visible light-induced photocatalysis, which harnesses solar energy, is particularly appealing for its environmental benefits. Semiconductors like TiO_2 and zinc oxide (ZnO) have been extensively explored to develop visible light-active photocatalysts. However, photocatalysis has notable limitations. Its reliance on light means that only surface plastics are effectively treated, leaving particles buried in sediments unaffected [54, 55]. Additionally, catalyst deactivation over time reduces long-term efficiency, and the process can generate toxic by-products depending on the type of photocatalyst and plastic. For instance, the degradation of polyethylene using polypyrrole/TiO_2 photocatalysts has been shown to produce harmful carbonyl compounds. These challenges highlight the need for thorough evaluation and optimization before adopting photocatalysis in large-scale treatment systems [8].

Fenton-based processes are widely employed due to their cost-effectiveness, as they operate efficiently under room temperature and atmospheric pressure conditions [39]. These methods rely on electron transfer reactions between hydrogen peroxide, potassium mono-persulfate, or potassium persulfate, and ferrous ions (Fe^{2+}), producing highly reactive hydroxyl radicals (•OH). These radicals are highly effective in breaking down persistent organic pollutants in water. In a Fenton-like reaction, ferrous ions (Fe^{2+}) are replaced with ferric ions (Fe^{3+}), enabling similar oxidative processes [56]. The photo-Fenton reaction enhances this mechanism by using UV light to accelerate the reduction of Fe^{3+} back to Fe^{2+}, increasing the efficiency of the system. Alternatively, the electro-Fenton process utilizes electrochemical methods to generate one or both of the Fenton reagents, further expanding the versatility of this approach. These variations make Fenton-based processes highly adaptable for water treatment applications [57].

Bioremediation presents a promising strategy for degrading microplastics and nanoparticles due to its sustainability, cost-effectiveness, and low energy requirements [44]. This process involves bacterial, fungal, and enzymatic degradation techniques, often combined with chemical methods to improve efficacy. It occurs in stages, beginning with biodeterioration, where biological agents alter the polymers, followed by bio-fragmentation, where enzymes transform them into simpler molecules. Microorganisms then assimilate these fragments, mineralizing them into products like CO_2 and H_2O [43]. Microbial plastic degradation employs two main mechanisms: intracellular, where microbes break down polymers on plastic surfaces, and extracellular, where hydrolase enzymes are secreted to dismantle complex structures. Despite great potential, challenges remain, especially with high-molecular-weight plastics [42]. Fungi and microalgae significantly contribute to this process through enzymatic activity and exopolysaccharide production, aiding in the colonization and breakdown of plastics. Enhancements through UV radiation and heat pretreatment have been effective for certain plastics, while key enzymes like *cutinases* and *lipases* facilitate the breakdown of plastic fragments. Bioremediation, integrating these biological processes, offers innovative solutions to address environmental plastic pollution sustainably [45, 46].

11.3 Nano- and Microplastics Degradation

11.3.1 Defective Photocatalysts

The sustainable conversion of solar energy into chemical energy presents a promising solution to environmental challenges and the global energy crisis. Among various photochemical reactions, photocatalysis stands out due to its wide-ranging applications, including organic synthesis and the treatment of water and air [58]. However, the broader adoption of photocatalytic reactions is hindered by challenges such as inefficient charge separation and limited charge transport. Given that photocatalysis involves complex processes under nonequilibrium conditions, a thorough understanding of the dynamics of photoexcited carriers is essential. Single photocatalytic materials like TiO_2, ZnO, CeO_2, and ternary compounds such as $BiVO_4$ have garnered significant attention for their high activity [59]. Nonetheless, these materials face inherent limitations, including poor absorption in the visible light spectrum, rapid recombination of photoexcited charge carriers, and sluggish surface reaction kinetics. To address these issues, optimizing the optoelectronic and interfacial properties of photocatalysts has become a major focus of research [60].

Factors such as facets, phases, and morphologies play a critical role in determining photocatalytic performance, as demonstrated by numerous studies [11]. Additionally, the overall efficiency of photocatalytic processes is influenced by various parameters, including the choice of precursor materials, synthesis methods, heat treatment conditions, and the type and concentration of dopants. This multifaceted nature of photocatalysis makes the optimization of photocatalysts a complex yet critical endeavor [61]. Defect engineering involves intentionally introducing defects into a material to improve its performance, particularly in photocatalytic applications. Defect sites play a pivotal role in the architecture of photocatalysts by altering surface reactivity and physicochemical properties, thereby enhancing catalytic activity [13]. These defects, such as vacancies or dislocations, create highly reactive unsaturated coordination sites that act as active centers for chemical reactions. For instance, vacancy defects on metal oxide surfaces serve as effective adsorption sites, facilitating the interaction of reactant molecules with the catalyst [12].

The catalytic effectiveness of defects primarily stems from their influence on the electronic structure of the material. Surface defects, such as oxygen vacancies, can reduce the bandgap of wide-bandgap semiconductors like TiO_2 and ZnO, improving light absorption and charge carrier separation [9]. These defects trap excited electrons, minimizing recombination and boosting redox reaction rates, ultimately enhancing photocatalytic activity. Conversely, bulk defects often have a detrimental effect on performance, as they increase electron–hole recombination, thereby reducing the efficiency of the photocatalytic process. Understanding and strategically controlling defect types and distributions is critical for optimizing photocatalyst design and performance [62].

Defect density, carrier concentration, and interfacial properties are key factors influencing the efficiency of photocatalytic processes, even for identical reactions and catalysts. Among these, defect concentration plays a central role, as it directly impacts all three parameters [63]. The relationship between defect density and catalytic performance is complex. While surface defects enhance reactivity by facilitating the adsorption of reactants like O_2 and H_2O, excessive defects, especially in the bulk, can reduce charge carrier lifetimes and increase recombination rates [64]. The balance between bulk and surface defects is critical for optimizing photocatalytic performance. High surface defect densities generally boost catalytic activity by promoting photocatalytic reactions, whereas bulk defects, which do not contribute to surface reactivity, can hinder charge transport. Photocatalysts with lower bulk-to-surface defect ratios have been shown to generate higher photocurrents and exhibit superior photocatalytic efficiency. Additionally, defects influence the material's electronic states, creating localized states within the bandgap [65]. These localized states enhance charge transfer and catalytic reactivity by enabling more efficient movement of charge carriers. Dislocations and grain boundaries can further modify electronic properties, creating regions with enhanced catalytic activity [66].

Theoretical studies using density functional theory provide valuable insights into the role of defects. For example, oxygen vacancies can act as shallow donors, while metal vacancies behave as shallow acceptors, influencing the material's conductivity [11, 67]. However, at high defect concentrations, defect aggregation can introduce disorder, negatively impacting overall performance. Achieving an optimal balance in defect density and distribution is therefore crucial for maximizing the efficiency of photocatalytic materials. Surface defects with dipole moments play a significant role in improving charge separation and enhancing selectivity in catalytic processes [62]. Recent advancements in defect-rich semiconductor photocatalysts have shown great promise, demonstrating impressive catalytic performance. However, critical aspects such as local surface microstructure, band structure, carrier transport, carrier concentration, and interface properties often remain insufficiently characterized, limiting a comprehensive understanding of these materials [65].

While defects can boost catalytic activity, excessive defect concentrations can lead to adverse effects, including poor charge transfer and reduced charge carrier lifetimes. Therefore, the rational design of photocatalysts requires careful control of defect densities and their spatial distribution to balance their beneficial and detrimental effects [68]. To achieve this, advanced modeling and characterization techniques are crucial. These tools provide insights into the intricate relationship between defects and photocatalytic efficiency, enabling the development of optimized photocatalysts. Such efforts are particularly valuable for enhancing visible-light photocatalysis, paving the way for more efficient and targeted applications [69].

11.3.2 Classification of Defects

Defects in photocatalytic materials exert varied effects on their activity, making it essential to classify and distinguish them. These defects can be categorized based on their atomic structures or their locations within the material. Structural irregularities are typically divided into four types: point defects, line defects, planar defects, and volume defects, as illustrated in Figure 11.2. A detailed understanding of these defects is critical for bridging the gap between theoretical models and real-world structures. This knowledge facilitates the development of structure-performance and activity-defect relationships, enabling the design of photocatalysts with optimized performance tailored to specific applications [11, 60].

Structural distortions are often responsible for the formation of cation and anion vacancies, which have been extensively studied in photocatalytic materials such as TiO_2 [70], BiO_2 [71], WO_3 [72], ZnO [73], and CeO_2 [74] due to their inherent photocatalytic activity and stability. Bai et al. provided statistical insights into various defects in photocatalysts, highlighting that significant lattice disorder is

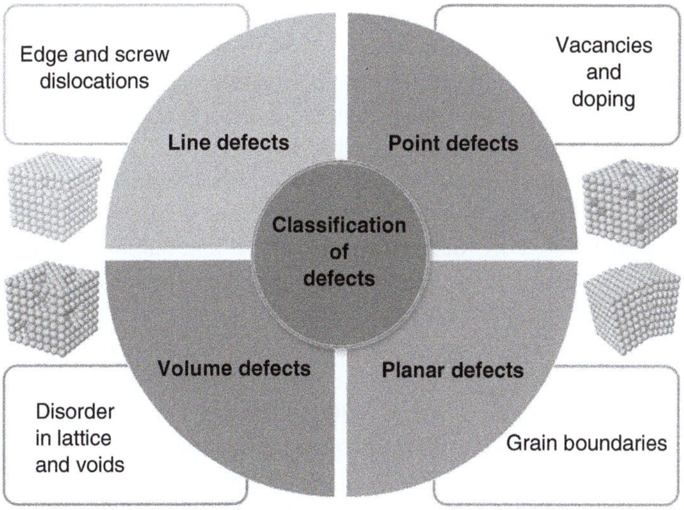

Figure 11.2 Classification of different types of defects with their examples.

frequently accompanied by point defects, particularly oxygen vacancies. Among these, point defects are of particular importance in semiconductor photocatalysts and are classified into cation and anion vacancies [11].

Cation vacancies influence the electronic structure of photocatalysts by introducing acceptor states within the bandgap, contributing to p-type conductivity [75]. During synthesis, processes like the high-temperature decomposition of interlayer organic compounds can lead to oxygen-rich structures, which in turn create cation vacancies through lattice distortion [76]. Commonly observed cation vacancies include those of titanium, cobalt, zinc, indium, nickel, and bismuth [77, 78]. Similarly, anion vacancies, such as sulfur, nitrogen, and iodine vacancies, have been explored in semiconductor photocatalysts based on ionic compounds. Examples include sulfur vacancies in ZnS [79], Zn_xCd_yS [80], and $ZnIn_2S_4$ [81], nitrogen vacancies in Ta_3N_5 [82], and iodine vacancies in $Bi_7O_9I_3$ [83]. The simultaneous presence of cation and anion vacancies can form a p–n homojunction, generating a built-in electric field that enhances charge carrier separation [61].

Bai et al. demonstrated these effects using TiO_2 as a prototype. Their study showed that differences in band bending at the junction interface vary with particle size [84]. The homo-phase junction formed by the coexistence of cation and anion vacancies significantly improves spatial charge separation and reduces recombination [85]. This approach often results in higher photocatalytic yields compared to heterojunction-based strategies, offering a promising avenue for enhancing photocatalytic performance [86]. Foreign element doping introduces another critical type of point defect in semiconductor photocatalysts by incorporating impurities that modify the material's composition. These dopant atoms or ions can either replace existing atoms/ions in the host lattice (substitutional doping) or occupy interstitial positions within the lattice. Substitutional dopants generally have atomic/ionic radii comparable to the atoms they replace, while interstitial dopants require significantly smaller radii [87]. This impurity doping has been widely applied to nearly all photocatalytic semiconductors. A notable example is nitrogen-doped TiO_2, where nitrogen can occupy either interstitial positions or substitute oxygen atoms. Similarly, anion doping in TiO_2 with elements such as carbon, boron, sulfur, phosphorus, and fluorine has been extensively studied [88].

Line defects, another category of defects in photocatalytic semiconductors, primarily involve dislocations, areas where the crystal lattice is misaligned. These dislocations are classified into edge dislocations, where a half-plane of atoms terminates within the crystal (observed in TiO_2 [89] and ZnO [90]), and screw dislocations, which guide the growth of semiconductors like Cu_2O [91], CdS [92], and Ag_3PO_4 [93]. Screw dislocations are characterized by defect lines that move perpendicularly to the direction of applied stress and atomic displacement. Unlike point defects, line defects do not alter the composition of photocatalytic materials. However, they play a significant role in influencing the crystal growth, mechanical properties, and surface reactivity of these materials, thereby affecting their photocatalytic performance [11]. Planar defects, like line defects, do not significantly alter the stoichiometric composition of semiconductor photocatalysts but can influence their structural and functional properties [61]. Among planar defects, grain boundaries are the most common and occur at the interface between two contacting grains in polycrystalline photocatalytic materials. These types of defects have been surveyed in photocatalytic materials like TiO_2 [94], ZnO [95], and CdS [96]. Another notable planar defect is the twin boundary, which introduces a plane of mirror symmetry within twinned crystals. Twin boundaries have been reported in photocatalytic materials like CdS [97], $Cd_{1-x}Zn_xS$ [98], and $AgBr_{1-x}I_x$ [99].

Volume defects, such as voids and lattice disorders, are also prominent in photocatalytic materials. Voids, which are small regions devoid of atoms, are essentially clusters of vacancies within the bulk phase of a material [11]. They have been identified in photocatalysts, including C_3N_4 [100], ZnO [101], TiO_2 [54], $ZnSnO_3$ and Zn_2SnO_4 [102], and $BiVO_4$. In contrast, lattice disorders involve regions where the periodic atomic arrangement is disrupted, forming disordered lattices. Disorder engineering has been widely applied, particularly on the surface of black TiO_2, and has also been introduced in other materials like Fe_2O_3 [103], In_2O_3 [104], ZnO [103], and ZrO_2 [103]. Disorder engineering is often associated with a high density of point defects. For instance, the hydrogenated black TiO_2 showcases a lattice-disordered shell, marked by a blend of oxygen vacancies and hydrogen dopants [105]. These defects and disorders play a critical role in modifying the electronic structure and catalytic performance of semiconductor photocatalysts.

11.3.3 Synthesis of Defective Photocatalysts

Defect engineering in photocatalytic feedstock is crucial to generate effective and selective photocatalysts. Establishing a clear connection between specific defects and the photocatalytic performance requires obtaining materials with precisely defined defects. The techniques used to create defects vary, depending on the types of defects desired. To date, numerous synthetic techniques have emerged for fabricating defect-laden photocatalytic materials, which can generally be divided into two primary approaches: one involves adding defects to preexisting photocatalytic substances, while the other focuses on inducing defects during the material's synthesis phase [11].

The technique of adding defects to already established foundational nanocrystals is a popular method for embedding lattice irregularities, voids, dopants, and vacancies within the surfaces or interiors of photocatalytic materials. This strategy enables the introduction of defects without relying on the development dynamics of the nanocrystals themselves. Commonly, cation vacancies in semiconductor lattices are created by reducing the foundational nanocrystals, for example, by treating pristine TiO_2 [106], ZnO [107], WO_3 [108], Bi_2WO_6 [109], and NaH [106]. Oxygen atoms can be eliminated from the lattice, leading to the creation of oxygen vacancies. During this process, the lattice oxygen may react with hydrogen from the reductants, facilitating the formation of these vacancies; however, most chemical reduction techniques primarily act on the surface of the photocatalysts [11]. For instance, Zhao et al. [110] demonstrated the introduction of oxygen vacancies into rutile TiO_2 nanorods by treating them with solid $NaBH_4$ at 300 °C, resulting in surface-localized vacancies. To achieve a uniform distribution throughout the nanorods, a precursor with bulk vacancies was first synthesized, followed by a similar surface reduction treatment. Additionally, photoreduction methods also have the ability to generate oxygen vacancies in specific semiconductor materials, like BiOCl [53]. Alternative anion vacancies can be generated through reduction processes, as demonstrated by the rich in nitrogen vacancy melon photocatalytic materials, which were produced by heating up melon in a hydrogen environment to convert lattice nitrogen into ammonia [111]. Additionally, lattice disorder can be introduced in preformed photocatalytic nanocrystals via chemical reduction processes, such as producing black TiO_2 by reducing it in a hydrogen atmosphere or utilizing hydrogen plasma at elevated temperatures [112].

Voids in semiconductor feedstocks can form via vacancy condensation [89]. For instance, Scheu et al. demonstrated that voids develop in rutile TiO_2 nanowires following annealing at 500 °C in air, resulting from the aggregation of oxygen vacancies that yield voids with an inverse Wulff shape [113]. Additionally, foreign atoms or ions can be incorporated into the lattice of semiconductor materials via diffusion doping, where dopants penetrate the crystal lattice of pre-synthesized host nanocrystals [86]. This diffusion process can occur at elevated temperatures using gaseous, liquid, or solid precursors. As an illustration, nitrogen can be integrated inside the lattice structure of TiO_2 [114], CeO_2 [115], and ZnS [116] through two main methods. One involves directly heating the materials in an ammonia-rich environment, while the other consists of combining the undoped substances with solid urea, hydrazine, and ammonium hydroxide, accompanied by a high-temperature processing step [11].

Ion exchange represents an effective approach for vicarious doping of foreign ions into preexisting ionic compound photocatalytic materials. This process typically involves the immersion of ion-exchangeable semiconductor materials in a solution containing the dopant ions at a specified temperature [86]. Layered oxides and hydroxides are particularly amenable to ion exchange, as their interlayer cations or anions can be readily substituted with foreign ions. For example, Sn^{2+} can be introduced into layered titanates and niobates, such as $K_2Ti_4O_9$, $K_4Nb_6O_{17}$, and $CsTi_2NbO_7$, by stirring these materials in a mixed solution of $SnCl_2$ and HCl, leading to the exchange of some alkaline metal ions with Sn^{2+} [117]. Since foreign ions can readily replace the interlayer cations or anions, multilayer oxide and hydroxide materials have been extensively investigated for ion exchange. By agitating the materials in a mixture of $SnCl_2$ and HCl, for example, Sn^{2+} can be doped into layered titanates and niobates, such as $K_2Ti_4O_9$, $K_4Nb_6O_{17}$, and $CsTi_2NbO_7$, where some of the alkaline metal ions were swapped for Sn^{2+} [118].

Defect engineering in photocatalytic feedstock is more commonly accomplished by creating defects during formation as opposed to introducing faults into preformed host nanocrystals. Because preformed nanocrystals have a limited reactive depth, posttreatment often only introduces surface flaws in pre-synthesized photocatalytic materials. In contrast, it is easier to create bulk flaws while photocatalytic materials are being formed. During the creation of photocatalytic materials, a variety of defects, including dislocations and boundaries that heavily rely on the growth kinetics of nanocrystals, can arise in addition to point defects [11].

To achieve defect formation in nanocrystals, precise control of their growth kinetics is essential. For example, ZnO nanostructures with abundant grain boundaries were synthesized by calcining electrospun polyvinyl alcohol zinc acetate composite nanofibers. The grain boundaries emerged from the merging of adjacent grains during growth, with the calcination temperature influencing the growth rate of ZnO grains and resulting in structures with varying grain boundary areas [12]. In another instance, pit-structured ZnO nanorods were produced by heating zinc nitrate hexahydrate and polyethylene glycol in an aqueous ammonia solution, followed by high-temperature calcination. Initially, $Zn(NH_3)_4{}^{2+}$ precursors formed and partially decomposed to create ZnO nuclei, while some precursors were incorporated into ZnO crystals, forming the $ZnO(NH_3)_n$ complex. Upon calcination at 300 °C, these units decomposed, releasing NH_3 and leading to the development of numerous surface pits on the ZnO nanorods [101].

11.3.4 Characterization

Characterizing defects is crucial for assessing the impact of defect engineering in photocatalysis. To effectively guide defect engineering within this field, it is essential to understand the atomic structures of defects and quantify their concentrations in photocatalytic materials, thereby establishing a structure–activity relationship [119, 120]. The methods used for defect characterization vary according to the type of defect, as shown in Figure 11.3. Microscopic techniques are the primary and most prevalent means for analyzing defects [121]. Various microscopes exist for observing different defect types. For example, transmission electron microscopy (TEM) [113] enables the observation of both bulk and surface defects due to the electron beam's high penetration capacity. However, distinguishing between

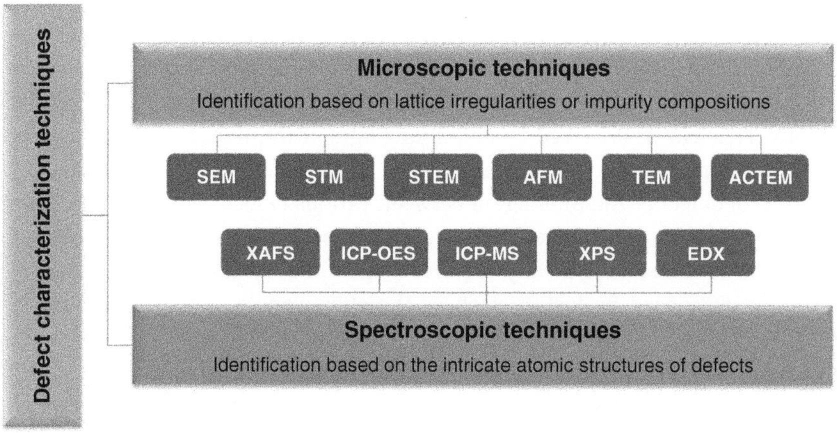

Figure 11.3 Various characterization methods utilized for defect identification in photocatalytic materials.

surface and bulk defects in solid nanocrystals can be challenging with TEM [11]. To focus on surface defects, scanning electron microscopy (SEM) and atomic force microscopy (AFM) are effective, particularly for two-dimensional materials [122, 123]. Alternatively, scanning transmission electron microscopy (STEM) has the capability to distinguish between defects found on the surface and those existing within the bulk of materials. For identifying point defects, scanning tunneling microscopy (STM) [124] and atomic-resolution spherical aberration-corrected transmission electron microscopy (ACTEM) [125] are vital. Higher-dimensional line defects are more easily detected using high-resolution transmission electron microscopy (HRTEM) by analyzing lattice fringes [126].

While microscopic techniques can identify defects based on lattice irregularities or impurity compositions, spectroscopic methods are essential for elucidating the intricate atomic structures of defects, particularly point defects, in photocatalytic materials [127, 128]. One of the key techniques is synchrotron radiation-based X-ray absorption fine structure (XAFS) spectroscopy, which is vital for investigating local atomic arrangements around defects [70]. Spectroscopic methods facilitate both qualitative and quantitative analyses of defects present in photocatalytic materials. Quantifying impurity defects, such as foreign dopants, is relatively straightforward [11, 129]. For instance, the concentration of dopants in bulk materials can be determined through techniques like inductively coupled plasma optical emission spectrometry (ICP-OES), inductively coupled plasma mass spectrometry (ICP-MS), and energy dispersive X-ray spectroscopy (EDX) [130, 131]. In contrast, the surface dopant content can be analyzed using X-ray photoelectron spectroscopy (XPS) [132]. The quantification of vacancies is linked to self-doping ions; for example, the concentration of oxygen vacancies in CeO_2 can be inferred from the measured Ce^{3+} content, based on the relationship that two Ce^{3+} ions correspond to each oxygen vacancy [11, 133].

11.3.5 Degradation of Nano- and Microplastics

The ability to create and characterize defects in a controlled manner allows for defect engineering to be harnessed to optimize the characteristics of photocatalytic materials. This can be achieved by amplifying light absorption, facilitating charge convey and separation, or adjusting surface reactions. In certain instances, a single type of defect or the interaction between various defects can simultaneously fine-tune multiple stages of the photocatalysis process [11]. The total breakdown of these plastic materials typically requires a span of several days to weeks. This duration is influenced by factors such as the effectiveness of the photocatalyst, the conditions under which the reaction occurs, and the specific type of plastic involved. This section explores recent advancements in the application of defect engineering in the photocatalytic degradation of plastic wastes, with a focus on specific examples to illustrate the effectiveness of various approaches [52].

Zhang et al. synthesized titanium dioxide with oxygen vacancies via a hydrothermal method to degrade polystyrene microplastics under UV light. The photocatalyst exhibited a removal efficiency of 98% after 12 hours of irradiation. The enhanced performance was attributed to the increased charge carrier separation facilitated by the oxygen vacancies [38]. Another example involves the work of Pumera and colleagues, who developed innovative autonomous microrobots for the photocatalytic breakdown of microplastics within intricate environments. Utilizing sunlight and magnetic fields for power, these robots are equipped with photocatalytic ($BiVO_4$) and magnetic (Fe_3O_4) components, allowing for autonomous movement and exact navigation. They proficiently decompose a variety of synthetic microplastics, including polylactic acid and polycaprolactone, by leveraging a localized self-stirring mechanism that improves interactions without relying on external mechanical devices [134].

In a recent study conducted by Qin et al., a method involving partial calcination was employed to create a heterojunction of ZnO/UiO-66-NH_2, in which ZnO particles are encased within the porous framework of UiO-66-NH_2 for the photocatalytic transformation of plastics. The synergistic interaction within the ZnO/UiO-66-NH_2 heterojunction fosters efficient pathways for mass transfer and carrier mobility while ensuring structural integrity. Consequently, this heterojunction demonstrates an impressive performance in converting polylactic acid and polyvinyl chloride into acetic acid, alongside hydrogen production [135].

Tofa et al. developed zinc oxide nanorods enhanced with the noble metal platinum as a cocatalyst and evaluated their effectiveness in photocatalytic degradation of low-density polyethylene. The Fourier transform infrared analysis indicated the presence of various oxygen-containing functional groups, including carbonyls and peroxide linkages. Evaluations based on specific indices demonstrated that the incorporation of platinum onto zinc oxide facilitated the separation of photogenerated charge carriers and broadened the spectrum of light absorption, which in turn significantly improved the photocatalytic activity for breaking down the plastic [136]. Further instances of defect-engineered photocatalysts and their applications in degrading NPs and MPs are summarized in Table 11.3.

Table 11.3 Degradation of plastics using defect-engineered photocatalysts.

Photocatalyst	Defect type	Pollutant	Removal efficiency (%)	References
$BiVO_4/Fe_3O_4$	Heterostructure construction	Polylactic acid	76	[9]
MXene/$Zn_xCd_{1-x}S$	Cocatalyst loading	Polyethylene terephthalate	—	[137]
Pt@N-TiO_2	Noble metal deposition	Polyethylene terephthalate	29	[138]
N-TiO_2	Nitrogen doping	Low-density polyethylene	5	[139]
Ag/TiO_2/rGO	Silver doping	Microplastic	76	[140]
$CoNi_{0.25}P$	Nickel modification	Polyethylene terephthalate	97	[141]
CuPc–ZnO	Copper phthalocyanine modified	Polystyrene	12	[142]
Au@Ni@TiO_2	Cocatalyst loading	Microplastics	67	[143]
Ag-TiO_2	Noble metal deposition	Polyethylene	18	[144]
C,N-TiO_2	C,N-doping	High-density polyethylene	71	[145]

11.4 Future Prospects

Defect engineering in photocatalytic materials offers a promising strategy for significantly enhancing the performance of wastewater treatment applications, specifically micro- and nanoplastic degradation. Defects contribute to photocatalytic efficiency by improving light absorption, facilitating charge transfer and separation, and fine-tuning surface reactions. The specific impacts of defects are influenced by their type, location, and concentration, as revealed by advanced characterization methods. By understanding these factors, researchers can strategically engineer photocatalytic systems to meet diverse needs, often improving the overall efficiency through the cooperative action of various defects at different stages of the photocatalytic process. Despite promising advancements, defect engineering in photocatalysts remains in its early stages and faces several challenges. One key issue is the complex task of designing defect structures that enhance rather than detract from photocatalytic performance, as one type of defect may benefit one reaction step while hindering another. Additionally, the complexity increases in multicomponent hybrid photocatalysts, where interactions between defects across different components can lead to unforeseen difficulties. Current methods for creating defects primarily address mono-component systems and struggle to maintain structural integrity across multiple components. In addition, the current characterization techniques have not evolved alongside advancements in synthesis, often failing to provide detailed, high-resolution information about defect structures and their functions in photocatalytic reactions. Overcoming these hurdles through improved methods and in-depth observation techniques could unveil new mechanisms and insights into defect behavior, paving the way for innovative designs in photocatalysis for micro- and nanoplastic degradation.

11.5 Conclusion

In summary, the increasing plastic waste crisis, particularly the prevalence of micro- and nanoplastics, poses significant ecological threats. These particles negatively impact various ecosystems, including marine and agricultural systems, and also human bodies. Recent research highlights the potential of defect-engineered photocatalytic materials for degrading micro- and nanoplastics, owing to their enhanced light absorption, charge separation, and interfacial reaction properties. However, the specific effects of defects on photocatalytic efficiency are not yet fully understood and depend on their type, location, and concentration. A deeper understanding of these factors will enable researchers to optimize defect-engineered photocatalytic systems for more effective plastic degradation, ultimately contributing to strategies for mitigating plastic pollution.

Conflict of Interest

The authors affirm that they have no conflict of interest.

List of Abbreviations

ACTEM	Aberration-Corrected Transmission Electron Microscopy
AFM	Atomic Force Microscopy
AOPs	Advanced Oxidation Processes
CFS	Coagulation, Flocculation, and Sedimentation
EDX	Energy Dispersive X-ray Spectroscopy
HDPE	High-Density Polyethylene
HRTEM	High-Resolution Transmission Electron Microscopy
ICP-MS	Inductively Coupled Plasma Mass Spectrometry
ICP-OES	Inductively Coupled Plasma Optical Emission Spectrometry
LDPE	Low-Density Polyethylene
MPs	Microplastics
NPs	Nanoplastics
PAM	Polyacrylamide
PE	Polyethylene

PET	Polyethylene Terephthalate
PP	Polypropylene
PS	Polystyrene
PS	Polysulfone
PU	Polyurethane
ROS	Reactive Oxygen Species
SEM	Scanning Electron Microscopy
STEM	Scanning Transmission Electron Microscopy
TEM	Transmission Electron Microscopy
UV	Ultraviolet Radiation
XAFS	X-ray Absorption Fine Structure
XPS	X-ray Photoelectron Spectroscopy

References

1 Jalilnejad, E., Alizadeh, M., and Jabbari, B. (2024). Ion-exchange membranes in microbial fuel cell systems. In: *Current Trends and Future Developments on (Bio-) Membranes: Recent Achievements for Ion-Exchange Membranes* (ed. A. Basile and K. Ghasemzadeh), 229–263. Elsevier.

2 Winiarska, E., Jutel, M., and Zemelka-Wiacek, M. (2024). The potential impact of nano- and microplastics on human health: understanding human health risks. *Environ. Res.* 251: 118535.

3 Ramsperger, A.F.R.M., Bergamaschi, E., Panizzolo, M. et al. (2023). Nano- and microplastics: a comprehensive review on their exposure routes, translocation, and fate in humans. *NanoImpact* 29: 100441.

4 Huang, Z., Weng, Y., Shen, Q. et al. (2021). Microplastic: a potential threat to human and animal health by interfering with the intestinal barrier function and changing the intestinal microenvironment. *Sci. Total Environ.* 785: 147365.

5 Ebrahimbabaie, P., Yousefi, K., and Pichtel, J. (2022). Photocatalytic and biological technologies for elimination of microplastics in water: current status. *Sci. Total Environ.* 806: 150603.

6 Paiman, S.H., Md Noor, S.F., Ngadi, N. et al. (2023). Insight into photocatalysis technology as a promising approach to tackle microplastics pollution through degradation and upcycling. *Chem. Eng. J.* 467: 143534.

7 Jeyavani, J., Al-Ghanim, K.A., Govindarajan, M. et al. (2024). A convenient strategy for mitigating microplastics in wastewater treatment using natural light and ZnO nanoparticles as photocatalysts: a mechanistic study. *J. Contam. Hydrol.* 267: 104436.

8 He, Y., Rehman, A.U., Xu, M. et al. (2023). Photocatalytic degradation of different types of microplastics by TiO_x/ZnO tetrapod photocatalysts. *Heliyon* 9 (11): e22562.

9 Zhou, G., Xu, H., Song, H. et al. (2024). Photocatalysis toward microplastics conversion: a critical review. *ACS Catal.* 14 (11): 8694–8719.

10 Wang, Z., Xiao, M., You, J. et al. (2022). Defect engineering in photocatalysts and photoelectrodes: from small to big. *Acc. Mater. Res.* 3 (11): 1127–1136.

11 Bai, S., Zhang, N., Gao, C., and Xiong, Y. (2018). Defect engineering in photocatalytic materials. *Nano Energy* 53: 296–336.

12 Dong, B., Liu, T., Li, C., and Zhang, F. (2018). Species, engineering and characterizations of defects in TiO_2-based photocatalyst. *Chin. Chem. Lett.* 29 (5): 671–680.

13 Xiong, J., Di, J., Xia, J. et al. (2018). Surface defect engineering in 2D nanomaterials for photocatalysis. *Adv. Funct. Mater.* 28 (39): 1801983.

14 Liu, J., Zhao, D., Wu, X. et al. (2025). Synergistic dual-defect band engineering for highly-efficient photocatalytic degradation of microplastics via Nb-induced oxygen vacancies in SnO_2 quantum dots. *J. Mater. Chem. A Mater.* 13: 4429 4443.

15 Morreale, M. and La Mantia, F.P. (2024). Current concerns about microplastics and nanoplastics: a brief overview. *Polym. (Basel)* 16 (11): 1525.

16 Zhou, J., Chen, M., Li, Y. et al. (2024). Microbial bioremediation techniques of microplastics and nanoplastics in the marine environment. *TrAC Trends Anal. Chem.* 180: 117971.

17 Gao, B., Chen, L., Wu, L. et al. (2025). Association between microplastics and the functionalities of human gut microbiome. *Ecotoxicol. Environ. Saf.* 290: 117497.

18 Arif, Y., Mir, A.R., Zieliński, P. et al. (2024). Microplastics and nanoplastics: source, behavior, remediation, and multi-level environmental impact. *J. Environ. Manag.* 356: 120618.

19 Dube, E. and Okuthe, G.E. (2023). Plastics and micro/nano-plastics (MNPs) in the environment: occurrence, impact, and toxicity. *Int. J. Environ. Res. Public Health* 20 (17): 6667.

20 Ghasemi, S., Parastesh, A., Ramezani, H. et al. (2024). Microplastics controlling approaches: removal reaction pathways, kinetics, and toxicity evaluation of the products. *ACS Symp. Ser.* 1482: 213–237.

21 Jaikumar, I.M., Tomson, M., Meyyazhagan, A. et al. (2025). A comprehensive review of microplastic pollution in freshwater and marine environments. *GAC* 12: 100202.

22 Chen, Q., Wang, P., Liu, R. et al. (2025). Experimental inconsistencies undermine accurate characterization of microplastics and identification of environmental drivers: a metadata analysis in Chinese aquatic environment. *Ecol. Indic.* 170: 113053.

23 Lu, N., Su, Q., Li, Y. et al. (2025). Characterization of microplastic distribution, sources and potential ecological risk assessment of domestic sewage from ships. *Environ. Res.* 268: 120755.

24 Feisal, N.A.S., Tengku Ibrahim, T.N.B., Ahmad, M.A. et al. (2025). The potential of using microalgae for microplastic degradation in aquatic ecosystem. *Algal Res.* 85: 103825.

25 Zhang, S., Wang, J., Yan, P. et al. (2021). Non-biodegradable microplastics in soils: a brief review and challenge. *J. Hazard. Mater.* 409: 124525.

26 Zhang, S., Wang, J., Liu, X. et al. (2019). Microplastics in the environment: a review of analytical methods, distribution, and biological effects. *TrAC Trends Anal. Chem.* 111: 62–72.

27 Chang, X., Fang, Y., Wang, Y. et al. (2022). Microplastic pollution in soils, plants, and animals: a review of distributions, effects and potential mechanisms. *Sci. Total Environ.* 850: 157857.

28 Ya, H., Jiang, B., Xing, Y. et al. (2021). Recent advances on ecological effects of microplastics on soil environment. *Sci. Total Environ.* 798: 149338.

29 Li, Z., Yang, Y., Chen, X. et al. (2023). A discussion of microplastics in soil and risks for ecosystems and food chains. *Chemosphere* 313: 137637.

30 Lu, S., Liu, L., Yang, Q. et al. (2021). Removal characteristics and mechanism of microplastics and tetracycline composite pollutants by coagulation process. *Sci. Total Environ.* 786: 147508.

31 Sembiring, E., Fajar, M., and Handajani, M. (2021). Performance of rapid sand filter – single media to remove microplastics. *Water Supply* 21 (5): 2273–2284.

32 Jiang, S., Li, Y., and Ladewig, B.P. (2017). A review of reverse osmosis membrane fouling and control strategies. *Sci. Total Environ.* 595: 567–583.

33 Ma, B., Xue, W., Hu, C. et al. (2019). Characteristics of microplastic removal via coagulation and ultrafiltration during drinking water treatment. *Chem. Eng. J.* 359: 159–167.

34 Wilken, R., Holländer, A., and Behnisch, J. (2002). Vacuum ultraviolet photolysis of polyethylene, polypropylene, and polystyrene. *Plasma Polym.* 7 (2): 185–205.

35 Wang, Z., Lin, T., and Chen, W. (2020). Occurrence and removal of microplastics in an advanced drinking water treatment plant (ADWTP). *Sci. Total Environ.* 700: 134520.

36 Li, L., Xu, G., Yu, H., and Xing, J. (2018). Dynamic membrane for micro-particle removal in wastewater treatment: performance and influencing factors. *Sci. Total Environ.* 627: 332–340.

37 Suhrhoff, T.J. and Scholz-Böttcher, B.M. (2016). Qualitative impact of salinity, UV radiation and turbulence on leaching of organic plastic additives from four common plastics – a lab experiment. *Mar. Pollut. Bull.* 102 (1): 84–94.

38 Nabi, I., Bacha, A.U.R., Li, K. et al. (2020). Complete photocatalytic mineralization of microplastic on TiO_2 nanoparticle film. *iScience* 23 (7): 101326.

39 Miao, F., Liu, Y., Gao, M. et al. (2020). Degradation of polyvinyl chloride microplastics via an electro-Fenton-like system with a TiO_2/graphite cathode. *J. Hazard. Mater.* 399: 123023.

40 Chen, R., Qi, M., Zhang, G., and Yi, C. (2018). Comparative experiments on polymer degradation technique of produced water of polymer flooding oilfield. *IOP Conf. Ser. Earth. Environ. Sci.* 113: 012208.

41 Zafar, R., Park, S.Y., and Kim, C.G. (2020). Surface modification of polyethylene microplastic particles during the aqueous-phase ozonation process. *Environ. Eng. Res.* 26 (5): 200412.

42 Espinosa, M.J.C., Blanco, A.C., Schmidgall, T. et al. (2020). Toward biorecycling: isolation of a soil bacterium that grows on a polyurethane oligomer and monomer. *Front. Microbiol.* 11: 404.

43 Ndahebwa Muhonja, C., Magoma, G., Imbuga, M., and Makonde, H.M. (2018). Molecular characterization of low-density polyethene (LDPE) degrading bacteria and fungi from Dandora dumpsite, Nairobi, Kenya. *Int. J. Microbiol.* 2018: 1–10.

44 Yadav, V., Dhanger, S., and Sharma, J. (2022). Microplastics accumulation in agricultural soil: evidence for the presence, potential effects, extraction, and current bioremediation approaches. *J. Appl. Biol. Biotechnol.* 10: 38–47.

45 Yoshida, S., Hiraga, K., Taniguchi, I., and Oda, K. (2021). Ideonella sakaiensis, PETase, and MHETase: from identification of microbial PET degradation to enzyme characterization. *Methods Enzymol.* 648: 187–205.

46 Raaman, N., Rajitha, N., Jayshree, A., and Jegadeesh, R. (2014). Biodegradation of plastic by *Aspergillus* spp. isolated from polythene polluted sites around Chennai. *J. Acad. Indus. Res.* 1 (6): 313–316.

47 Miranda Zoppas, F., Sacco, N., Soffietti, J. et al. (2023). Catalytic approaches for the removal of microplastics from water: recent advances and future opportunities. *Chem. Eng. J. Adv.* 16: 100529.

48 Pandey, B., Pathak, J., Singh, P. et al. (2022). Microplastics in the ecosystem: an overview on detection, removal, toxicity assessment, and control release. *Water (Basel)* 15 (1): 51.

49 Venkataramana, C., Botsa, S.M., Shyamala, P., and Muralikrishna, R. (2021). Photocatalytic degradation of polyethylene plastics by $NiAl_2O_4$ spinels-synthesis and characterization. *Chemosphere* 265: 129021.

50 Kamalian, P., Khorasani, S.N., Abdolmaleki, A. et al. (2020). Toward the development of polyethylene photocatalytic degradation. *J. Polym. Eng.* 40 (2): 181–191.

51 Alhussain, H., Ferjani, H., Oyewo, O.A. et al. (2024). Photocatalytic degradation of methyl orange using CuO/Al_2O_3 nanocomposite obtained by microwave-assisted thermal decomposition route. *Inorg. Chem. Commun.* 162: 112272.

52 Chu, S., Zhang, B., Zhao, X. et al. (2022). Photocatalytic conversion of plastic waste: from photodegradation to photosynthesis. *Adv. Energy Mater.* 12 (22): 2200435.

53 Zhang, L., Wang, W., Jiang, D. et al. (2015). Photoreduction of CO_2 on BiOCl nanoplates with the assistance of photoinduced oxygen vacancies. *Nano Res* 8 (3): 821–831.

54 Banerjee, B., Amoli, V., Maurya, A. et al. (2015). Green synthesis of Pt-doped TiO_2 nanocrystals with exposed (001) facets and mesoscopic void space for photo-splitting of water under solar irradiation. *Nanoscale* 7 (23): 10504–10512.

55 Haghighi, A., Haghighi, M., Shabani, M., and Fard, S.G. (2021). Oxygen-rich bismuth oxybromide nanosheets coupled with Ag_2O as Z-scheme nano-heterostructured plasmonic photocatalyst: solar light-activated photodegradation of dye pollutants. *J. Hazard. Mater.* 408: 124406.

56 Wang, Y., Zhong, H., Xu, Q. et al. (2025). Vacancy-rich NiFe-LDH/carbon paper as a novel self-supporting electrode for the electro-Fenton degradation of polyvinyl chloride microplastics. *J. Hazard. Mater.* 485: 136797.

57 Ma, S., Tang, S., and Zhao, Y. (2024). Upcycling polystyrene microplastics to Fe/Mo-doped sponge-carbon: Mo^{5+} enhanced electron transfer for boosting Fenton-like performance. *Chem. Eng. J.* 498: 155460.

58 Spasiano, D., Marotta, R., Malato, S. et al. (2015). Solar photocatalysis: materials, reactors, some commercial, and pre-industrialized applications. A comprehensive approach. *Appl Catal B* 170–171: 90–123.

59 Zhang, Y., Afzal, N., Pan, L. et al. (2019). Structure-activity relationship of defective metal-based photocatalysts for water slitting: experimental and theoretical perspectives. *Adv. Sci.* 6 (10): 1900053.

60 Maarisetty, D. and Baral, S.S. (2020). Defect engineering in photocatalysis: formation, chemistry, optoelectronics, and interface studies. *J. Mater. Chem. A Mater.* 8 (36): 18560–18604.

61 Liu, J., Wei, Z., and Shangguan, W. (2019). Defects engineering in photocatalytic water splitting materials. *ChemCatChem* 11 (24): 6177–6189.

62 Muhammad, P., Zada, A., Rashid, J. et al. (2024). Defect engineering in nanocatalysts: from design and synthesis to applications. *Adv. Funct. Mater.* 34 (29): 2314686.

63 Naldoni, A., Allieta, M., Santangelo, S. et al. (2012). Effect of nature and location of defects on bandgap narrowing in black TiO_2 nanoparticles. *J. Am. Chem. Soc.* 134 (18): 7600–7603.

64 Fu, B., Wu, W., Gan, L., and Zhang, Z. (2019). Bulk/surface defects engineered TiO_2 nanotube photonic crystals coupled with plasmonic gold nanoparticles for effective *in vivo* near-infrared light photoelectrochemical detection. *Anal. Chem.* 91 (22): 14611–14617.

65 Xu, C., Ravi Anusuyadevi, P., Aymonier, C. et al. (2019). Nanostructured materials for photocatalysis. *Chem. Soc. Rev.* 48 (14): 3868–3902.

66 Appavoo, K., Liu, M., Black, C.T., and Sfeir, M.Y. (2015). Quantifying bulk and surface recombination processes in nanostructured water splitting photocatalysts via in situ ultrafast spectroscopy. *Nano Lett.* 15 (2): 1076–1082.

67 Cushing, S.K., Meng, F., Zhang, J. et al. (2017). Effects of defects on photocatalytic activity of hydrogen-treated titanium oxide nanobelts. *ACS Catal.* 7 (3): 1742–1748.

68 Nowotny, J., Alim, M.A., Bak, T. et al. (2015). Defect chemistry and defect engineering of TiO_2-based semiconductors for solar energy conversion. *Chem. Soc. Rev.* 44 (23): 8424–8442.

69 Yu, Y., Yan, W., Wang, X. et al. (2018). Surface engineering for extremely enhanced charge separation and photocatalytic hydrogen evolution on g-C_3N_4. *Adv. Mater.* 30 (9): 1705060.

70 Zhao, Y., Chen, G., Bian, T. et al. (2015). Defect-rich ultrathin ZnAl-layered double hydroxide nanosheets for efficient photoreduction of CO_2 to CO with water. *Adv. Mater.* 27 (47): 7824–7831.

71 Li, J., Wu, X., Pan, W. et al. (2018). Vacancy-rich monolayer BiO_2-x as a highly efficient UV, visible, and near-infrared responsive photocatalyst. *Angew. Chem. Int. Ed.* 57 (2): 491–495.

72 Park, M., Seo, J.H., Song, H., and Nam, K.M. (2016). Enhanced visible light activity of single-crystalline WO_3 microplates for photoelectrochemical water oxidation. *J. Phys. Chem. C* 120 (17): 9192–9199.

73 Chen, D., Wang, Z., Ren, T. et al. (2014). Influence of defects on the photocatalytic activity of ZnO. *J. Phys. Chem. C* 118 (28): 15300–15307.

74 Amoresi, R.A.C., Oliveira, R.C., Marana, N.L. et al. (2019). CeO_2 nanoparticle morphologies and their corresponding crystalline planes for the photocatalytic degradation of organic pollutants. *ACS Appl. Nano Mater.* 2 (10): 6513–6526.

75 Zhang, R., Zhang, Y.-C., Pan, L. et al. (2018). Engineering cobalt defects in cobalt oxide for highly efficient electrocatalytic oxygen evolution. *ACS Catal.* 8 (5): 3803–3811.

76 Long, M. and Zheng, L. (2017). Engineering vacancies for solar photocatalytic applications. *Chin. J. Catl.* 38 (4): 617–624.

77 Fu, X., Wang, X., Chen, Z. et al. (2010). Photocatalytic performance of tetragonal and cubic β-In_2S_3 for the water splitting under visible light irradiation. *Appl Catal B* 95 (3–4): 393–399.

78 Zhao, Y., Jia, X., Chen, G. et al. (2016). Ultrafine NiO nanosheets stabilized by TiO_2 from monolayer NiTi-LDH precursors: an active water oxidation electrocatalyst. *J. Am. Chem. Soc.* 138 (20): 6517–6524.

79 Fang, Z., Weng, S., Ye, X. et al. (2015). Defect engineering and phase junction architecture of wide-bandgap ZnS for conflicting visible light activity in photocatalytic H_2 evolution. *ACS Appl. Mater. Interfaces* 7 (25): 13915–13924.

80 Zhang, X., Zhao, Z., Zhang, W. et al. (2016). Surface defects enhanced visible light photocatalytic H_2 production for Zn-Cd-S solid solution. *Small* 12 (6): 793–801.

81 Wang, J., Chen, Y., Zhou, W. et al. (2017). Cubic quantum dot/hexagonal microsphere $ZnIn_2S_4$ heterophase junctions for exceptional visible-light-driven photocatalytic H_2 evolution. *J. Mater. Chem. A Mater.* 5 (18): 8451–8460.

82 Dabirian, A. and van de Krol, R. (2013). Resonant optical absorption and defect control in Ta_3N_5 photoanodes. *Appl. Phys. Lett.* 102 (3): 033905.

83 Long, M., Hu, P., Wu, H. et al. (2015). Understanding the composition and electronic structure dependent photocatalytic performance of bismuth oxyiodides. *J. Mater. Chem. A Mater.* 3 (10): 5592–5598.

84 Bai, Y., Zhou, Y., Zhang, J. et al. (2019). Homophase junction for promoting spatial charge separation in photocatalytic water splitting. *ACS Catal.* 9 (4): 3242–3252.

85 Guo, L., Yang, Z., Marcus, K. et al. (2018). MoS_2/TiO_2 heterostructures as nonmetal plasmonic photocatalysts for highly efficient hydrogen evolution. *Energy Environ. Sci.* 11 (1): 106–114.

86 Pradhan, N., Das Adhikari, S., Nag, A., and Sarma, D.D. (2017). Luminescence, plasmonic, and magnetic properties of doped semiconductor nanocrystals. *Angew. Chem. Int. Ed.* 56 (25): 7038–7054.

87 Livraghi, S., Paganini, M.C., Giamello, E. et al. (2006). Origin of photoactivity of nitrogen-doped titanium dioxide under visible lght. *J. Am. Chem. Soc.* 128 (49): 15666–15671.

88 Wang, J., Tafen, D.N., Lewis, J.P. et al. (2009). Origin of photocatalytic activity of nitrogen-doped TiO_2 nanobelts. *J. Am. Chem. Soc.* 131 (34): 12290–12297.

89 Zhao, B., Chen, F., Qu, W., and Zhang, J. (2009). The evolvement of pits and dislocations on TiO_2-B nanowires via oriented attachment growth. *J. Solid State Chem.* 182 (8): 2225–2230.

90 Ohno, Y., Tokumoto, Y., Yonenaga, I. et al. (2012). Optical properties of edge dislocations on (1100) prismatic planes in wurtzite ZnO introduced at elevated temperatures. *J. Appl. Phys.* 111 (11): 113514.

91 Hacialioglu, S., Meng, F., and Jin, S. (2012). Facile and mild solution synthesis of Cu_2O nanowires and nanotubes driven by screw dislocations. *Chem. Commun.* 48 (8): 1174–1176.

92 Wu, H., Meng, F., Li, L. et al. (2012). Dislocation-driven CdS and CdSe nanowire growth. *ACS Nano* 6 (5): 4461–4468.

93 Wang, J.-D., Liu, J.-K., Luo, C.-X. et al. (2013). Silver phosphate crystal growth by screw dislocation driven of dynamic-template. *Cryst. Growth Des.* 13 (11): 4837–4843.

94 Kamei, M. (2008). Localization of the photocatalytic reaction on the grain boundary of bicrystalline TiO_2. *Appl. Phys. Express* 1: 101201.

95 Senthamizhan, A., Balusamy, B., Aytac, Z., and Uyar, T. (2016). Grain boundary engineering in electrospun ZnO nanostructures as promising photocatalysts. *CrystEngComm* 18 (34): 6341–6351.

96 Liang, Y.C., Chung, C.C., Lin, T.Y., and Cheng, Y.R. (2016). Synthesis and microstructure-dependent photoactivated properties of three-dimensional cadmium sulfide crystals. *J. Alloys Compd.* 688: 769–775.

97 He, K., Wang, M., and Guo, L. (2015). Novel-CdS-nanorod with stacking fault structures: preparation and properties of visible-light-driven photocatalytic hydrogen production from water. *Chem. Eng. J.* 279: 747–756.

98 Liu, M., Wang, L., and (Max) Lu, G., Yao, X., and Guo, L. (2011). Twins in Cd1−xZnxS solid solution: highly efficient photocatalyst for hydrogen generation from water. *Energy Environ. Sci.* 4 (4): 1372.

99 Yin, B., Huang, X., Mishra, R., and Sadtler, B. (2017). Compositionally induced twin defects control the shape of ternary silver halide nanocrystals. *Chem. Mater.* 29 (3): 1014–1021.

100 Dong, G., Chen, D., Luo, J. et al. (2017). Voids padding induced further enhancement in photocatalytic performance of porous graphene-like carbon nitride. *J. Hazard. Mater.* 335: 66–74.

101 Wu, D., Wang, W., Tan, F. et al. (2013). Fabrication of pit-structured ZnO nanorods and their enhanced photocatalytic performance. *RSC Adv.* 3 (43): 20054.

102 Kumari, V., Patra, A.K., and Bhaumik, A. (2014). Self-assembled ultra-small zinc stannate nanocrystals with mesoscopic voids via a salicylate templating pathway and their photocatalytic properties. *RSC Adv.* 4 (26): 13626–13634.

103 Fan, C., Yu, S., Qian, G., and Wang, Z. (2017). Ultrasonic-induced disorder engineering on ZnO, ZrO_2, Fe_2O_3 and SnO_2 nanocrystals. *RSC Adv.* 7 (30): 18785–18792.

104 Lin, L., Huang, J., Li, X. et al. (2017). Effective surface disorder engineering of metal oxide nanocrystals for improved photocatalysis. *Appl Catal B* 203: 615–624.

105 Hu, Y.H. (2012). A highly efficient photocatalyst—hydrogenated black TiO_2 for the photocatalytic splitting of water. *Angew. Chem. Int. Ed.* 51 (50): 12410–12412.

106 Wang, M., Nie, B., Yee, K.-K. et al. (2016). Low-temperature fabrication of brown TiO_2 with enhanced photocatalytic activities under visible light. *Chem. Commun.* 52 (14): 2988–2991.

107 Wang, J., Wang, Z., Huang, B. et al. (2012). Oxygen vacancy induced band-gap narrowing and enhanced visible light photocatalytic activity of ZnO. *ACS Appl. Mater. Interfaces* 4 (8): 4024–4030.

108 Meng, J., Lin, Q., Chen, T. et al. (2018). Oxygen vacancy regulation on tungsten oxides with specific exposed facets for enhanced visible-light-driven photocatalytic oxidation. *Nanoscale* 10 (6): 2908–2915.

109 Tian, J., Sang, Y., Yu, G. et al. (2013). A Bi_2WO_6-based hybrid photocatalyst with broad spectrum photocatalytic properties under UV, visible, and near-infrared irradiation. *Adv. Mater.* 25 (36): 5075–5080.

110 Zhao, Z., Zhang, X., Zhang, G. et al. (2015). Effect of defects on photocatalytic activity of rutile TiO_2 nanorods. *Nano Res* 8 (12): 4061–4071.

111 Niu, P., Yin, L., Yang, Y. et al. (2014). Increasing the visible light absorption of graphitic carbon nitride (melon) photocatalysts by homogeneous self-modification with nitrogen vacancies. *Adv. Mater.* 26 (47): 8046–8052.

112 Chen, X., Liu, L., Yu, P.Y., and Mao, S.S. (2011). Increasing solar absorption for photocatalysis with black hydrogenated titanium dioxide nanocrystals. *Science (1979)* 331 (6018): 746–750.

113 Folger, A., Ebbinghaus, P., Erbe, A., and Scheu, C. (2017). Role of vacancy condensation in the formation of voids in rutile TiO_2 nanowires. *ACS Appl. Mater. Interfaces* 9 (15): 13471–13479.

114 Kachina, A., Puzenat, E., Ould-Chikh, S. et al. (2012). A new approach to the preparation of nitrogen-doped titania visible light photocatalyst. *Chem. Mater.* 24 (4): 636–642.

115 Jorge, A.B., Sakatani, Y., Boissière, C. et al. (2012). Nanocrystalline N-doped ceria porous thin films as efficient visible-active photocatalysts. *J. Mater. Chem.* 22 (7): 3220.

116 Zhou, Y., Chen, G., Yu, Y. et al. (2015). An efficient method to enhance the stability of sulphide semiconductor photocatalysts: a case study of N-doped ZnS. *Phys. Chem. Chem. Phys.* 17 (3): 1870–1876.

117 Zong, X. and Wang, L. (2014). Ion-exchangeable semiconductor materials for visible light-induced photocatalysis. *J. Photobio. C Photochem. Rev.* 18: 32–49.

118 Hosogi, Y., Kato, H., and Kudo, A. (2008). Photocatalytic activities of layered titanates and niobates ion-exchanged with Sn^{2+} under visible light irradiation. *J. Phys. Chem. C* 112 (45): 17678–17682.

119 Sun, Y., Dong, S., Zhang, H., and Shi, Y. (2015). Numerical investigation of the geodynamic mechanism for the late Jurassic deformation of the Ordos block and surrounding orogenic belts. *J. Asian Earth Sci.* 114: 623–633.

120 Gordon, T.R., Cargnello, M., Paik, T. et al. (2012). Nonaqueous synthesis of TiO_2 nanocrystals using TiF4 to engineer morphology, oxygen vacancy concentration, and photocatalytic activity. *J. Am. Chem. Soc.* 134 (15): 6751–6761.

121 Gasparotto, A., Barreca, D., Bekermann, D. et al. (2011). F-Doped Co_3O_4 photocatalysts for sustainable H_2 generation from water/ethanol. *J. Am. Chem. Soc.* 133 (48): 19362–19365.

122 Sun, Y., Liu, Q., Gao, S. et al. (2013). Pits confined in ultrathin cerium(IV) oxide for studying catalytic centers in carbon monoxide oxidation. *Nat. Commun.* 4 (1): 2899.

123 Lei, F., Sun, Y., Liu, K. et al. (2014). Oxygen vacancies confined in ultrathin indium oxide porous sheets for promoted visible-light water splitting. *J. Am. Chem. Soc.* 136 (19): 6826–6829.

124 Zhou, M., Zu, X.T., Sun, K. et al. (2017). Enhanced photocatalytic hydrogen generation of nano-sized mesoporous $InNbO_4$ crystals synthesized via a polyacrylamide gel route. *Chem. Eng. J.* 313: 99–108.

125 Esch, F., Fabris, S., Zhou, L. et al. (2005). Electron localization determines defect formation on ceria substrates. *Science (1979)* 309 (5735): 752–755.

126 Hu, S., Ma, L., You, J. et al. (2014). Enhanced visible light photocatalytic performance of g-C_3N_4 photocatalysts co-doped with iron and phosphorus. *Appl. Surf. Sci.* 311: 164–171.

127 Zhang, N., Li, X., Ye, H. et al. (2016). Oxide defect engineering enables to couple solar energy into oxygen activation. *J. Am. Chem. Soc.* 138 (28): 8928–8935.

128 Li, H., Shang, J., Zhu, H. et al. (2016). Oxygen vacancy structure associated photocatalytic water oxidation of BiOCl. *ACS Catal.* 6 (12): 8276–8285.

129 Yang, W., Zhang, L., Xie, J. et al. (2016). Enhanced photoexcited carrier separation in oxygen-doped ZnIn2S4 nanosheets for hydrogen evolution. *Angew. Chem. Int. Ed.* 55 (23): 6716–6720.

130 Meksi, M., Kochkar, H., Berhault, G., and Guillard, C. (2015). Effect of cerium content and post-thermal treatment on doped anisotropic TiO_2 nanomaterials and kinetic study of the photodegradation of formic acid. *J. Mol. Catal. A Chem.* 409: 162–170.

131 Zhang, N., Jalil, A., Wu, D. et al. (2018). Refining defect states in W18O49 by Mo doping: a strategy for tuning N2 activation towards solar-driven nitrogen fixation. *J. Am. Chem. Soc.* 140 (30): 9434–9443.

132 Sun, S., Gao, P., Yang, Y. et al. (2016). N-Doped TiO_2 nanobelts with coexposed (001) and (101) facets and their highly efficient visible-light-driven photocatalytic hydrogen production. *ACS Appl. Mater. Interfaces* 8 (28): 18126–18131.

133 Liu, Z.Z., Shang, X., Dong, B., and Chai, Y.M. (2018). Triple Ni-Co-Mo metal sulfides with one-dimensional and hierarchical nanostructures towards highly efficient hydrogen evolution reaction. *J. Catal.* 361: 204–213.

134 Qin, J., Dou, Y., Zhou, J. et al. (2023). Photocatalytic valorization of plastic waste over zinc oxide encapsulated in a metal–organic framework. *Adv. Funct. Mater.* 33 (28): 2214839.

135 Tofa, T.S., Ye, F., Kunjali, K.L., and Dutta, J. (2019). Enhanced visible light photodegradation of microplastic fragments with plasmonic platinum/zinc oxide nanorod photocatalysts. *Catalysts* 9 (10): 819.

136 Beladi-Mousavi, S.M., Hermanová, S., Ying, Y. et al. (2021). A maze in plastic wastes: autonomous motile photocatalytic microrobots against microplastics. *ACS Appl. Mater. Interfaces* 13 (21): 25102–25110.

137 Cao, B., Wan, S., Wang, Y. et al. (2022). Highly-efficient visible-light-driven photocatalytic H_2 evolution integrated with microplastic degradation over MXene/$Zn_xCd_{1-x}S$ photocatalyst. *J. Colloid Interface Sci.* 605: 311–319.

138 Zhou, D., Luo, H., Zhang, F. et al. (2022). Efficient photocatalytic degradation of the persistent PET fiber-based microplastics over Pt nanoparticles decorated N-doped TiO_2 nanoflowers. *Adv. Fiber Mater.* 4 (5): 1094–1107.

139 Llorente-García, B.E., Hernández-López, J.M., Zaldívar-Cadena, A.A. et al. (2020). First insights into photocatalytic degradation of HDPE and LDPE microplastics by a mesoporous N–TiO_2 coating: effect of size and shape of microplastics. *Coat.* 10 (7): 658.

140 Fadli, M.H., Ibadurrohman, M., and Slamet, S. (2021). Microplastic pollutant degradation in water using modified TiO_2 photocatalyst under UV-irradiation. *IOP Conf. Ser. Mater. Sci. Eng.* 1011 (1): 012055.

141 Zhou, H., Ren, Y., Li, Z. et al. (2021). Electrocatalytic upcycling of polyethylene terephthalate to commodity chemicals and H_2 fuel. *Nat. Commun.* 12 (1): 4679.

142 Nakatani, H., Motokucho, S., and Miyazaki, K. (2015). Difference in polystyrene oxo-biodegradation behavior between copper phthalocyanine modified TiO_2 and ZnO paint photocatalyst systems. *Polym. Degrad. Stab.* 120: 1–9.

143 Wang, L., Kaeppler, A., Fischer, D., and Simmchen, J. (2019). Photocatalytic TiO_2 micromotors for removal of microplastics and suspended matter. *ACS Appl. Mater. Interfaces* 11 (36): 32937–32944.

144 Yuwendi, Y., Ibadurrohman, M., Setiadi, S., and Slamet, S. (2022). Photocatalytic degradation of polyethylene microplastics and disinfection of E. coli in water over Fe- and Ag-modified TiO_2 nanotubes. *Bull. Chem. React. Eng.* 17 (2): 263–277.

145 Ariza-Tarazona, M.C., Villarreal-Chiu, J.F., Hernández-López, J.M. et al. (2020). Microplastic pollution reduction by a carbon and nitrogen-doped TiO_2: effect of pH and temperature in the photocatalytic degradation process. *J. Hazard. Mater.* 395: 122632.

12

Defective Photocatalytic Materials for Desalination: Synthesis Strategies, Types of Defects, and Mechanisms

Fangdi Wu[1] *and Yongqian Cui*[2]

[1] *Wuyi University, College of Ecology and Resources Engineering, Fujian Provincial Key Laboratory of Eco-Industrial Green Technology, Key Laboratory of Green Chemical Technology of Fujian Province University, Baihua Road, Wuyishan 354300, PR China*
[2] *Shaanxi University of Science and Technology, School of Environmental Science and Engineering, Weiyang University Park, Xi'an 710021, PR China*

12.1 Introduction

Due to the rapid population growth and increasing demand for clean water from industrial development, it is crucial for humanity to explore and utilize eco-friendly energy sources to alleviate freshwater scarcity [1–3]. Among the common clean energy sources, such as wind, tidal, geothermal, and solar energy, solar energy stands out as one of the cleanest, most abundant, and most easily accessible energy resources on Earth [3–5]. Under sunlight, seawater can be evaporated into water vapor and recycled as rain and dew. This process of solar-driven water evaporation occurs naturally every day [6]. However, the efficiency of water evaporation is influenced by weather conditions, making this natural evaporation process inefficient [7]. Therefore, the search for low-cost, low-toxicity, reusable photothermal conversion materials with high solar-driven water evaporation efficiency is the key to solving the problem of freshwater scarcity.

Photothermal materials can achieve energy conversion by collecting solar energy. These materials absorb solar radiation and convert it into heat (Figure 12.1) [8]. The essence of achieving efficient solar energy conversion lies in maximizing the utilization of sunlight by enhancing absorption while concurrently reducing light reflectance to a minimum. Photothermal materials can generally be categorized into three broad groups based on their diverse photothermal conversion processes (as shown in Figure 12.2) [6, 8]. The first type is plasma metallic materials that exhibit localized surface plasmon resonance (LSPR) effects. These metal nanoparticles are capable of absorbing light at their resonant wavelengths (Figure 12.2a). LSPR is a phenomenon of resonant photon-induced coherent charge oscillations that occurs when the frequency of photons matches the inherent frequency of electrons on the metal surface. This results in fluctuations within the electrons, prompting a transition from occupied to unoccupied energy states, generating hot electrons, and

Defect Engineering in Photocatalysis: From Fundamentals to Applications,
First Edition. Edited by Chuanyi Wang, Abdelkader Labidi, and Qiuhui Zhu.

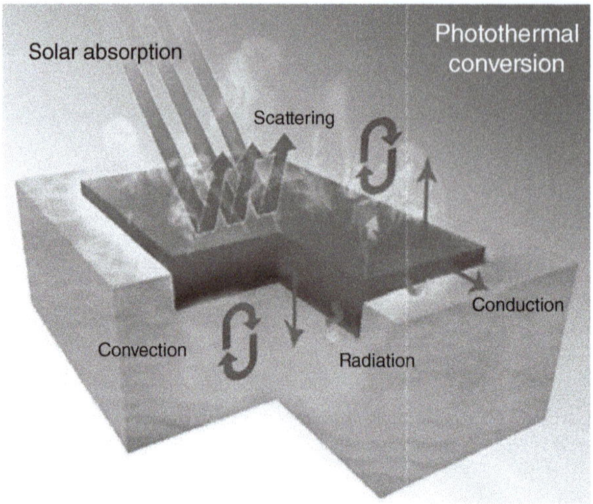

Figure 12.1 Schematic diagram of the photothermal conversion process. Reproduced from Gao et al. [8] / with permission of Royal Society of Chemistry.

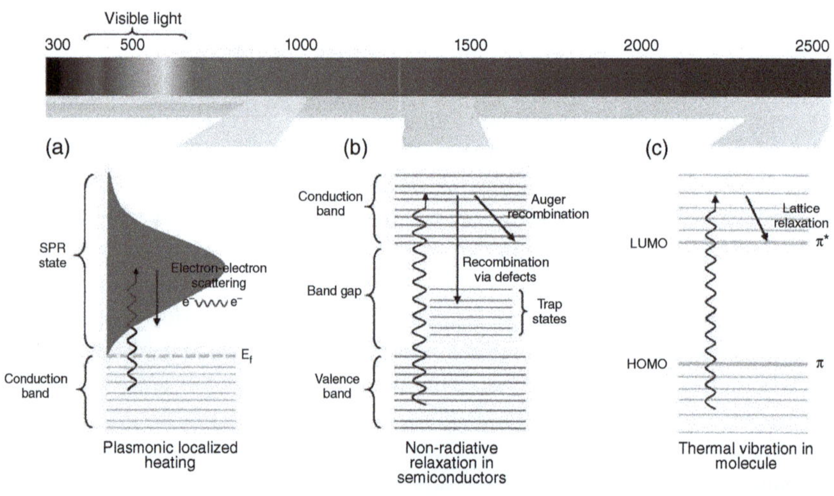

Figure 12.2 Three different types of photothermal conversion. Reproduced from Gao et al. [8] / with permission of Royal Society of Chemistry.

ultimately establishing a hot charge distribution. The decay of these hot electrons can result in the emission of visible radiation or electron–electron interactions, leading to a surge in carriers. The decay through electron–electron scattering can redistribute the energy of hot electrons, causing a rapid increase in the local surface temperature of the metal. After the local temperature alteration, equilibrium cooling occurs subsequently as energy is transferred from electrons to lattice phonons. The lattice then undergoes cooling via phonon–phonon interactions and dissipates heat into the surrounding environment, resulting in a localized temperature increase in the system. Zhou et al. exhibit a solar desalination system enhanced by plasmons, which is constructed through the three-dimensional (3D) self-assembly of aluminum nanoparticles into a 3D, porous membrane structure. It achieves high efficiency (about 90%) and effective desalination [9]. The second type of semiconductor materials releases energy through radiative relaxation (Figure 12.2b). The photothermal effect in semiconductor materials is exhibited through the process of excited electrons returning to their low-energy states, accompanied by local heating of the lattice. In semiconductor materials, the band that can be completely filled by electrons is known as the valence band (VB). The band that cannot be completely filled by electrons is the conduction band (CB). The local region between the CB and the VB is the bandgap of the semiconductor material. The optical absorption spectrum of semiconductor materials depends on their bandgap energy. When semiconductor materials receive photons with energy greater than the bandgap energy, electrons can be excited from the VB to the CB, leaving holes in the VB and free electrons in the CB. These electrons and holes ultimately release energy to return to their ground states through either nonradiative relaxation via phonons or radiative relaxation via photons. Consequently, the diffusion of photoexcited carriers and surface/bulk recombination leads to a temperature distribution and an increase in temperature. Ghim et al. prepared chemically exfoliated MoS_2 nanosheets, which achieved high solar evaporation efficiencies of approximately 76% and 82% under light intensities of 0.76 and 5.35 $kW\,m^{-2}$, respectively [10]. The third type is composite carbon-based photothermal materials that rely on molecular thermal vibrations. In some materials, the internal lattice vibrates due to the absorption of solar energy, generating thermal energy through optical transitions of π–π^* bonds (Figure 12.2c). There exists a large energy band between σ–σ^* bonds, making σ–σ^* transitions difficult to achieve under solar irradiation. Due to the weaker bonding electrons, π bonds are generally weaker than σ bonds, allowing these electrons to be excited from π orbitals to π^* orbitals with lower energy input. Furthermore, the presence of conjugated π bonds leads to a redshift in the absorption spectrum. With an increasing number of π bonds, the energy difference between the highest occupied molecular orbital (HOMO) and the lowest unoccupied molecular orbital (LUMO) narrows. Carbon-based materials that inherently appear black possess good photothermal conversion capabilities. When exposed to sunlight, excited electrons relax through electron–electron or electron–phonon scattering, generating heat and leading to an overall increase in temperature. Li et al. fixed a wrapped cotton core onto polystyrene thermal insulation foam wrapped in black nonwoven cotton fabric and studied its effect for seawater desalination under different light intensities. The study found that under one-sun irradiation, the water evaporation rate of this solar evaporator was 1.47 $kg\,m^{-2}\,h^{-1}$ [11].

Besides salt impurities, seawater also contains various other pollutants, including inorganic and organic contaminants such as volatile organic compounds (VOCs), microorganisms, and heavy metals. During solar distillation, heavy metals and most organic pollutants remain in the seawater [12]. However, VOCs tend to evaporate along with water vapor and are incorporated into the collected freshwater during condensation, thereby affecting the quality of the freshwater. Photocatalysis can directly utilize solar energy as a light source to facilitate deep reactions at room temperature, leveraging its unique properties to drive the reactions. Photocatalytic technology has demonstrated significant advantages in the fields of environmental pollution treatment and clean energy production, making it an ideal and highly regarded production technology. Due to the unique advantages of photocatalysis, photocatalytic oxidation technology has developed into an environmentally friendly technology with great application prospects and is considered an effective method for degrading VOCs.

In the application of semiconductor photocatalytic materials in seawater desalination projects, their photothermal properties (as shown in Figure 12.2b) are utilized to convert sunlight into thermal energy, which effectively evaporates seawater. Simultaneously, the photocatalytic materials efficiently degrade the VOCs both in the seawater that is about to evaporate and those that have already evaporated, yielding clean distilled water that meets drinking water standards. Furthermore, the heat generated by the photothermal properties of the materials can warm the materials themselves, enhancing their photocatalytic performance. This achieves a simultaneous action of both light and thermal catalysis, further improving the photothermal–photocatalytic performance.

The original semiconductor photocatalytic materials often suffer from limitations such as limited light absorption range, easy electron–hole recombination, and poor adsorption and activation capabilities for surface reactant molecules, which greatly reduce their photocatalytic activity [13–15]. Over the past few decades, various strategies have been developed to enhance photocatalytic performance, including adjusting the morphology and structure of photocatalysts, employing element doping techniques, introducing cocatalysts, and constructing heterojunctions [14, 15]. Among the numerous improvement strategies, defect engineering can indirectly control the properties of materials by manipulating their defect states, crystalline phases, exposed facets, and other characteristics. Defects not only introduce new energy levels, narrowing the bandgap of semiconductor materials, but they can also serve as adsorption sites for charge transfer, hindering the recombination of photogenerated carriers and thus promoting charge separation and enhancing photocatalytic activity [16–21]. This chapter mainly introduces the synthesis methods of defective photocatalytic materials, the reaction mechanisms and progress of defective photocatalytic materials in photocatalytic desalination, and finally outlines the challenges and opportunities related to defect engineering in photocatalytic desalination.

12.2 Synthesis Methodologies for Defective Photocatalytic Materials

On one hand, defect engineering can promote the migration of photogenerated carriers, causing significant electron delocalization and providing more active sites for the catalytic process. On the other hand, it can also induce nonradiative relaxation, releasing energy in the form of phonons and increasing the temperature of the semiconductor, which in turn raises the temperature of the catalytically active sites.

This not only effectively enhances the photocatalytic efficiency but also achieves photothermal coupling, thereby benefiting the seawater desalination process. In this section, we discuss various methods that can be employed to manufacture defective photocatalytic materials.

12.2.1 Chemical Reduction Method

Chemical reduction methods can introduce defects into semiconductors. For instance, heat treatment in a reducing atmosphere and chemical reduction using reducing agents are considered promising strategies for generating surface defects in various photocatalysts. Employing reducing atmospheres, such as H_2, CO, NH_3, and H_2S, can produce a variety of defects in photocatalysts. Song et al. [22] reported the use of hydrogen as a reducing agent to prepare tungsten oxide with oxygen defects, such as $WO_{2.72}$, which exhibits high catalytic activity for the hydrogenation of linear alkenes, cyclic alkenes, and aryl nitros. Compared to non-noble metal nickel catalysts, it demonstrates significant advantages in both activity and selectivity. Computational analyses based on density functional theory demonstrate that oxygen-deficient surfaces exhibit enhanced kinetic and thermodynamic favorability for H_2 activation. Examinations of various oxygen-deficient tungsten oxides unveil a direct correlation between their hydrogenation activity and the concentration of oxygen vacancies (OVs). Given tungsten's abundance and the feasibility of large-scale synthesis of $WO_{2.72}$ through cost-effective methods, it presents itself as an optimal catalyst for industrial applications. Yu et al. [23] chose glyoxal as a reducing agent to prepare $Sr_2Bi_2Nb_2TiO_{12}$ with adjustable surface OV concentration through a simple low-temperature water bath method. The introduction of OV greatly enhances the light absorption of $Sr_2Bi_2Nb_2TiO_{12}$, covering the entire visible light region, and significantly promotes the separation of photoinduced charge carriers and improves CO_2 adsorption capacity. $Sr_2Bi_2Nb_2TiO_{12}$ with the optimal concentration of OVs exhibits a remarkable CO production rate of 17.11 $\mu mol\, g^{-1}\, h^{-1}$, which is approximately 58 times higher than that of the bulk phase ($Sr_2Bi_2Nb_2TiO_{12}$ without OVs) prepared by high-temperature solid-state reaction (SBNT-SSR). Table 12.1 summarizes the reductants used for forming various types of defects in different photocatalysts, as well as the impact of these defects on photocatalytic activity.

12.2.2 Thermal Treatment Method

Thermal annealing treatment can provide higher energy to atoms and accelerate their transport, thereby allowing surface atoms to escape and produce defects. Sun et al. [41] reported the rapid preparation of MoO_{3-x}-rGO composites with synergistic photothermal and photocatalytic properties through a solid-phase microwave thermal shock method. The formation of MoO_{3-x} containing OVs was induced by the hotspots generated on the surface of microwave-reduced rGO. Owing to the LSPR effect, the photothermal and photocatalytic performance of the near-infrared MoO_{3-x}-rGO catalyst was significantly enhanced. Under one-sun

Table 12.1 Description of various reducing agents for defect formation, defect types, and their corresponding roles in improving photocatalytic performance.

Photocatalyst	Preparation method	Type of defect	Properties and application	References
Black TiO_2	Hydrogenation; Al chemical reduction	Ti^{3+}, oxygen vacancy, Ti-H bonds	Valence band shift, enhancing solar absorption; Photocatalytic hydrogen production, photodegradation, photothermic effect	[24–29]
TiO_2	N_2H_4 reduction	Ti^{3+}	Enhanced activity as photoanode, higher electronic conductivity; Water oxidation	[30]
Sulfur-doped g-C_3N_4	NH_3 chemical reduction	Carbon vacancy	Provide many active sites, increase the charge–carrier separation rate; Photocatalytic nitrogen fixation	[31]
Amorphous α-Zn–Ge–O	$NaBH_4$ chemical reduction	Oxygen vacancy	Valence band shift, enhancing solar absorption, provide many active sites; Splitting of CO_2	[32]
Zn–Cd–S solid solution	Hydrazine hydrate	Sulfur vacancy	Wider visible light absorption range, promote photogenerated charge transfer; Photocatalytic hydrogen production	[33]
ZnO	Propanetriol reduction	Zn vacancy	Switch n-type ZnO into stable p-type, induce room-temperature ferromagnetism; High photocatalytic activity	[34]
BiOCl	Propanetriol reduction	Oxygen vacancy	Enhancing solar absorption, reduce the recombination of photogenerated electron–hole pairs; Photocatalytic hydrogen production	[35]
WO_3	Vacuum/hydrogen Post-treatment	Oxygen vacancy	Enhancing solar absorption, enhanced photocurrent responses; Photocatalytic water oxidation	[36]
MoO_{3-x}	Ethanol reduction	Oxygen vacancy	Enhancing visible light absorption, Photocatalytic hydrogen production	[37]
Bi_2WO_6	Ethylene glycol reduction	Oxygen vacancy	Enhancing visible light absorption, Reduce bandgap; photocatalytic CO_2 reduction	[38]
$BiPO_4$	Ethylene glycol reduction, calcination	Oxygen vacancy	Enhancing visible light absorption, increases the migration ability of photogenerated charges and separation ability of photogenerated carriers; Photodegradation	[39]
$W_{18}O_{49}$	Ethanol reduction	Oxygen vacancy	Enhancing visible light absorption, Inhibit electron–hole pair recombination; Photocatalytic CO_2 reduction	[40]

irradiation of the optimal composite catalyst (GMW-3), the seawater evaporation rate and photothermal conversion efficiency can reach 1.58 kg m^{-2} h^{-1} and 87.2%, respectively. Additionally, the GMW-3 composite evaporator exhibited excellent antibacterial performance during the evaporation process under the synergistic effect of photothermal and photocatalytic activities. These findings provide new insights and a platform for designing efficient catalysts for large-scale seawater desalination and purification.

Zheng et al. [42] successfully prepared black TiO_2 with Ti^{3+}/OV tunable defects using a simple method of NaN_3 detonation, which allows for ultrafast and efficient formation of defective TiO_2 at 450 °C. With the incorporation of defective Ti^{3+}/OV sites, the bandgap-tuned black TiO_2 (specifically, sample DP-3) experiences improved light harvesting across the entire solar spectrum. Notably, DP-3 demonstrates a water evaporation rate of 1.624 kg m^{-2} h^{-1} and achieves a conversion efficiency of 85% under standard solar irradiation conditions. Hu et al. [43] successfully synthesized iodine-doped g-C_3N_4 with carbon vacancies through an in situ decomposition-thermal polymerization method using dicyandiamide and KIO_3 as precursors at 823 K for four hours. Experimental results showed that the bandgap of g-C_3N_4 was significantly reduced, and the average fluorescence lifetime increased after iodine doping. The spectral response range of iodine-doped g-C_3N_4 was effectively expanded, and the separation efficiency of photogenerated carriers was also significantly improved. g-C_3N_4 with 10 wt% iodine doping and carbon vacancies exhibited excellent photocatalytic nitrogen fixation and ammonia synthesis performance under simulated sunlight (reaching 200.8 mg L^{-1} g^{-1} in 3 hours, which is 2.8 times that of bulk g-C_3N_4) and maintained good stability. Lei et al. [44] obtained ultrathin In_2O_3 porous nanosheets with different concentrations of OVs by rapidly heating ultrathin $In(OH)_3$ nanosheets in various atmospheres, where the phase transition from $In(OH)_3$ to In_2O_3 introduced OVs. Theoretical and experimental results indicated that the OVs imparted new donor levels and increased the density of states in the five-atomic-layer-thick In_2O_3 nanosheets, thereby narrowing the bandgap from the ultraviolet (UV) to visible light range and enhancing the separation efficiency of photogenerated carriers.

12.2.3 External Force Induction Method

Apart from the aforementioned strategies for defect generation, researchers have also developed various externally induced methods, such as exfoliation, plasma etching, lithium-induced conversion, UV irradiation, and ball milling, to synthesize defective photocatalysts for solar energy conversion applications. Li et al. [45] employed n-butyllithium (under an argon atmosphere) to exfoliate multilayered MoS_2 and $Bi_{12}O_{17}Cl_2$, followed by ultrasonication to produce monolayer MoS_2 and $Bi_{12}O_{17}Cl_2$ nanosheets. During the ultrasound-mediated exfoliation process, organic ligands escaped from the interior of $Bi_{12}O_{17}Cl_2$ (or MoS_2), inevitably removing some oxygen atoms from the lattice, resulting in the formation of surface OVs in the monolayer $Bi_{12}O_{17}Cl_2$ (or MoS_2) nanosheets. Zhang et al. [46] added the synthesized bulk $ZnIn_2S_4$ to a hexane solution of *n*-butyllithium (under a nitrogen atmosphere). The mixture was then heated at 60 °C for 72 hours while stirring to obtain Li-intercalated $ZnIn_2S_4$ ($LixZnIn_2S_4$). The obtained $LixZnIn_2S_4$ powder was dispersed in ice-water under ultrasonic conditions and centrifuged to produce $ZnIn_2S_4$ nanosheets containing sulfur vacancies (Vs). The Vs confined within the monolayer $ZnIn_2S_4$ established an important link between the electronic manipulation and activity of $ZnIn_2S_4$. Its photocatalytic hydrogen production activity was 11 times higher than that of bulk $ZnIn_2S_4$. Ye et al. [47] prepared black BiOCl containing OVs through argon bubbling and UV light irradiation. The prepared black BiOCl samples exhibited a 20-fold higher visible-light photocatalytic activity for RhB degradation compared to white BiOCl. Zhu et al. [48] employed a simple ball-milling process to synthesize $BiPO_4$ with dual defects of Bi and O. Shen et al. [49] used a ball-milling method to prepare graphite and carbon nanotube materials with more exposed edges. Electrochemical measurements indicated that for the oxygen reduction reaction (ORR), edge carbon atoms are more active than basal plane carbon atoms.

12.3 Mechanisms and Advancements

12.3.1 Desalination Mechanisms

The process of photocatalytic reaction primarily depends on the wavelength or photon energy and the catalyst. Semiconductor materials are the primary photocatalysts, and due to their unique electronic structure characterized by the VB and CB, they can serve as sensitizers for light-stimulated redox processes. The semiconductor photocatalytic oxidation–reduction mechanism is shown in Figure 12.3 [50]. When the energy of incident photons equals or exceeds the bandgap of a semiconductor photocatalyst, VB electrons may be excited and transferred to the CB of the semiconductor photocatalyst, thereby generating holes (h^+) in VB and electrons (e^-) in CB. The holes produced in the VB can oxidize pollutant molecules and react with water molecules through advanced oxidation processes to produce hydroxyl radicals. The hydroxyl radicals generated through this oxidation process serve as strong oxidizing agents, leading to the degradation of pollutants in the water purification process. The CB electrons can react with dissolved oxygen species to form superoxide anions and induce redox reactions [51, 52].

As mentioned in the introduction, the photothermal effect of semiconductor materials is also an effective means for seawater desalination. It primarily utilizes the energy released through radiative relaxation when excited electrons in semiconductors return to their low-energy states (Figure 12.2b), thereby enabling the evaporation of seawater. Additionally, semiconductor photocatalysts harness their redox capabilities under sunlight to disinfect and decompose harmful substances produced during the desalination process. For instance, their integration with membrane devices can help address the prevalent issue of biofouling in desalination membranes today. Furthermore, the combination of parallel filtration units with photocatalytic degradation represents an ideal setup for desalinating seawater

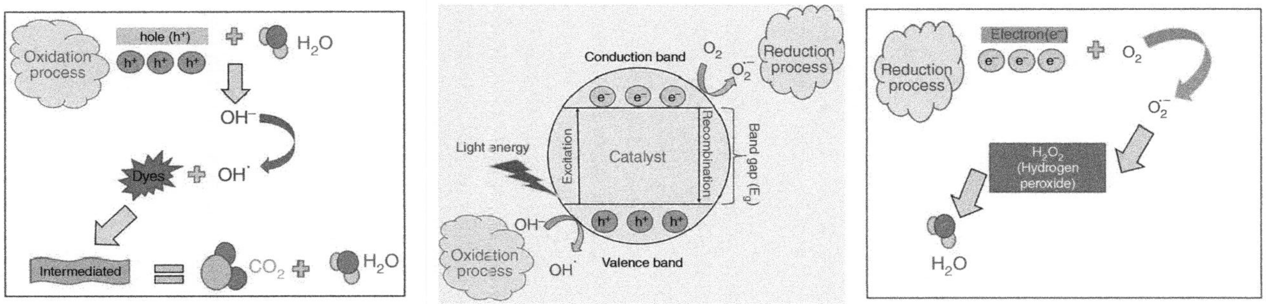

Figure 12.3 Schematic diagram of semiconductor photocatalytic oxidation–reduction mechanism. Reproduced from Saravanan et al. [50] / with permission of Springer Nature.

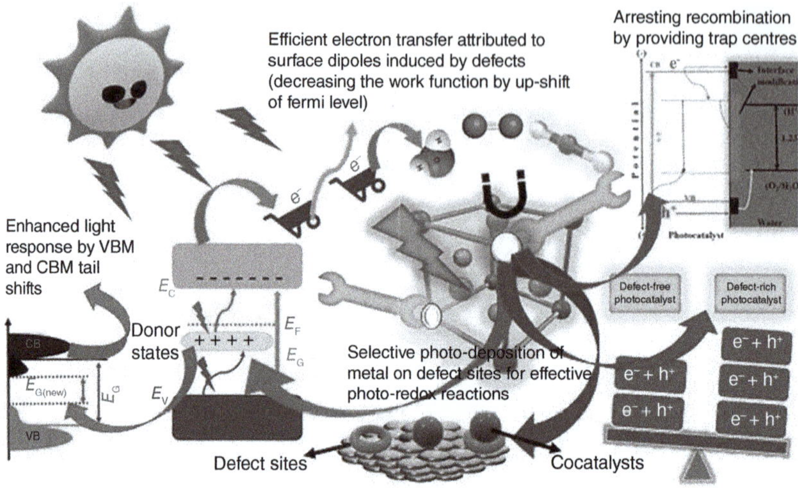

Figure 12.4 The effect of defects involving modifications in the band structure, electronic properties, interface properties, and photocatalytic mechanism. Reproduced from Maarisetty and Baral [54] / with permission of The Royal Society of Chemistry.

contaminated with various organic and inorganic pollutants [51]. Therefore, semiconductor photocatalysts can effectively couple the processes of photothermal seawater desalination and catalytic pollutant degradation.

The Shockley–Read–Hall (SRH) recombination theory indicates that defect levels in semiconductors serve as effective nonradiative recombination centers, generating a photothermal effect under light illumination. The resulting thermoelectric potential can overcome external voltages, promote carrier separation, suppress recombination, extend carrier lifetime, and enhance the performance of photocatalysts [53]. The influence of defects on band structure, electronic properties, interface properties, and photocatalytic mechanisms is illustrated in Figure 12.4 [54]. Defect engineering has emerged as a promising strategy to improve the photocatalytic performance of catalysts by forming surface dipoles, adjusting the position of the Fermi level, improving dissociation and adsorption, providing defect sites, and modulating band edges.

12.3.2 Advancements of Defective Photocatalysts in Desalination

12.3.2.1 Oxygen Vacancy

Yang et al. [55], inspired by plants' selective utilization of the solar spectrum, designed a spectrally tailored aerogel for solar energy utilization, as illustrated in Figure 12.5. This aerogel consists of semiconductor HNb_3O_8 (D-HNb_3O_8) nanosheets rich in OV defects and a polyacrylamide (PAM) framework, enabling integrated photochemical and photothermal conversion of the entire solar spectrum. Specifically, D-HNb_3O_8 absorbs the full range of solar radiation to perform two distinct spectral-selective functions: UV-driven photocatalytic excitation and visible-near-infrared induced photothermal conversion. Interconnected thermally

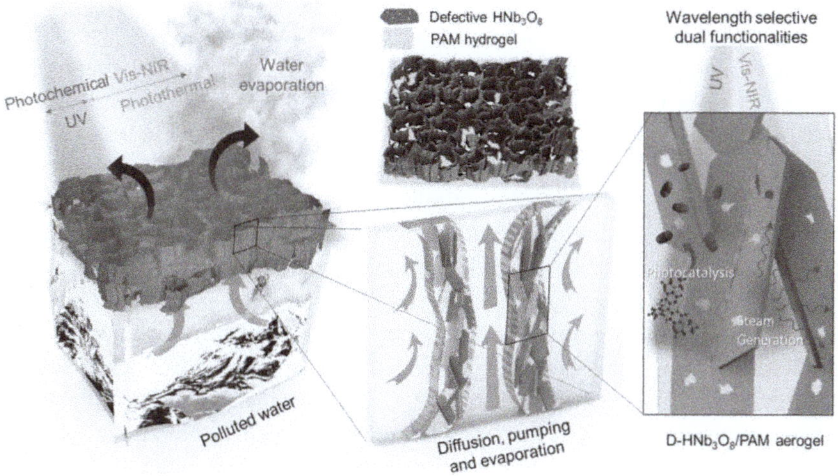

Figure 12.5 A schematic depiction of the concurrent photothermal and photochemical solar energy conversion in a defective D-HNb_3O_8/PAM aerogel. Reproduced from Yang et al. [55] / with permission of John Wiley & Sons.

insulating porous networks are constructed using hydrophilic PAM chains, which possess capillary pumping capabilities to minimize heat loss and provide sufficient water for continuous solar steam generation. Furthermore, the microporous PAM framework exhibits strong adsorption capacity for reactant molecules. More importantly, due to the uniform infiltration of D-HNb_3O_8 within the PAM polymer matrix, the solar energy collected and converted by D-HNb_3O_8 can be directly transferred to the reactant molecules or water adsorbed by the PAM network. In addition, the local thermal energy caused by the insulation effect of PAM increases the near-field temperature around the D-HNb_3O_8 reactant, which accelerates the photochemical reaction and synergistically promotes the integrated conversion of photochemistry and photothermal reactions. D-HNb_3O_8/PAM aerogel achieved a high solar steam generation rate of 1.401 $kg\,m^{-2}\,h^{-1}$, with a solar thermal conversion efficiency of up to 91%.

Cao et al. [56] have constructed a photothermal interfacial evaporator, utilizing OV-containing BiOBr (BiOBr-OVs) as the photothermal material. Except for the photothermal material and sodium alginate coating, all other raw materials for this evaporator come from daily life (the composition materials of the water-absorbing layer and floating layer are daily towels and recycled hydrophobic sponges, respectively), thus reducing its cost. Significantly, the optimal evaporator achieved a water vapor generation rate of 1.63 $kg\,m^{-2}\,h^{-1}$ under one-sun irradiation, marking a 317.95% increase compared to pure water evaporation (0.39 $kg\,m^{-2}\,h^{-1}$). This enhancement is attributed to the efficient interfacial evaporation and the combined effects of the photothermal, water-absorbing, and floating layers. Notably, during steam production under one-sun irradiation, the evaporator effectively photodegrades RhB in bulk water with a remarkable efficiency of 99.9%. The result is due to the nonradiative relaxation effect of OVs in BiOBr, holes oxidation, and the synergistic effect among the layers of the optimal evaporator, which indicates the evaporator's ability to prevent dye aggregation during water evaporation. Additionally, compared to pure water, the steam generation rate of this evaporator decreases by only 1.23% and 3.07% when using 3.5 and 10 wt% NaCl solutions, respectively. Meanwhile, the evaporator achieves a removal rate of 99.98% for K^+, Na^+, and Mg^{2+} in simulated seawater. Therefore, high-salinity simulated seawater desalination is achieved through the optimization of the evaporator.

Tan et al. [57] utilized indium oxide as a prototype to explore how point defects affect its physicochemical characteristics and performance in interfacial solar steam generation (ISSG). Their results indicated that OVs played a pivotal role in light absorption by introducing multiple deep energy levels, which in turn augmented sub-bandgap photon capture. This improved light absorption led to a higher carrier concentration and a substantial reduction in carrier lifetime. Consequently, energy was predominantly dissipated through nonradiative carrier recombination in the form of heat, enhancing photothermal conversion efficiency. Simultaneously, hydrophilicity was enhanced due to the high surface energy induced by OVs. Remarkably, defect-laden black In_2O_3 (b-In_2O_3) demonstrated an impressive photothermal conversion efficiency of 74% and robust ISSG performance with an evaporation rate of 2.3 kg m^{-2} h^{-1}, maintaining exceptional stability for over 60 hours under one-sun conditions. Furthermore, the practical application b-In_2O_3 in wastewater purification was highlighted, achieving a significant reduction in metal ion concentration in the treated water that adhered to World Health Organization (WHO) standards. These discoveries offer profound insights into the development of oxide-based photothermal materials and underscore the promise of defect-engineered b-In_2O_3 as an innovative material for efficient solar-powered water purification, presenting a sustainable approach to addressing global water scarcity.

Li et al. [58] synthesized a gray-black TiO_2 material rich in OVs via high-frequency low-temperature plasma. They combined this material with a modified PVDF sponge and polystyrene foam to form a 3D solar evaporator. The evaporator showed broad spectral absorption, low thermal conductivity, and high evaporation efficiency. Experimental data confirmed that OVs narrowed the TiO_2 bandgap, enhancing light absorption and photothermal performance. In desalination tests, the system achieved a notable evaporation rate of 2.91 kg m^{-2} h^{-1} and an evaporation efficiency of 75.52% efficiency under one-sun illumination, seven times higher than natural evaporation. The desalinated water met WHO standards and even approached soft water levels. This research contributes to the development of high-performance TiO_2 photothermal materials and seawater evaporators, promoting sustainable and efficient desalination technologies.

12.3.2.2 Sulfur Vacancy

He et al. [59] introduced a 3D wood membrane coated internally with sulfur-defective MoS_2 (WM-H) for effective seawater desalination. This design leverages the vertically aligned mesoporous wood channels to increase the water–MoS_2 interface. Due to sulfur defects, the membrane's surface temperature reaches about 53 °C under one-sun illumination. Notably, the sulfur-defective MoS_2-coated wood boasts a high evaporation flux of 1.46 kg m^{-2} h^{-1} (under one sun) and a solar-thermal conversion efficiency of 82.5%. After desalination, the ion concentration of the evaporated water is far below WHO standards. Moreover, defective MoS_2 exhibits enhanced Na^+ adsorption peaks, being 2.17 and 1.49 times higher on both sides compared to defect-free MoS_2. WM-H also demonstrates robust structural stability and salt tolerance, affirming its reliability for prolonged use in solar seawater desalination.

Yuan et al. [60] have developed a high-efficiency solar energy utilization system by utilizing vertically aligned MoS_{2-x} nanosheet arrays (MoS_{2-x} NSAs) with S vacancies, grown in situ on an Mo mesh, as the solar absorber. In this system, interfacial heat is harnessed to drive water evaporation and photothermally enhanced photodegradation, while the generated low-grade waste heat is simultaneously converted into electrical energy. The MoS_{2-x} NSAs exhibit a solar absorption rate of 94.2%, along with excellent photothermal conversion and thermal localization properties. The localized heat and S vacancies synergistically enhance the photodegradation performance by promoting the separation of photogenerated carriers, thus preventing the accumulation of dyes on the surface of the MoS_{2-x} NSAs over extended operation periods. The water evaporation rates achieved by the MoS_{2-x} NSAs and 3D U-shaped MoS_{2-x} NSAs under one-sun illumination are 1.32 and 1.77 kg m^{-2} h^{-1}, respectively.

Shridharan et al. [61] synthesized uniform, submicron-sized red phosphorus (RP) nanosheets enriched with defects through a simple ball-milling technique, showcasing their superior and long-lasting solar steam generation properties. By adjusting the milling duration, they obtained three RP sizes, with RP60 standing out for its smallest lateral particle size, nanosheet shape, and defect-rich surface. RP60 also exhibited distinctive characteristics, including a narrow bandgap (1.44 eV), mini mal thermal conductivity (0.07 W m^{-1} K^{-1}), and low heat capacity (0.66 J g^{-1} K^{-1}), paired with exceptional wettability. When exposed to simulated one-sun conditions (100 mW cm^{-2}), the RP60 photothermal absorber delivered a high-water evaporation rate of 1.34 kg $(m^2{\cdot}h)^{-1}$ and maintained a consistent solar steam generation efficiency of 74.1% for over 10 hours. The design featuring one-dimensional water channels and a porous polyurethane support optimized water delivery, ensured a large contact area, localized heat, and facilitated steam release. By incorporating plasmonic heating using silver nanoparticles on RP60, they achieved a notable increase in solar steam generation efficiency, surpassing 96.0%, and a water evaporation rate of 1.75 kg $(m^2{\cdot}h)^{-1}$. This research underscores the importance of morphology, particle size, and defect management in augmenting photothermal performance for efficient and enduring interfacial seawater desalination.

Huang et al. [62] reported for the first time on a solar water evaporator based on Cu_xS. This was achieved by assembling and designing biphasic Cu_xS composite nanorods into a 3D hierarchical structured foam with efficient light harvesting, super-hydrophilic, and underwater super-oleophobic properties. The preparation of the Cu_xS/Cu foam (SCF) and its application in multifunctional solar distillation are illustrated in Figure 12.6. Due to enhanced synergistic photothermal conversion

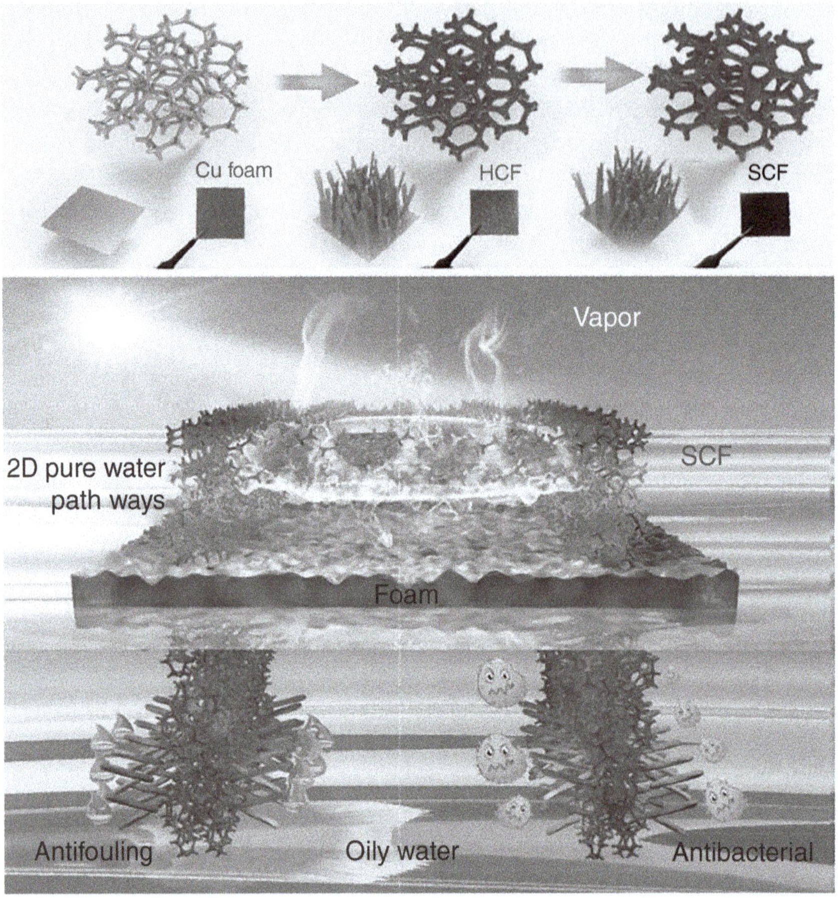

Figure 12.6 Schematic illustration for the fabrication of SCF and its application for multifunctional solar distillation. Reproduced from Huang et al. [62] / reprinted with permission of Elsevier.

and confined two-dimensional (2D) water pathways, the manufactured SCF foam evaporator with copper vacancies can achieve an extremely high-water evaporation rate of 1.96 kg m^{-2} h^{-1}, with an energy efficiency of 94.5%, making it one of the most performant inorganic semiconductor-based evaporators. The developed device also maintained a high evaporation rate of 1.73 kg m^{-2} h^{-1} and an average evaporation rate of 5.2 L m^{-2} d^{-1} over 30 consecutive days of outdoor testing, producing clean water with excellent antifouling, antibacterial, and ion exclusion properties from oil-in-water emulsions. As a proof-of-concept application, the Cu_xS-based evaporator demonstrated remarkable ability to purify clean water and organic solvents from dyed solutions through multiphase processes.

Zhao et al. [63] prepared an $Ag/MoS_2/TiO_{2-x}$ ternary heterojunction through a combination of hydrothermal and photodeposition processes, combined with an in situ solid-state chemical reduction method. Their findings revealed that this heterojunction featured Ti^{3+} doping, with Ag nanoparticles and MoS_2 nanosheets seamlessly adhered to the TiO_2 nanobelt surface. Notably, the $Ag/MoS_2/TiO_{2-x}$ system exhibited remarkable performance, achieving a bisphenol A degradation rate of 96.7% in low-temperature water and a hydrogen production rate of 1.98 mmol h^{-1} g^{-1}, both significantly surpassing pristine TiO_2. Furthermore, the photothermal properties of $Ag/MoS_2/TiO_{2-x}$ were surprisingly impressive. The exceptional photocatalytic activity and photothermal efficiency stemmed from the combined effects of the heterojunction formation, Ti^{3+} and surface OV defects, and the surface plasmon resonance of silver nanoparticles. These factors collectively broadened the photoresponse into the visible-infrared range and facilitated effective spatial separation of photogenerated charge carriers.

Xu et al. [64] developed a 2D/2D cesium tungsten bronze/copper sulfide (2D/2D $Cs_{0.32}WO_3/CuS$) nanoheterojunction and incorporated it into a cellulose-based hybrid hydrogel to construct a multifunctional evaporator. Thanks to the increased nonradiative recombination centers in the deep-level defects (W^{5+}) and OVs of $Cs_{0.32}WO_3$, as well as the shorter carrier migration distance and higher redox potential in the $Cs_{0.32}WO_3/CuS$ nanoheterojunction, the evaporator exhibited significantly enhanced thermal output capabilities, achieving ultra-efficient seawater evaporation (4.22 kg m^{-2} h^{-1}) and photodegradation of organic pollutants (methylene blue and rhodamine B) with removal rates approaching 99%. Furthermore, the evaporator demonstrated long-term stability and sustainable self-cleaning properties, preventing the accumulation of salt.

12.4 Conclusions and Future Perspectives

To overcome the issue of freshwater scarcity, semiconductor photocatalytic materials rich in defects offer an innovative solution for seawater desalination. Defect engineering in photocatalytic materials represents a promising and attractive approach to significantly improve the performance of energy-related applications. The primary functions of defects of photocatalysts include expanding the absorption range of sunlight, facilitating charge transfer and separation, and modulating surface catalytic reactions, which are crucial for enhancing photocatalytic activity [14]. In many cases, while a defect can promote a photocatalytic reaction, it may also have negative impacts [58]. Therefore, it poses a significant challenge to eliminate the disadvantages of defects in photocatalytic materials and maximize their advantages by custom-designing defective photocatalysts with the desired defect type, location, and concentration [65].

Despite the confirmed recent advancements in semiconductor-based photothermal catalysis, understanding is needed to optimize the photothermal catalytic process. More importantly, understanding the photothermal effect and identifying the catalytic mechanisms, including the relaxation and transfer of hot carriers on the femtosecond timescale, will be crucial [66]. Meanwhile, the preparation and design of semiconductor photocatalytic materials for solar-driven seawater desalination involve multiple theories, methods, and technologies. Especially when combining the fields of photothermal water evaporation and photocatalysis, many new challenges await resolution, requiring continuous accumulation of practical experience and refinement of the preparation system. Further research can still be conducted in the following aspects:

1) As the understanding of defect engineering deepens, researchers will explore new material systems that exhibit promising properties for desalination. This may include metals, metal oxides, sulfides, selenides, and other inorganic compounds that can be engineered to have defects that enhance their photocatalytic activity.
2) Photogenerated carriers (electrons–holes) play a significant role in the synergistic process of photothermal conversion and catalytic conversion for semiconductor photocatalysts. The proportion of carriers' contributions to these two processes, the influencing factors, and the microscopic interactions between the photothermal and photocatalytic processes all remain to be further studied.
3) The ability to accurately characterize defective photocatalytic materials is crucial for understanding their properties and optimizing their performance. Future research will rely on advanced characterization techniques, such as high-resolution electron microscopy, spectroscopy, and computational modeling, to provide insights into the structure, composition, and behavior of defects within the materials.
4) Due to the differences in active sites for solar energy utilization among various types of photothermal catalysts, elucidating the exact distribution of different forms of energy conversion during the photothermal catalytic seawater desalination process using semiconductor photocatalysts will be a key research question in the future.
5) One of the challenges in the adoption of defective photocatalytic materials for desalination is their scalability and cost-effectiveness. Research will focus on developing scalable synthesis methods that can produce large quantities of material with consistent properties. Additionally, efforts will be made to reduce the cost of production by exploring low-cost raw materials, efficient processing techniques, and recycling strategies.
6) The rate of photocatalytic reactions is not only limited by the activity of the catalyst but also influenced by the mass transfer process of pollutant molecules to the catalyst surface. Therefore, further research and exploration into the mass transfer behavior during the photothermal catalytic seawater desalination process are needed in the future.
7) Insufficient binding force between photothermal materials and photocatalysts can affect the efficiency of seawater desalination. Direct mixing not only greatly reduces the extent to which the catalyst is irradiated by sunlight but also easily leads to the detachment of the photocatalyst. Therefore, new strategies for combining photothermal water evaporation with photocatalytic materials need to be further developed. The interaction between photocatalysts and photothermal materials can be enhanced through methods such as chemical grafting or in situ growth.
8) Currently, solar evaporators lack sufficient salt resistance when used in large-scale seawater desalination processes. Therefore, it is essential to develop

new types of seawater desalination evaporators that can operate under high salt concentrations while maintaining high photothermal conversion efficiency.

In summary, the realm of desalination employing defective semiconductor materials offers immense potential in tackling the pressing issue of global water scarcity. As researchers continue to evolve, the innovation of novel material compositions, sophisticated analytical tools, and scalable, economically viable manufacturing processes will be pivotal in propelling this technology forward. Furthermore, efforts will concentrate on refining defect engineering strategies, bolstering material durability, enhancing charge separation and transport efficiencies, and seamless integration with existing desalination techniques, marking these as central areas of focus for future advancements.

Acknowledgments

This work was supported by the project of National Science Foundation for Fujian Province, China (2025J011072), and the Introduced Talent Start-up Fund Project of Wuyi University (No. YJ202323).

Conflict of Interest

The authors declare that they have no conflict of interest.

References

1 Liu, P., Hu, Y., Li, X. et al. (2022). Enhanced solar evaporation using a scalable MoS_2-based hydrogel for highly efficient solar desalination. *Angew. Chem. Int. Ed.* 61 (37): e202208587.

2 Ni, G.W., Zandavi, S.H., Javid, S.M. et al. (2018). A salt-rejecting floating solar still for low-cost desalination. *Energy Environ. Sci.* 11: 1510–1519.

3 Wei, Z., Wang, J., Guo, S., and Tan, S.C. (2022). Towards highly salt-rejecting solar interfacial evaporation: photothermal materials selection, structural desings, and energy management. *Nano Res. Energy* 1: e9120014.

4 Jiang, H.L., Ai, L.H., Chen, M., and Jiang, J. (2020). Broadband nickel sulfide/nickel foam-based solar evaporator for highly efficient water purification and electricity generation. *ACS Sustain. Chem. Eng.* 8: 10833–10841.

5 Wang, X., Liu, Q., Wu, S. et al. (2019). Multilayer polypyrrole nanosheets with self-organized surface structures for flexible and efficient solar-thermal energy conversion. *Adv. Mater.* 31: 1807716.

6 Bai, H., Zhao, T., and Cao, M. (2020). Interfacial solar evaporation for water production: from structure design to reliable performance. *Mol. Syst. Des. Eng.* 5 (2): 419–432.

7 Zheng, Z., Li, H., Zhang, X. et al. (2020). High-absorption solar steam device comprising Au@Bi_2MoO_6-CDs: extraordinary desalination and electricity generation. *Nano Energy* 68: 104298.

8 Gao, M., Zhu, L., Peh, C.K., and Ho, G.W. (2019). Solar absorber material and system designs for photothermal water vaporization towards clean water and energy production. *Energy Environ. Sci.* 12: 841–864.

9 Zhou, L., Tan, Y., Wang, J. et al. (2016). 3D self-assembly of aluminium nanoparticles for plasmon-enhanced solar desalination. *Nat. Photonics* 10: 393–398.

10 Ghim, D., Jiang, Q., Cao, S. et al. (2018). Mechanically interlocked 1T/2H phases of MoS_2 nanosheets for solar thermal water purification. *Nano Energy* 53: 949–957.

11 Li, X., Li, J., Lu, J. et al. (2018). Enhancement of interfacial solar vapor generation by environmental energy. *Joule* 2 (7): 1331–1338.

12 Shi, L., Shi, Y., Zhuo, S. et al. (2019). Multi-functional 3D honeycomb ceramic plate for clean water production by heterogeneous photo-Fenton reaction and solar-driven water evaporation. *Nano Energy* 60: 222–230.

13 Tong, H., Ouyang, S., Bi, Y. et al. (2012). Nano-photocatalytic materials: possibilities and challenges. *Adv. Mater.* 24: 229.

14 Bai, S., Zhang, N., Gao, C., and Xiong, Y. (2018). Defect engineering in photocatalytic materials. *Nano Energy* 53: 296–336.

15 Li, H., Li, J., Ai, Z. et al. (2018). Oxygen vacancy-mediated photocatalysis of BiOCl: reactivity, selectivity and perspective. *Angew. Chem. Int. Ed.* 57 (1): 122–138.

16 Fang, W. and Shangguan, W. (2018). A review on bismuth-based composite oxides for photocatalytic hydrogen generation. *Int. J. Hydrogen Energ.* 4 (2): 895–912.

17 Nowotny, J., Alim, M.A., Bak, T. et al. (2015). Defect chemistry and defect engineering of TiO_2-based semiconductors for solar energy conversion. *Chem. Soc. Rev.* 44: 8424–8442.

18 Zhang, N., Li, X., Ye, H. et al. (2016). Oxide defect engineering enables to couple solar energy into oxygen activation. *J. Am. Chem. Soc.* 138: 8928.

19 Chen, X., Liu, L., and Huang, F. (2015). Black titanium dioxide (TiO_2) nanomaterials. *Chem. Soc. Rev.* 44: 1861–1885.

20 Guan, M., Xiao, C., Zhang, J. et al. (2013). Vacancy associates promoting solar-driven photocatalytic activity of ultrathin bismuth oxychloride nanosheets. *J. Am. Chem. Soc.* 135: 10411.

21 Kong, M., Li, Y., Chen, X. et al. (2011). Tuning the relative concentration ratio of bulk defects to surface defects in TiO_2 nanocrystals leads to high photocatalytic efficiency. *J. Am. Chem. Soc.* 133: 16414–16417.

22 Song, J., Huang, Z.F., Pan, L. et al. (2015). Oxygen-deficient tungsten oxide as versatile and efficient hydrogenation catalyst. *ACS Catal.* 5 (11): 6594–6599.

23 Yu, H., Li, J., Zhang, Y. et al. (2019). Three-in-one oxygen vacancies: whole visible-spectrum absorption, efficient charge separation, and surface site activation for robust CO_2 photoreduction. *Angew. Chem. Int. Ed.* 58 (12): 3880–3884.

24 Chen, X., Liu, L., Yu, P.Y., and Mao, S.S. (2011). Increasing solar absorption for photocatalysis with black hydrogenated titanium dioxide nanocrystals. *Science* 331: 746–750.

25 Liu, N., Schneider, C., Freitag, D. et al. (2014). Black TiO_2 nanotubes: cocatalyst-free open-circuit hydrogen generation. *Nano Lett.* 14: 3309–3313.

26 Teng, F., Li, M., Gao, C. et al. (2014). Preparation of black TiO_2 by hydrogen plasma assisted chemical vapor deposition and its photocatalytic activity. *Appl. Catal. B Environ.* 148–149: 339–343.

27 Yu, X., Kim, B., and Kim, Y.K. (2013). Highly enhanced photoactivity of anatase TiO_2 nanocrystals by controlled hydrogenation-induced surface defects. *ACS Catal.* 3: 2479–2486.

28 Wang, Z., Yang, C., Lin, T. et al. (2013). Visible-light photocatalytic, solar thermal and photoelectrochemical properties of aluminium-reduced black titania. *Energy Environ. Sci.* 6: 3007–3014.

29 Cui, H., Zhao, W., Yang, C. et al. (2014). Black TiO_2 nanotube arrays for high-efficiency photoelectrochemical water-splitting. *J Mater Chem A* 2: 8612–8616.

30 Mao, C., Zuo, F., Hou, Y. et al. (2014). In situ preparation of a Ti^{3+}self-doped TiO_2 film with enhanced activity as photoanode by N_2H_4 reduction. *Angew. Chem. Int. Ed.* 53: 10485.

31 Cao, S., Fan, B., Feng, Y. et al. (2018). Sulfur-doped g-C_3N_4 nanosheets with carbon vacancies: general synthesis and improved activity for simulated solar-light photocatalytic nitrogen fixation. *Chem. Eng. J.* 353: 147–156.

32 Wang, B., Wang, X., Lu, L. et al. (2018). Oxygen-vacancy-activated CO_2 splitting over amorphous oxide semiconductor photocatalyst. *ACS Catal.* 8 (1): 516–525.

33 Zhang, X., Zhao, Z., Zhang, W. et al. (2016). Surface defects enhanced visible light photocatalytic H_2 production for Zn-Cd-S solid solution. *Small* 12 (6): 793–801.

34 Pan, L., Wang, S., Mi, W. et al. (2014). Undoped ZnO abundant with metal vacancies. *Nano Energy* 9: 71–79.

35 Ye, L., Jin, X., Leng, Y. et al. (2015). Synthesis of black ultrathin BiOCl nanosheets for efficient photocatalytic H_2 production under visible light irradiation. *J. Power Sources* 293: 409–415.

36 Yan, J., Wang, T., Wu, G. et al. (2015). Tungsten oxide single crystal nanosheets for enhanced multichannel solar light harvesting. *Adv. Mater.* 27: 1580.

37 Cheng, H., Kamegawa, T., Mori, K., and Yamashita, H. (2014). Surfactant-free nonaqueous synthesis of plasmonic molybdenum oxide nanosheets with enhanced catalytic activity for hydrogen generation from ammonia borane under visible light. *Angew. Chem. Int. Ed.* 53: 2910.

38 Kong, X.Y., Choo, Y.Y., Chai, S.P. et al. (2016). Oxygen vacancies induced Bi_2WO_6 for realization of full solar spectrum photocatalytic CO_2 reduction: from UV to NIR region. *Chem. Commun.* 52: 14242.

39 Wei, Z., Liu, Y., Wang, J. et al. (2015). Controlled synthesis of a highly dispersed $BiPO_4$ photocatalyst with surface oxygen vacancies. *Nanoscale* 7: 13943.

40 Xi, G., Ouyang, S., Li, P. et al. (2012). Ultrathin $W_{18}O_{49}$ nanowires with diameters below 1 nm: synthesis, near-infrared absorption, photoluminescence, and photochemical reduction of carbon dioxide. *Angew. Chem. Int. Ed.* 51: 2395.
41 Sun, Y., Qiu, S., Fang, Z. et al. (2023). Rapid synthesis of oxygen-deficient MoO_{3-x}-rGO composites for synergistic photothermal seawater desalination and photocatalytic sterilization. *ACS Sustain. Chem. Eng.* 11: 3359–3369.
42 Zheng, P., Tang, J., Zhou, Z. et al. (2021). Ultrafast synthesis of defective black TiO_2 via one-step NaN_3 deflagration for high-efficiency solar water evaporation. *Surf Interfaces* 22: 100901.
43 Hu, X.L., Zhang, W.J., Yong, Y.W. et al. (2020). One-step synthesis of iodine-doped g-C3N4 with enhanced photocatalytic nitrogen fixation performance. *Appl. Surf. Sci.* 510: 145413.
44 Lei, F., Sun, Y., Liu, K. et al. (2014). Oxygen vacancies confined in ultrathin indium oxide porous sheets for promoted visible-light water splitting. *J. Am. Chem. Soc.* 136: 6826.
45 Li, J., Zhan, G., Yu, Y., and Zhang, L. (2016). Superior visible light hydrogen evolution of Janus bilayer junctions via atomic-level charge flow steering. *Nat. Commun.* 7: 11480.
46 Zhang, S., Liu, X., Liu, C. et al. (2018). MoS_2 quantum dot growth induced by S vacancies in a $ZnIn_2S_4$ monolayer: atomic-level heterostructure for photocatalytic hydrogen production. *ACS Nano* 12: 751.
47 Ye, L., Deng, K., Xu, F. et al. (2012). Increasing visible-light absorption for photocatalysis with black BiOCl. *Phys. Chem. Chem. Phys.* 14: 82.
48 Zhu, Y., Ling, Q., Liu, Y. et al. (2016). Photocatalytic performance of $BiPO_4$ nanorods adjusted via defects. *Appl. Catal. B Environ.* 187: 204–211.
49 Shen, A., Zou, Y., Wang, Q. et al. (2014). Oxygen reduction reaction in a droplet on graphite: direct evidence that the edge is more active than the basal plane. *Angew. Chem. Int. Ed.* 53 (40): 10804–10808.
50 Saravanan, R., Gracia, F., and Stephen, A. (2017). *Basic Principles, Mechanism, and Challenges of Photocatalysis, Nanocomposites for Visible Light-Induced Photocatalysis*, 19–40. Springer.
51 Anis, S.F., Hashaikeh, R., and Hilal, N. (2019). Functional materials in desalination: a review. *Desalination* 468: 114077.
52 Kumar, K. and Chowdhury, A. (2020). Use of novel nanostructured photocatalysts for the environmental sustainability of wastewater treatments. *Encycl. Renewable and Sustainable Mater.* 1: 949–964.
53 Yang, Z., Huang, T., Li, M. et al. (2024). Unveiling the synergistic role of frustrated lewis pairs in carbon-encapsulated Ni/NiO_x photothermal cocatalyst for enhanced photocatalytic hydrogen production. *Adv. Mater.* 36 (24): 2313513.
54 Maarisetty, D. and Baral, S.S. (2020). Defect engineering in photocatalysis: formation, chemistry, optoelectronics, and interface studies. *J Mater Chem A* 8: 18560–18604.
55 Yang, M., Tan, C.F., Lu, W. et al. (2020). Spectrum tailored defective 2D semiconductor nanosheets aerogel for full-spectrum-driven photothermal water evaporation and photochemical degradation. *Adv. Funct. Mater.* 30 (43): 2004460.

56 Cao, Y., Zhou, Z., Zhang, T. et al. (2023). Vacancy defects assist solar interfacial evaporator to realize simultaneously clean water production and pollutant removal. *Sep. Purif. Technol.* 326: 124803.

57 Tan, R., Shridharan, T.S., Lee, J.H. et al. (2025). Defect-engineered black indium oxide: a high-performance photothermal material for solar-driven water purification. *Desalination* 599: 118440.

58 Li, F., Liu, C., Dong, Y. et al. (2024). Thermal plasma-synthesized gray-black TiO_2 with abundant oxygen vacancies for high-efficiency solar desalination. *Environ. Sci.: Water Res. Technol.* 10: 1271.

59 He, X., Zhang, L., Hu, X., and Zhou, Q. (2021). Formation of S defects in MoS_2-coated wood for high-efficiency seawater desalination. *Environ. Sci. Nano* 8: 2069.

60 Yuan, B.H., Yang, L.X., Yang, H.W. et al. (2022). Flexible vacancy-mediated MoS_{2-x} nanosheet arrays for solar-driven interfacial water evaporation, photothermal-enhanced photodegradation, and thermoelectric generation. *Energy Convers. Manag.* 252: 115070.

61 Shridharan, T.S., Sivanantham, A., Tan, R. et al. (2023). Defect-enriched red phosphorus nanosheets as efficient and stable photothermal absorber material for interfacial solar desalination. *Desalination* 561: 116700.

62 Huang, W., Su, P., Cao, Y. et al. (2020). Three-dimensional hierarchical Cu_xS-based evaporator for high-efficiency multifunctional solar distillation. *Nano Energy* 69: 104465.

63 Zhao, T., Xing, Z., Xiu, Z. et al. (2019). Synergistic effect of surface plasmon resonance, Ti^{3+} and oxygen vacancy defects on $Ag/MoS_2/TiO_{2-x}$ ternary heterojunctions with enhancing photothermal catalysis for low-temperature wastewater degradation. *J. Hazard. Mater.* 364: 117–124.

64 Xu, J., Zhuo, S., Luo, Y. et al. (2023). 2D/2D $Cs_{0.32}WO_3/CuS$ nano-heterojunctions for simultaneous high-efficiency solar desalination, photocatalytic decontamination, and electricity generation. *Small Methods* 7 (9): 2300558.

65 Xie, Q., Cai, Z., Li, P. et al. (2018). Layered double hydroxides with atomic-scale defects for superior electrocatalysis. *Nano Res* 11 (9): 4524–4534.

66 Song, C., Wang, Z., Yin, Z. et al. (2022). Principles and applications of photothermal catalysis. *Chem. Catal.* 2: 52–83.

13

Defective Photocatalytic Materials for Organic Pollutant Degradation

Jiapeng Hu

Wuyi University, Fujian Provincial Key Laboratory of Eco-Industrial Green Technology, College of Ecology and Resources Engineering, Baihua Road, Wuyishan, 354300, PR China

13.1 Introduction

Water is one of the fundamental elements of life on Earth, serving as the primary material basis for the survival and activities of all living organisms. Although water covers approximately 71% of the Earth's surface, only about 2.5% of it is freshwater. Freshwater is an essential resource for sustaining life and the environment, encompassing rivers, lakes, ponds, and groundwater [1]. The quality of freshwater directly impacts industrial development and the progress of human society. However, of all freshwater resources, only about 1% is readily accessible for use, while the remainder is largely locked in inaccessible groundwater and glaciers [2].

Due to explosive population growth and accelerated industrialization, freshwater resources are increasingly being polluted. Organic pollutants are generally classified into four categories based on their biological toxicity and biodegradability [3]: (i) nontoxic and easily degradable, which includes most organic compounds; (ii) nontoxic but difficult to degrade, such as lignin, cellulose, sodium alkylbenzene sulfonate, and polyvinyl alcohol; (iii) toxic but biodegradable, including compounds like chlorophenol, formaldehyde, phenol, and nitro compounds; and (iv) toxic and difficult to degrade, such as polychlorinated biphenyls (PCBs), quinoline, and pyridine. Typically, the term "refractory organic pollutants" refers to the substances in categories 2, 3, and 4. With the development of agriculture and animal husbandry, surface water pollution caused by eutrophication and the accumulation of persistent organic pollutants (POPs) like pesticides in organisms, as well as the discharge of industrial wastewater containing large amounts of toxic and harmful POPs during industrialization, have led to global pollution issues. These organic pollutants accumulate in the environment and are transmitted through the food chain, ultimately posing risks to human health. Industrial wastewater from industries such as dyes, printing, coking, pesticides, pharmaceuticals, fine chemicals, and petroleum is characterized by containing large quantities of toxic, harmful, and biologically refractory organic pollutants [4]. Table 13.1 summarizes the main refractory organic pollutants and their associated hazards.

Defect Engineering in Photocatalysis: From Fundamentals to Applications,
First Edition. Edited by Chuanyi Wang, Abdelkader Labidi, and Qiuhui Zhu.

Table 13.1 Major refractory organic pollutants and their hazards.

Industry	Main difficulty in degrading pollutants contained	Characteristics and hazards	Representative pollutants
Coking and petrochemical wastewater	Polycyclic aromatic hydrocarbons	Stable in nature and highly carcinogenic	Benzo (a) pyrene, 7,12-dimethylanthracene
	Heterocyclic compounds	Stable in nature, biologically enriched, with mutagenic and carcinogenic effects	Benzo (a) acridine, Pyridine, Pyrrole
	Organic cyanide compounds	Highly toxic substances	Acetonitrile, nitrile, and acrylonitrile
Dyes and printing and dyeing wastewater	Heterocyclic compounds	Same as mentioned earlier	—
	Polychlorinated biphenyls (PCBs)	Enters the human body through the food chain, causing acute poisoning and carcinogenic effects on the body	—
	Synthetic dye	High chromaticity, toxic, and carcinogenic	Rhodamine B (RhB), Methyl orange (MO), Methylene blue (MB), Acid orange (AO)
Pesticide wastewater	Heterocyclic compounds	Same as mentioned earlier	—
	Synthetic pesticides	Toxic and carcinogenic to the human body	Dimethylamine, Glyphosate
Textile fiber industry, paper industry, and food industry	Synthetic detergent	Foaming affects the biological treatment effect and has a solubilizing effect on polycyclic aromatic hydrocarbons	Anionic surfactants (LAS), unsaturated fatty acids, protein adhesives
	Synthetic dye	Same as mentioned earlier	—
Pharmaceutical wastewater	Heterocyclic compounds	Same as mentioned earlier	—
	PCBs	Same as mentioned earlier	—
Synthetic plastic	Plasticizer	Strong stability and inhibitory effect on the central nervous system of the human body	Di (2-ethylhexyl) phthalate, Di (2-ethylhexyl) oxalate, Di (isononyl) phthalate

The reasons for the difficulty in biodegrading compounds can be mainly attributed to internal and external factors. The internal factor lies in the chemical composition and structure of the compound itself, which render it resistant to degradation. The external factor pertains to environmental conditions that are unfavorable for the growth and reproduction of microorganisms. Generally, the biorefractoriness of organic pollutants is primarily due to internal factors [5]. In summary, aromatic hydrocarbons are more difficult to degrade than aliphatic hydrocarbons; cyclic hydrocarbons are more refractory than chain hydrocarbons; branched-chain hydrocarbons are more resistant to degradation than straight-chain hydrocarbons; saturated hydrocarbons are harder to biodegrade than unsaturated ones; short-chain hydrocarbons (*n*-alkanes shorter than C9 are generally difficult to biodegrade) are more persistent than long-chain hydrocarbons; the introduction of electron-withdrawing groups and large substituents reduces the biodegradability of organic compounds; and organic compounds with large molecular weights are more challenging to degrade [6].

Upon structural analysis, it can be observed that most refractory organic pollutants contain benzene rings, saturated long-chain or branched alkyl groups [7]. The presence of these groups renders these organics highly hydrophobic or endowed with stable hydrophobic ends. This characteristic hinders effective contact between the organics and microorganisms, thereby making these organics refractory to biodegradation or increasing the difficulty of biochemical treatment of wastewater [8]. Developing targeted wastewater treatment technologies based on this characteristic will be an effective approach to remove these pollutants.

Currently, water pollution treatment technologies primarily encompass physical methods [9], chemical methods [10], and biological methods [11]. Physical methods include adsorption, membrane filtration, and ultrasonic treatment, while biological methods involve the use of microorganisms, constructed wetlands, and bioenzymes. Chemical methods encompass chemical oxidants, the Fenton process, electrocatalysis, and photocatalysis, among others. These methods all exhibit certain efficacy in wastewater treatment. However, in practical applications, single physical methods cannot meet the water quality standards for effluent, while membrane treatment and electrocatalysis are costly to operate and difficult to implement in large-flow wastewater treatment processes. The Fenton process requires harsh operating conditions, and chemical oxidants can easily cause secondary pollution to water bodies. Despite their effectiveness in wastewater treatment, these methods are often constrained in practical applications. Therefore, the development of economical and efficient water treatment technologies to address water pollution is urgent. Solar energy is an inexhaustible green energy source, and photocatalysis has been a popular research direction in recent years, widely applied in energy conversion and pollutant degradation. Utilizing semiconductor materials to generate electron–hole pairs via sunlight is currently an economical and efficient means for treating contaminated wastewater. Constructing stable semiconductor photocatalysts to produce highly oxidative reactive radicals for synergistic degradation of water pollutants provides an effective solution to water pollution and wastewater treatment. Numerous traditional photocatalytic materials, such as TiO_2, ZnO, and g-C_3N_4, have also been used for the degradation of organic pollutants in water, but they suffer from drawbacks like a narrow light response range and low photon quantum efficiency [12].

To enhance the degradation activity of photocatalytic materials, researchers have employed various strategies such as the construction of heterojunctions, noble metal loading, and sensitization [13]. The formation of defects plays a crucial role in the photocatalytic degradation of organic pollutants, exerting profound and intricate influences. During the design and preparation of photocatalytic materials, the intelligent introduction and control of defects can substantially elevate the material's performance in photocatalytic degradation of organic pollutants. These defects, whether they reside within TiO_2, ZnO, g-C_3N_4, or exist in the form of vacancies, interstitial atoms, dislocations, and other configurations, all exert significant impacts on the photocatalytic process [14]. Specifically, the introduction of defects aids in optimizing the energy band structure of photocatalytic materials, enabling them to better absorb and utilize light energy. Additionally, defects facilitate the effective separation and transport of photogenerated carriers, reducing the recombination probability of electrons and holes, and thus enhancing the efficiency of photocatalytic reactions. Furthermore, an increase in defects translates to an increase in surface active sites on the photocatalytic material, which can more effectively adsorb and degrade organic pollutants, further augmenting the effectiveness of photocatalytic degradation [15].

This chapter systematically reviews the crucial and multifaceted roles that defective photocatalytic materials play in the degradation of organic compounds. We specifically focus on four major types of defects: point defects, line defects, planar defects, and volume defects, and delve into their profound impacts on photocatalytic performance. These diverse types of defects not only significantly enrich the microstructure and macroscopic properties of photocatalytic materials but also provide a solid theoretical foundation and practical guidance for optimizing the performance of photocatalytic materials in practical applications.

13.2 Point Defects

Defects hold significant importance in the development of photocatalysis, including point defects, line defects, planar defects, and volume defects (Figure 13.1) [17]. First, defects such as vacancies, dopant atoms, or lattice distortions can trap photogenerated electrons or holes, suppress the recombination of electron–hole pairs, thereby prolonging carrier lifetime and enhancing photocatalytic efficiency [18]. Second, defects introduce defect levels into the material's band structure, broadening the light absorption range and enabling photocatalysts to utilize visible light, thus improving solar energy utilization [18]. Additionally, defects serve as active sites, facilitating the adsorption and activation of pollutants and inducing redox reactions through photogenerated carriers to degrade pollutants. Simultaneously, defects alter the electronic structure and chemical properties of the material's surface, enhancing the generation of reactive oxygen species (such as $\cdot OH, \cdot \cdot O_2^-$, etc.), further accelerating pollutant degradation. By regulating the type

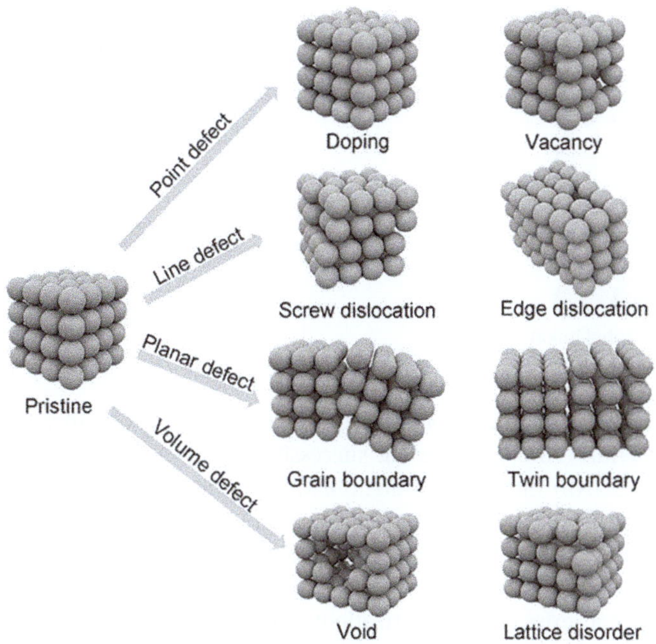

Figure 13.1 Schematic diagram of the original crystal. *Source:* Reproduced from Ma et al. [16] / with permission of John Wiley & Sons.

and concentration of defects, the performance of photocatalysts can be optimized to play a more efficient role in the degradation of water pollutants. Therefore, defects possess important application value in photocatalysis technology.

In the development of photocatalysis, point defects play a crucial role, encompassing vacancies and doping, among others [18]. In photocatalytic materials, point defects refer to localized imperfections at the atomic scale within the crystal structure, often arising from atomic size mismatches, impurity atoms, or random fluctuations during crystal growth. These defects significantly impact the photocatalytic activity of the materials. As shown in Figure 13.2, the main types of point defects include: Schottky defects, where atoms migrate from the interior of the crystal to the surface or edges, leaving internal vacancies, commonly observed in ionic crystals and aiding in maintaining charge balance; Frenkel defects, where atoms move from their lattice sites to interstitial positions, leaving behind vacancies; and impurity atom defects, where nonintrinsic atoms or ions exist within the crystal, potentially introducing additional energy levels and affecting the electronic structure [19]. Furthermore, under certain energetic conditions, two single vacancies can combine to form a vacancy pair or divacancy, and three vacancies combining together are known as a trivacancy, both of which can enhance photocatalytic activity [20]. The presence of point defects notably alters the band structure of photocatalytic materials, as well as the distribution and number of surface-active sites. Therefore, understanding and controlling point defects is key to improving photocatalytic activity.

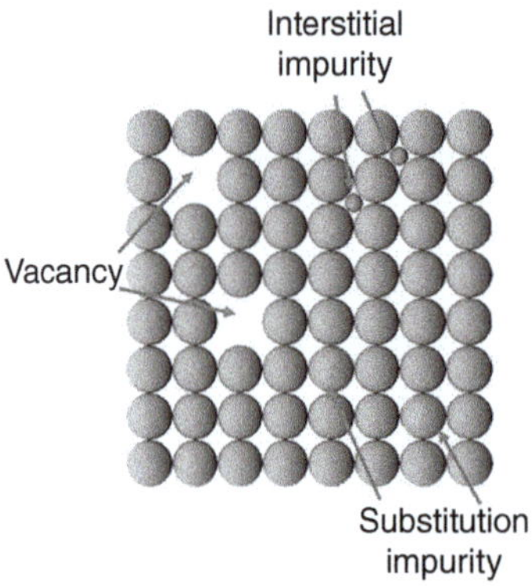

Figure 13.2 Different types of defects. *Source:* Reproduced from Ma et al. [16] / with permission of John Wiley & Sons.

In order to further enhance the photocatalytic degradation performance of g-C_3N_4 toward amoxicillin (AMX), Luo et al. successfully prepared single-atom Nb-anchored g-C_3N_4 (CNNb) through a straightforward thermal polymerization method. Compared to pure g-C_3N_4, the optimized Nb-anchoring samples exhibited superior photocatalytic oxidation efficiency in degrading AMX. Furthermore, the underlying reasons for the improvement in CNNb's photocatalytic activity were uncovered through a series of characterizations and theoretical calculations. Initially, the crystal structures of both g-C_3N_4 and CNNb systems were examined by XRD. As illustrated in Figure 13.3a, each catalyst displayed two distinct diffraction peaks at 2θ values of 13.0° and 27.4°, corresponding to the (100) and (002) crystal planes, respectively. However, an increase in the Nb anchoring amount resulted in a decrease in the intensity of these diffraction peaks, suggesting damage to the crystal phase of the system and the formation of more defects with reduced crystallinity. The influence of different Nb anchoring positions on photocatalytic activity varied, prompting the use of DFT calculations to investigate the energy required for various anchoring sites. The results, summarized in Figure 13.3b, indicated that replacing C atoms with Nb required a significant amount of external heat (433.074 eV), making this process challenging. In contrast, the four anchoring methods—interlayer intercalation, cavity padding, replacement of bridging N, and triazine N—required energies of approximately −6.81464, −8.8682, −444.155, and −442.273 eV, respectively. Here, exothermic processes are indicated by negative energy values, which imply the practicality of various anchoring techniques. Among them, the substitution of bridging N stands out as the preferred method for anchoring Nb. As depicted in Figure 13.3c, pure g-C_3N_4 exhibited low pollutant removal rates through photocatalysis. Furthermore, the photocatalytic performance of CNNb composites was closely linked to the Nb content. For example, CNNb0.06 boosted AMX removal efficiency from 15.3% to 69.1% compared to pure g-C_3N_4. To elucidate the roles of reactive species in AMX degradation over CNNb0.06 (Figure 13.3d), specific masking agents like triethanolamine (TEOA), potassium dichromate ($K_2Cr_2O_7$), p-benzoquinone (PBQ), and isopropanol (IPA) were used to target h^+, e^-, $\bullet O_2^-$, and •OH, respectively. Notably, KI and PBQ additions significantly hindered AMX removal in CNNb0.06, pinpointing $\bullet O_2^-$ as the primary species driving AMX removal. Based on theoretical and experimental findings, three

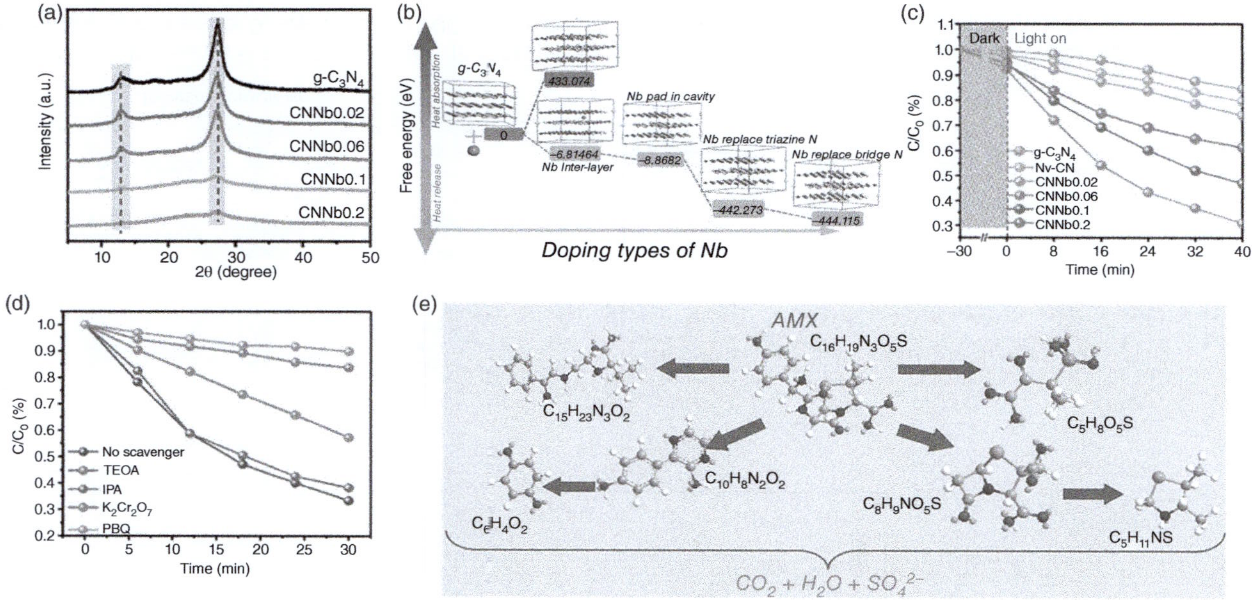

Figure 13.3 (a) X-ray diffraction (XRD) patterns of prepared samples; (b) free energy analysis of various single-atom Nb configurations in DFT calculations; (c) photocatalytic degradation efficiency of AMX CNNb0.06; (d) photocatalytic efficiency with diverse scavengers over CNNb0.06; and (e) primary mechanisms of AMX removal by CNNb0.06. *Source:* Reproduced from Luo et al. [21] / with permission of Elsevier.

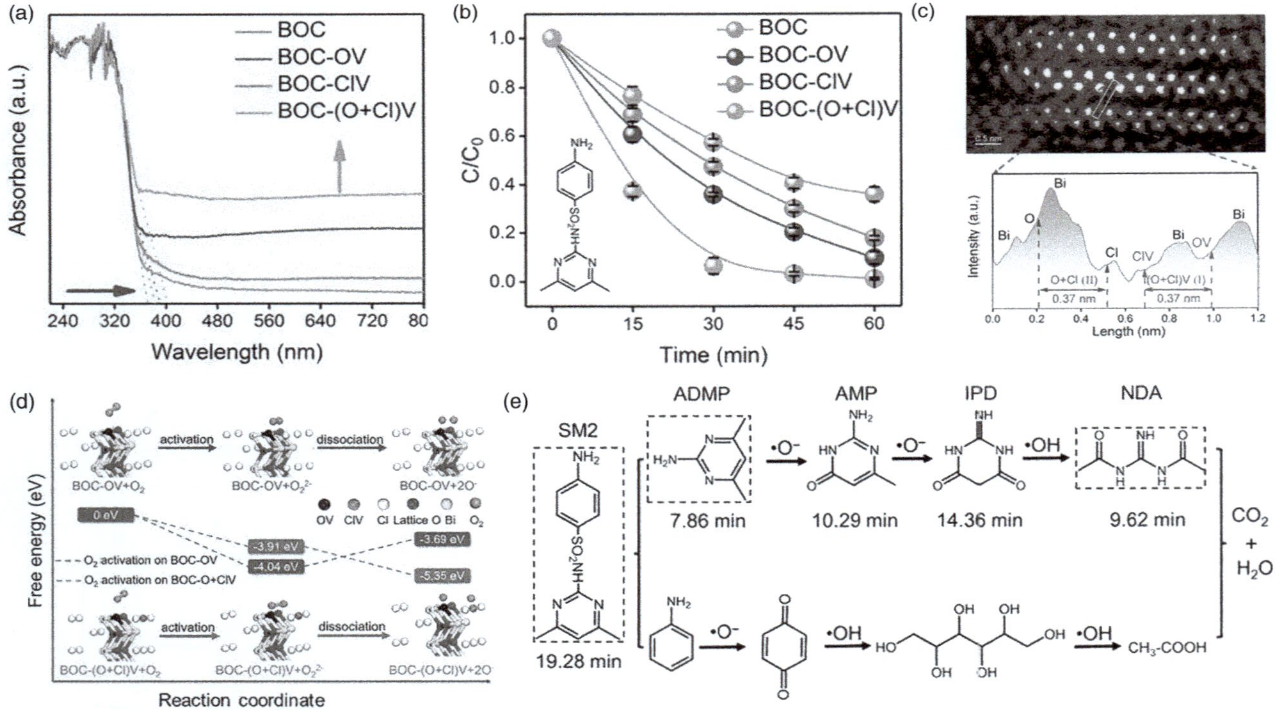

Figure 13.4 (a) UV-vis DRS spectra of BOC-010 samples; (b) STEM images of BOC-(O + Cl)V structure; (c) degradation profiles of SM2 under photocatalysis; (d) energy barriers for O_2 separation on BOC-OV and BOC-(O + Cl)V surfaces; and (e) hypothesized pathway for sulfadimidine photodegradation. *Source:* Reproduced from Yang et al. [22] / with permission of American Chemical Society.

potential AMX removal pathways were proposed, with key intermediates illustrated in Figure 13.3e.

Vacancies also play a crucial role in enhancing photocatalytic activity. Zhang et al. [22] prepared dual (O + Cl) vacancies on the BOC-010 surface, designated as BOC-(O + Cl)V, for the photocatalytic degradation of sulfamethazine (SM2). As illustrated in Figure 13.4a, the UV-vis absorption spectra revealed remarkable light absorption by BOC-(O + Cl)V, attributed to the impurity energy levels associated with the presence of these dual vacancies. As shown in Figure 13.4b, BOC-(O + Cl)V demonstrated the capability to fully eliminate SM2 within 30 minutes under simulated solar light. In contrast, BOC, BOC-OV, and BOC-ClV required an hour to complete the degradation process. To further ascertain the types of vacancies, HAADF-STEM analysis was employed, confirming the existence of (O + Cl) vacancies in BiOCl (Figure 13.4c). DFT calculations (Figure 13.4d) indicated that the introduction of dual vacancies enhanced BOC-(O + Cl)V's capacity for adsorbing and dissociating molecular O_2 into $\bullet O_2^-$ radicals. Based on GC-MS analysis, a mechanism for the ring-opening and deep oxidation of SM2, catalyzed by BOC-(O + Cl)V, was proposed (Figure 13.4e). The essential role of $\bullet O_2^-$ in breaking down persistent aromatic heterocyclic structures, thereby enhancing the complete mineralization of SM2 into CO_2 and H_2O. Therefore, BiOCl with double vacancies not only helps broaden the light response range but also improves the production of reactive oxygen species, thereby enhancing the photocatalytic degradation activity toward organic pollutants.

13.3 Line Defects

Line defects, primarily manifested as dislocations, arise from the misalignment of atomic arrangements within the crystal structure [23]. In photocatalysis, dislocations exert a significant influence on the material's properties. They can alter the local electronic structure of the material, giving rise to additional energy levels, which in turn affect the generation and separation of photogenerated carriers. Furthermore, dislocation regions typically possess higher surface energy, enabling them to serve as active sites that facilitate the adsorption and activation of reactants, thereby enhancing photocatalytic activity. However, an excessive number of dislocations may also act as recombination centers for electron–hole pairs, diminishing photocatalytic efficiency. Consequently, rational modulation of dislocation density and distribution emerges as one of the critical factors in optimizing the performance of photocatalysts [18].

The generation of highly reactive oxygen species also contributes to enhancing the performance of photocatalytic degradation of pollutants. Li et al. [24] investigated two bulk metal oxide single crystals, introducing dislocations with specific orientations to explore the influence of linear defects on the photocatalytic properties of metal oxides and their underlying mechanisms. Unlike metals, ceramics are typically viewed as brittle materials with limited dislocation activity. However, uniaxial deformation of ceramic single crystals at high temperatures can activate specific slip systems. Figure 13.5a,b illustrate tetragonal $BaTiO_3$ and rutile TiO_2 samples

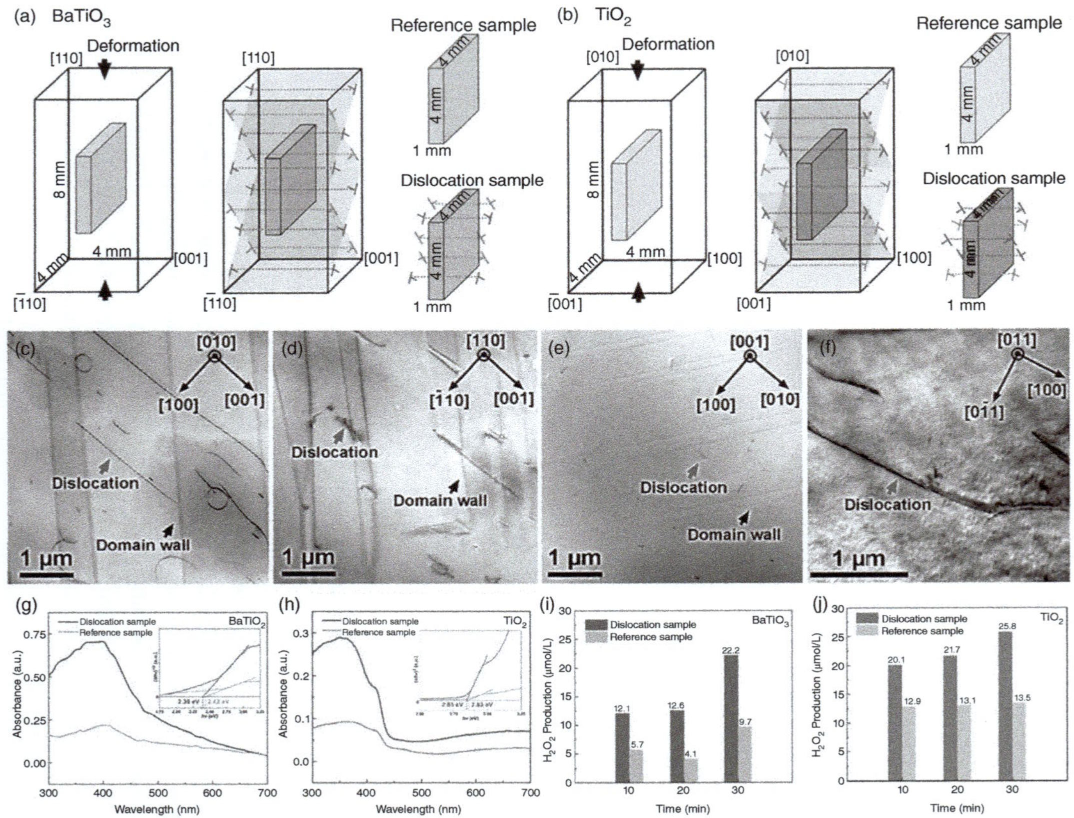

Figure 13.5 Schematic diagrams of (a) pristine $BaTiO_3$ and (b) pristine TiO_2 before and after uniaxial deformation; (c–e) bright-field STEM images of dislocations (blue arrows) in $BaTiO_3$ on (010), (110), and (001) planes; (f) bright-field TEM image of dislocations in TiO_2 on (011) plane; and (g–h) DRS spectra and (i–g) photocatalytic H_2O_2 production of $BaTiO_3$ and TiO_2, with and without deformation. *Source:* Reproduced from Zhang et al. [24] / with permission of Royal Society of Chemistry.

before and after deformation. $BaTiO_3$ was deformed at 1150 °C, while TiO_2 required a slightly lower temperature of 1050 °C. High-temperature compression activated {100}⟨001⟩ and {110}⟨110⟩ slip systems, generating dislocations primarily along [100] and [001] in $BaTiO_3$ and TiO_2, respectively. Figure 13.5c–e show bright-field STEM images of dislocations in $BaTiO_3$ on (010), (110), and (001) planes, revealing their spatial distribution. Dislocations along [001] in $BaTiO_3$ were parallel to (010) and (110) planes and perpendicular to (001). Figure 13.5f highlights dislocations in TiO_2 on the (011) plane, consistent with reported slip planes for rutile TiO_2. Figure 13.5g,h showcase the light absorbance spectra for $BaTiO_3$ and TiO_2 crystals, with a focus on samples both with and without dislocations. The findings indicate that the plastic deformation and high dislocation density in these samples did not significantly alter their light absorption ranges. Specifically, the absorbance edges were observed at approximately 498 nm for $BaTiO_3$ and 438 nm for TiO_2. Figure 13.5i,j show the H_2O_2 production catalyzed by $BaTiO_3$ and TiO_2 in O_2-enriched pure water, revealing that dislocations in these single crystals significantly enhance their photocatalytic efficiency. In conclusion, dislocations introduced through high-temperature deformation can serve as fast pathways for the migration of photogenerated electrons, thereby increasing the number of photogenerated electrons on the surface of the photocatalyst. This enhancement benefits electron-dominated photocatalytic reactions. Using photocatalytic hydrogen peroxide synthesis as a benchmark reaction, single-crystal metal oxides with high dislocation densities achieved a significant improvement in hydrogen peroxide production compared to dislocation-free samples, even without the assistance of additional sacrificial agents.

13.4 Planar Defects

Surface defects primarily encompass grain boundaries and twin boundaries, which arise from the interfaces between differing crystalline structures [25]. In the realm of photocatalysis, these surface defects exert a profound influence on the material's properties. Grain boundaries and twin boundaries can serve as rapid transmission pathways for photogenerated carriers, facilitating the separation of electrons and holes. Consequently, by judiciously controlling the type, density, and distribution of surface defects, it is possible to optimize the performance of photocatalysts and enhance their efficiency in photocatalytic degradation of pollutants.

Zhang et al. [26] synthesized a variety of $Zn_{0.5}Cd_{0.5}S$ materials, encompassing single-crystal $Zn_{0.5}Cd_{0.5}S$ without sulfur vacancies and featuring a twin structure (ZCS), single-crystal $Zn_{0.5}Cd_{0.5}S$ abundant in sulfur vacancies (ZCS-V), twin-structured $Zn_{0.5}Cd_{0.5}S$ devoid of sulfur vacancies (T-ZCS), and twin-crystal $Zn_{0.5}Cd_{0.5}S$ rich in sulfur vacancies (T-ZCSv), all through a straightforward and practical approach. T-ZCS demonstrated the most superior photocatalytic activity, and then a series of characterizations were conducted to uncover the underlying reasons. The microscopic morphology and crystalline lattice structure of T-ZCSv and ZIF-8@T-ZCSv were analyzed using TEM. Figure 13.6a shows a rounded nanoparticle with visible twin boundaries. HRTEM images and Fast Fourier Transform (FFT)

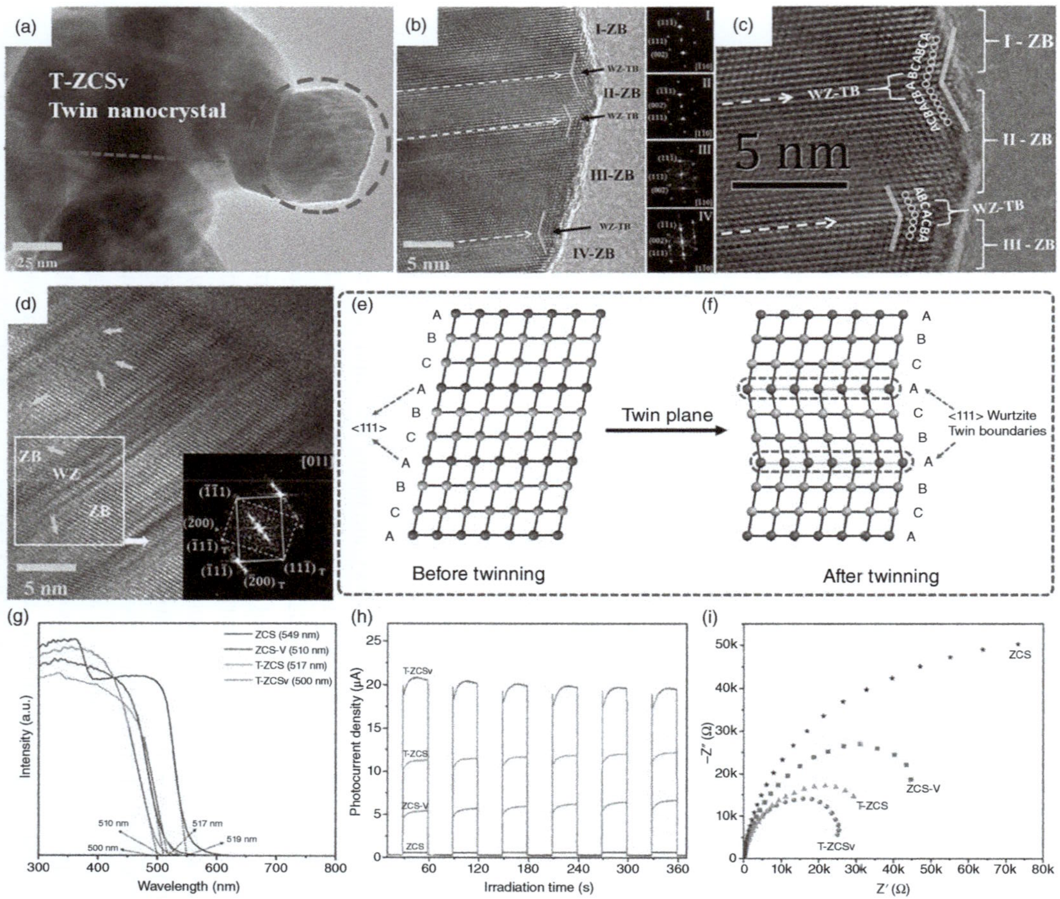

Figure 13.6 (a–c) TEM images of T-ZCSv and (d) ZIF-8@T-ZCSv; (e–f) the corresponding structural model of the twinning nanocrystal; (g) UV-vis diffuse reflectance spectra; (h) Photocurrent response; and (i) electrochemical impedance spectroscopy of ZCS, ZCS-V, T-ZCS, and T-ZCSv. *Source:* Reproduced from Zhang et al. [26] / with permission of Royal Society of Chemistry.

analysis confirmed the presence of sphalerite-Zinc Blende (ZB) and Wurtzite-Twin Boundaries (WZ-TBs) in the ZnS crystalline phase. Regions I, II, III, and IV were identified as ZB structures, indexed along $[\bar{1}10]$ and $[1\bar{1}0]$ zone axes, with mirrored symmetry. Atomic stacking faults (SFs) formed WZ-TBs or dislocation boundaries, indicating $(111)/[11\bar{2}]$ slips that disrupted the ZB structure, creating twin crystals and homojunctions from WZ-TBs. These boundaries produced twinned superlattices, as shown in Figure 13.6b,c, with mirror-symmetric lattice stripes and alternating twin arrangements. The ZB/WZ twin boundaries formed interfacial homojunctions, enhancing charge carrier separation and transport. Figure 13.6d reveals twinned ZB structures with WZ-TBs in ZIF-8@T-ZCSv (1300 mg), confirmed by FFT patterns indexed along the [011] zone axis. The presence of $(111)/[11\bar{2}]$-slips confirmed that ZIF-8 did not alter the twinning characteristics of T-ZCSv. Lattice defects, marked by yellow arrows, were attributed to S vacancies, which are expected to enhance photocatalytic activity. As shown in Figure 13.6e, the ZB phase has an FCC structure ("···ABCABC···" order) and the WZ phase has a hexagonal structure ("···ABABAB···" order). Periodic SFs in the ZB structure introduced discontinuities, forming dislocations with "···CAC···" or "···BAB···" stacking, creating coherent twin boundaries (WZ-TBs) (Figure 13.6f). Furthermore, the absorption edges of ZCS, T-ZCS, ZCS-V, and T-ZCSv were observed at approximately 549, 517, 510, and 500 nm, respectively (Figure 13.6g), indicating their visible light absorption capabilities. T-ZCSv exhibited the largest blueshift (49 nm), attributed to its twinned structure and S vacancies, which introduce defective state energy levels. Transient photocurrent (TPC) measurements under intermittent irradiation ($\lambda \geq 420$ nm) revealed that T-ZCSv had the highest photocurrent density, followed by T-ZCS, ZCS-V, and ZCS (Figure 13.6h), highlighting enhanced charge separation and migration due to the twinned structure and S vacancies. Electrochemical Impedance Spectra (EIS) further confirmed T-ZCSv's superior performance, showing the smallest curvature radius in Nyquist plots (Figure 13.6i), indicating the lowest charge transfer resistance and improved electrical conductivity. These results demonstrate that T-ZCSv's twinned structure and S vacancies facilitate efficient charge transfer, reduce recombination, and significantly enhance photocatalytic hydrogen production. In conclusion, the formation of planar defects can also broaden the light response range and increase the number of surface-active sites, thereby enhancing the photoelectric activity and improving the performance of photocatalytic degradation of pollutants.

13.5 Volume Defects

The role of bulk defects in photocatalysis is predominantly reflected in their dual impact on the structure and performance of materials [27]. An optimal level of bulk defects, such as pores and cracks, can enhance photocatalytic performance by boosting light scattering and absorption, providing more reactive sites, and increasing the specific surface area, thereby elevating the activity of photocatalytic degradation of pollutants in water [28]. Additionally, bulk defects often coexist with other types of defects.

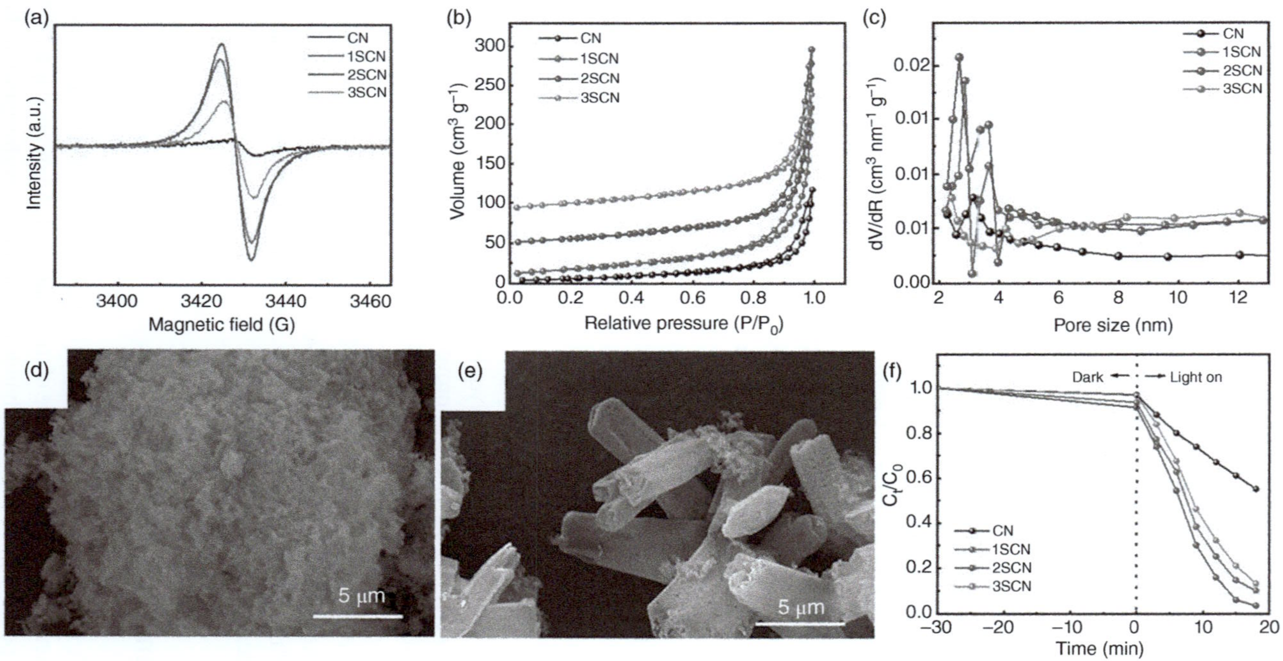

Figure 13.7 (a) EPR spectra; (b) BET isotherms; and (c) pore size distribution of CN and SCN; SEM images of (d) CN and (e) 2SCN; and (f) photocatalytic degradation efficiency of RhB under visible light. *Source:* Li et al. [29] / The Royal Society of Chemistry / CC BY 3.0.

Zhang et al. synthesized g-C_3N_4 with nitrogen defects and porous structures (SCN) via thermal condensation of melamine and trithiocyanuric acid. EPR analysis (Figure 13.7a) revealed a symmetrical peak at g = 2.0038, indicating nitrogen defects. Pure g-C_3N_4 (CN) showed a weak signal from unpaired C atoms in aromatic heterocycles, while SCN exhibited a stronger signal, reflecting increased nitrogen defects. Defect density rose with trithiocyanuric acid content, peaking at 2SCN, where edge N-bridge atoms were predominantly missing. The generation of point defects often leads to the formation of bulk defects. N_2 adsorption–desorption isotherms (Figure 13.7b) confirmed mesoporous structures (Type IV with H3 hysteresis). Specific surface areas were 23.2, 65.3, 70.8, and 54.6 $m^2\ g^{-1}$ for CN, 1SCN, 2SCN, and 3SCN, respectively, with 2SCN having the highest area. Pore volumes were 0.16, 0.32, 0.38, and 0.22 $m^3\ g^{-1}$ for CN, 1SCN, 2SCN, and 3SCN, respectively. Mesopore sizes ranged from 2 to 6 nm (Figure 13.7c), enhancing light scattering and photocatalytic activity. SEM images (Figure 13.7d,e) showed bulk structures for pure g-C_3N_4, while SCN formed irregular nanorods (~5 μm long, hundreds of nm wide) with microporous features and cracks. Photocatalytic degradation of RhB under visible light ($\lambda \geq 420$ nm) reached 45%, 90%, 97%, and 87% in 18 minutes for CN, 1SCN, 2SCN, and 3SCN, respectively (Figure 13.7f). 2SCN's degradation rate was 2.5 times that of CN. The enhanced photocatalytic degradation activity is attributed to the co-effect of point defects and volume defects.

13.6 Conclusion

In addressing the global challenge of water pollution, defective photocatalytic materials have emerged as an innovative strategy, demonstrating unique advantages in the degradation of organic pollutants. By introducing various types of defects (such as point defects, line defects, planar defects, and volume defects), these materials effectively enhance photocatalytic performance, providing new solutions for water purification. But there are also disadvantages that cannot be ignored.

Advantages:

1) **Enhanced photocatalytic activity.** The introduction of defects can regulate the light absorption, charge separation, and transport properties of photocatalytic materials, thereby improving their photocatalytic activity. This allows defective photocatalytic materials to achieve efficient degradation of organic pollutants even under low light intensities.
2) **Broadened light response range.** Some defects can broaden the light response range of photocatalytic materials, enabling them to utilize more sunlight for catalytic reactions. This not only improves the utilization rate of solar energy but also reduces the energy consumption of the photocatalytic process.
3) **Environmentally friendly.** Defective photocatalytic materials are typically synthesized using nontoxic or low-toxic raw materials and theoretically do not produce harmful by-products during use. This aligns with environmental protection principles and contributes to the protection of human health and the ecological environment.

Disadvantages:

1) **Complexity in synthesis.** The synthesis process of defective photocatalytic materials is often complex, requiring precise control of reaction conditions and raw material ratios. This increases production difficulty and cost, limiting their large-scale application.
2) **Instability in performance.** Different types of defects have varying effects on photocatalytic performance and are susceptible to environmental factors. This may lead to unstable performance of the materials in practical applications, making it difficult to achieve the expected results.
3) **Photocorrosion issues.** Some defective photocatalytic materials are prone to photocorrosion under prolonged illumination, resulting in structural damage and performance degradation. This limits the lifespan and catalytic efficiency of the materials.
4) **Challenges in recycling and reuse.** The recycling and reuse of defective photocatalytic materials after degrading organic pollutants may pose certain challenges. This increases treatment costs and potentially causes secondary pollution to the environment.
5) **Toxicity concerns.** Release of Harmful By-products: Although defective photocatalytic materials are theoretically environmentally friendly and harmless, they may produce intermediate compounds that are more toxic than the original pollutants during the actual degradation process. The toxicity of these by-products must be carefully monitored and assessed to ensure environmental safety. This finding emphasizes the importance of conducting rigorous toxicological testing and long-term environmental impact assessments when using these materials.

In summary, defective photocatalytic materials exhibit significant advantages in the degradation of organic pollutants but still face some urgent issues that need to be addressed. Future research should focus on optimizing synthesis processes, improving material stability and performance, exploring effective recycling and reuse methods, and comprehensively assessing the ecological toxicity and environmental impact of the materials to promote the further development of defective photocatalytic materials in practical applications.

Conflict of Interest

The authors declare that they have no conflict of interest.

References

1 Suresh, K., Tang, T., Van Vliet, M.T. et al. (2023). Recent advancement in water quality indicators for eutrophication in global freshwater lakes. *Environ. Res. Lett.* 18: 063004.

2 Gaaloul, N. and Eslamian, S. (2022). Groundwater quality in arid environments. *Clean Water and Sanitation* 260–272.

3 Singh, A.K., Bilal, M., Iqbal, H.M., and Raj, A. (2021). Trends in predictive biodegradation for sustainable mitigation of environmental pollutants: Recent progress and future outlook. *Sci. Total Environ.* 770: 144561.

4 Jones, K.C. (2021). Persistent organic pollutants (POPs) and related chemicals in the global environment: some personal reflections. *Environ. Sci. Technol.* 55: 9400–9412.

5 Kong, W., Li, Y., Zhang, Y., and Liu, H. (2023). Enhanced degradation of refractory organics by bioelectrochemical systems: a review. *J. Clean. Prod.* 138675.

6 Hu, J., Li, R., Zhang, J. et al. (2025). Deciphering the N1-substituent effects on biodegradation of sulfonamides: Novel insights revealed from molecular biology and computational chemistry approaches. *Water Res.* 274: 123037.

7 Wang, H., Wang, X., Wang, M. et al. (2023). Degradation and transformation of linear alkyl-benzene sulfonates (LAS) in integrated constructed wetland–microbial fuel cell systems. *Chemosphere* 321: 138135.

8 Ma, D., Yi, H., Lai, C. et al. (2021). Critical review of advanced oxidation processes in organic wastewater treatment. *Chemosphere* 275: 130104.

9 Ahmed, S., Mofijur, M., Nuzhat, S. et al. (2021). Recent developments in physical, biological, chemical, and hybrid treatment techniques for removing emerging contaminants from wastewater. *J. Hazard. Mater.* 416: 125912.

10 Yan, H., Lai, C., Wang, D. et al. (2021). In situ chemical oxidation: peroxide or persulfate coupled with membrane technology for wastewater treatment. *J Mater Chem A* 9: 11944–11960.

11 Rathi, B.S., Kumar, P.S., Vo, D.-V.N. et al. (2021). Critical review on hazardous pollutants in water environment: occurrence, monitoring, fate, removal technologies and risk assessment. *Sci. Total Environ.* 797: 149134.

12 Das, S. and Chowdhury, A. (2021). Recent advancements of g-C_3N_4-based magnetic photocatalysts towards the degradation of organic pollutants: a review. *Nanotechnology* 33: 072004.

13 Hills-Kimball, K., Yang, H., Cai, T. et al. (2021). *Adv. Sci.* 8: 2100214.

14 Muhammad, P., Zada, A., Rashid, J. et al. (2024). Recent advances in ligand design and engineering in lead halide perovskite nanocrystals. *Adv. Funct. Mater.* 2314686.

15 Ahmad, I., Aftab, M.A., Fatima, A. et al. (2024). A comprehensive review on the advancement of transition metals incorporated on functional magnetic nanocomposites for the catalytic reduction and photocatalytic degradation of organic pollutants. *Coord. Chem. Rev.* 514: 215904.

16 Ma, J., Long, R., Liu, D. et al. (2022). Defect engineering in photocatalytic methane conversion. *Small Struct.* 3: 2100147.

17 Wang, Z., Xiao, M., You, J. et al. (2022). Defect engineering in photocatalysts and photoelectrodes: from small to big. *Acc. Mater. Res.* 3: 1127–1136.

18 Zafar, Z., Yi, S., Li, J. et al. (2022). Recent development in defects engineered photocatalysts: an overview of the experimental and theoretical strategies. *Energy Environ. Mater.* 5: 68–114.

19 Yu, X., Ng, S.F., Putri, L.K. et al. (2021). Point-defect engineering: leveraging imperfections in graphitic carbon nitride (g-C_3N_4) photocatalysts toward artificial photosynthesis. *Small* 17: 2006851.

20 Shi, T., Qiu, X., Zhou, Y. et al. (2023). Unconventional energetics of small vacancy clusters in BCC high-entropy alloy $Nb_{0.75}ZrTiV_{0.5}$. *J. Mater. Sci. Technol.* 146: 61–71.

21 Luo, J., Han, H., Wang, X. et al. (2023). *Appl Catal B* 328: 122495.

22 Yang, Z., Shi, Y., Li, H. et al. (2022). Single-atom Nb anchored on graphitic carbon nitride for boosting electron transfer towards improved photocatalytic performance. *Environ. Sci. Technol.* 56: 3587–3595.

23 Wang, J., Zhao, X., Zou, G. et al. (2023). Oxygen and chlorine dual vacancies enable photocatalytic O_2 dissociation into monatomic reactive oxygen on BiOCl for refractory aromatic pollutant removal. *Mater. Today Nano* 22: 100336.

24 Zhang, H., Gao, S., Wang, H. et al. (2024). Crystal-defect engineering of electrode materials for energy storage and conversion. *J Mater Chem A* 12: 23910–23919.

25 Mei, Y., Liu, J., Cui, T. et al. (2024). Enhancing the photoelectric performance of metal oxide semiconductors by introduction of dislocations. *Adv. Mater.* 2411311.

26 Zhang, Y., Lu, D., Wang, Z. et al. (2023). Synergetic microstructure engineering by induced ZB/WZ twin boundaries and S vacancies in a $Zn_{0.5}Cd_{0.5}S$-based S-scheme photocatalyst for highly efficient photocatalytic hydrogen production. *Inorg. Chem. Front.* 10: 6683–6700.

27 Zhu, J., Zhu, Y., Zhou, Y. et al. (2023). Synergistic promotion of photocatalytic degradation of Methyl orange by fluorine-and silicon-doped TiO_2/AC composite material. *Molecules* 28: 5170.

28 Pandeya, S., Ding, R., Ma, Y. et al. (2024). Self-standing CdS/TiO_2 Janus nanofiberous membrane: COD removal, antibacterial activity and photocatalytic degradation of organic pollutants. *J. Environ. Chem. Eng.* 12: 112521.

29 Li, M., Bai, X., Rao, X. et al. (2022). Nitrogen defect-containing polymeric carbon nitride for efficient photocatalytic H_2 evolution and RhB degradation under visible light irradiation. *RSC Adv.* 12: 24713–24723.

14

NO_x Removal Using Defective Photocatalytic Materials

Qiuhui Zhu[1], Yongqian Cui[2], Xue Xiao[1,2], Juiyeh Rau[1], Yin Shen[3], and Chuanyi Wang[2]

[1] *Wuyi University, College of Ecology and Resources Engineering, Fujian Provincial Key Laboratory of Eco-Industrial Green Technology, Baihua Road, Wuyishan, 354300, PR China*
[2] *Shaanxi University of Science and Technology, School of Environmental Science and Engineering, Weiyang University Park, Xi'an 710021, PR China*
[3] *Wuyi University, College of Tea and Food Science, Baihua Road, Wuyishan, 354300, PR China*

14.1 Introduction

The rapid development of the global economy and the substantial elevation of industrialization levels have been heavily reliant on the extensive consumption of fossil fuels, leading to increasingly severe environmental pollution issues [1]. Nitrogen oxides (NO_x) constitute a primary component of air pollutants, and their excessive emissions contribute to acid rain, photochemical smog, haze, and other phenomena [2]. Furthermore, under intense sunlight, NO_x can undergo a series of chemical reactions with primary pollutants such as volatile organic compounds (VOCs), resulting in the formation of secondary pollutants including ozone (O_3), highly reactive radicals, peroxyacyl nitrates, aldehydes, and more [3]. Consequently, the over-emission of NO_x has imposed severe impacts on the production and living conditions of people worldwide.

To address the environmental pollution resulting from excessive NO_x emissions, researchers from various countries have developed numerous methods to reduce NO_x concentrations while implementing a series of NO_x reduction measures [4]. Currently, the widely applied NO_x prevention and control technologies include catalytic reduction, sorbent absorption, liquid absorption, and others. The above methods are primarily suitable for treating high concentrations of NO_x, yet the NO_x concentration in exhaust gases remains excessively high [5]. Lu et al. [6] employed a wet oxidation and absorption method for NO removal. When the NO concentration was 500 ppm, the flue gas flow rate was 1.0 L min^{-1}, and the O_2 molar fraction was 3.0%, a $NaClO_2$ solution with a pH of 5-6 was used as the oxidant, and 15% $Ca(OH)_2$ served as the absorbent. Under the aforementioned conditions, the denitrification efficiency process reached 80%, but the NO_x content in the exhaust gas remained

Defect Engineering in Photocatalysis: From Fundamentals to Applications,
First Edition. Edited by Chuanyi Wang, Abdelkader Labidi, and Qiuhui Zhu.

relatively high. Therefore, it is of utmost importance to further reduce the NO_x concentration in exhaust gases based on the aforementioned methods.

Photocatalytic conversion of NO_x utilizes semiconductor-based photocatalytic technology, where electrons (e^-) and holes (h^+) generated under illumination interact with H_2O and O_2 adsorbed on the surface of the photocatalytic material to produce reactive oxygen species [7]. These species, along with the photogenerated carriers, convert ppb-level NO_x into nitrite (NO_2^-), nitrate (NO_3^-), N_2, and O_2, achieving the purpose of air purification [8]. Due to the limitation of photon quantum efficiency in the photocatalytic conversion of NO, photocatalysis technology is currently only at the laboratory stage, with small-scale pilot studies and exploratory applications. Introducing defects into photocatalytic materials can broaden their light response range and improve the utilization rate of photogenerated carriers, thereby enhancing thc activity of photocatalytic NO_x conversion [9]. In this chapter, we will introduce the role of common photocatalytic materials with defects, including Ti-based, Zn-based, tungsten-based, and bismuth-based materials, in the photocatalytic conversion of NO_x, and clarify the role played by these defects.

14.2 Photocatalytic Mechanism of NO Removal

14.2.1 Photocatalytic Oxidation of NO

During the photocatalytic reaction, semiconductor materials are excited by light of sufficient energy, causing electrons in the valence band to transition to the conduction band, while h^+ remain in the valence band. Consequently, h^+ react with water or hydroxyl groups to generate highly oxidative hydroxyl radicals (•OH), and the photogenerated electrons react with oxygen molecules to produce superoxide radicals ($\bullet O_2^-$). Additionally, h^+, due to their oxidizing capacity, can directly react with NO [11]. The primary reaction processes are as follows:

$$TiO_2 + h\nu \rightarrow h^+ + e^- \quad (14\text{-}1)$$

$$h^+ + OH^- \rightarrow \bullet OH \quad (14\text{-}2)$$

$$h^+ + H_2O \rightarrow \bullet OH + H^+ \quad (14\text{-}3)$$

$$e^- + O_2 \rightarrow \bullet O_2^- \quad (14\text{-}4)$$

$$\bullet O_2^- + N(II)O + h^+ \rightarrow N(V)O_3^- \quad (14\text{-}5)$$

$$N(II)O + \bullet OH + h^+ \rightarrow N(III)O_2^- + H^+ \quad (14\text{-}6)$$

$$N(III)O_2^- + \bullet OH + h^+ \rightarrow N(IV)O_2 + H_2O \quad (14\text{-}7)$$

$$N(IV)O_2 + \bullet OH + h^+ \rightarrow N(V)O_3^- + H^+ \quad (14\text{-}8)$$

In terms of treating NO_x using photocatalytic oxidation technology, while it is capable of converting low-concentration NO into NO_2^- and NO_3^-, the oxidation products tend to adsorb onto the active sites on the surface of the photocatalytic material, leading to catalyst deactivation (Figure 14.1) [12]. Furthermore, during the process of photocatalytic oxidation of NO, harmful by-products such as O_3 may also be produced, posing a risk of secondary pollution [13].

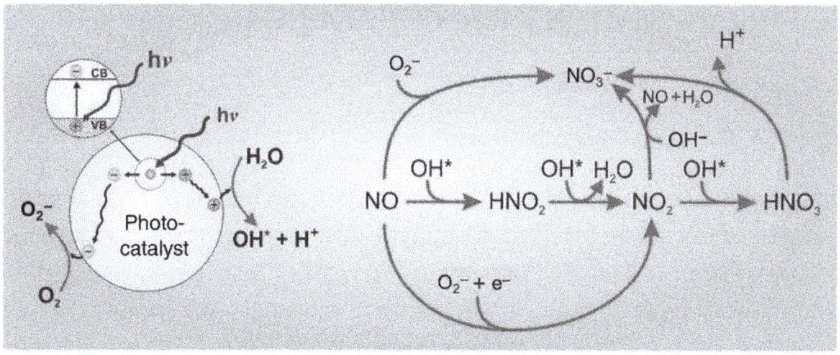

Figure 14.1 Possible reaction pathways for photocatalytic oxidation of NO. *Source:* Reproduced from Nguyen et al. [10] / with permission of Elsevier.

14.2.2 Photocatalytic Reduction of NO

The technology of photocatalytic reduction of NO is capable of converting NO into harmless N_2 and O_2. The process is not only environmentally friendly, reducing reliance on traditional chemical reduction methods but also consumes less energy compared to other treatment methods that require high temperatures and pressures, aligning more closely with the environmental concept of energy conservation and emission reduction. The primary reaction pathway is illustrated in Figure 14.2. However, given that the bond energy of NO is $632\,kJ\,mol^{-1}$, during the process of photocatalytic oxidation of NO, NO is predominantly oxidized into NO_2^- and NO_3^-, with only a small fraction being reduced to N_2 and O_2 [14]. To enhance the proportion of NO undergoing photocatalytic conversion, researchers have employed various strategies: (i) introducing reducing gases such as CO and NH_3 into the gas composition, and (ii) increasing the reaction temperature.

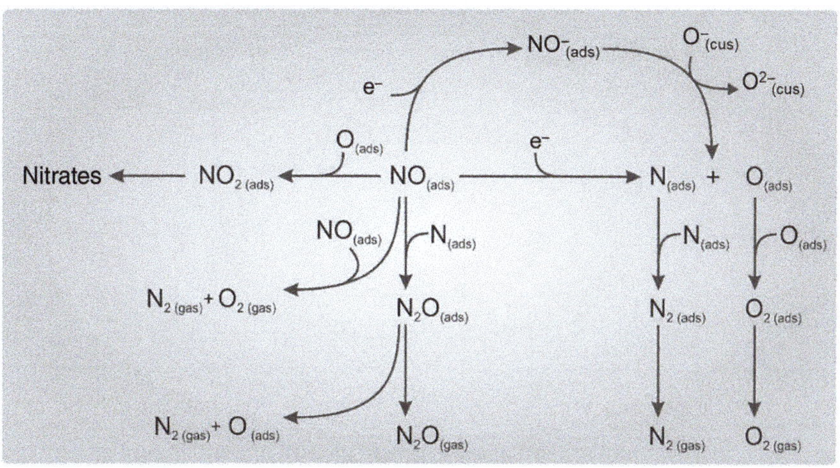

Figure 14.2 Possible reaction pathways for photocatalytic reduction of NO. *Source:* Reproduced from Nguyen et al. [10] / with permission of Elsevier.

14.3 The Photocatalytic NO Removal Over Defective Catalysts

14.3.1 Defective Ti-based Photocatalysts

Ti-based photocatalytic materials, notably the star photocatalyst titanium dioxide (TiO_2), are renowned for their high photocatalytic activity, exceptional chemical stability, superb corrosion resistance, and environmentally friendly nature. Despite these numerous advantages, Ti-based photocatalytic materials also possess some potential drawbacks, such as the narrow light response range of TiO_2, which only absorbs ultraviolet light, thereby limiting further enhancement of its photocatalytic activity [15].

To elevate the photocatalytic activity of Ti-based materials, researchers have introduced defects to expand the spectrum of light responsiveness and optimize the dynamics of photogenerated e^- and h^+, thereby improving the photocatalytic activity. Dai et al. [16] synthesized N-doped TiO_2 (N–TiO_2) through a co-deposition method. As depicted in Figure 14.3a, TiO_2 demonstrated a prominent absorption cutoff for ultraviolet radiation near 390 nm, stemming from its intrinsic absorption properties. In contrast, N–TiO_2 displayed a notable red shift in the

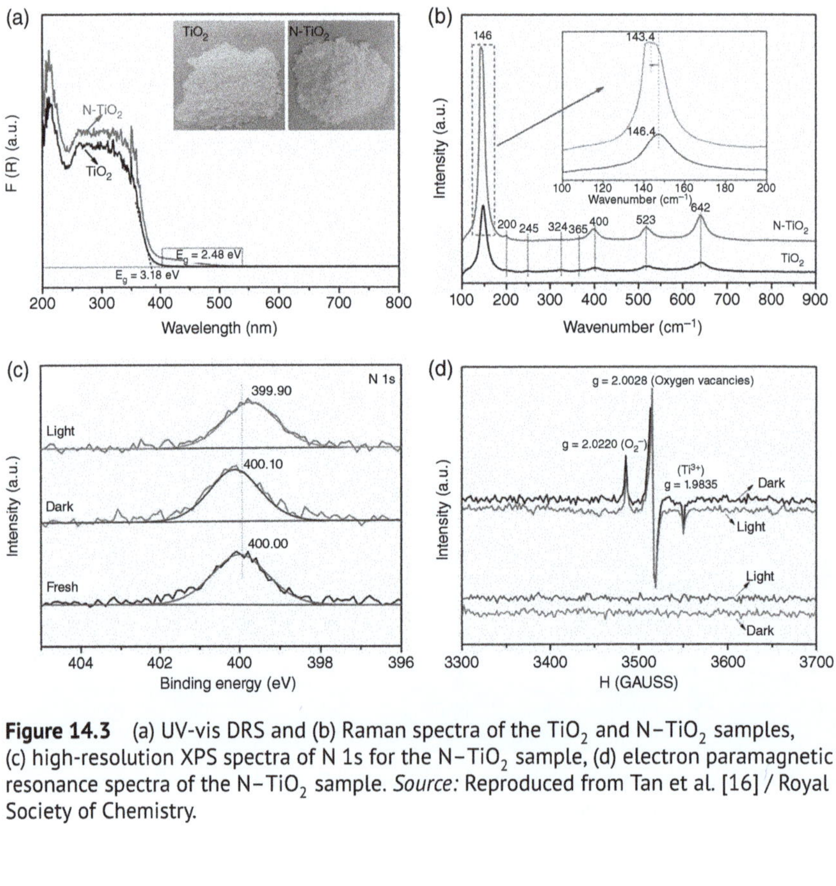

Figure 14.3 (a) UV-vis DRS and (b) Raman spectra of the TiO_2 and N–TiO_2 samples, (c) high-resolution XPS spectra of N 1s for the N–TiO_2 sample, (d) electron paramagnetic resonance spectra of the N–TiO_2 sample. *Source:* Reproduced from Tan et al. [16] / Royal Society of Chemistry.

absorption edge and an enhanced visible light-harvesting capacity around 500 nm, accompanied by a color change from white to pale yellow. To further elucidate the broadening of the light absorption range in N–TiO_2, a distinct peak was identified in the TiO_2 samples, positioned approximately at 146 cm^{-1} and associated with the O–Ti–O bending vibration. Notably, the above-mentioned peak underwent a shift toward a lower wavenumber range, specifically from 146.4 to 143.4 cm^{-1}, in the N-doped TiO_2 (N–TiO_2), as illustrated in the magnified portion of Figure 14.3b. The observed Raman frequency shift is tentatively attributed to the presence of surface imperfections, notably oxygen vacancies, potentially resulting from the incorporation of the doped N element. The high-resolution X-ray photoelectron spectroscopy (XPS) spectra of N 1s confirmed the successful incorporation of N into TiO_2 (Figure 14.3c). Figure 14.3d demonstrates that the N–TiO_2 samples display a prominent paramagnetic resonance signal, corresponding to surface oxygen vacancies, at a g-value of 2.0028. Meanwhile, the peak observed at $g = 1.9835$ is attributed to Ti^{3+}. Additionally,$\cdot O_2^-$ species were also detected, corresponding to the signal peak at $g = 2.022$. Notably, the oxygen vacancies displayed an intensified signal at the g-value of 2.0028 under visible light irradiation, suggesting that visible light facilitated the formation of surface oxygen vacancies. These findings demonstrate that the doping of N induced the formation of oxygen vacancies, thereby broadening the light response range of Ti-based photocatalytic materials.

TiO_2 and N–TiO_2 were employed for photocatalytic NO removal under visible light. As illustrated in Figure 14.4a, N–TiO_2 demonstrated superior photocatalytic performance for NO removal and catalytic stability, maintaining nearly 100% conversion. Compared to TiO_2, N–TiO_2 showed a decreasing trend and reached only 40% conversion after four cycles (Figure 14.4b). Additionally, N–TiO_2 was more effective in inhibiting NO_2 production. The O_2-TPD of N–TiO_2 in Figure 14.4c revealed a new peak at 350 °C under oxygen feed and visible light, indicating that defective sites enhanced O_2 adsorption. Figure 14.4d presents the in-situ DRIFTS spectra of NO and O_2 co-adsorption onto N–TiO_2 samples, captured at different reaction times under visible light illumination. Notably, the NO adsorption bands underwent a shift toward higher wavenumbers, ranging from 1904 to 2021 cm^{-1}, in an oxygen-rich environment, predominantly at Ti^{4+} sites. As the photocatalytic NO oxidation reaction progressed, the band gradually disappeared, suggesting that NO chemical adsorption easily transitioned into activated species under visible light. Simultaneously, small peaks at 2021 cm^{-1} were observed, attributed to the NO activation product Ti^{4+}–NO^+ species. Notably, new NO adsorption bands emerged at 1715 cm^{-1} as the reaction continued, assigned to NO adsorbed at Ti^{3+} sites. The enhanced band intensity suggested an increase in oxygen vacancies within the N–TiO_2 under visible light exposure, which correlated well with electron paramagnetic resonance (EPR) findings and facilitated the activation and dissociation of molecular O_2. These results underscore the significance of oxygen vacancies in the adsorption and activation of molecular oxygen. Furthermore, visible light stimulation fostered the creation of oxygen vacancies and novel NO adsorption sites, such as Ti^{3+}, during the photocatalytic oxidation of NO. The effective circulation of oxygen between molecular and lattice forms within the catalysts elevated both

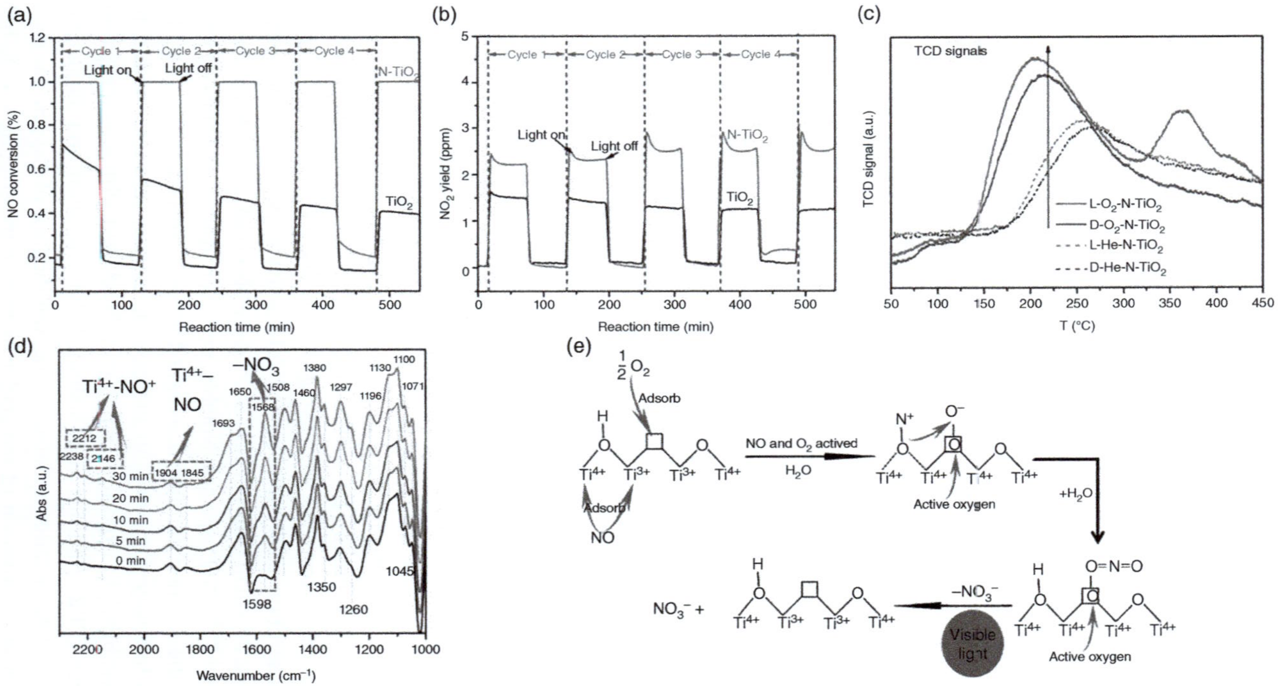

Figure 14.4 (a) Photocatalytic NO removal; (b) NO_2 yield in four cycle tests over TiO_2 and $N-TiO_2$ samples. (c) O_2-TPD of TiO_2 and $N-TiO_2$, (d) FT-IR spectra of the co-adsorbing $NO+O_2$ over $N-TiO_2$, (e) mechanism of photocatalytic NO oxidation under visible light over the $N-TiO_2$ samples. *Source:* Reproduced from Tan et al. [16] / with permission of Royal Society of Chemistry.

catalytic efficiency and product selectivity for NO oxidation under visible light. Figure 14.4e illustrates the mechanism of NO oxidation over N–TiO_2 under visible light conditions.

Ti-based polymetallic oxides with defects are also widely utilized in photocatalytic NO removal. Among them, $Bi_2Ti_2O_7$ stands out as a typical representative. To further enhance the photocatalytic activity of $Bi_2Ti_2O_7$, Wang et al. [17] synthesized full-spectrum Bi@$Bi_2Ti_2O_7$ enriched with oxygen vacancies (OVs) (Bi@BT-OV) using a one-pot hydrothermal method. The study revealed that the incorporation of Bi and OVs significantly broadens the light absorption range of $Bi_2Ti_2O_7$ (BT) to cover the full spectrum (Figure 14.5a). With an increase in the reducing agent, the absorption capacity also improved continuously. As the concentration of OVs rose, the band gap narrowed, with direct band gaps of 2.93, 2.61, 2.61, 2.5, and 2.39 eV for BT, Bi@BT-OV-1, Bi@BT-OV-2, Bi@BT-OV-4, and Bi@BT-OV-5, respectively (Figure 14.5b). In Bi@BT-OV, a symmetrical signal was observed at $g = 2.003$, which was absent in BT (as shown in Figure 14.5c). Furthermore, as the amount of glucose added increases, the intensity of the OV signal gradually intensifies, indicating that the formation of OVs in BT can be controlled by adjusting the glucose concentration. To investigate how OVs influence BT's band structure, density functional theory (DFT) calculations were performed to determine the phonon density of states (PDOS) for BT, BT-OV, Bi@BT, and Bi@BT-OV-4. As illustrated in Figure 14.5d,e, the bandgap values for BT and BT-OV-5% (with 5% oxygen vacancy concentration) are 2.85 and 2.51 eV, respectively. The result suggests that incorporating OVs into BT narrows its bandgap by refining the CB and VB edges, thereby reducing the energy required for electron excitation and transition. Meanwhile, the impact of deposited Bi on tuning the bandgap of both BT and BT-OV is minimal, as evident from Figure 14.5f,g. Moreover, the concentration of OVs plays a significant role in modifying the direct bandgap of BT. According to XPS analysis, Bi@BT-OV-4 contains 16% OVs, which is substantially higher than that in BT-OV. Integrating experimental data, the PDOS for Bi@BT-OV-4 has been calculated to be 0.76 eV. The above results indicate that oxygen vacancies play a decisive role in optimizing the band structure of BT.

Figure 14.6a illustrates the photocatalytic NO removal efficiency of BT and Bi@BT-OV variants (1, 2, 4, and 5), with efficiencies of 31.79%, 66.1%, 68.24%, 79%, and 74.62%, respectively. Notably, Bi@BT-OV-4 emerges as the superior photocatalyst for NO removal. The DeNO*x* index was employed to further evaluate the photocatalytic performance. In contrast, Figure 14.6b highlights the inefficiency of BT due to its high NO_2 yield. And, the introduction of Bi and OVs enhances the photocatalytic activity of BT and suppresses NO_2 generation. Bi@BT-OV-4 boasts a positive $DeNO_x$ index (+0.87), attributed to its optimal Bi and OV content. The photocatalytic activity over Bi@BT-OV-4 was evaluated under monochromatic light (spanning UV to near infrared region (NIR), including 365, 420, 520, 590, 660, 740, and 850 nm) (Figure 14.3c). The apparent quantum yield (AQY) for NO removal correlates strongly with light wavelength in the visible-NIR region. Based on a series of characterizations, the mechanism of charge transfer has been clarified. An ohmic contact between metallic Bi and semiconductor BT-OV creates

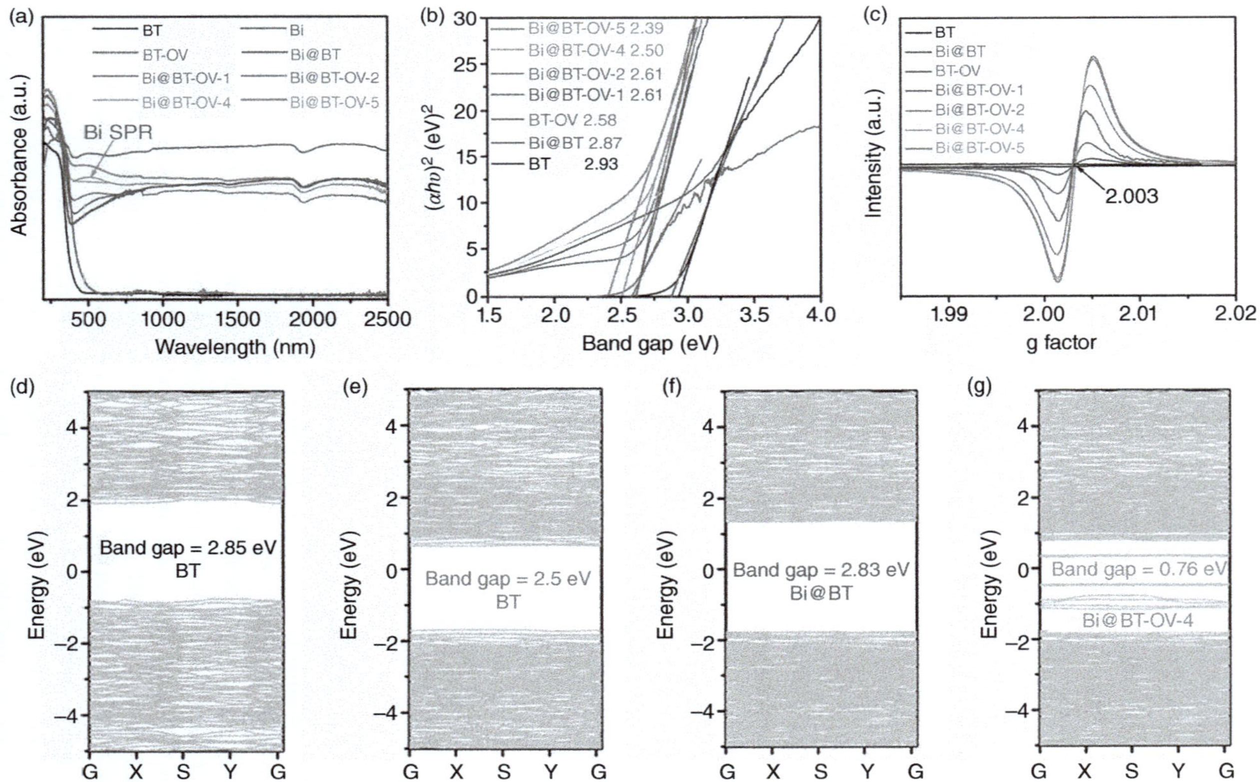

Figure 14.5 (a) UV–vis–NIR spectra, (b) band gap and (c) EPR spectra of the prepared samples, and calculated band structures of (d) BT, (e) BT-OV, (f) Bi@BT, and (g) Bi@BT-OV-4. *Source:* Reproduced from Zhu et al. [17] / with permission of Elsevier.

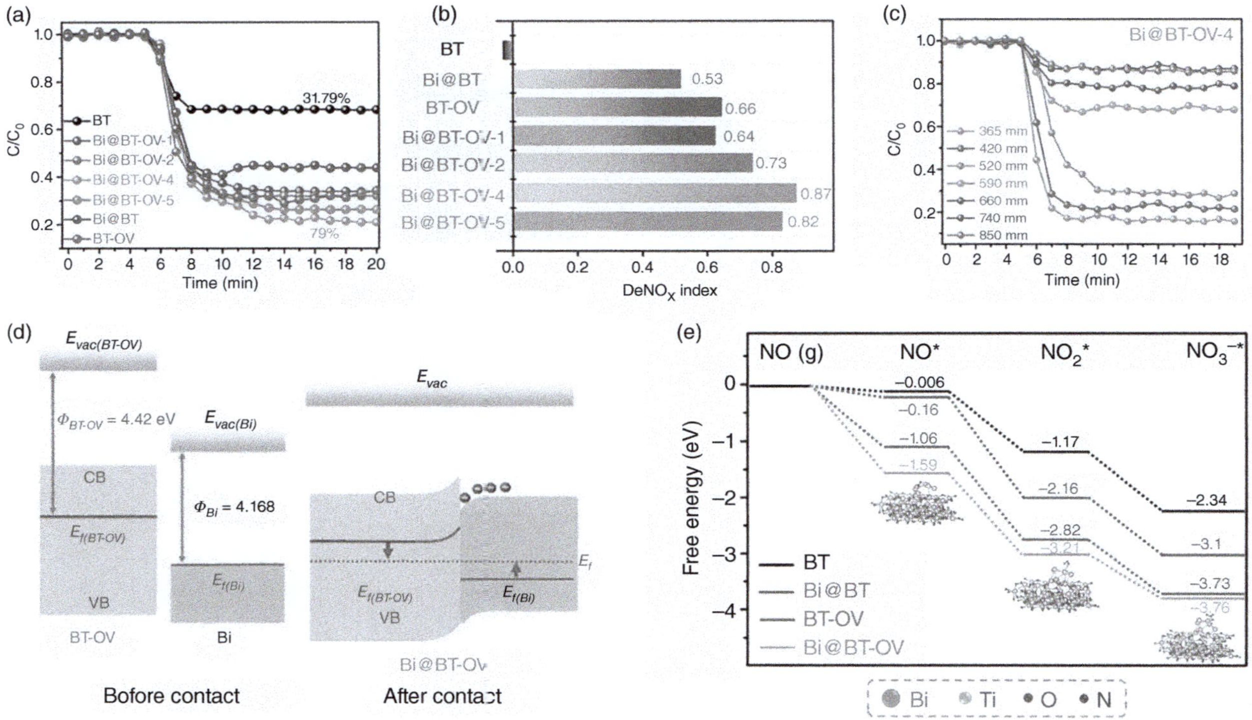

Figure 14.6 (a) NO removal efficiency of prepared samples via photocatalysis, (b) $DeNO_x$ indices for various photocatalysts, (c) photocatalytic performance of Bi@BT-OV-4 under monochromatic light, (d) schematic of electron transfer dynamics before and after Bi-BT-OV interaction, and (e) DFT-predicted free energy for NO oxidation on BT, Bi@BT, BT-OV, and Bi@BT-OV surfaces. *Source:* Reproduced from Zhu et al. [17] / with permission of Elsevier.

a built-in electric field, facilitating hot electron transfer from Bi to BT-OV until Fermi level equilibrium is reached (Figure 14.6d). DFT calculations (Figure 14.6e) further investigate the synergistic effect of Bi and OVs on Bi@BT-OV-4 during photocatalytic removal of NO. The energy barriers for NO conversion (NO(g)→NO_3) are calculated for BT, BT-OV, Bi@BT, and Bi@BT-OV. The combined presence of Bi and OVs makes the photocatalytic removal of NO to NO_3^- on Bi@BT-OV more thermodynamically favorable, explaining its higher activity compared to BT. Importantly, the synergistic effect of Bi and OVs surpasses their individual contributions, significantly enhancing photocatalytic NO removal performance and stability while inhibiting intermediate NO_2 generation on BT.

However, during the photocatalytic oxidation of NO, the oxidation products of NO can adsorb onto the surface of the photocatalytic material, leading to the deactivation of the material. Therefore, the exploration of green and sustainable technologies for the photocatalytic conversion of low-concentration NO holds significant environmental importance and urgency. It is worth noting that the oxidation and reduction pathways often coexist during the photocatalytic conversion of NO. Wonyong Choi et al. [18] prepared the mesoporous TiO_{2-x} (*mp*-TiO_{2-x}) for photocatalytic NO to N_2 and O_2 by Ti^{3+} self-doping method. When exposed to visible light, Ti^{3+}-enriched *mp*-TiO_{2-x} successfully decomposed 81.8% of NO into nearly equal quantities of N_2 and O_2, achieving a remarkable N_2 selectivity of 93.8% (Figure 14.7a,b). Based on the in situ Fourier transform infrared spectroscopy

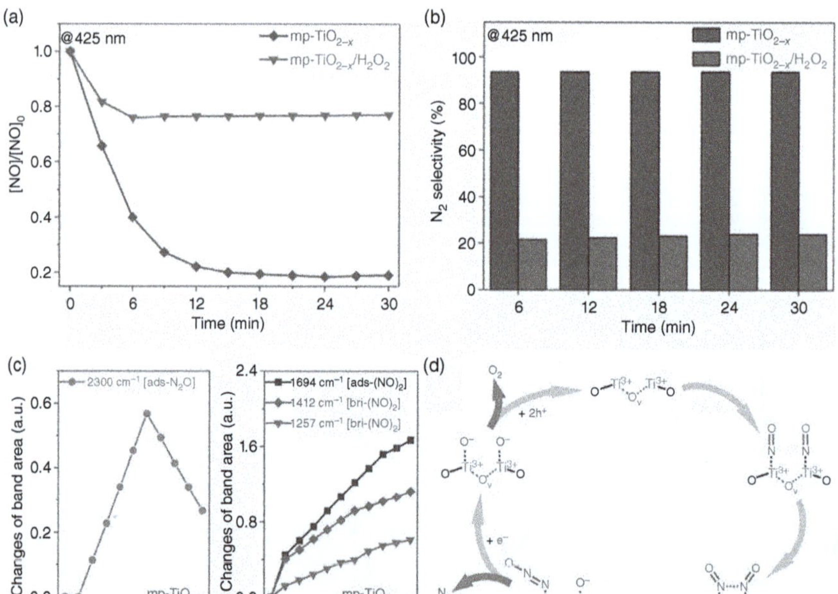

Figure 14.7 (a) The evolution of photocatalytic NO removal, (b) N_2 selectivity in photocatalytic NO decomposition on *mp*-TiO_{2-x}, (c) in situ FTIR spectra for photocatalytic NO photoreduction of *mp*-TiO_{2-x}, and (d) mechanism hypothesis for photocatalytic NO conversion to N_2 and O_2 on *mp*-TiO_{2-x}. *Source:* Reproduced from Zhao and Choi [18] / with permission of Springer Nature.

(FTIR) (Figure 14.7c), the significance of Ti^{3+} sites was established in facilitating the N–N coupling of adjacent NO molecules and enhancing the transformation of N_2O to N_2 (Figure 14.7d).

14.3.2 Defective Zn-based Material

In addition to Ti-based materials, Zn-based photocatalysts also demonstrate exceptional performance in photocatalytic reactions. Furthermore, Zn-based photocatalysts are not only environmentally friendly and cost-effective but also readily modifiable through techniques such as doping and oxygen vacancy engineering, which broaden their response range in the visible light spectrum and enhance their efficiency [19]. These advantages render Zn-based photocatalysts as materials with extensive application potential and promising prospects in fields such as environmental remediation and energy conversion [20].

Wang et al. [21] prepared ZnO nanoparticles with abundant OVs using ZIF-8 as a precursor. UV-Vis DRS analysis revealed that ZnO samples exhibited an additional absorption peak in the visible light region compared to pure ZnO (P-ZnO) and ZIF-8 due to the formation of color centers caused by O atom loss (Figure 14.8a). Furthermore, ZIF-8-derived ZnO samples exhibited significant signal peaks at $g = 2.001 \pm 0.002$, which was attributed to the presence of OVs, with the signal intensity decreasing as the calcination temperature increased (Figure 14.8b). The

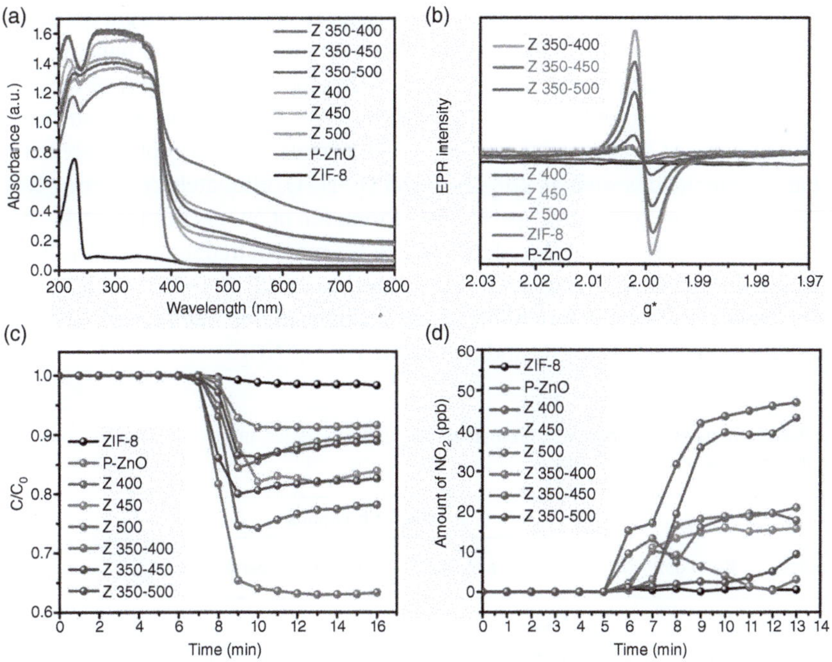

Figure 14.8 (a) UV-vis DRS (b) EPR spectra, (c) photocatalytic performance for NO removal over different photocatalysts under visible light irradiation, and (d) corresponding amount of NO_2 generated. *Source:* Reproduced from Zhu et al. [21] / with permission of Elsevier.

ZnO photocatalysts derived from ZIF-8 and treated at different temperatures demonstrated enhanced photocatalytic performance for NO removal compared to ZIF-8 and P-ZnO. Among them, Z 350–400 exhibited the highest photocatalytic NO removal activity and the lowest NO_2 yield (Figure 14.8c,d). In summary, the introduction of oxygen vacancies in ZnO can broaden the light response range, enhance the photocatalytic activity for NO conversion, and inhibit the production of NO_2. To further elucidate the role of defects in the photocatalytic conversion of NO by ZnO, Reshalaiti Hailili et al. [14] prepared oxygen vacancy-modified ZnO and achieved precise control over the concentration of oxygen vacancies by utilizing flux agents. As depicted in Figure 14.9a, the VO signal on ZnO−1 is barely detectable, whereas the signal intensity on ZnO−3 is three times greater than that on ZnO−2. The photocatalytic efficiency for NO removal over V_O-poor ZnO and V_O-rich ZnO are 26.11% and 42.67%, which are 3.3-fold and 5.4-fold increments than that of V_O-free ZnO (Figure 14.9b). Moreover, incorporating a gradient V_O on the sample surfaces enhances the carrier lifetime, increasing from 2.504 ns in the pristine state to 3.575 ns in the V_O-rich variant, as illustrated in Figure 14.9c. Additionally, Figure 14.9d presents a band energy diagram revealing a notable upward shift in both CB and VB edges due to alterations in optical characteristics and surface imperfections. Notably, the CB of the VO-rich ZnO shifts upwards by approximately 0.29 eV compared to pristine ZnO, highlighting the significant influence of surface defects on the energy band structure. Table 14.1 provides a comprehensive overview of the microstructure, optical properties, and corresponding band edge positions of ZnO. Therefore, OVs aid in optimizing the energy band structure and the photogenerated carrier dynamics of ZnO, thus enhancing the activity for photocatalytic conversion of NO. Why does the formation of oxygen vacancies contribute to reducing the yield of NO_2? For the defective ZnO surfaces with more exposed polar facets and atomic characters, NO adsorption prevails, facilitating interactions such as Zn···O—N and O···N—O, ultimately converting NO to NO_3^- with the help of active species. The formation of NO_2 is influenced by the concentrations of OH^- and •OH radicals, which are more abundant in pristine ZnO due to its more positive VB edges (2.54 eV) compared to $E°$ (•OH/OH^- = 1.99 eV) (Figure 14.9e). Overall, the formation of defects helps to improve the efficiency of photocatalytic conversion of NO and reduce the yield of NO_2.

Table 14.1 Synthesis conditions, microstructure characteristics, and optical attributes of ZnO samples.

		Band edge positions	
Samples	**Band gap (eV)**	**Valence band (eV)**	**Conduction band (eV)**
Pristine	3.26	2.54	−0.72
VO-poor	3.24	2.25	−0.99
VO-rich	3.19	2.18	−1.01

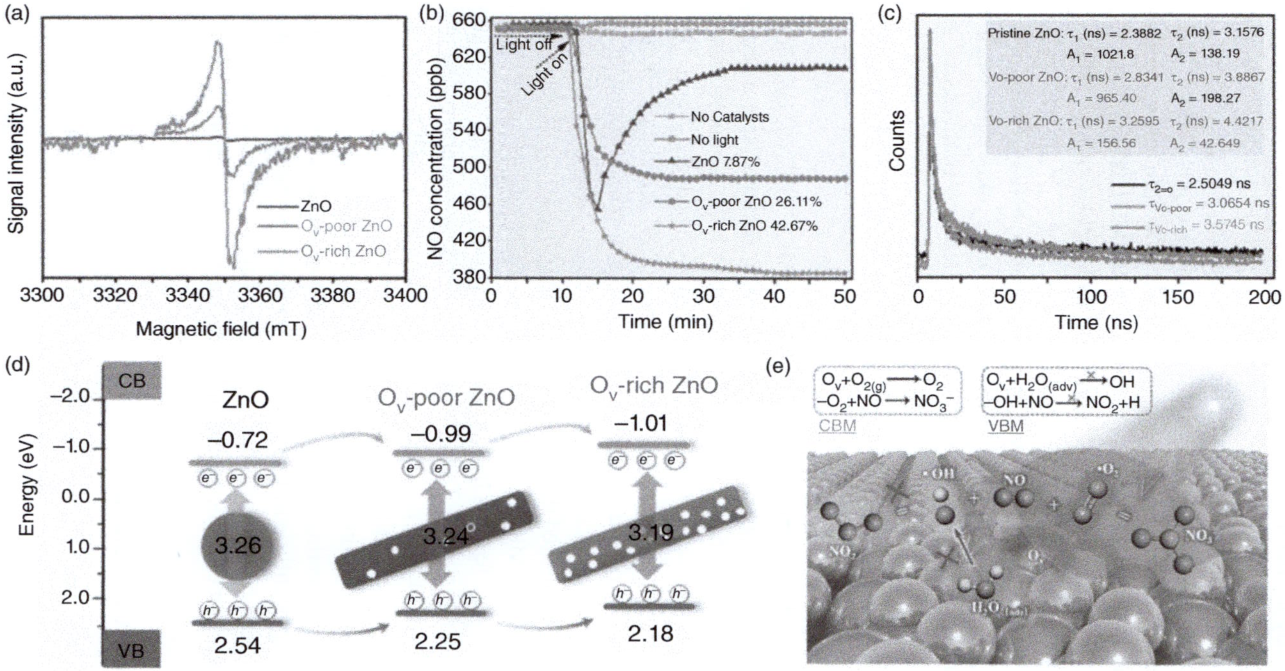

Figure 14.9 (a) EPR spectra, (b) photocatalytic NO conversion on pristine, V_O-poor, and V_O-rich ZnO, (c) nanosecond-level fluorescence spectrum, (d) energy diagram of ZnO band edges of pristine, V_O-poor, and V_O-rich ZnO, and (e) defective ZnO enhances photocatalytic NO conversion via improved charge-carrier dynamics and molecular oxygen activation by defects. *Source:* Reproduced from Hailili et al. [14] / with permission of American Chemical Society.

14.3.3 Defective g-C_3N_4

Carbon nitride (g-C_3N_4) has emerged in recent years as a novel non-metal semiconductor photocatalytic material, widely applied in the field of photocatalysis due to its exceptional light absorption capabilities, low production costs, high stability, and high tunability of its electronic structure [22]. g-C_3N_4 stands out as a typical polymeric semiconductor, wherein C and N atoms form a highly delocalized π-conjugated system through sp^2 hybridization. Within this system, the N_{pz} orbitals constitute the highest occupied molecular orbital (HOMO) of g-C_3N_4, while the C_{pz} orbitals form the lowest unoccupied molecular orbital (LUMO). The bandgap between them, approximately 2.7 eV, enables g-C_3N_4 to absorb light with wavelengths less than 475 nm from the solar spectrum [23]. Nevertheless, g-C_3N_4 materials still possess several drawbacks that need addressing, such as high photogenerated carrier recombination rates, low electrical conductivity, and a lack of absorption for light with wavelengths above 460 nm, which limit their practical applications. To overcome these shortcomings, researchers have systematically optimized g-C_3N_4 in terms of synthesis methods, electronic structure, and nanomorphology [24].

Yu et al. [25] prepared N vacancies modified g-C_3N_4 with K^+ (KCNN) by thermal polymerization. The KCNN sample, after optimization, demonstrates a significantly improved photocatalytic NO removal efficiency of 70.5%, outperforming bulk g-C_3N_4 at 38.1%, porous g-C_3N_4 at 54.5%, and K-doped g-C_3N_4 at 58.6% (Figure 14.10a). As shown in Figure 14.10b and c, theoretical calculations indicate that the incorporation of defects facilitates the adsorption and activation of NO, enabling a more favorable reaction pathway ($NO \rightarrow NO^+ \rightarrow NO_3^-$). Furthermore, Lv et al. [26] further investigated the role of defects in the photocatalytic conversion of NO by preparing g-C_3N_4 with surface hydroxylation (DCN-180), focusing on two aspects: (i) the impact of defects on the energy band structure, and (ii) the influence of defects on the activation of O_2. The introduction of hydroxyl improved the photocatalytic performance of pure g-C_3N_4 for NO removal (65%), which is twice that of pure g-C_3N_4 (32.5%), and inhibited the generation of NO_2 (Figures 14.10d and e). In addition, hydroxyl groups onto the surface of g-C_3N_4 results in the formation of defect states, adversely affecting the segregation of photogenerated charge carriers (Figures 14.10 f). Importantly, surface hydroxyl groups can promote the activation of O_2 (Figures 14.10 g) and improve the formation of superoxide radicals, thereby increasing the conversion efficiency of NO and reducing the production of NO_2.

14.3.4 Other Defective Materials

In the field of photocatalytic NO conversion, in addition to Ti-based, Zn-based, and g-C_3N_4, there exist various representative materials with unique defect characteristics. Each of these materials, such as bismuth-based photocatalysts [27] and sulfide semiconductors [28], face challenges such as high photorecombination rates of charge carriers, poor stability, or limited light absorption ranges. However, by skillfully introducing defects through methods like vacancy creation and

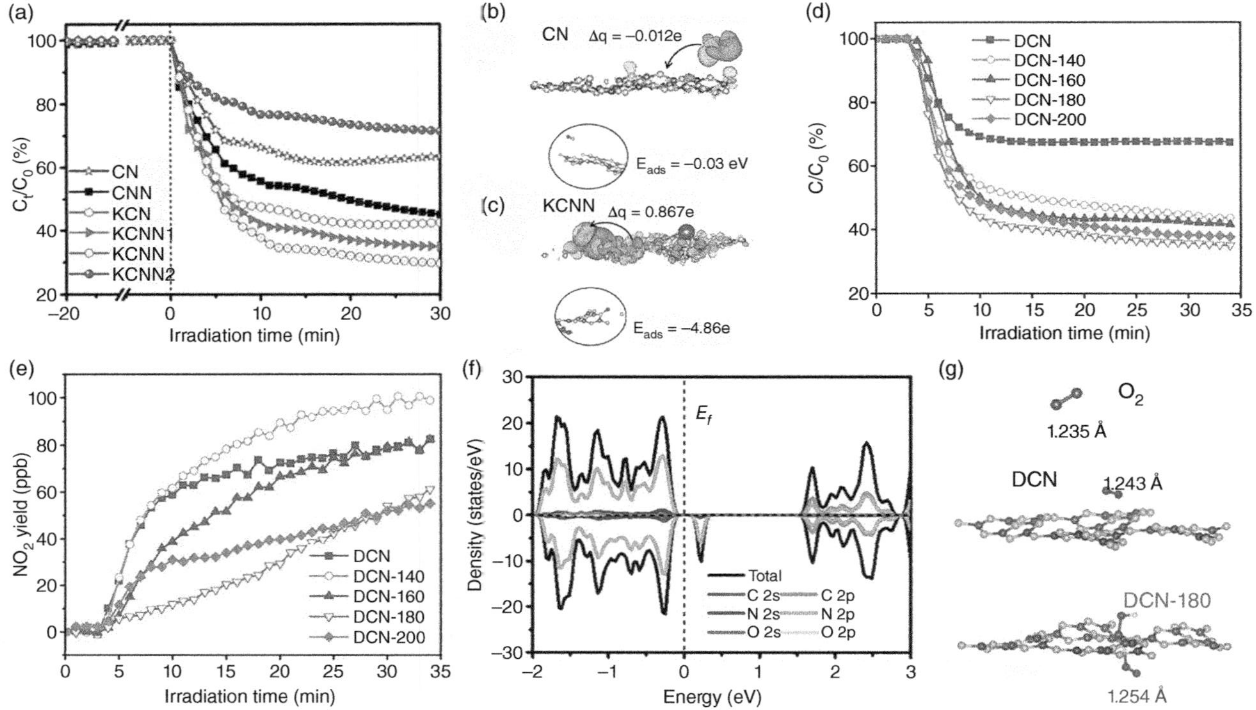

Figure 14.10 (a) Photocatalytic NO removal for prepared samples. (b) Charge density difference in CN with adsorbed NO. *Source:* Reproduced from Xia et al. [25] / with permission of Elsevier. Copyright (2024) Elsevier. (c) Charge density difference in KCNN with adsorbed NO, (d) photocatalytic NO oxidation activities, (e) NO_2 yield over as-prepared samples, (f) DOS of DCN-180, and (g) O_2 bond length variation before and after adsorption. *Source:* Reproduced from Yu et al. [26] / with permission of Elsevier.

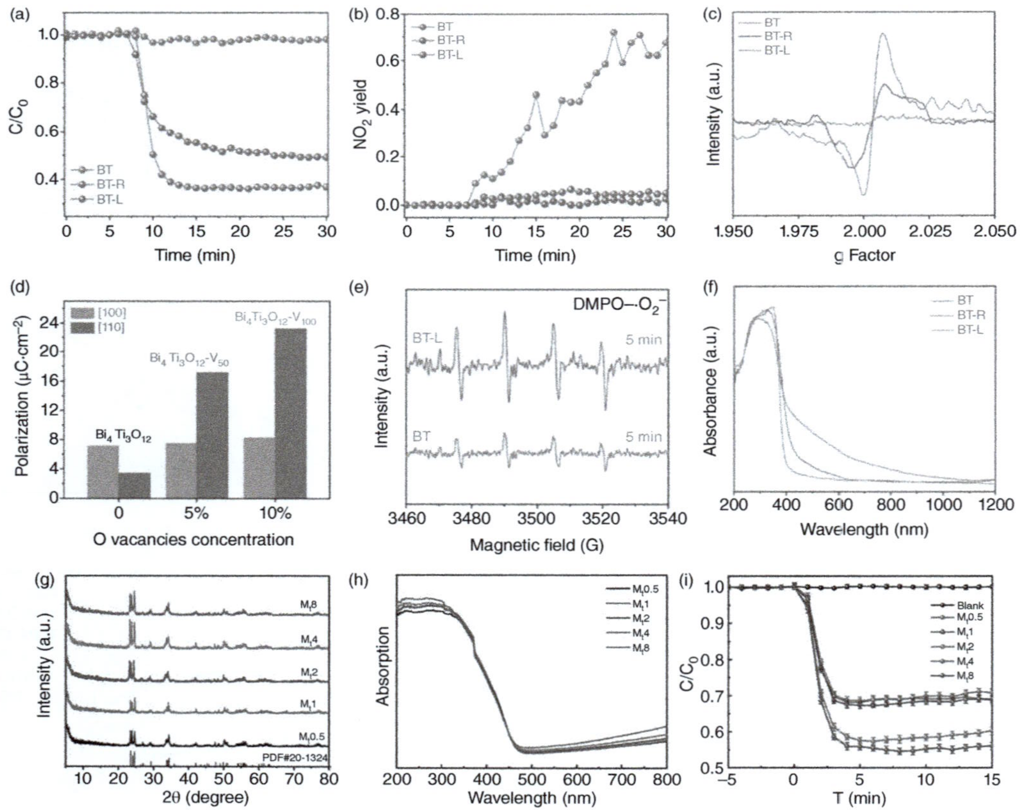

Figure 14.11 (a) Photocatalytic efficiencies for NO removal, (b) NO_2 yields, (c) EPR spectra, (d) calculated polarization of BT, BT-R, and BT-L, (e) DMPO-•O_2^- of BT and BT-L under visible light, and (f) UV–vis DRS of BT, BT-R, and BT-L. *Source:* Reproduced from Zhu et al. [29] / with permission of Elsevier. Copyright (2024) Elsevier. (g) XRD patterns, (h) UV-Vis DRS, and (i) photocatalytic efficiencies for NO removal using the M_t series. *Source:* Reproduced from Wang et al. [30] / with permission of Taylor & Francis Group.

doping, the electronic structure and optical properties of these semiconductor materials can be modulated and improved, thereby enhancing their photocatalytic performance for NO conversion. For example, in bismuth-based materials, the introduction of defects can suppress the recombination of photogenerated electrons and holes, boosting photocatalytic efficiency. Sulfide semiconductors, on the other hand, improve their stability and modulate their electronic structure through defects, enhancing photocatalytic activity. Therefore, in the research and development of novel photocatalytic materials, the design and regulation of defects have become a critical direction, aiming to continually optimize and elevate the photocatalytic performance of these materials for NO conversion.

Wang et al. [29] prepared $Bi_4Ti_3O_{12}$ nanorods (BT-R) with preferential [010] orientation growth via a hydrothermal method. The addition of ethylene glycol was utilized to enhance the growth of $Bi_4Ti_3O_{12}$ along the [010] direction, resulting in the synthesis of oxygen-vacancy-modified $Bi_4Ti_3O_{12}$ nanowires (BT-L). BT-L was employed for the photocatalytic conversion of NO_x at the ppb level, achieving a conversion efficiency of 67.5% under visible light ($\lambda > 420$ nm) with a low NO_2 yield (Figure 14.11a,b). Based on EPR (Figure 14.11c) and theoretical calculations (Figure 14.11d), the polarization intensity of $Bi_4Ti_3O_{12}$ is primarily concentrated in the [010] direction and positively correlates with the oxygen vacancy concentration. Consequently, the formation of oxygen vacancies facilitates the adsorption and activation of O_2 (Figure 14.11e) and contributes to the expanded light response range (Figure 14.11f), thereby enhancing the photocatalytic activity for NO conversion. To improve the preparation efficiency of defective photocatalytic materials, Chen et al. [30] prepared the defective WO_3 by a ball milling method (Figure 14.11g). As shown in Figure 14.11h, the absorption in the visible light region is gradually enhanced from $M_t0.5$ to M_t8, mainly associated with the increased surface defects. M_t1 exhibited the best performance for photocatalytic NO removal, which was attributed to the low recombination of charge carriers and the size of the as-prepared samples (Figure 14.11i).

14.4 Conclusion

In conclusion, the investigation into defect-engineered photocatalytic materials has revealed their promising potential for enhancing the photocatalytic conversion of NO. By introducing defects into the material structure, the optical and electronic properties were tailored to optimize the absorption of visible light and facilitate efficient charge separation. These modifications led to improved photocatalytic activity, as evidenced by the enhanced removal rates of NO. Furthermore, the defect sites acted as active centers for the reduction of NO, promoting the conversion process. Overall, the study underscores the significance of defect engineering in developing advanced photocatalytic materials for effective NO conversion, paving the way for innovative applications in environmental remediation.

Acknowledgements

This work was supported by Fujian Provincial Natural Science Foundation of China (No. 2024J08084) and the Introduced Talent Start-up Fund Project of Wuyi University (No. YJ202325).

Conflict of Interest

The authors declare that they have no conflict of interest.

References

1 Skakkebæk, N.E., Lindahl-Jacobsen, R., Levine, H. et al. (2022). Environmental factors in declining human fertility. *Nat. Rev. Endocrinol.* 18: 139–157.
2 Shaw, S. and Van Heyst, B. (2022). Nitrogen oxide (NO_x) emissions as an indicator for sustainability. *Env. Sustain. Indicat.* 15: 100188.
3 Hui, K., Yuan, Y., Xi, B., and Tan, W. (2023). A review of the factors affecting the emission of the ozone chemical precursors VOCs and NO_x from the soil. *Environ. Int.* 172: 107799.
4 Asghar, U., Rafiq, S., Anwar, A. et al. (2021). Review on the progress in emission control technologies for the abatement of CO_2, SO_x and NO_x from fuel combustion. *J. Environ. Chem. Eng.* 9: 106064.
5 Doppalapudi, A., Azad, A., and Khan, M. (2023). Advanced strategies to reduce harmful nitrogen-oxide emissions from biodiesel fueled engine. *Renew. Sust. Energ. Rev.* 174: 113123.
6 Yan, J., Zhou, F., Zhou, Y. et al. (2018). Wet oxidation and absorption procedure for NO_x removal. *Environ. Technol. Innov.* 11: 41–48.
7 de Assis, G.C., Silva, I.M., dos Santos, T.G. et al. (2021). Photocatalytic processes for biomass conversion. *Cat. Sci. Technol.* 11: 2354–2360.
8 Rhimi, B., Padervand, M., Jouini, H. et al. (2022). Recent progress in NO_x photocatalytic removal: Surface/interface engineering and mechanistic understanding. *J. Environ. Chem. Eng.* 10: 108566.
9 Li, N., Wang, C., Zhang, K. et al. (2022). Progress and prospects of photocatalytic conversion of low-concentration NO_x. *Chin. J. Catal.* 43: 2363–2387.
10 Nguyen, V.-H., Nguyen, B.-S., Huang, C.-W. et al. (2020). Photocatalytic NO_x abatement: Recent advances and emerging trends in the development of photocatalysts. *J. Clean. Prod.* 270: 121912.
11 Li, F., Liu, G., Liu, F., and Yang, S. (2023). A review of self-cleaning photocatalytic surface: effect of surface characteristics on photocatalytic activity for NO. *Environ. Pollut.* 327: 121580.
12 Chang, F., Wang, X., Yang, C. et al. (2022). Enhanced photocatalytic NO removal with the superior selectivity for NO_2^-/NO_3^- species of $Bi_{12}GeO_{20}$-based

composites via a ball-milling treatment: Synergetic effect of surface oxygen vacancies and np heterojunctions. *Compos. Part B* 231: 109600.

13 Tang, Y., Cui, W., Wang, S., and Dong, F. (2024). Efficient photocatalytic NO removal with inhibited NO_2 formation and catalyst loss over sponge-supported and functionalized g-C_3N_4. *J. Hazard. Mater.* 465: 133323.

14 Hailili, R., Ji, H., Wang, K. et al. (2022). ZnO with controllable oxygen vacancies for photocatalytic nitrogen oxide removal. *ACS Catal.* 12: 10004–10017.

15 (a) Jiang, L., Zhou, S., Yang, J. et al. (2022). Near-infrared light responsive TiO_2 for efficient solar energy utilization. *Adv. Funct. Mater.* 32: 2108977. (b) Li, Z., Li, Z., Zuo, C., and Fang, X. (2022). *Adv. Mater.* 34: 2109083.

16 Tan, X., Qin, G., Cheng, G. et al. (2020). Application of nanostructured TiO_2 in UV photodetectors: a review. *Cat. Sci. Technol.* 10: 6923–6934.

17 Zhu, Q., Hailili, R., Xin, Y. et al. (2022). Efficient full spectrum responsive photocatalytic NO conversion at $Bi_2Ti_2O_7$: Co-effect of plasmonic Bi and oxygen vacancy. *Appl Catal B* 319: 121888.

18 Zhao, S. and Choi, W. (2024). Ti^{3+} sites on mesoporous titania enable selective photocatalytic decomposition of NO to N_2 and O_2 under visible light. *Chem Catal.* 4 (8): 101058.

19 Tran, D.P.-H., Pham, M.-T., Wang, Y.-F. et al. (2023). Tuning visible light-driven photocatalytic NO removal: Insights from glucose-derived CQDs/ZnO nanorods composite. *J. Environ. Chem. Eng.* 11: 111561.

20 Wan, Y., Li, J., Ni, J. et al. (2022). Crystal-facet and microstructure engineering in ZnO for photocatalytic NO oxidation. *J. Hazard. Mater.* 435: 129073.

21 Zhu, P., Yin, X., Gao, X. et al. (2021). Enhanced photocatalytic NO removal and toxic NO_2 production inhibition over ZIF-8-derived ZnO nanoparticles with controllable amount of oxygen vacancies. *Chin. J. Catal.* 42: 175–183.

22 Luo, Y., Zhu, Y., Han, Y. et al. (2023). g-C_3N_4-based photocatalysts for organic pollutant removal: a critical review. *Carbon Res.* 2: 14.

23 Ong, W.-J., Tan, L.-L., Ng, Y.H. et al. (2016). Graphitic carbon nitride (g-C_3N_4)-based photocatalysts for artificial photosynthesis and environmental remediation: are we a step closer to achieving sustainability? *Chem. Rev.* 116: 7159–7329.

24 Duan, L., Li, G., Zhang, S. et al. (2021). Preparation of S-doped g-C_3N_4 with C vacancies using the desulfurized waste liquid extracting salt and its application for NO_x removal. *Chem. Eng. J.* 411: 128551.

25 Xia, Y., Yang, H., Ho, W. et al. (2024). Promoting the photocatalytic NO oxidation activity of hierarchical porous g-C_3N_4 by introduction of nitrogen vacancies and charge channels. *Appl Catal B* 344: 123604.

26 Yu, M., Chang, S., Ma, L. et al. (2025). Remarkable improvement in photocatalytic activity of g-C_3N_4 for NO oxidation through surface hydroxylation. *Sep. Purif. Technol.* 354: 128695.

27 Lu, X., Wang, L., Li, Z. et al. (2024). 2D bismuth-based nanomaterials for photocatalytic nitrogen oxide removal: progress and prospects. *ACS Sustain. Chem. Eng.* 12: 11444–11466.

28 Zhao, J., Tian, S., Huang, Q. et al. (2022). Ultra-highly photocatalytic removal of pollutants by polypyrrole/cadmium sulfide/polyether sulfone hybrid porous membrane in single-pass mode. *Chem. Eng. J.* 432: 134300.

29 Zhu, Q., Wang, Y., Wang, J. et al. (2024). Synergistic polarization and oxygen vacancies engineering for enhancing photocatalytic NO removal over $Bi_4Ti_3O_{12}$ nanowires. *Appl Catal B* 346: 123734.

30 Wang, Y., Chang, F., Wei, Z. et al. (2022). Photocatalytic NO removal by WO_3 samples prepared via a ball milling treatment under different parameters. *Inorg. Nano-Metal Chem.* 54: 547–559.

15

Defective Photocatalytic Materials for Biomass Conversion

Nafiseh Pourshirband[1,2], Aydin Hassani[3,4], and Paria Sharafi-Badr[5]

[1] *Department of Chemistry, Shah. C., Islamic Azad University, 311-86145, Shahreza, Isfahan, Islamic Republic of Iran*
[2] *Department of Chemistry, Isf. C., Islamic Azad University, 81595-158, Isfahan, Islamic Republic of Iran*
[3] *Department of Materials Science and Nanotechnology Engineering, Faculty of Engineering, Near East University, 99138 Nicosia, Mersin 10, Turkey*
[4] *Research Center for Science, Technology and Engineering (BILTEM), Near East University, 99138 Nicosia, Mersin 10, Turkey*
[5] *Department of Pharmacognosy and Pharmaceutical Biotechnology, School of Pharmacy, Iran University of Medical Sciences, 1475886671, Tehran, Iran*

15.1 Introduction

Environmental concerns regarding resource scarcity, resulting from the substantial depletion of fossil fuels, have emerged as some of the most extensively discussed and formidable challenges of the past decade [1, 2]. Consequently, the urgent demands of these environmental issues drive individuals to seek new renewable resources that can substitute or partially replace existing fossil fuels. The most prevalent renewable carbon resource on Earth is not only easily accessible and sustainable but also economically viable, making its efficient utilization an appealing solution for bridging the gap between rising energy demand and insufficient energy sources [3].

Lignocellulosic biomass (LCB) constitutes the majority of plant raw materials and serves as an abundant and sustainable source of green carbon that does not compete with food crops [4, 5]. This biomass is made up three primary polymers: cellulose (40–60%), hemicellulose (10–40%), and lignin (15–30%). The proportions of these polymers vary depending on the type of substrate [6, 7]. Cellulose is a linear polymer that, along with branched carbohydrates known as hemicelluloses, forms a key structural component of plant cell walls. These polymers are interconnected and surrounded by phenolic compounds called lignin, forming a sturdy matrix. Collectively, cellulose, hemicelluloses, and lignin create microfibrils that demonstrate remarkable resistance to chemical oxidation [8]. Cellulose, hemicelluloses, and saccharide macromolecules are composed of C_5 or $C_5 + C_6$ sugars and can be hydrolyzed to provide monosaccharides such as glucose, arabinose, and xylose [9]. In contrast, lignin serves as a primary source of persistent aromatics, enabling

Defect Engineering in Photocatalysis: From Fundamentals to Applications,
First Edition. Edited by Chuanyi Wang, Abdelkader Labidi, and Qiuhui Zhu.

the synthesis of monocyclic aromatic compounds. Biomass and its derivatives are unique renewable resources that provide a variety of organic carbons. These materials have significant potential for the production of high-value compounds and clean energy. By 2030, it is projected that 35% of chemicals and products will be derived from biomass [10, 11]. In the recent years, biomass photoreforming utilizing the photocatalysis process has attracted significant attention from researchers and industrialists due to its potential to generate high-value-added chemicals and fuels. However, some reported photocatalytic feedstocks still face challenges in biomass conversion technologies, such as charge carrier recombination, limited light absorption, and low efficiency in converting biomass into valuable products.

To address these issues, the creation/formation of defects in photocatalytic materials plays a critical role in enhancing the performance of photocatalysts, offering a promising strategy for advancing biomass conversion technologies. For example, crystallographic defects, which are disruptions in the ideal periodic arrangement of atoms or molecules within crystalline materials, are universally present in photocatalytic materials and significantly influence their performance. However, the impact of these structural defects is not yet fully understood. While defects are often considered detrimental to photocatalytic activity [12, 13], they can also modify the electronic structure of an otherwise-perfect periodic crystal. Additionally, defects act as scattering centers for electrons and holes, impeding the diffusion of charge carriers [14, 15]. Recently, the beneficial role of defects in enhancing the photocatalytic performance has gained recognition, driven by advancements in defect-controlled photocatalysts and a deeper understanding of photocatalytic mechanisms [16, 17]. Structural imperfections in solid materials can be broadly categorized into four types based on their dimensions: point defects (e.g., vacancies and doping), one-dimensional defects (e.g., screw and edge dislocations), two-dimensional defects (e.g., grain boundaries and twin boundaries), and three-dimensional defects (e.g., lattice disorder and voids) [18]. Among these, point defects, particularly vacancies, are the most common type in photocatalytic materials [19, 20]. Oxygen vacancies (O_V) are widely studied in metal oxide semiconductor photocatalysts, including TiO_2 [21], ZnO [22], WO_3 [23], Bi_2WO_6 [24], BiOCl [25], $BiPO_4$ [26], BiO_2 [27], and $SrTiO_3$ [28]. Similarly, sulfur vacancies have been identified in materials such as ZnS [29], Zn_xCd_yS [30], and $ZnIn_2S_4$ [31], while nitrogen vacancies are observed in Ta_3N_5 [32]; iodine vacancies have also been reported in $Bi_7O_9I_3$, all of which are examples of anion vacancies in ionic compound-based photocatalysts [33]. Additionally, cation vacancies, such as bismuth [34] and titanium [35] vacancies, have been documented in photocatalytic materials alongside anion vacancies. One effective approach to controlling the non-stoichiometric composition of photocatalysts without introducing contaminants is through vacancy creation or self-doping. In contrast, foreign element doping introduces significant point defects into the semiconductor host, altering its composition by incorporating impurities [36]. Dopant ions or atoms can either replace existing ions or atoms within the host or occupy interstitial spaces. For substitutional doping, the radii of the dopant atoms or ions should be relatively close to those of the atoms they replace, whereas interstitial dopants typically have much smaller radii

[37, 38]. The formation or creation of defects in photocatalytic feedstocks provides a promising platform for the synthesis of novel photocatalysts and advancement of their applications in biomass conversion technologies. This chapter explores defect formation in various photocatalytic feedstocks, highlighting synthetic protocols and characterization techniques for these defective materials. Additionally, it discusses the advantages and disadvantages of using defective photocatalytic materials for biomass photo-reforming into valuable chemicals and fuels.

15.2 Types of Defects

Understanding the various types of defects in photocatalytic materials is crucial for optimizing their performance in applications such as energy conversion and environmental remediation. These defects significantly influence charge carrier dynamics, recombination rates, and overall photocatalytic efficiency [39, 40]. Defects in photocatalytic materials can be broadly categorized into intrinsic, extrinsic, surface, and structural defects. Intrinsic defects are natural irregularities within the crystal lattice, including vacancies (missing atoms), interstitials (extra atoms within the lattice), and antisite defects (atoms occupying incorrect positions) [40, 41]. For example, O_V in ZnO photocatalysts are common intrinsic defects that can enhance photocatalytic activity by providing active sites for reactions and facilitating charge–carrier separation. Extrinsic defects are intentionally introduced through doping, where foreign elements are added to modify the material's properties [41, 42]. For instance, incorporating transition metals into semiconductor photocatalysts creates new energy levels within the bandgap, optimizing light absorption and charge separation. Surface defects refer to imperfections at the material's surface, such as dangling bonds or reconstructed surface atoms, which often play a critical role in the catalytic reactions [41, 42]. For example, surface O_V on TiO_2 can enhance reactant adsorption and promote charge transfer, facilitating photocatalytic reactions. Structural defects involve alterations in the material's overall structure, such as phase transitions, amorphization, or the formation of heterogeneous structures [41]. A notable example is the creation of heterojunctions, where two distinct semiconductor materials are combined to improve charge separation and reactivity by leveraging their differing band structures. Defects influence photocatalytic efficiency through several key mechanisms, including charge generation, separation, and recombination [41]. They contribute to charge generation by creating localized states within the bandgap, facilitating the formation of electron–hole pairs upon light excitation. Additionally, defects enhance charge separation by reducing recombination rates, particularly in dual-defect systems where multiple defects synergistically improve efficiency [41, 43]. However, defects can also act as recombination centers, underscoring the importance of maintaining an optimal defect concentration to ensure effective participation of charge carriers in photocatalytic reactions [44]. Recent advancements in defect engineering emphasize the controlled introduction and manipulation of defects to optimize photocatalytic materials [40, 41]. Techniques such as chemical vapor deposition (CVD) and plasma treatments are increasingly used to create specific defects at precise locations within the crystal lattice [45, 46]. These methods aim to establish dual-defect systems, where coexisting defects enhance light absorption and chemical reactivity. Such fine-tuning of electronic properties holds a significant promise for developing highly efficient photocatalysts, addressing the challenges in energy sustainability and environmental remediation [42]. It is worth noting that the role of defect engineering in enhancing photocatalytic performance has been extensively reviewed [39–41].

15.3 Preparation and Characterization

15.3.1 Fabrication Techniques

Defect engineering aims to enhance the performance and properties of materials and systems by systematically studying and modifying their imperfections. Nanomaterials, through their unique surface reactivities, significantly influence catalytic processes. Native defects in materials can profoundly alter their chemical and physical properties, with surface defects playing a particularly critical role in catalysis. The adsorption of reactant molecules onto the catalyst surface is a fundamental step in catalysis, initiating chemical transformations that lead to product formation. Surface defects create active sites that enhance reactivity by facilitating the binding of reactant molecules and promoting chemical reactions [47]. Unlike fully coordinated sites, defects generate unsaturated sites that are inherently more reactive. Understanding the relationship between the macroscopic features of reaction kinetics and adsorption states of reaction intermediates on these active sites is crucial for catalysis. These sites serve as locations where reactants can bind and undergo chemical transformations. By introducing dopants and defects, nanocatalysts can reduce their bandgaps, enhance light absorption, and modify electrical conductivity, thereby promoting charge separation and migration [48]. The type and concentration of these defects are key factors in determining the material's magnetic and luminescent properties. Additionally, the correlation between the macroscopic features of reaction kinetics and adsorption states of reaction intermediates on active sites is essential for advancing catalytic studies [49].

Defects can create unsaturated coordination sites that exhibit higher reactivity compared to fully coordinated sites. For instance, a vacancy defect on a metal oxide surface, characterized by the absence of an atom, serves as an effective adsorption site for molecules involved in chemical reactions. The introduction of O_v in wide-bandgap semiconductors, such as TiO_2 and ZnO, reduces the bandgap and enhances the kinetics of redox reactions. These O_v traps capture excited electrons and improve charge separation, ultimately boosting the photocatalytic activity of semiconductors [50]. Defects can also modify the electronic structure of the catalyst, significantly influencing their catalytic activity. Substitutional defects, where one atom is replaced by another, can alter the electronic states of the catalyst, affecting its reactivity toward specific reactants. Similarly, the dislocation of atoms at semiconductor edges creates localized defect states that facilitate redox processes, as these dislocated centers act as charge-scattering clusters. These localized states within the bandgap can enhance charge transfer and catalytic reactivity. Additionally, dislocations and grain boundaries can create regions with unique electronic properties, potentially leading to enhanced catalytic activity in these specific areas. A comprehensive understanding of the role of defects in catalysis is essential for developing catalysts with improved activity, selectivity, and durability. By regulating the introduction of defects, the physical properties of materials can be precisely tailored for advanced applications [51]. Optimizing defect architecture in materials provides an effective strategy for customizing their properties, thereby enhancing their performance in various catalytic applications.

15.3.2 Characterization Tools

Defect characterization technology is essential for understanding, applying, and managing defects. These techniques are used to analyze defect morphology, location, density, and chemical properties. By enhancing the understanding and regulation of how defects influence material and catalytic activity, these methodologies play a critical role in optimizing and advancing defect engineering solutions [52]. Figure 15.1 summarizes the commonly used methods for defect characterization.

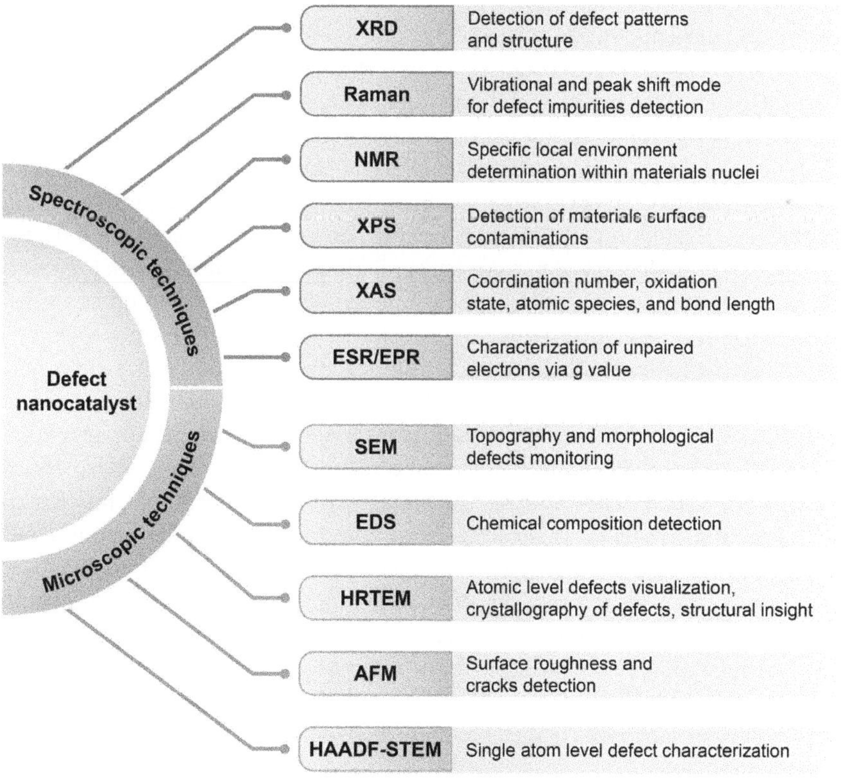

Figure 15.1 The latest and most advanced methods for characterizing defect structures. *Source:* Adapted from Muhammad et al. [52].

15.4 Biomass Conversion

15.4.1 Biomass Feedstocks

Biomass feedstocks are primarily composed of proteins, lipids, and carbohydrates, including lignocellulosic materials, sugars, starch, and polysaccharides. The composition of biomass varies significantly, leading to its classification into several categories. LCB, which constitutes the majority of plant fibers, includes materials such as grasses, straw, and wood. It is primarily composed of lignin, hemicelluloses, and cellulose. This category can be further divided into herbaceous biomass (non-woody plants that wilt and die during the growing season) and woody biomass (forest-based plants, including trees, bushes, and shrubs) [53, 54]. Carbohydrate-rich biomass, found in sugar beets and sugar cane, contains sugars such as glucose and fructose, as well as disaccharides like sucrose. Starch-rich biomass, including crops like wheat, corn, and potatoes, is characterized by a high concentration of reserve polysaccharides such as starch and inulin. Oil-rich biomass, such as rapeseed and various micro- and macroalgae, contains a high lipid content. High-protein biomass sources include oilseeds like sunflower and soybean, as well as legumes such as peas and lentils. However, animal-based biomass sources, such as fish and pigs, are not considered viable bioenergy feedstocks due to their high cost, nitrogen concentration, and the food-first principle, which limits their long-term financial and practical feasibility.

Biomass can also be classified based on its physical state into two: wet biomass and dry biomass. Wet biomass, with a water content ranging from 15% to over 90% (e.g., algae), requires additional processing steps, such as drying or ensiling, similar to procedures used for biogas and feed. Dry biomass, with a moisture content of less than 13%, is easier to handle in terms of collection, transportation, and storage [53]. Additionally, biomass used for energy can be categorized based on its source or production sector. Agricultural biomass includes materials grown on agricultural land, such as lignocellulosic materials, starches, and oil crops, as well as both edible and non-edible crops used for food and energy purposes. Forest biomass refers to wood sourced from forests not primarily used for farming or urban development, including tree plantations in forested areas intended for energy production, and woody biomass resulting from forests management practices [55]. By-products, residues, and waste encompass biomass from agricultural, forestry, and industrial side streams, including organic residues from municipal solid waste. Aquatic biomass includes plant and animal materials produced in aquatic environments, such as microalgae, seaweed, and aquatic plants. Biomass classifications often combine two criteria: the source and characteristics of the biomass. For example, agricultural biomass includes both herbaceous and woody elements. Short-rotation coppice (SRC), the cultivation of woody biomass for raw material and/or fuel, involves harvesting on a cycle of two to five years. SRC is not limited to specific land types and can be cultivated in both forested and agricultural environments [54, 55].

15.4.2 Photocatalytic Biomass Conversion

The photocatalytic conversion of LCB has received significant attention as a sustainable and efficient method for biomass conversion, presenting an appealing alternative to classical catalysis, which often requires substantial energy input [56–58]. LCB exhibits limited biodegradability due to its intrinsic characteristics when untreated. Pretreatment methods disrupt the lignin barrier, facilitating microbial and enzymatic degradation of the carbohydrate components. Certain pretreatment techniques can partially or completely fractionate biomass into hemicelluloses, cellulose, and lignin. The integrated biorefinery model enhances efficiency by segregating LCB into distinct component streams, enabling the on-site synthesis of fuels, chemicals, and other products [59]. This biorefinery concept improves the economic viability of fuel production by allowing simultaneous production within a single facility, similar to petroleum refineries. Streams subjected to hydrothermal pretreatment demonstrate elevated levels of hemicellulose and xylose, while concentrations of glucose, arabinose, and galactose remain low. Isolating this stream from cellulose-rich particles enhance its quality due to its high content of hemicellulose and xylose. Enzymatic hydrolysis of the liquid xylose stream, along with the cellulose- and lignin-rich solids, facilitates the separation of lignin-enriched solids from the xylose- and glucose-enriched streams. Recent research indicates that an inhibitor-tolerant strain of *Actinobacillus succinogenes* can produce succinic acid ($42.8\,g\,L^{-1}$) from xylose-enriched biorefinery streams, demonstrating resistance to furfural and hydroxymethylfurfural [60].

The production of succinic acid from biomass-derived streams is relatively underexplored but holds potential as a co-product in biorefineries, offering a sustainable alternative to petroleum-derived chemicals for various applications. The enzymatic action of endoxylanases from glycoside hydrolase families GH10 and GH11 extracts high-value products, such as xylo oligosaccharides (XOS), from the insoluble arabinoxylan fraction of pretreated wheat bran and brewers' spent grain. XOS, the novel prebiotics used in the food industry, can be produced economically on a large-scale [61–63]. Previous findings indicate that xylanases generated by *Pseudozyma hubeiensis*, isolated in the laboratory, can produce XOS with the desired degree of polymerization (DP, X_3–X_7) without requiring additional downstream processing [64]. Xu et al. employed cost-effective chromium phosphate as an efficient catalyst, achieving furfural yields of 88% from xylose and 67% from wheat bran, demonstrating the potential for integrating these processes to utilize the xylose stream from LCB [65].

The substantial loss of hemicelluloses in the liquid stream poses a critical challenge for applying this pretreatment to the fractionation of LCB components. Lignin-enriched liquid streams are preferable for lignin valorization compared to the solid residual lignin produced by dilute acid pretreatment, as these alkaline streams have been utilized to synthesize precursors such as polyhydroxyalkanoates and muconic acid [66]. An editorial article in Green Chemistry emphasizes that multidisciplinary approaches are necessary to address the challenges associated with lignin valorization, unlocking the true potential of this underutilized resource (LCB) [67].

15.4.3 Role of Defects

Single-atom catalysts (SACs) can significantly enhance their surface catalytic performance through defect engineering. A controlled number of defects on the catalyst surface can boost the catalytic activity. Understanding and analyzing defects in catalysts is essential for their synthesis and characterization. In SACs, defects play a critical role in anchoring metal atoms, as the elevated energy of these defect sites makes them highly reactive. During reactions, defects adjacent to and surrounding the active atoms significantly influence the adsorption or activation of the reaction substrate, rather than the single atom alone [68, 69]. Figure 15.2 illustrates the synergistic effects of single atoms and defects, highlighting key mechanisms, such as the reaction adsorption mechanism [70], local promotion mechanism, active site renewal mechanism, and selective adsorption mechanism. These mechanisms provide a foundation developing advanced defect-based SACs with enhanced activity and additional functionalities [6, 69, 71–74].

Defect chemistry suggests that imperfections in catalysts can create additional vacant coordination sites, improving the adsorption of reactants or intermediates [75]. In SACs, the mechanism of adsorption and activation involving individual atoms and defects is crucial. Defect sites provide more active adsorption sites on the surface compared to individual atoms, and their interaction with these atoms lowers the energy barrier for adsorption, thereby enhancing the reaction's activity.

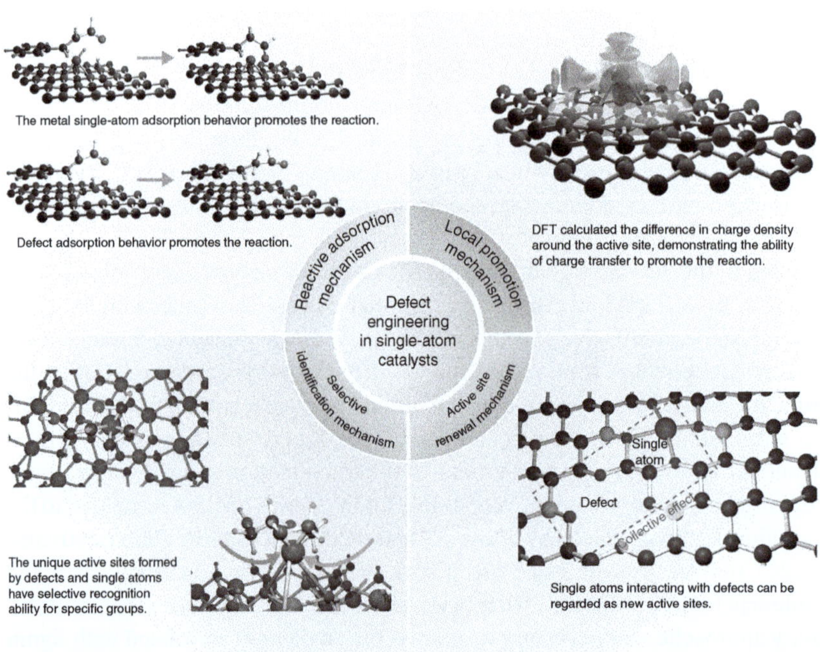

Figure 15.2 The reaction mechanisms of defect engineering in single-atom catalysts. *Source:* Reproduced from Wang et al. [70] / with permission of Elsevier.

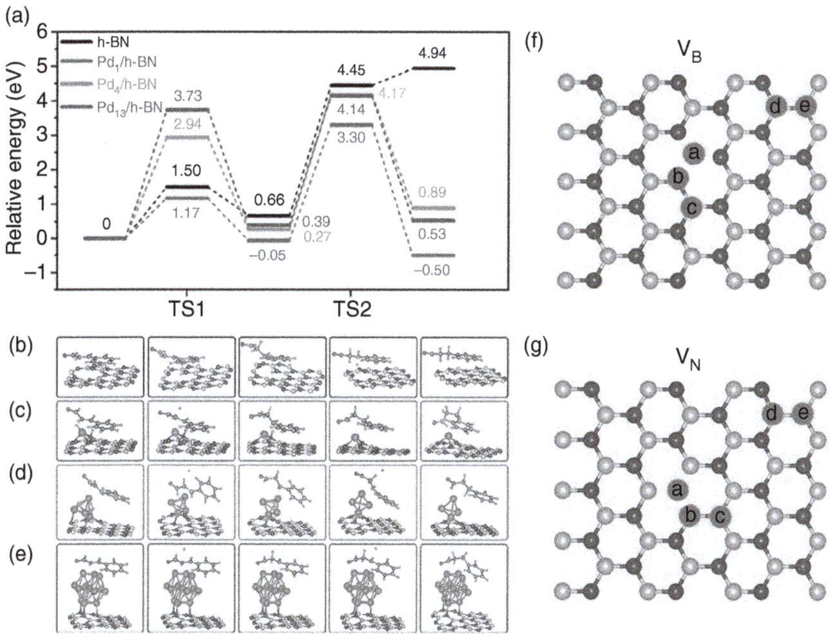

Figure 15.3 (a) Reaction pathways and the corresponding energies of defective h-BN, Pd_1/h-BN, Pd_4/h-BN, and Pd_{13}/h-BN in the hydrogenation of cinnamaldehyde. (b–e) Corresponding structures of intermediates and transition states for the key elementary steps. The atomic structure of hexagonal boron nitride (f) with a B vacancy and (g) an N vacancy. B and N atoms are represented by light grey and dark grey spheres. The site labeled 'a' is the defect site, while sites labeled 'a–e' are all possible sites for Pd atom adsorption. *Source:* Adapted from Li et al. [78].

Additionally, defect sites boost catalytic activity by serving as direct adsorption sites. In contrast, surfaces without defects exhibit higher activation energy for adsorption, often preventing the reaction from occurring [74, 76]. Furthermore, individual atoms located at defect sites can activate the reaction substrate, enhancing catalytic reactions, including biomass conversion processes [77]. For example, Li et al. developed SACs on hexagonal boron nitride (h-BN) to improve the selective hydrogenation of cinnamaldehyde to hydrocinnamaldehyde [78]. As illustrated in Figure 15.3, defects in nitrogen-doped h-BN effectively stabilize Pd atoms. Surface imperfections in h-BN substrates, which are typically inert, facilitate the attachment of a single Pd atom to H_2 molecules, leading to the hetero-fission of H_2 and formation of two H atoms. This unique interaction between defect sites and individual atoms significantly influences the catalytic activity of SACs. While specific reactions may not focus solely on particular defect sites, single atoms interacting with defects can exhibit varying catalytic activities. In the Pd–N structure, a hydrogen atom is situated at an adjacent anchoring point. The interaction between the cinnamaldehyde and Pd atoms occurs via the C=C bond and vacant *d* orbital of the Pd, while the aromatic ring remains unchanged during the catalytic process. The Pd/h-BN catalyst demonstrated remarkable effectiveness in the selective hydrogenation of cinnamaldehyde. Similarly, Pt–Cu alloy SACs have shown high efficiency in hydrogenolysis of glycerol to 1,2-propanediol (1,2-PDO). Glycerol molecules adsorb and activate at the Pt–Cu interface sites, with oxygen atoms attached to primary carbon adsorbing onto Cu sites and hydrogen atoms on primary carbon adsorbing onto Pt sites [71]. The primary C—O bond and secondary C—H bond are then cleaved, forming 2,3-enol, which rearranges into acetol. A hydrogen atom interacts with acetol to produce the target molecule, 1,2-PDO. The introduction of Cu atoms alters the electronic configuration of the carrier, demonstrating that defect sites can selectively interact with specific atoms through chemical bonding or surface attachment. This interaction influences the electronic structure of the atom, altering its reactivity and selectivity. When Cu and isolated Pt work together as catalysts, they utilize different atomic-scale surface-active sites, enhancing both activity and selectivity. Experimental findings show that single-atom synergistic defects act as direct adsorption sites for active compounds, strengthening the adsorption process. However, an excess of adsorption sites can compromise the structural integrity of SACs, negatively affecting their stability.

In 2024, Huang et al. reported the synthesis of TiO_2 nanoparticles with tunable O_V (O_V-TiO_2) using a facile hydrothermal method for the photocatalytic conversion of woody biomass to green H_2 production (Figure 15.4) [79]. They highlighted that O_V can reduce the energy barrier for the formation of •OH radicals, which are crucial for oxidizing woody biomass. The as-prepared O_V-TiO_2 demonstrated excellent photocatalytic conversion of α-cellulose and poplar wood chips to H_2 when paired with Pt nanoclusters as a cocatalyst. They observed maximum H_2 production rates of 1146 and 59 μmol h^{-1} g^{-1} for α-cellulose and poplar wood chips, respectively, with an apparent quantum yield of 4.89% at 380 nm in α-cellulose aqueous solution.

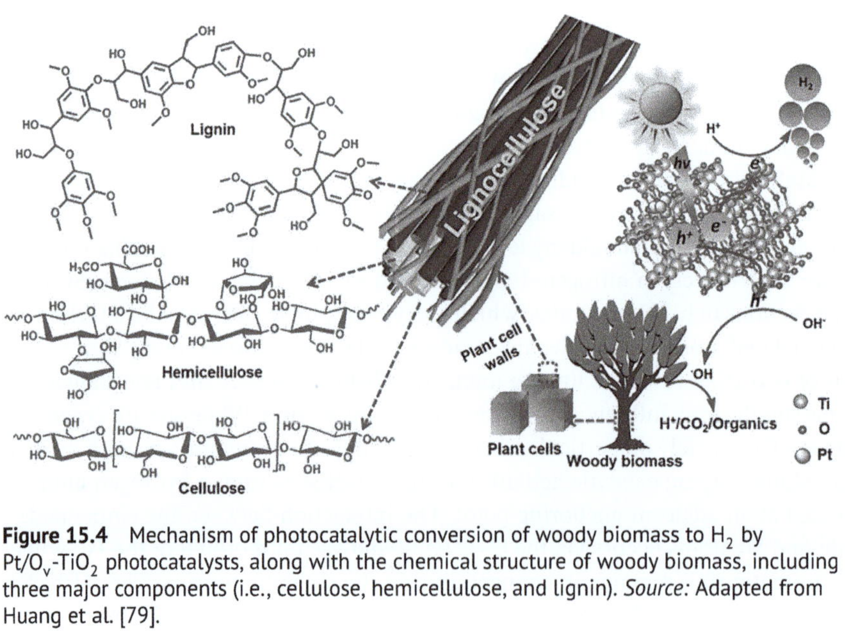

Figure 15.4 Mechanism of photocatalytic conversion of woody biomass to H_2 by Pt/O_V-TiO_2 photocatalysts, along with the chemical structure of woody biomass, including three major components (i.e., cellulose, hemicellulose, and lignin). *Source:* Adapted from Huang et al. [79].

Regarding the mechanism of biomass conversion, they proposed a three-step process: (i) the formation of photogenerated charges and their transfer to the O_V-TiO_2 surface; (ii) the oxidation of water by photogenerated holes to produce •OH radicals, which subsequently oxidize woody biomass; and (iii) the reduction of absorbed protons to H_2 by the Pt cocatalyst, which accepts photogenerated electrons from the O_V-TiO_2 catalyst. The study emphasized that the superior photocatalytic performance of the Pt/O_V-TiO_2 photocatalyst for woody biomass conversion to is attributed to the presence of O_V. These vacancies reduce the energy barrier for •OH formation and accelerate the kinetic process of •OH production, providing the driving force for the efficient oxygenolysis of woody biomass.

In another study, Dai et al. reported the development of ZIF-8-NH_2@Bi/Bi_2MoO_6, a material containing dual junctions (i.e., heterojunction and Schottky junction) and a surface plasmon resonance (SPR) effect, for the effective cleavage of the Cβ–O bond in a typical lignin model, 2-phenoxy-1-phenylethanol (PP-ol), and alkali lignin [80]. They indicated that the O_V significantly boost O_2 capture and activation, generating singlet oxygen ($^1O_2$), which plays a crucial role in facilizing Cβ–O bond cleavage. Meanwhile, photogenerated holes convert PP-ol into 2-phenoxy-1-acetophenone (Figure 15.5a,b). In their experiments, ZIF-8-NH_2@Bi/Bi_2MoO_6 achieved 93% conversion of PP-ol under air conditions, yielding 57% phenol and 48% acetophenone (AP). These results outperformed those of ZIF-8-NH_2@/Bi_2MoO_6, which showed only 46% PP-ol conversion, with 8.7% phenol yield and 6.2% AP yield. The study revealed two reaction pathways: (i) In the absence of $^1O_2$, hydroxyl groups in PP-ol are deprotonated by holes and oxidized into PP-one (pathway I) (Figure 15.5b,c). In the presence of $^1O_2$, Cα is preferentially deprotonated into carbon-centered radicals by $^1O_2$ (pathway II). Additionally, they highlighted that adsorbed O_2 on the catalyst's surface, O_V, and holes play complementary roles in lignin depolymerization (Figure 15.5d).

15.5 Challenges and Limitations

Recent studies have focused on enhancing photocatalysts through methods such as doping, coupling with various nanomaterials, precipitation with metal particles, and manipulating crystal growth patterns. These approaches aim to provide deeper insights into the photocatalyst design and photocatalytic valorization of biomass. Defective photocatalysts have been shown to improve the efficiency of biomass transformation into valuable chemicals and fuels. However, challenges, such as insufficient activity, reduced stability, and high costs, limit the practical implementation of photocatalysis for biomass reforming. Photocatalysis can degrade biomass into simpler molecules or mineralize it into harmless by-products. The presence of defects in photocatalytic materials significantly enhances the photocatalytic degradation of biomass under various conditions (e.g., air and solution). Defective photocatalysts can improve reaction rates and selectivity toward desired products during biomass photoreforming. However, while defects can boost charge separation, they may also act as recombination centers for charge carriers, potentially

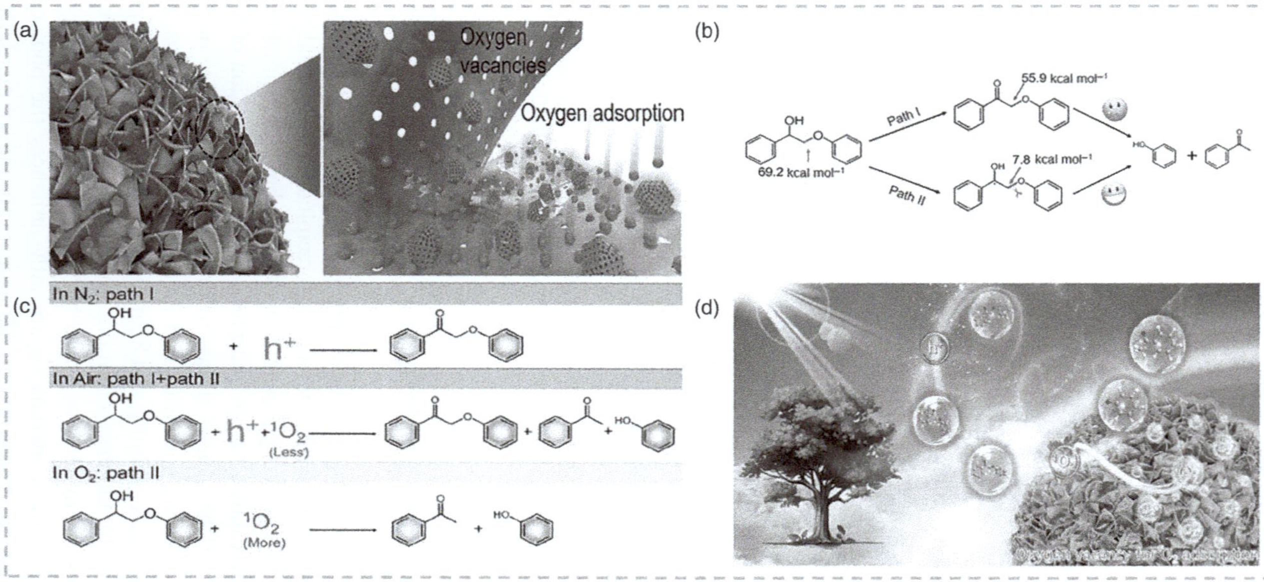

Figure 15.5 (a) Schematic presentation of oxygen adsorption on ZB-3-400, (b) proposed pathways for photocatalytic depolymerization of phenoxy-1-phenylethanol, (c) phenoxy-1-phenylethanol conversion pathways in different gas atmospheres, and (d) reactive species and oxygen vacancies for driving lignin depolymerization over ZIF-8-NH_2@Bi/Bi_2MoO_6. *Source:* Reproduced from Dai et al. [80] / with permission of American Chemical Society.

limiting the overall efficiency of photocatalytic materials for biomass conversion. Biomass feedstocks are often complex mixtures of different compounds, leading to competition between reaction pathways and complicating the photocatalytic process, even when using defective photocatalysts. The precise mechanisms by which defects enhance photocatalytic biomass degradation remain poorly understood. Additionally, the preparation of defective photocatalysts can generate waste products, and their use in biomass photoreforming may sometimes result in the formation of toxic by-products. Incorporating green chemistry principles into the large-scale synthesis of defective photocatalysts presents notable challenges, including managing energy consumption and mitigating the formation of toxic products during biomass conversion. The scalability of defective photocatalysts for biomass conversion is still unsatisfactory, and advanced preparation methods are urgently needed to overcome these challenges. Optimizing the synthesis and application of defective photocatalysts in biomass conversion technologies remains a critical area for future research.

15.6 Future Perspectives

The integration of artificial intelligence (AI) and machine learning (ML) techniques into biomass-to-biofuel processes holds significant potential for advancing the field. These technologies provide powerful tools for predictive modeling and optimization, enabling researchers to identify optimal conditions, forecast yield outcomes, and design efficient catalytic systems. By leveraging data-driven approaches, AI and ML can analyze vast amounts of experimental data on the synthesis of defective photocatalysts, which can be effectively utilized in biomass photoreforming. This can lead to more efficient process designs, reduced experimentation time, and improved overall performance of biomass conversion technologies. Furthermore, AI and ML can facilitate real-time monitoring and control of biofuel production processes, enabling adaptive optimization and enhanced productivity. As the field evolves, the synergy between AI, ML, and biomass-to-biofuel research is expected to play a crucial role in accelerating the development and implementation of sustainable and economically viable biomass waste-conversion technologies.

A major challenge for future research lies in developing innovative and reliable characterization methods for defective photocatalysts. Current limitations hinder our understanding of how defects affect photocatalytic efficiency. The relationship between structure and activity in photocatalytic processes is complex, as existing techniques often only provide indirect information and low-resolution data about the defect arrangements in photocatalytic materials.

Defect engineering is a critical strategy for advancing photocatalysts with improved photocatalytic activity. Introducing defects, such as dislocations or vacancies, enhances the activity and selectivity of catalytic sites, leading to better performance in biomass photoreforming. These defects create active sites on catalyst surfaces, significantly expanding the potential applications, particularly for biomass conversion. By precisely manipulating defects, researchers can tailor catalyst selectivity and optimizing photocatalytic processes, enhancing efficiency and promoting sustainability

Advancements in nanotechnology have enabled the design of nanocatalysts with atomic-scale precision, particularly SACs, which offer enhanced functionalities and efficiency for biomass photoreforming. However, challenges such as site deactivation, electron–hole recombination, and structural instability impede their performance in converting biomass into useful chemicals and fuels. Future research should focus on developing advanced defective materials and employing computational modeling to address these challenges. Overcoming these barriers will unlock the potential of SACs for biomass conversion applications, driving sustainable technological advancement in biomass transformation.

Integrating green chemistry principles into large-scale synthesis of defective photocatalysts presents a significant challenge, but addressing these challenges could greatly benefit biomass conversion technologies. Moreover, understanding the interactions between defective photocatalysts and biomass feedstocks could enable the tailoring of the specific types of defective photocatalysts to specific biomass types, leading to more efficient photoreforming process. Advanced characterization techniques could also provide valuable insights into the behavior of defects during photocatalytic reactions, optimizing their design for efficient biomass conversion technologies.

15.7 Conclusion

This chapter aims to elucidate the potential impact of incorporating defects in photocatalytic materials, emphasizing their critical role in the efficient conversion of biomass into value-added chemicals and fuels. Defective photocatalysts have shown distinct advantage, such as enhanced light absorption, improved charge separation efficiency, and increased surface reaction activity, all of which contribute to more effective biomass conversion. As highlighted by the examples discussed in this chapter, defective photocatalysts represent a promising approach for advancing biomass conversion processes. By utilizing defect engineering, researchers and industrialists can improve the efficiency and selectivity of photocatalytic reactions, offering a sustainable and attractive method for transforming biomass into valuable chemicals and fuels. Despite significant progress in understanding and applying defects for the synthesis of photocatalysts over recent decades, challenges remain in developing efficient and stable defective photocatalysts. Key obstacles include the precise control of defect types and concentrations, unexplored regulatory effects of defects on the interfacial properties of photocatalysts, and mechanisms by which defects affect charge separation and transport. Moving forward, advanced techniques for the precise engineering of defects are essential to optimize photocatalytic properties of materials and enable the efficient conversion of biomass into high-yield, useful products.

Conflict of Interest

The authors declare that they have no conflict of interest.

References

1 Tan, H., Li, J., He, M. et al. (2021). Global evolution of research on green energy and environmental technologies: a bibliometric study. *J. Environ. Manag.* 297: 113382.

2 Islam, A., Hossain, M.B., Mondal, M.A.H. et al. (2021). Energy challenges for a clean environment: Bangladesh's experience. *Energy Rep.* 7: 3373–3389.

3 Antar, M., Lyu, D., Nazari, M. et al. (2021). Biomass for a sustainable bioeconomy: an overview of world biomass production and utilization. *Renew. Sust. Energ. Rev.* 139: 110691.

4 Malik, P., Awasthi, M., and Sinha, S. (2021). Biomass-based gaseous fuel for hybrid renewable energy systems: an overview and future research opportunities. *Int. J. Energy Res.* 45: 3464–3494.

5 Lin, C.-Y. and Lu, C. (2021). Development perspectives of promising lignocellulose feedstocks for production of advanced generation biofuels: a review. *Renew. Sust. Energ. Rev.* 136: 110445.

6 He, J., Li, N., Li, Z.G. et al. (2021). Strategic defect engineering of metal–organic frameworks for optimizing the fabrication of single-atom catalysts. *Adv. Funct. Mater.* 31: 2103597.

7 Kumar, M., Ambika, S., Hassani, A., and Nidheesh, P.V. (2023). Waste to catalyst: role of agricultural waste in water and wastewater treatment. *Sci. Total Environ.* 858: 159762.

8 Roy, R., Rahman, M.D.S., and Raynie, D.E. (2020). Recent advances of greener pretreatment technologies of lignocellulose. *Curr. Res. Green Sustain. Chem.* 3: 100035.

9 Li, T., Chen, C., Brozena, A.H. et al. (2021). Developing fibrillated cellulose as a sustainable technological material. *Nature* 590: 47–56.

10 Chheda, J.N., Huber, G.W., and Dumesic, J.A. (2007). Liquid-phase catalytic processing of biomass-derived oxygenated hydrocarbons to fuels and chemicals. *Angew. Chem. Int. Ed.* 46: 7164–7183.

11 Yin, Y., Liu, Q., Zhao, Y. et al. (2023). Recent progress and future directions of biomass-derived hierarchical porous carbon: designing, preparation, and supercapacitor applications. *Energy Fuel* 37: 3523–3554.

12 Eghbali, P., Hassani, A., Wacławek, S. et al. (2024). Recent advances in design and engineering of MXene-based catalysts for photocatalysis and persulfate-based advanced oxidation processes: a state-of-the-art review. *Chem. Eng. J.* 480: 147920.

13 Zu, D., Wei, H., Lin, Z. et al. (2024). The role of point defects in heterojunction photocatalysts: perspectives and outlooks. *Adv. Funct. Mater.* 34: 2408213.

14 Serpone, N., Emeline, A.V., Ryabchuk, V.K. et al. (2016). Why do hydrogen and oxygen yields from semiconductor-based photocatalyzed water splitting remain disappointingly low? Intrinsic and extrinsic factors impacting surface redox reactions. *ACS Energy Lett.* 1: 931–948.

15 Lordi, V., Erhart, P., and Åberg, D. (2010). Charge carrier scattering by defects in semiconductors. *Phys. Rev. B - Condens. Matter Mater. Phys.* 81: 235204.

16 Guan, M., Xiao, C., Zhang, J. et al. (2013). Vacancy associates promoting solar-driven photocatalytic activity of ultrathin bismuth oxychloride nanosheets. *J. Am. Chem. Soc.* 135: 10411–10417.

17 Chen, X., Liu, L., Yu, P.Y., and Mao, S.S. (2011). Increasing solar absorption for photocatalysis with black hydrogenated titanium dioxide nanocrystals. *Science* 331: 746–750.

18 Fang, Z., Bueken, B., De Vos, D.E., and Fischer, R.A. (2015). Defect-engineered metal organic frameworks. *Angew. Chem. Int. Ed.* 54: 7234–7254.

19 Nowotny, J. (2008). Titanium dioxide-based semiconductors for solar-driven environmentally friendly applications: impact of point defects on performance. *Energy Environ. Sci.* 1: 565–572.

20 Long, M. and Zheng, L. (2017). Engineering vacancies for solar photocatalytic applications. *Chin. J. Catal.* 38: 617–624.

21 Pan, X., Yang, M.Q., Fu, X. et al. (2013). Defective TiO_2 with oxygen vacancies: synthesis, properties and photocatalytic applications. *Nanoscale* 5: 3601–3614.

22 Guo, H.L., Zhu, Q., Wu, X.L. et al. (2015). Oxygen deficient ZnO_{1-x} nanosheets with high visible light photocatalytic activity. *Nanoscale* 7: 7216–7223.

23 Li, Y., Tang, Z., Zhang, J., and Zhang, Z. (2016). Defect engineering of air-treated WO_3 and its enhanced visible-light-driven photocatalytic and electrochemical performance. *J. Phys. Chem. C* 120: 9750–9763.

24 Kong, X.Y., Choo, Y.Y., Chai, S.P. et al. (2016). Oxygen vacancy induced Bi_2WO_6 for the realization of photocatalytic CO_2 reduction over the full solar spectrum: from the UV to the NIR region. *Chem. Commun.* 52: 14242–14245.

25 Ye, L., Deng, K., Xu, F. et al. (2012). Increasing visible-light absorption for photocatalysis with black BiOCl. *Phys. Chem. Chem. Phys.* 14: 82–85.

26 Lv, Y., Zhu, Y., and Zhu, Y. (2013). Enhanced photocatalytic performance for the $BiPO_{4-x}$ nanorod induced by surface oxygen vacancy. *J. Phys. Chem. C* 117: 18520–18528.

27 Li, J., Wu, X., Pan, W. et al. (2018). Vacancy-rich monolayer BiO_{2-x} as a highly efficient UV, visible, and near-infrared responsive photocatalyst. *Angew. Chem. Int. Ed.* 57: 491–495.

28 Tan, H., Zhao, Z., Zhu, W.B. et al. (2014). Oxygen vacancy enhanced photocatalytic activity of pervoskite $SrTiO_3$. *ACS Appl. Mater. Interfaces* 6: 19184–19190.

29 Fang, Z., Weng, S., Ye, X. et al. (2015). Defect engineering and phase junction architecture of wide-bandgap ZnS for conflicting visible light activity in photocatalytic H_2 evolution. *ACS Appl. Mater. Interfaces* 7: 13915–13924.

30 Zhang, X., Zhao, Z., Zhang, W. et al. (2016). Surface defects enhanced visible light photocatalytic H_2 production for Zn-Cd-S solid solution. *Small* 12: 793–801.

31 Wang, J., Chen, Y., Zhou, W. et al. (2017). Cubic quantum dot/hexagonal microsphere $ZnIn_2S_4$ heterophase junctions for exceptional visible-light-driven photocatalytic H_2 evolution. *J Mater Chem A* 5: 8451–8460.

32 Dabirian, A. and Van De Krol, R. (2013). Resonant optical absorption and defect control in Ta_3N_5 photoanodes. *Appl. Phys. Lett.* 102: 033905–033905.4.

33 Long, M., Hu, P., Wu, H. et al. (2015). Understanding the composition and electronic structure dependent photocatalytic performance of bismuth oxyiodides. *J Mater Chem A* 3: 5592–5598.

34 Wang, J., Yuan, Y., Li, C. et al. (2015). Thermodynamic assessments of the Yb–Bi and the Yb–Te systems. *Calphad* 51: 306–313.

35 Wu, Q., Huang, F., Zhao, M. et al. (2016). Ultra-small yellow defective TiO_2 nanoparticles for co-catalyst free photocatalytic hydrogen production. *Nano Energy* 24: 63–71.

36 Pradhan, N., Das Adhikari, S., Nag, A., and Sarma, D.D. (2017). Luminescence, plasmonic, and magnetic properties of doped semiconductor nanocrystals. *Angew. Chem. Int. Ed.* 56: 7038–7054.

37 Zhang, J., Wu, Y., Xing, M. et al. (2010). Development of modified N doped TiO_2 photocatalyst with metals, nonmetals and metal oxides. *Energy Environ. Sci.* 3: 715–726.

38 Wang, J., Tafen, D.N., Lewis, J.P. et al. (2009). Origin of photocatalytic activity of nitrogen-doped TiO_2 nanobelts. *J. Am. Chem. Soc.* 131: 12290–12297.

39 Bai, S., Zhang, N., Gao, C., and Xiong, Y. (2018). Defect engineering in photocatalytic materials. *Nano Energy* 53: 296–336.

40 Zafar, Z., Yi, S., Li, J. et al. (2022). Recent development in defects engineered photocatalysts: an overview of the experimental and theoretical strategies. *Energy Environ. Mater.* 5: 68–114.

41 Sharma, M., Sajwan, D., Gouda, A. et al. (2024). Recent progress in defect-engineered metal oxides for photocatalytic environmental remediation. *Photochem. Photobiol.* 100 (4): 830–896.

42 Li, T., Li, Y., Guo, C., and Hu, Y. (2024). Advances and challenges of dual-defect semiconductor photocatalysts for solar-to-chemical conversion. *Chem. Commun.* 60 (17): 2320–2348.

43 Raciti, E., Gali, S.M., Melchionna, M. et al. (2022). Radical defects modulate the photocatalytic response in 2D-graphitic carbon nitride. *Chem. Sci.* 13: 9927–9939.

44 Liu, J., Zhao, D., Wu, X. et al. (2025). Synergistic dual-defect band engineering for highly-efficient photocatalytic degradation of microplastics via Nb-induced oxygen vacancies in SnO_2 quantum dots. *J Mater Chem A* 13 (3): 4429–4443.

45 Sovizi, S., Angizi, S., Ahmad Alem, S.A. et al. (2023). Plasma processing and treatment of 2D transition metal dichalcogenides: tuning properties and defect engineering. *Chem. Rev.* 123: 13869–13951.

46 Yi, K., Liu, D., Chen, X. et al. (2021). Plasma-enhanced chemical vapor deposition of two-dimensional materials for applications. *Acc. Chem. Res.* 54: 1011–1022.

47 Xie, C., Yan, D., Li, H. et al. (2020). Defect chemistry in heterogeneous catalysis: recognition, understanding, and utilization. *ACS Catal.* 10: 11082–11098.

48 Li, H., Jia, X., Zhang, Q., and Wang, X. (2018). Metallic transition-metal dichalcogenide nanocatalysts for energy conversion. *Chem* 4: 1510–1537.

49 Swedha, M., Balasurya, S., Syed, A. et al. (2022). Continuous photocatalysis via Z-scheme based nanocatalyst system for environmental remediation of pharmaceutically active compound: modification, reaction site, defect engineering and challenges on the nanocatalyst. *J. Mol. Liq.* 353: 118745.

50 Yang, A., Hou, Q., Qi, M., and Zhang, Y. (2023). Effects of Sm and point defects (V_{Zn}, V_O, H_i) at different valence states on the electronic structure and photocatalytic properties of ZnO. *Mater Today Commun* 34: 105451.

51 Afrid, S.M.T.-S. (2024). Defect engineered magnetism induction and electronic structure modulation in monolayer MoS_2. *Heliyon* 10 (1): e23384.

52 Muhammad, P., Zada, A., Rashid, J. et al. (2024). Defect engineering in nanocatalysts: from design and synthesis to applications. *Adv. Funct. Mater.* 34 (29): 2314686.

53 Vasu, D., Srivastava, R., Patil, N.G. et al. (2018). A comparative assessment of land suitability evaluation methods for agricultural land use planning at village level. *Land Use Policy* 79: 146–163.

54 FAO Unified bioenergy terminology, in, food and agriculture organization of the United Nations Rome, 2004.

55 FRA, F. (2012). Forest resources assessment 2015: terms and definitions. *FAO Report* 36.

56 Liang, J., Labidi, A., Lu, B. et al. (2025). Nanoarchitectonics of S-scheme CdS-SH QDs/Bi_2MoO_6 heterojunction for boosted photocatalytic lignin biomass conversion: interplay of singlet oxygen and interfacial engineering. *Appl. Catal. B-Environ. Energy* 367: 125092.

57 Labidi, A., Ren, H., Zhu, Q. et al. (2024). Coal fly ash and bottom ash low-cost feedstocks for CO_2 reduction using the adsorption and catalysis processes. *Sci. Total Environ.* 912: 169179.

58 Wang, H., Cheng, X., Li, Z. et al. (2024). Photothermal catalytic enhancement of lignocellulosic biomass conversion: a more efficient way to produce high-value products and fuels. *Chem. Eng. J.* 496: 153772.

59 Li, Z., Li, W., Wang, J. et al. (2024). Roles of cocatalysts in biomass photo(electro)refining. *Adv. Energy Mater.* 14: 2401838.

60 Salvachúa, D., Mohagheghi, A., Smith, H. et al. (2016). Succinic acid production on xylose-enriched biorefinery streams by *Actinobacillus succinogenes* in batch fermentation. *Biotechnol. Biofuels* 9: 1–15.

61 Mathew, S., Aronsson, A., Karlsson, E.N., and Adlercreutz, P. (2018). Xylo-and arabinoxylooligosaccharides from wheat bran by endoxylanases, utilisation by probiotic bacteria, and structural studies of the enzymes. *Appl. Microbiol. Biotechnol.* 102: 3105–3120.

62 Amorim, C., Silvério, S.C., and Rodrigues, L.R. (2019). One-step process for producing prebiotic arabino-xylooligosaccharides from brewer's spent grain employing *Trichoderma* species. *Food Chem.* 270: 86–94.

63 Mhetras, N., Mapre, V., and Gokhale, D. (2019). Xylooligosaccharides (XOS) as emerging prebiotics: its production from lignocellulosic material. *Adv. Microbiol.* 9: 14–20.

64 Adsul, M.G., Bastawde, K.B., and Gokhale, D.V. (2009). Biochemical characterization of two xylanases from yeast *Pseudozyma* hubeiensis producing only xylooligosaccharides. *Bioresour. Technol.* 100: 6488–6495.

65 Xu, S., Pan, D., Wu, Y. et al. (2018). Efficient production of furfural from xylose and wheat straw by bifunctional chromium phosphate catalyst in biphasic systems. *Fuel Process. Technol.* 175: 90–96.

66 Chen, X., Wang, W., Ciesielski, P. et al. (2016). Improving sugar yields and reducing enzyme loadings in the deacetylation and mechanical refining (DMR) process through multistage disk and szego refining and corresponding techno-economic analysis. *ACS Sustain. Chem. Eng.* 4: 324–333.

67 Bruijnincx, P.C.A., Rinaldi, R., and Weckhuysen, B.M. (2015). Unlocking the potential of a sleeping giant: lignins as sustainable raw materials for renewable fuels, chemicals and materials. *Green Chem.* 17: 4860–4861.

68 Zhang, G., Li, W., Fan, G. et al. (2019). Controlling product selectivity by surface defects over MoO_x-decorated Ni-based nanocatalysts for γ-valerolactone hydrogenolysis. *J. Catal.* 379: 100–111.

69 Yuan, S., Zhang, J., Hu, L. et al. (2021). Decarboxylation-induced defects in MOF-derived single cobalt atom@carbon electrocatalysts for efficient oxygen reduction. *Angew. Chem. Int. Ed.* 60: 21685–21690.

70 Wang, D., Shan, H., Yin, W., and Li, H. (2024). Defect engineering of single-atom catalysts in biomass conversion. *Fuel* 355: 129439.

71 Zhang, X., Cui, G., Feng, H. et al. (2019). Platinum–copper single atom alloy catalysts with high performance towards glycerol hydrogenolysis. *Nat. Commun.* 10: 5812.

72 Jia, Y., Jiang, K., Wang, H., and Yao, X. (2019). The role of defect sites in nanomaterials for electrocatalytic energy conversion. *Chem* 5: 1371–1397.

73 Liu, G., Robertson, A.W., Li, M.M.-J. et al. (2017). MoS_2 monolayer catalyst doped with isolated Co atoms for the hydrodeoxygenation reaction. *Nat. Chem.* 9: 810–816.

74 Zhang, L., Zhao, X., Yuan, Z. et al. (2021). Oxygen defect-stabilized heterogeneous single atom catalysts: preparation, properties and catalytic application. *J Mater Chem A* 9: 3855–3879.

75 Hosseini, P., Wolkersdörfer, K., Wark, M. et al. (2020). Morphology and conductivity of copper hexacyanoferrate films. *J. Phys. Chem. C* 124: 16849–16859.

76 Blanco, E., Carrales-Alvarado, D., Belen Dongil, A., and Escalona, N.s. (2021). Effect of the support functionalization of mono-and bimetallic Ni/Co supported on graphene in hydrodeoxygenation of guaiacol. *Ind. Eng. Chem. Res.* 60: 18870–18879.

77 Xu, W., Zhang, Y., Wang, J. et al. (2022). Defects engineering simultaneously enhances activity and recyclability of MOFs in selective hydrogenation of biomass. *Nat. Commun.* 13: 2068.

78 Li, Z., Wei, W., Li, H. et al. (2021). Low-temperature synthesis of single palladium atoms supported on defective hexagonal boron nitride nanosheet for chemoselective hydrogenation of cinnamaldehyde. *ACS Nano* 15: 10175–10184.

79 Huang, Z.-X., Wang, Z.-Y., Yuan, Y.-J. et al. (2024). Defect engineering of ultrasmall TiO_2 nanoparticles enables highly efficient photocatalysts for solar H_2 production from woody biomass. *Nano Lett.* 24: 11968–11975.

80 Dai, D., Qiu, J., Xia, G. et al. (2023). Defect engineering promoted photocatalysis for lignin depolymerization: performance and mechanism insight. *ACS Catal.* 13: 14987–14995.

16

Photocatalytic Reduction of CO_2 on Defect-Rich Catalysts

Wen Chen[1], Bao Pan[1], Jiani Qin[2], Jui-Yeh Rau[3], and Chuanyi Wang[2,4]

[1] *Shaanxi University of Science & Technology, Shaanxi Key Laboratory of Chemical Additives for Industry, College of Chemistry and Chemical Engineering, No. 3, Xuefu Middle Road, Xi'an, 710021, PR China*
[2] *Shaanxi University of Science & Technology, College of Environmental Science and Engineering, No. 3, Xuefu Middle Road, Xi'an, 710021, PR China*
[3] *Wuyi University, Fujian Provincial Key Laboratory of Eco-Industrial Green Technology, College of Ecology and Resources Engineering, No. 358, Baihua Road, Wuyishan, 354300, PR China*
[4] *Department of Environmental and Sustainable Engineering, Faculty of Engineering, Chulalongkorn University, 254 Phayathai Road, Pathumwan, Bangkok, 10330, Thailand*

16.1 Introduction

With the acceleration of global industrialization and excessive population growth, anthropogenic activities have resulted in a significant increase in greenhouse gas emissions, exacerbating the issue of global warming [1]. The primary cause of global warming is the increase in carbon dioxide (CO_2) emissions. To mitigate this challenge, researchers are exploring converting CO_2 into valuable chemicals and fuels through CO_2 capture, utilization, and storage (CCUS). Photocatalytic conversion of CO_2 into hydrocarbon fuels is considered one of the most promising approaches to address both energy demands and environmental pollution [2, 3]. The key to achieving large-scale industrial application of photocatalytic CO_2 reduction technology in the future hinges on the design and development of highly efficient semiconductor photocatalysts.

Defect engineering strategy has been proven as an effective way to elevate the CO_2 reduction performance of photocatalysts. Notably, advancements in characterization techniques have facilitated the exploration of the potential advantages of defects in catalytic processes. Even minor defects can greatly alter the properties of crystalline materials and enhance catalytic performance. In semiconductor materials, the influence of defects on electronic structure modulation is particularly pronounced. If defects in semiconductor materials can be systematically engineered and precisely controlled, the electronic structure of photocatalysts can be optimized to broaden light absorption spectra and dramatically enhance carrier separation efficiency. Moreover, the ligand-unsaturated active sites generated by these defects can markedly promote the adsorption and activation of CO_2 molecules, thereby greatly enhancing the activity of the photocatalytic CO_2 reduction reaction.

Defect Engineering in Photocatalysis: From Fundamentals to Applications,
First Edition. Edited by Chuanyi Wang, Abdelkader Labidi, and Qiuhui Zhu.

This advancement can lead to new breakthroughs and development opportunities in relevant scientific research and application fields [4].

16.2 Thermodynamics and Kinetics of CO_2 Reduction

CO_2 is a stable, nonpolar, linear molecule, a characteristic that significantly influences its activation and reduction. This is particularly evident given the high bond energy of C=O, which reaches 799 kJ mol^{-1} [5]. Figure 16.1 presents the potential pathways for the photocatalytic CO_2 reduction reaction, including electron and proton transfer, hydrogenation and deoxygenation processes, cleavage of the C—O bond, and formation of the C—H bond. The photocatalytic CO_2 reduction reaction typically commences with the adsorption of CO_2 on the photocatalyst surface. The modes of CO_2 adsorption involve carbon coordination, oxygen coordination, and mixed coordination with surface atoms of the photocatalyst. The affinity of C, H, and

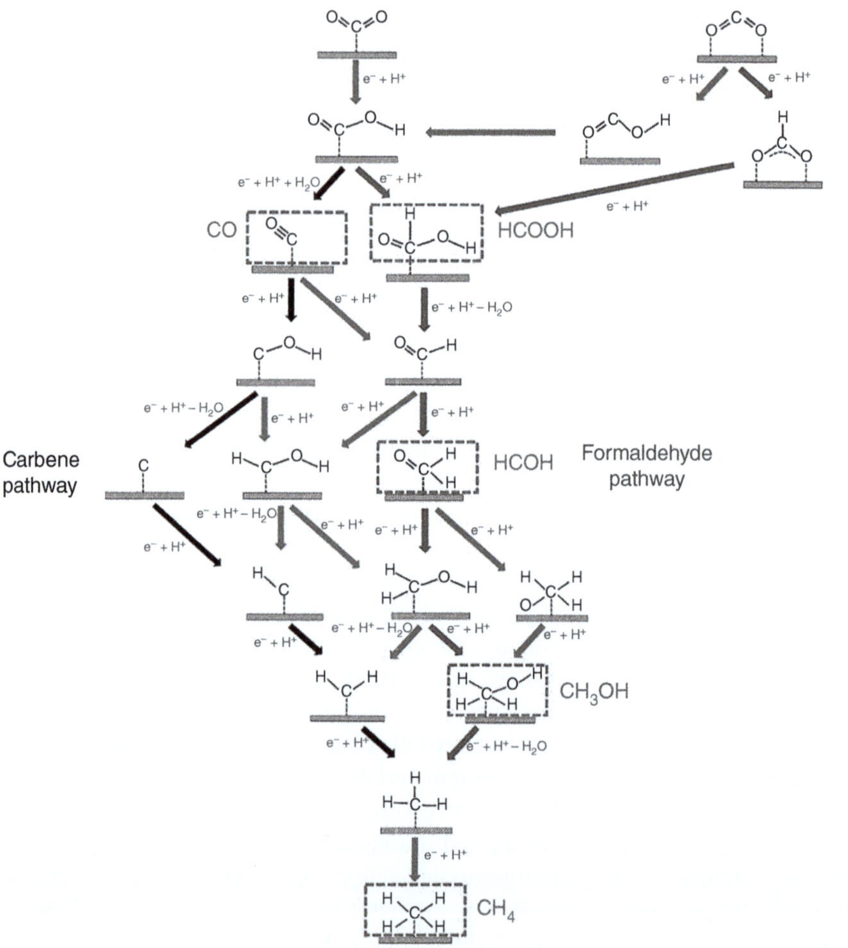

Figure 16.1 Possible pathways for photocatalytic CO_2 reduction on photocatalyst surfaces. *Source:* Reproduced from Fu et al. [6] / with permission of Elsevier.

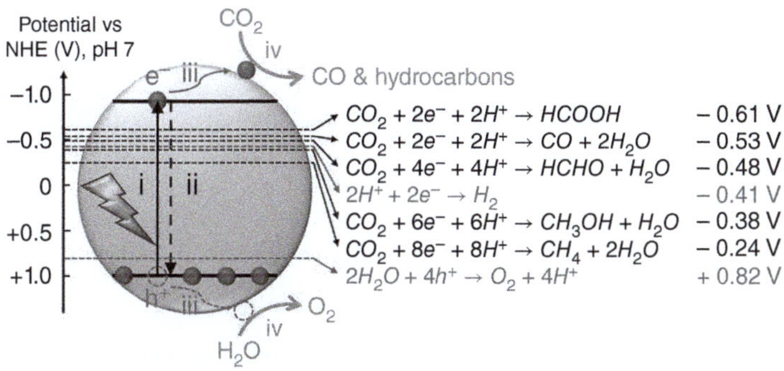

Figure 16.2 Steps in the photocatalytic CO_2 reduction: (i) photoexcitation and generation of electron–hole pairs, (ii) recombination of electron–hole pairs, (iii) transfer of electron–holes to the surface active sites, and (iv) photocatalytic reduction of CO_2 to CO and/or hydrocarbons with concomitant oxidation of H_2O to O_2. Chemical equations and the corresponding standard electrode potentials with respect to the normal hydrogen electrode (NHE) are listed. *Source:* Reproduced from Wang et al. [7] / with permission of Elsevier.

O for the photocatalyst greatly influences the sequence of adsorption/desorption, hydrogenation, and deoxygenation of reactants and intermediates, thereby determining the product pathways and selectivities. Notably, the adsorption of CO_2 can change its linear molecular configuration, activating the CO_2 molecule and promoting the photocatalytic CO_2 reduction [6].

However, in the current context, photocatalytic CO_2 reduction faces challenges due to low light harvesting efficiency, rapid recombination of photogenerated carriers, and sluggish CO_2 reduction kinetics, leading to limited productivity. First, when the photon energy exceeds the bandgap, the photocatalyst can be excited to produce electron–hole pairs (Figure 16.2), step (i). The efficiency of this process is determined by the photocatalyst's absorbance and exposed surface area. Second, a substantial portion of these electron–hole pairs recombine rapidly due to strong coulombic attraction (Figure 16.2) step (ii). Third, the carriers must migrate to the surface active sites for redox reactions (Figure 16.2), step (iii), during which they are prone to trapping and recombination. Finally, the remaining electrons and holes participate in the photocatalytic CO_2 reduction and oxygen evolution reactions (Figure 16.2), step (iv), respectively, resulting in the conversion of CO_2 into various products [7].

16.3 Classification of Defects

Defects in crystalline materials arise from the irregular arrangement of atoms. Based on their spatial dimensions, as shown in Figure 16.3, defects can be classified into point defects, line defects, planar defects, and volume defects [8]. Specifically, point defects result from the localized anomalies in the atomic structure, such as vacancies or the presence of heteroatoms [9]. At the microscopic level, point defects manifest as vacant lattice sites where atoms are missing. Analogous to an orderly square array, a few specific positions may exhibit vacancies, which represent a typical form of point defects. Additionally, the doping or substitution of heteroatoms,

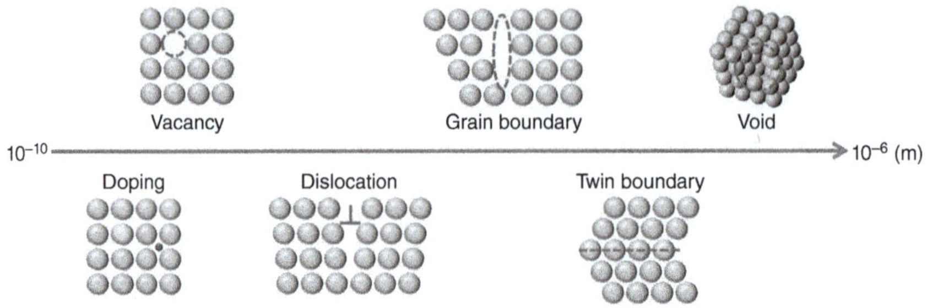

Figure 16.3 Schematic structure of different types of defects in semiconductor photocatalysts. *Source:* Reproduced from Zhong et al. [8] / with permission of John Wiley & Sons.

which alters the local atomic composition, also falls under the category of point defects. Line defects primarily manifest as dislocations within the crystal lattice. The crystal can be visualized as an assembly of numerous small units arranged according to specific rules. When the arrangement of these units becomes dislocated along a particular line, a line defect is created. This type of defect corresponds to the boundary between the portion of the crystal that has experienced relative sliding and the portion that remains unaffected. Planar defects are boundaries that partition a material into regions where the crystal structure remains consistent within each region but exhibits different orientation across regions [10]. Common types of planar defects include grain boundaries, twin boundaries, and stacking faults. Volumetric defects primarily encompass lattice disorder and voids in crystalline materials. Lattice disorder refers to the loss of regular atomic arrangement within a specific spatial range, leading to disorganization. Voids denote empty spaces within the crystal lattice, which compromise its structural integrity and constitute a significant aspect of volumetric defects.

Defects can also be classified based on their causes and their effects on material properties. The primary causes include growth-related defects and processing-induced defects. Growth-related defects arise during the crystal growth process due to factors such as non-uniform growth conditions or impurity incorporation. Processing-induced defects occur during subsequent material processing steps, including cutting, grinding, polishing, and others. Furthermore, from the perspective of their impact on material properties, defects can be categorized into detrimental defects and beneficial defects.

16.4 Creation of Defects

The preparation method of defects is closely dependent on the type of defects, which can be broadly classified into two categories: introducing defects into pre-synthesized photocatalysts and generating defects during the synthesis process. Introducing defects into pre-formed host nanocrystals is a widely adopted approach that facilitates the incorporation of vacancies, dopants, voids, and lattice disorder into either the surface or bulk of photocatalysts.

Reduced host nanocrystals can introduce cationic and anionic vacancies into the lattice of semiconductor material. For instance, treatment of pristine TiO_2, ZnO, WO_3, Bi_2WO_6, $BiPO_4$, $SrTiO_3$, and $K_4Nb_6O_{17}$ with reducing agents such as $NaBH_4$ [11], KBH_4 [12], N_2H_4 [13], H_2 [14], and NaH [15] effectively removes oxygen atoms from the lattice of these metal oxide photocatalysts, thereby forming oxygen vacancies. It is important to note that most chemical reduction methods primarily affect the surface of the photocatalyst, limiting the extent of oxygen atom removal to this region. In addition to reduction methods, heat treatment is another crucial approach for generating vacancies. Specifically, thermal treatment can induce various types of vacancies in photocatalysts. However, the use of a suitable etchant facilitates the selective removal of targeted atoms. For example, Yang's team synthesized porous graphitic carbon nitride (HGCN) with carbon vacancies by thermally treating bulk graphitic carbon nitride (BGCN) under an NH_3 atmosphere [16]. The NH_3 etching process not only promotes the exfoliation of stacked layers into porous nanosheets but also generates carbon vacancies through the removal of graphitic carbon species. Exogenous atoms or ions can be incorporated into the lattice of semiconductor material through diffusive doping, where the dopant is introduced into the pre-formed host nanocrystal lattice via diffusion [17]. This process can be facilitated at high temperatures using gaseous, liquid, or solid precursors. Our group [18] constructed an ultrathin $ZnIn_2S_4$ (ZIS) nanosheet by polymer polyvinylpyrrolidone (PVP) to achieve oxygen doping (Figure 16.4), and the photoreactivity of CO_2 reduction under visible illumination was 7.0 times higher than that of the pristine ZIS. O doping of ZIS nanosheets facilitated the migration and efficiently suppressed the recombination of the photogenerated carriers, which demonstrated the effect of oxygen doping on the crystal structure of the process of CO_2 reduction under the visible-light condition, electronic properties, and photocatalytic activity. Ion exchange represents another approach for fabricating ionic compound

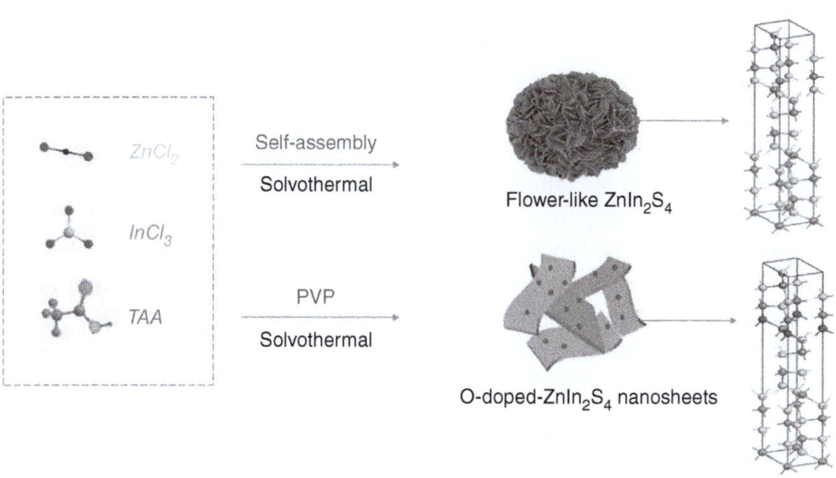

Figure 16.4 Schematic illustration of the synthesis process of ZIS and O-ZIS samples. *Source:* Reproduced from Pan et al. [18] / with permission of Elsevier.

photocatalysts, typically involving the mixing of semiconductor materials with dopant ion solutions at a specified temperature to induce ion exchange [19]. Layered oxide and hydroxide materials have been extensively investigated for ion exchange applications, as their interlayer cations or anions can be readily replaced by foreign ions. The extent of substitution is significantly influenced by the particle size of the oxide. Additionally, lattice disorder engineering can be performed on pre-synthesized photocatalytic nanocrystals using chemical reduction techniques.

Generating defects during synthesis is a more prevalent method in defect engineering compared to introducing defects into pre-formed host nanocrystals. Due to the limited reaction depth within pre-formed nanocrystals, post-processing treatment typically only introduces surface defects in most cases. In contrast, bulk defects are more likely to be generated during the formation of photocatalysts. A wide range of defects, including dislocations and grain boundaries, can form during the growth of photocatalysts, which are strongly influenced by the kinetics of nanocrystal growth.

In the preparation of photocatalysts, the formation of vacancies is more favorable in environments deficient in specific elements or enriched with vacancy-inducing elements. Notably, oxygen-deficient environments are commonly employed to generate oxygen vacancies during photocatalyst synthesis [20]. Additionally, a reducing atmosphere can also promote the formation of oxygen vacancies [21]. Similarly, other types of vacancies can be created by depleting specific elements during the synthesis process. The incorporation of precursors containing dopant elements into the synthesized system is the most widely adopted method for doping in photocatalysts. This doping process is typically integrated with the nucleation and growth of nanocrystals or the transformation of solid precursors into nanocrystals. By adjusting the ratio of dopant elements in the starting precursor relative to the host nanocrystals, the degree of doping in photocatalysts can be precisely controlled, which represents a significant advantage.

16.5 Application of Defects in CO_2 Reduction

In the research and application field of photocatalytic materials, vacancies are widely used and studied as common and easy-to-prepare defects. For metal oxides, vacancies are typically categorized into two types: oxygen vacancies (OVs) resulting from oxygen deficiency and metal vacancies arising from metal deficiency [22]. In early studies, OVs were considered to be detrimental to the photocatalytic process as recombination centers of charge carriers [23]. But with the controlled synthesis of OVs and the rapid development of advanced characterization techniques, it has been gradually recognized that OVs can potentially influence the fundamental steps of semiconductor photocatalytic CO_2 reduction. This discovery provides a plethora of opportunities for performance optimization of OV engineering.

OVs were introduced in the 1960s by Tompkins et al. [24]. Subsequently, it was discovered that OVs are ubiquitous in metal oxides and are formed through the removal of oxygen atoms from the crystal lattice, significantly influencing the electronic and chemical properties of the material surface [25]. The absence of oxygen atoms can result in a neutral state, leading to an accumulation of electrons on nearby metal cations, which critically affects catalytic reactions involving electron transfer. Specifically, in photocatalytic CO_2 reduction, OVs enhance photocatalyst performance by adjusting light absorption spectra, facilitating photogenerated charge migration, and improving CO_2 molecule adsorption [26]. The presence of OVs increases the Fermi energy level of the oxides and introduces defect-induced energy levels within the bandgap, thereby narrowing the bandgap. The OVs are present in both the surface and bulk phases of photocatalysts, and their positional differences lead to distinct effects. Surface OVs facilitate the separation of photogenerated electrons, while bulk OVs act as charge trapping centers, hindering electron transfer. Li et al. [27] prepared TiO_2 photocatalysts with varying OVs positions: bulk OVs (TiO_2–BO), surface OVs (TiO_2–SO), and a combination of both (TiO_2–SBO). They investigated the role of OVs in photocatalytic CO_2 reduction. Despite the different impacts on photogenerated carrier separation, both types of OVs enhance the photocatalyst's response to visible light [28]. Consequently, TiO_2–SBO exhibits two synergistic advantages, leading to superior photocatalytic CO_2 reduction activity (Figure 16.5a).

The structure and concentration of active sites in photocatalysts significantly influence CO_2 conversion and selectivity. The presence of OVs in metal oxide photocatalysts increases the number of electron-rich active sites, thereby enhancing the catalyst's affinity for CO_2 molecules [30]. Uncoordinated metal atom sites surrounding the OVs synergistically interact with these active sites and OVs during the reaction process, modulating the conformation of CO_2 molecules and consequently affecting their adsorption and activation. During photocatalytic CO_2 reduction, OVs are readily populated and consumed by oxygen atoms in the CO_2 molecule, leading to catalyst deactivation. Preventing the deactivation of OVs is a big challenge in defect engineering. Zhao et al. employed a rapid flame reduction technique to construct bridged OVs on $SnTa_2O_6$ nanosheets, which altered the adsorption conformation of CO_2 from monodentate to bidentate, with oxygen atoms binding to Sn and Ta atoms, respectively [31]. The bridged structure of OVs not only facilitated the adsorption and activation of CO_2 molecules but also effectively prevented the filling and depletion of OVs. As the flame treatment time increased, the morphology of $SnTa_2O_{6-x}$ evolved from nanosheets to nanoparticles. Under the synergistic effect of the bridged OVs and morphological transformation, the photocatalytic CO_2 reduction performance of $SnTa_2O_{6-x}$ was enhanced by 7.5-fold.

In addition to common OVs, metal oxides also exhibit metal cation vacancies. Compared with OVs, the formation energy of metal cation vacancies is much higher, which poses a greater challenge for their design and manipulation in photocatalysts [32]. Therefore, fewer studies have focused on metal cation vacancies in photocatalytic materials. Metal cation vacancies possess distinct orbital distributions and characteristic electronic configurations that substantially influence the physicochemical and electronic properties of metal oxides. Gao et al. [29] designed V_V-rich and V_V-poor o-$BiVO_4$ samples, investigating the relationship between metal cation vacancy structures and photocatalytic performance through the modulation of vanadium (V) defects. Additionally, the V_V-rich o-$BiVO_4$ exhibited stronger surface

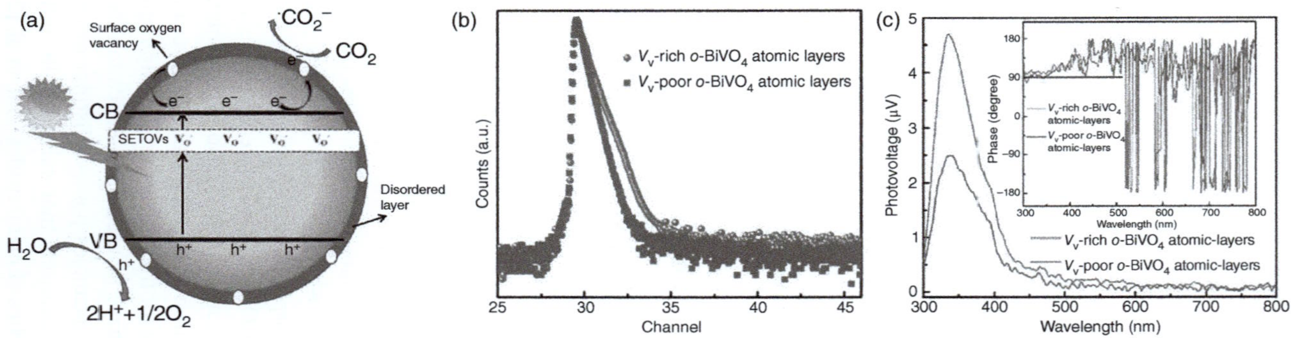

Figure 16.5 Schematic diagram of the CO_2 photoreduction mechanism on TiO_2–SBO (a), positron lifetime spectra (b), and surface photovoltage spectra (c) of V_V-rich and V_V-poor o-BiVO$_4$ atomic layers. *Source:* Li et al. [27] and Gao et al. [29].

photovoltage responses and longer fluorescence lifetimes (Figure 16.5b,c), suggesting that the presence of a defect level enhances photogenerated charge separation efficiency, thereby improving solar conversion efficiency and CO_2 reduction activity.

16.6 Conclusions and Outlook

Defect engineering is an exceptionally effective strategy for enhancing the performance of metal oxide photocatalysts in CO_2 reduction. Extensive efforts have been devoted to developing advanced catalysts and elucidating the underlying reaction mechanism. Defects inherent in metal oxides significantly influence the generation and separation of photogenerated charges, and future exploration of novel semiconductor materials with tailored defects for CO_2 reduction should focus on the following aspects:

1) Achieving simultaneous optimization of defect structure, concentration, location, and other parameters through precise tuning of defects in photocatalysts.
2) Realizing the rational design and synergistic integration of two or more types of defects to effectively address diverse defect-related challenges and collectively enhance the photocatalytic CO_2 reduction reaction.
3) Designing multicomponent photocatalysts while thoroughly considering the interactions between different component catalysts.
4) Developing characterization techniques with enhanced spatial, temporal, and spectral resolution and establishing more realistic in situ detection methods.

In conclusion, photocatalytic CO_2 reduction technology holds the potential to simultaneously alleviate the energy crisis and mitigate the greenhouse effect. However, numerous technical challenges must be addressed before this technology can be scaled up for industrial applications. As various characterization techniques and analytical methods continue to advance, our understanding of material defects deepens, leading to the discovery of additional functionalities and improved performance in the future.

Conflict of Interest

The authors declare that they have no conflict of interest.

References

1 Gillett, N.P., Kirchmeier-Young, M., Ribes, A. et al. (2021). Constraining human contributions to observed warming since the pre-industrial period. *Nat. Clim. Chang.* 11 (3): 207–212.

2 Lin, H., Luo, S., Zhang, H., and Ye, J. (2022). Toward solar-driven carbon recycling. *Joule* 6 (2): 294–314.

3 Pan, B., Qin, J., Wang, X., and Su, W. (2020). Efficient self-assembly synthesis of $LaPO_4$/CdS hierarchical heterostructure with enhanced visible-light photocatalytic CO_2 reduction. *Appl. Surf. Sci.* 504: 144379.

4 Pastor, E., Sachs, M., Selim, S. et al. (2022). Electronic defects in metal oxide photocatalysts. *Nat. Rev. Mater.* 7 (7): 503–521.

5 Pan, B., Lv, Y., Dong, Y. et al. (2024). Hydrochar-supported $NiFe_2O_4$ nanosheets with a tailored microstructure for enhanced CO_2 photoreduction to syngas. *Inorg. Chem.* 63 (4): 2148–2156.

6 Fu, J., Jiang, K., Qiu, X. et al. (2020). Product selectivity of photocatalytic CO_2 reduction reactions. *Mater. Today* 32: 222–243.

7 Wang, L., Zhu, B., Zhang, J. et al. (2022). S-scheme heterojunction photocatalysts for CO_2 reduction. *Matter* 5 (12): 4187–4211.

8 Zhong, K., Sun, P., and Xu, H. (2025). Advances in defect engineering of metal oxides for photocatalytic CO_2 reduction. *Small* 21: 2310677.

9 Bai, S., Zhang, N., Gao, C., and Xiong, Y. (2018). Defect engineering in photocatalytic materials. *Nano Energy* 53: 296–336.

10 Li, S., Lin, J., Chen, Y. et al. (2024). Growth anisotropy and morphology evolution of line defects in monolayer MoS_2: atomic-level observation, large-scale statistics, and mechanism understanding. *Small* 20 (4): 2303511.

11 Xie, Q., Qin, J., Gao, T. et al. (2024). Engineering $ZnIn_2S_4$ nanosheets with zinc vacancies: unleashing enhanced photocatalytic degradation of tetracycline. *Langmuir* 40 (47): 25327–25333.

12 Bi, W., Ye, C., Xiao, C. et al. (2014). Spatial location engineering of oxygen vacancies for optimized photocatalytic H_2 evolution activity. *Small (Weinheim an der Bergstrasse, Germany)* 10 (14): 2820–2825. 2742.

13 Mao, C., Zuo, F., Hou, Y. et al. (2014). In situ preparation of a Ti^{3+} self-doped TiO_2 film with enhanced activity as photoanode by N_2H_4 reduction. *Angew. Chem.* 126 (39): 10653–10657.

14 Liu, X., Yang, L., Huang, M. et al. (2022). Oxygen vacancy-regulated metallic semiconductor MoO_2 nanobelt photoelectron and hot electron self-coupling for photocatalytic CO_2 reduction in pure water. *Appl. Catal. B Environ.* 319: 121887.

15 Wang, M., Nie, B., Yee, K.-K. et al. (2016). Low-temperature fabrication of brown TiO_2 with enhanced photocatalytic activities under visible light. *Chem. Commun.* 52 (14): 2988–2991.

16 Liang, Q., Li, Z., Huang, Z.H. et al. (2015). Holey graphitic carbon nitride nanosheets with carbon vacancies for highly improved photocatalytic hydrogen production. *Adv. Funct. Mater.* 25 (44): 6885–6892.

17 Pradhan, N., Das Adhikari, S., Nag, A., and Sarma, D. (2017). Luminescence, plasmonic, and magnetic properties of doped semiconductor nanocrystals. *Angew. Chem. Int. Ed.* 56 (25): 7038–7054.

18 Pan, B., Wu, Y., Rhimi, B. et al. (2021). Oxygen-doping of $ZnIn_2S_4$ nanosheets towards boosted photocatalytic CO_2 reduction. *J. Energy Chem.* 57: 1–9.

19 Qin, J., Zhao, Q., Zhao, Y. et al. (2021). Metal-free phosphorus-doped $ZnIn_2S_4$ nanosheets for enhanced photocatalytic CO_2 reduction. *J. Phys. Chem. C* 125 (43): 23813–23820.

20 Guo, H.-L., Zhu, Q., Wu, X.-L. et al. (2015). Oxygen deficient ZnO_{1-x} nanosheets with high visible light photocatalytic activity. *Nanoscale* 7 (16): 7216–7223.

21 Zuo, F., Wang, L., Wu, T. et al. (2010). Self-doped Ti^{3+} enhanced photocatalyst for hydrogen production under visible light. *J. Am. Chem. Soc.* 132 (34): 11856–11857.

22 Campbell, C.T. and Peden, C.H. (2005). Oxygen vacancies and catalysis on ceria surfaces. *Science* 309 (5735): 713–714.

23 Qian, R., Zong, H., Schneider, J. et al. (2019). Charge carrier trapping, recombination and transfer during TiO_2 photocatalysis: an overview. *Catal. Today* 335: 78–90.

24 Tompkins, F.C. (1960). Superficial chemistry and solid imperfections. *Nature* 186 (4718): 3–6.

25 Zhang, X., Tang, F., Wang, M. et al. (2020). Femtosecond visualization of oxygen vacancies in metal oxides. *Sci. Adv.* 6 (10): eaax9427.

26 Yu, H., Li, J., Zhang, Y. et al. (2019). Three-in-one oxygen vacancies: whole visible-spectrum absorption, efficient charge separation, and surface site activation for robust CO_2 photoreduction. *Angew. Chem. Int. Ed.* 58 (12): 3880–3884.

27 Li, J., Zhang, M., Guan, Z. et al. (2017). Synergistic effect of surface and bulk single-electron-trapped oxygen vacancy of TiO_2 in the photocatalytic reduction of CO_2. *Appl. Catal. B Environ.* 206: 300–307.

28 Gao, W., Li, H., Hu, J. et al. (2024). Recent advances of metal active sites in photocatalytic CO_2 reduction. *Chem. Sci.* 15: 14081–14103.

29 Gao, S., Gu, B., Jiao, X. et al. (2017). Highly efficient and exceptionally durable CO_2 photoreduction to methanol over freestanding defective single-unit-cell bismuth vanadate layers. *J. Am. Chem. Soc.* 139 (9): 3438–3445.

30 Zhao, J., Xiong, Z., Wang, J. et al. (2023). Constructing bridging oxygen vacancies in $SnTa_2O_{6-x}$ for efficient and stable photocatalytic CO_2 reduction under visible light irradiation. *Appl. Surf. Sci.* 636: 157833.

31 Liu, W., Xiu, F., Sun, K. et al. (2010). Na-doped p-type ZnO microwires. *J. Am. Chem. Soc.* 132 (8): 2498–2499.

32 Pan, B., Zhou, L., Qin, J. et al. (2022). Modulating $CoFeO_X$ nanosheets towards enhanced CO_2 photoreduction to syngas: effect of calcination temperature and mixed-valence multi-metals. *Chem Eur J* 28 (72): e202201992.

17

H_2 Evolution Utilizing Defect-Engineered Photocatalytic Materials

Atif Sial[1], Abdelkader Labidi[1], Haitao Ren[2], Ting Gao[1], and Chuanyi Wang[1,3]

[1]*Shaanxi University of Science and Technology, School of Environmental Science and Engineering, Xi'an, 710021, PR China*
[2]*Xijing University, Xi'an Key Laboratory of Advanced Photo-Electronics Materials and Energy Conversion Device, Technological Institute of Materials & Energy Science (TIMES), Xi'an, 710123, China*
[3]*Department of Environmental and Sustainable Engineering, Faculty of Engineering, Chulalongkorn University, 254 Phayathai Road, Pathumwan, Bangkok, 10330, Thailand*

17.1 Introduction

The global energy landscape is undergoing a transformative shift as societies strive for sustainable alternatives to fossil fuels, which are increasingly being recognized as contributors to climate change and environmental degradation [1]. An essential player in this transition is H_2, a clean and versatile energy carrier that can beutilized across various applications, including fuel cells, energy storage systems, and feedstock for industrial processes [2]. The distinct advantage of H_2 lies in its ability to release only H_2O as a byproduct during combustion, making it an environmentally friendly option that aligns with the goals of reducing greenhouse gas emissions and achieving carbon neutrality [3]. However, the widespread adoption of H_2 as a viable energy source hinges on the development of efficient and cost-effective production methods.

Among the numerous approaches for H_2 generation, photocatalysis has emerged as a particularly promising method due to its ability to harness solar energy, which is the most abundant and renewable energy source on Earth. Photocatalysis involves the use of semiconductor materials that can absorb light and drive chemical reactions, specifically the splitting of H_2O into H_2 and O_2 through photolytic processes [4]. This technique not only capitalizes on solar energy but also offers the potential for decentralized H_2 production, thereby supporting energy security and accessibility in diverse geographical contexts.

Despite its promise, the efficiency of photocatalytic water-splitting is heavily influenced by the properties of the semiconductor materials employed. Key factors such as the material's bandgap, surface area, charge carrier mobility, and the dynamics of electron–hole pair generation and recombination all play critical roles in determining the overall photocatalytic activity [5]. Traditional semiconductor

Defect Engineering in Photocatalysis: From Fundamentals to Applications,
First Edition. Edited by Chuanyi Wang, Abdelkader Labidi, and Qiuhui Zhu.

material, titanium dioxide (TiO_2), has been extensively studied for photocatalytic applications [6]. However, it also needs to be stated that even those reports which have shown high and very high HER efficiencies are not free of ambiguities. On the other hand, the diverse range of other semiconductor-based photocatalysts includes oxides like ZnO, perovskites, and non-oxide counterparts such as CdS, CuS, and ZnS. Quantum dots (QDs), nitrides, conjugated polymers, graphene and graphitic carbon nitride (g-C_3N_4)-based photocatalytic materials, MOFs, and heterojunction-based photocatalysts have also been widely applied in photocatalytic H_2 evolution due to their unique and promising properties [7]. Not only can they be synthesized by simple methods but also they show stability and possess a crucial advantage of scalability. This is a significant practical advantage, making them a promising avenue for further exploration and application in various photocatalytic H_2 evolution reactions. However, it is essential to recognize that even these photocatalysts have challenges to be addressed, particularly in optimizing efficiency and environmental concerns associated with their production and use.

To address these limitations, researchers have increasingly turned to the concept of defect engineering. Defect engineering refers to the intentional introduction of imperfections in the crystal lattice of semiconductor materials, which can significantly enhance their photocatalytic properties [8]. Defects can take various forms, including vacancies, interstitials, antisite defects, and dopants, and can result from a range of synthesis methods, post-treatment processes, or even environmental conditions [9]. These imperfections can alter the electronic structure of the material, leading to increased light absorption, improved charge carrier separation, and enhanced catalytic activity [10]. The introduction of defects in photocatalytic materials can create localized energy levels within the bandgap [11]. In addition, these defect states can serve as electron or hole traps, which can facilitate the separation of charge carriers generated upon light absorption [10]. By minimizing the recombination of electron–hole pairs, defects can enhance the overall efficiency of H_2 production. For instance, recent studies have shown that certain defects can act as active sites for chemical reactions, promoting the formation of H_2 and intermediate species necessary for the water-splitting process [12]. Moreover, defect engineering can improve the stability and resilience of photocatalytic materials. Photocorrosion is a significant challenge in photocatalytic systems, where semiconductor materials degrade over time due to exposure to UV light and reactive species [13]. By strategically introducing defects, researchers can enhance the structural integrity of the photocatalysts and mitigate the adverse effects of photocorrosion. For example, certain defect types can effectively anchor the crystal lattice, preventing the migration of ions and the collapse of the material structure during prolonged exposure to environmental conditions [14]. This enhanced stability is crucial for maintaining high photocatalytic performance over extended periods and under varying operational conditions.

In the photocatalytic water-splitting system, activated electrons and holes go to the surface of the photocatalyst after light illumination and act as reducing and oxidizing agents, respectively [15]. Upon contact with water, whether on the surface of the photocatalyst or within the cocatalyst, the ensuing oxidation and reduction

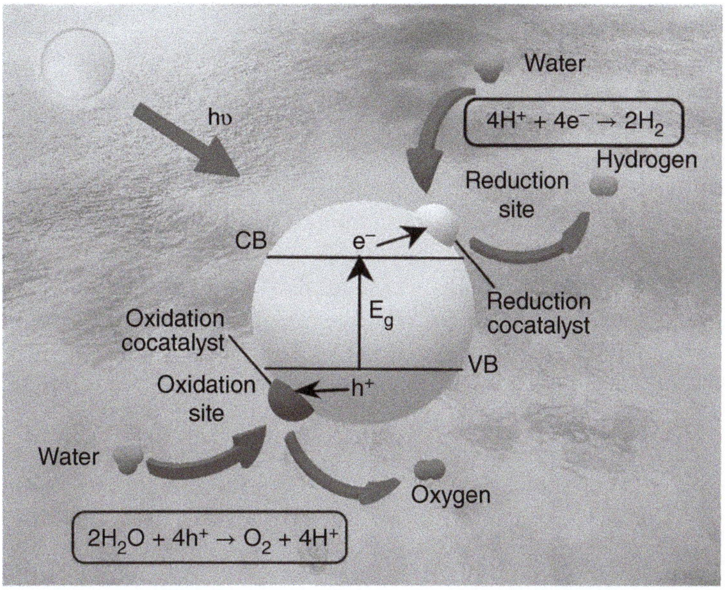

Figure 17.1 Graphical representation of the overall photocatalytic water-splitting reaction. Reproduced from [7] / with permission of Elsevier.

processes initiate promptly, as described in Eq. (17.1–17.4). In a photocatalyst, the effect of photons on the creation of pairs of electrons and holes is as follows:

$$\text{Photocatalyst} + h\nu \rightarrow 4e^- + 4h^+ \tag{17.1}$$

$$2H_2O + 4h^+ \rightarrow O_2 + 4H^+ \text{ (oxidation half-reaction)} \tag{17.2}$$

$$4H^+ + 4e^- \rightarrow 2H_2 \text{ (reduction half-reaction)} \tag{17.3}$$

$$2H_2O + 4e^- + 4h^+ \rightarrow 2H_2\,(g) + O_2\,(g)\ \text{(overall water-splitting reaction)} \tag{17.4}$$

The standard Gibbs free energy change (ΔG_0) essential for the uphill mechanism of separating water into H_2 and oxygen is 237 kJ mol^{-1} or 1.23 eV [15]. As such, to be able to achieve water-splitting, the bandgap of the photocatalyst must be more than 1.23 eV (~1000 nm). However, utilizing visible light necessitates the bandgap approaching 3.0 eV (>400 nm) [15]. Figure 17.1 is a schematic diagram outlining the overall photocatalytic H_2 generation reaction.

17.2 Synthesis Protocols

The process of defect engineering encompasses various techniques, each with its advantages and limitations. Common methods for introducing defects into semiconductor materials include ion implantation, high-energy irradiation, thermal treatment, and elemental doping. These techniques can manipulate the microstructure and electronic properties of the materials, allowing researchers to design tailored photocatalysts with optimized performance for H_2 evolution.

17.2.1 Ion Implantation

Ion implantation is a well-established technique that enables the controlled introduction of defects at specific depths and concentrations within the semiconductor material [16]. By bombarding the material with energetic ions, researchers can create vacancies and interstitials that enhance photocatalytic activity. The precision offered by ion implantation allows for the selective tuning of defect types, which can be critical for improving charge carrier dynamics. However, this method often requires subsequent annealing to repair the damage caused during implantation and to activate the defects [17]. In a recent study, Hu et al. synthesized defective photocatalytic materials, specifically $TiO_2/Co/PA$, using an ion implantation method (Figure 17.2a). This innovative hybrid material, formed by anchoring

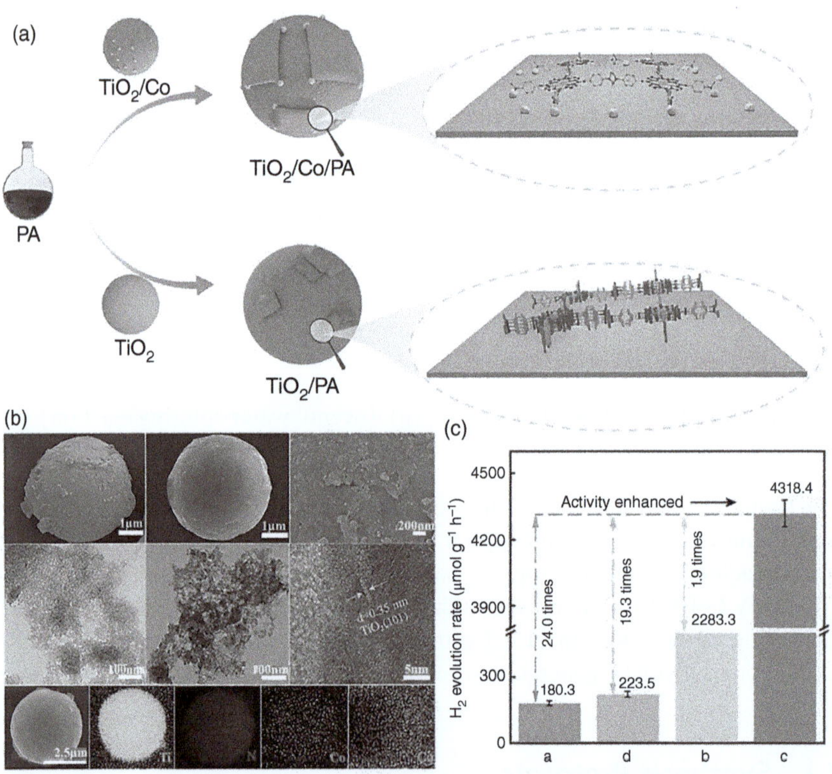

Figure 17.2 (a) The synthesis methodology for the $TiO_2/Co/PA$ and TiO_2/PA nano-micro hybrid materials. (b) SEM and TEM images, supplemented by corresponding elemental mapping analyses of the TiO_2/PA and $TiO_2/Co/PA$ composites. (c) Illustration of the comparative average photocatalytic H_2 production rates for the following samples: (a) TiO_2, (b) TiO_2/PA, (c) $TiO_2/Co/PA$, and (d) TiO_2/Co. Reproduced from [18] / with permission of Elsevier.

meso-tetra (4-carboxylphenyl) porphyrin copper (II) (PA) onto TiO_2 microspheres with cobalt (Co) ions (Figure 17.2b), exhibited outstanding photocatalytic performance for H_2 production [18]. The TiO_2/Co/PA achieved a H_2 evolution rate of 4318.4 $\mu mol\,g^{-1}\,h^{-1}$, outperforming pure TiO_2 by 24 times and TiO_2/PA by 1.9 times (Figure 17.2c). The enhanced performance is attributed to the improved light absorption, efficient charge separation, and reduced H-type aggregation of PA, highlighting the potential of ion implantation in developing efficient bifunctional photocatalysts for sustainable H_2 production.

17.2.2 High-Energy Irradiation

High-energy irradiation, such as gamma rays or electron beams, is another effective method for introducing defects into semiconductor materials [19]. This technique can induce a broad spectrum of defects throughout the materials, leading to significant modifications in their electronic and optical properties. The defects generated by high-energy irradiation can enhance light absorption by creating sub-bandgap states, which facilitate the capture of lower-energy photons. Furthermore, the random nature of defect formation can lead to a diverse range of active sites, potentially improving overall photocatalytic efficiency [20]. For instance, Mohapatra et al. investigated the synthesis of defective photocatalytic materials for H_2 production using high-energy electron beam (EB) irradiation [21]. The researchers prepared nitrogen-vacated g-C_3N_4 nanosheets through the thermal polycondensation of melamine at 550 °C, followed by EB irradiation with absorbed doses ranging from 100 to 500 kGy (Figure 17.3a). The resulting optimized photocatalyst, designated as gCN-300, exhibited exceptional photocatalytic activity, achieving a H_2 evolution rate of $750.57 \pm 9.9\,\mu mol\,g^{-1}\,h^{-1}$ (Figure 17.3b). This performance was further complemented by a significant reduction of Cr(VI) by 83.5% within just two hours, indicating the material's dual functionality in energy conversion and environmental remediation. The enhanced photocatalytic efficiency was attributed to several critical factors. The introduction of mid-bandgap states through EB irradiation increased visible-light absorption, while nitrogen vacancies created surface-active sites that facilitated reactant interactions.

Moreover, the modifications improved electronic conductivity, allowing for better charge carrier mobility and efficient charge separation, thereby reducing recombination rates (Figure 17.3c). Characterization techniques, including XRD, FTIR, and ESR, confirmed the presence of nitrogen vacancies and elucidated the structural and electronic changes that underpinned the enhanced performance of gCN-300 [21]. This study paves the way for the innovative use of high-energy EB irradiation as a scalable and cost-effective method for engineering defect-rich photocatalysts.

17.2.3 Thermal Treatment

Thermal treatment involves subjecting semiconductor materials to high temperatures in controlled atmospheres, resulting in the formation of defects through processes like oxidation, reduction, or volatilization [22]. This method can also

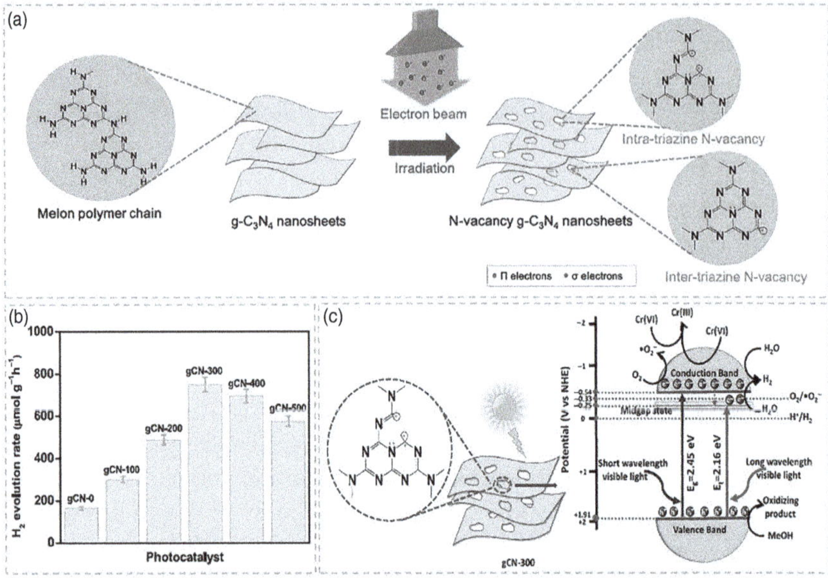

Figure 17.3 (a) An illustrative schematic detailing the fabrication approach for nitrogen-deficient g-C_3N_4 utilizing EB irradiation method. (b) The photocatalytic H_2 generation rates for as-prepared variants. (c) A conceptual framework depiction to elucidate the photocatalytic reduction mechanism of Cr(VI) alongside the H_2 generation process occurring over the g-CN-300 sample. Reproduced from [21] / with permission of Elsevier.

lead to changes in the phase and morphology of the materials, further enhancing their photocatalytic properties. For instance, calcination can induce the formation of oxygen vacancies (OVs), which significantly boost the photocatalytic activity of materials such as TiO_2 [23]. However, careful control over temperature and atmosphere is required to prevent unwanted phase transitions or sintering.

In this aspect, a recent study by Tian et al. reported the synthesis of sulfur-doped honeycomb porous g-C_3N_4 nanosheets utilizing a thermal method combined with a defective construction approach [24]. The melamine, cyanuric acid, and thiourea were utilized to create a supramolecular liquid precursor, which was subjected to high-temperature calcination. This process facilitated the selective volatilization of the precursor, resulting in the formation of a unique honeycomb porous structure (Figure 17.4a). As mentioned in the Figure 17.4b, the optimized photocatalyst, designated as S-PDCN-3, exhibited a remarkable H_2 production rate of 23.78 mmol g^{-1} h^{-1}, representing a significant enhancement of 21.8 times compared to the bulk g-C_3N_4 (BCN) sample, which produced only 1.1 mmol g^{-1} h^{-1}. The exceptional performance of S-PDCN-3 was attributed to its increased specific surface area of 356.60 m^2 g^{-1}, which provided ample active sites for photocatalytic reactions, and the successful incorporation of sulfur into the g-C_3N_4 lattice. Furthermore, the addition of the photosensitizer Erythrosin B (EB) further amplified the H_2 production rate to an impressive 63.8 mmol g^{-1} h^{-1} (Figure 17.4c) [24]. The porous nature of the S-PDCN-3 structure facilitated efficient charge carrier

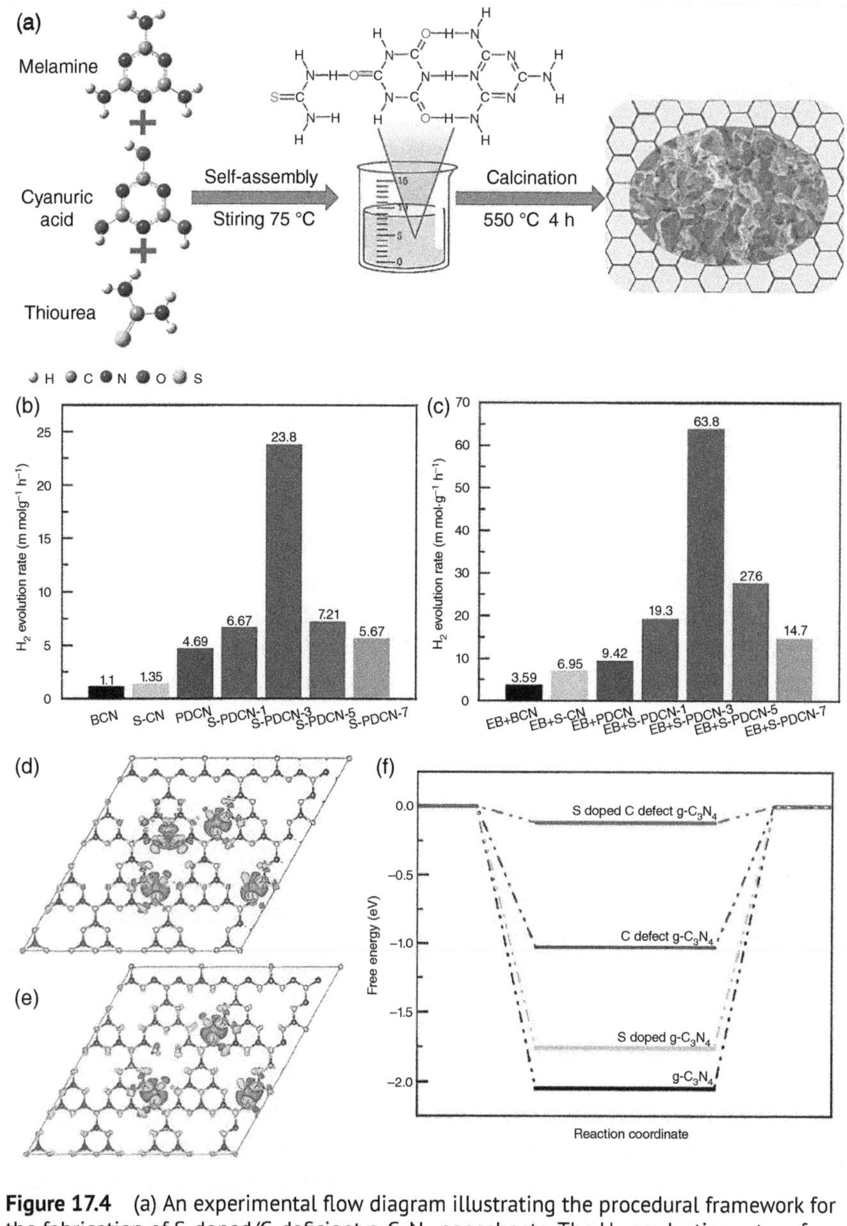

Figure 17.4 (a) An experimental flow diagram illustrating the procedural framework for the fabrication of S-doped/C-deficient g-C_3N_4 nanosheets. The H_2 production rates of as-prepared variants in the absence (b) and presence (c) of EB photosensitizer. The charge density variations, examined for (d) S-doped g-C_3N_4, (e) S-doped/C-deficient g-C_3N_4, and (f) the estimated free-energy profile pertaining to the H_2 evolution reaction. Reproduced from [24] / with permission of Elsevier.

separation and migration, while the sulfur doping enhanced electron transport properties and broadened the light absorption spectrum. Furthermore, the charge density difference analysis of S-doped g-C_3N_4 (Figure 17.4d), particularly with carbon defects (Figure 17.4e), revealed enhanced electron density at active sites, significantly improving photocatalytic efficiency. Additionally, Figure 17.4f presents the calculated free-energy diagram for the H_2 production reaction, which illustrates lowered energy barriers due to these modifications, indicating that S-doping facilitates more effective H_2 evolution. These insights highlight the potential of S-doped g-C_3N_4 as a highly efficient photocatalyst for sustainable energy applications.

17.2.4 Elemental Doping

Doping with other elements is a widely used strategy for defect engineering that can modify the electronic structure of semiconductor materials. By substituting specific cations or anions within the crystal lattice, researchers can create new energy levels that enhance light absorption and improve charge carrier dynamics [25]. For example, doping TiO_2 with nonmetals such as nitrogen or carbon can narrow the bandgap, enabling the material to absorb a broader spectrum of visible light [26]. This adjustment is crucial, as the majority of solar energy lies within the visible range, and enhancing light absorption can lead to improved photocatalytic efficiency. Additionally, elemental doping can generate additional defect sites that act as charge carrier traps, thereby facilitating the separation of photo-induced electrons and holes [27]. This separation is vital for minimizing recombination losses, which can significantly hinder the photocatalytic process.

For instance, studies have demonstrated that nitrogen-doped TiO_2 exhibits enhanced photocatalytic activity for H_2 evolution compared to its undoped counterpart [28]. The introduction of nitrogen not only narrows the bandgap but also creates localized energy states that support the formation of reactive species necessary for water-splitting. Similarly, the incorporation of transition metals such as iron [29], cobalt [30], or nickel [31] into semiconductor lattices can also enhance photocatalytic performance by introducing new electronic states that promote charge separation and increase the material's overall catalytic activity. In particular, Chen et al. investigated the enhancement of photocatalytic H_2 production through the elemental doping of InP/ZnSeS/ZnS QDs [32]. The researchers synthesized three different QDs: InP/ZnSeS/ZnS, InP/ZnSeS:Cu/ZnS, and InP/ZnSeS:Mn/ZnS, using a thermal injection approach (Figure 17.5a). This elemental doping strategy aimed to modulate the energy-level structure of the QDs to optimize their photocatalytic performance. Morphological characterization of the synthesized QDs revealed that the doping process preserved the original internal structure while significantly enhancing the visible-light absorption capabilities (Figure 17.5b–d). Notably, the Cu-doped QDs exhibited a remarkable 10-fold increase in absorption intensity compared to their undoped counterparts. The valence band positions were measured at 1.25 V for undoped, 1.11 V for Cu-doped, and 1.27 V for Mn-doped QDs, with corresponding bandgaps of 3.14, 2.74, and 2.95 eV, respectively. The photocatalytic performance of the materials was evaluated, demonstrating that the Cu-doped QDs

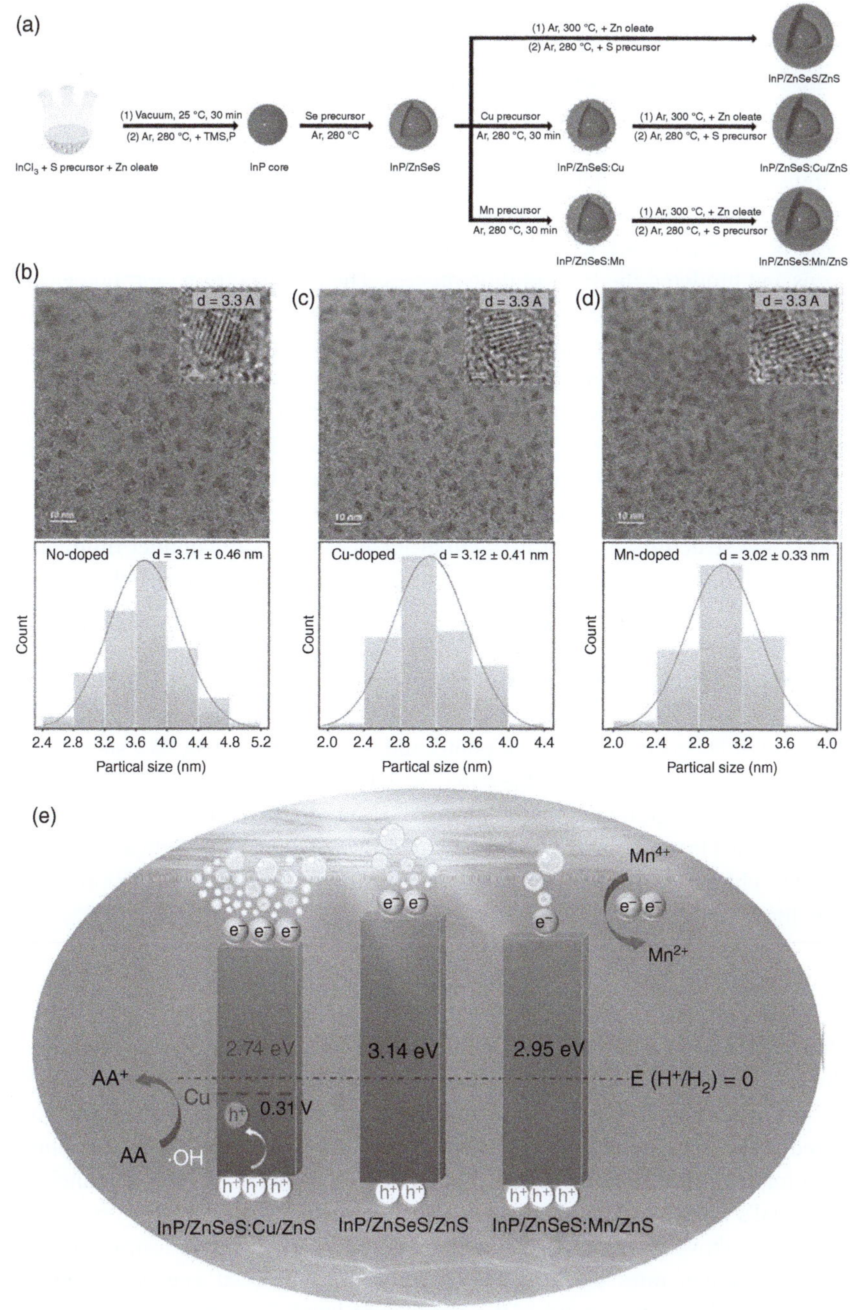

Figure 17.5 (a) A detailed synthesis strategy for the three QDs samples. TEM images of (b) undoped, (c) Cu-doped, and (d) Mn-doped QDs, accompanied by histograms quantifying the distribution of particle sizes. The insets provide HR-TEM images, elucidating the corresponding lattice interplanar spacings. (e) The schematic illustration of the photocatalytic H_2 generation mechanism within these three QD systems. Reproduced from [32] / with permission of Elsevier.

achieved an impressive H_2 production rate of 337.5 μmol g^{-1} h^{-1}, which is over six times higher than the undoped QDs (53.8 μmol g^{-1} h^{-1}) and significantly greater than the Mn-doped QDs (8.97 μmol g^{-1} h^{-1}). This enhancement in performance was attributed to the optimization of energy levels in the Cu-doped QDs, which efficiently introduced an additional energy level to trap holes and facilitate rapid hole transfer to electron donors, ultimately leading to improved electron–hole separation (Figure 17.5e). In contrast, the Mn-doped QDs exhibited inferior catalytic performance due to the reduction of Mn^{4+} to Mn^{2+}, which consumed electrons and hindered the H_2 evolution process. The findings from morphological characterization, band-structure analysis, and photoelectrochemical investigations supported the proposed mechanism of enhanced photocatalytic activity.

While elemental doping presents a powerful strategy for enhancing photocatalytic performance, it also comes with its challenges. The degree of doping must be carefully optimized, as excessive doping can lead to increased defect densities that may negatively affect the crystal structure and reduce material stability. Furthermore, the interaction between dopants and the host lattice can introduce complexities in the electronic structure, making it essential for researchers to conduct thorough characterizations to understand the impact of doping on photocatalytic behavior.

17.3 H_2 Evolution Reactions

The process of H_2 evolution through photocatalysis involves several key steps that are influenced by the properties of the photocatalytic materials. Understanding these steps is essential for optimizing the efficiency of H_2 production. The overall reaction can be simplified into a series of interconnected processes, including light absorption followed by charge carrier generation, charge separation, and surface reactions and product formation.

17.3.1 Light Absorption and Charge Carrier Generation

The first step in photocatalytic water-splitting is the absorption of light by the semiconductor material. The efficiency of this process is determined by the bandgap of the material, which dictates the wavelengths of light that can be effectively absorbed [33]. As mentioned earlier, defect-engineering strategies such as doping and ion implantation can be employed to adjust the bandgap, thereby enhancing the material's ability to absorb visible light. The overall water-splitting reaction is mentioned above in Eq. 17.1–17.4 and Figure 17.1. Upon light absorption, the semiconductor generates electron–hole pairs as the absorbed energy promotes electrons from the valence band to the conduction band. The effectiveness of this process is influenced by the material's electronic structure, which can be modified through defect-engineering techniques [34]. For instance, the presence of defects may create localized energy states within the bandgap. These states can act as intermediate energy levels that facilitate the excitation of electrons with lower energy, effectively broadening the range of the light spectrum that can be utilized for charge carrier generation. As a result, defect-engineered materials can achieve higher quantum efficiencies for H_2 evolution, even under visible-light illumination.

The research study conducted by Zheng et al. reported the successful synthesis of protonated C_3N_4 porous nanosheets with nitrogen-defects, specifically N1-vacancy and cyano (—C≡N) groups, utilizing a straightforward hydrothermal treatment of melamine in the presence of nitric acid, followed by thermal polymerization [35]. The resultant material was designated as C_3N_4-ND, which exhibited remarkable photocatalytic performance for H_2 production under visible-light irradiation. The incorporation of nitrogen-defects significantly enhanced the optical properties of the synthesized material. The presence of N1-vacancies and —C≡N groups effectively reduced the bandgap, thereby optimizing light absorption across the visible spectrum. These modifications fostered improved charge generation reactions, essential for the photocatalytic H_2 evolution process. The C_3N_4-ND1.7 sample demonstrated an astounding 39.18 mmol g^{-1} h^{-1} for H_2 evolution when employing 0.5 vol% triethanolamine (TEOA) as a sacrificial agent. Notably, these rates signify a 22.4-fold and 75.34-fold enhancement over the performance of BCN, respectively. The enhanced photocatalytic activity was attributed to the synergistic effects of the porous nanosheet morphology and the nitrogen defects, which not only increased the specific surface area but also facilitated the efficient absorption of light and the optimal distribution of photogenerated charge carriers. The unique porous structure permitted increased interactions with incident light, allowing for the reflection and scattering of photons, which resulted in heightened light absorption efficiency.

The results of density functional theory (DFT) calculations corroborated these experimental findings, revealing that nitrogen-defects play a crucial role in enhancing the electronic structure and charge distribution within the C_3N_4 framework. The optimized structures, estimated bandgaps, and charge density differences of BCN and C_3N_4-ND1.7 are depicted in Figure 17.6a–c [35]. Moreover, by elucidating the charge generation mechanism the researchers stated that the nitrogen defects acted as active sites for the adsorption of H^+ ions, thereby promoting charge separation and reducing the recombination of photogenerated electron–hole pairs (Figure 17.6d). This study elucidated the critical interplay between light absorption and charge generation reactions in defect-engineered photocatalytic materials, emphasizing the potential of C_3N_4-ND for applications in the green H_2 energy sector. The detailed mechanism of photocatalytic H_2 production over C_3N_4-ND1.7 sample is illustrated in Figure 17.6e. Their work not only sheds light on the underlying mechanisms but also lays a foundation for further research into optimizing photocatalytic systems for enhanced efficiency and sustainability.

Additionally, the nature and concentration of defects can significantly influence the rate of charge carrier generation. Certain defects may enhance the absorption cross-section of the material, leading to an increased number of excitons produced per unit of incident light. However, it is crucial to balance defect concentrations because an excessive number of defects can lead to non-radiative recombination pathways, where the generated electron–hole pairs recombine before contributing to the photocatalytic reaction. Thus, the optimization of defect levels is key to maximizing charge carrier generation while minimizing recombination losses.

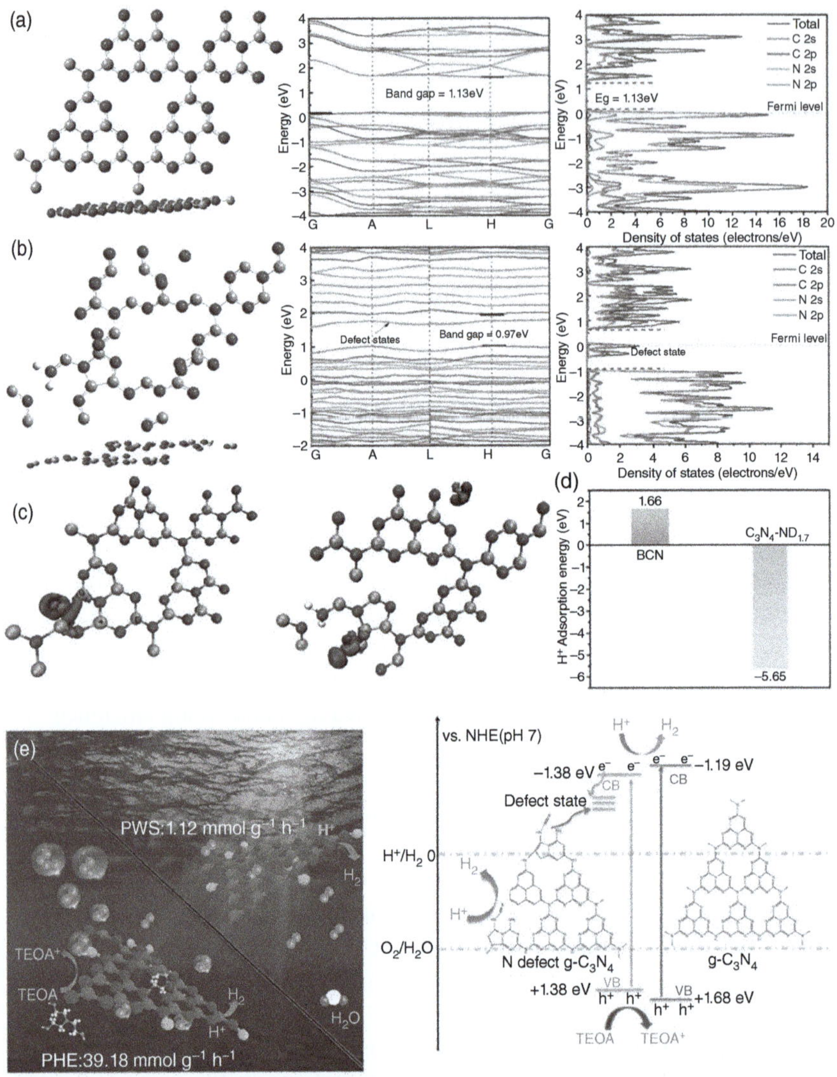

Figure 17.6 (a) Structural model, computed band topology, and associated DOS for BCN, along with (b) C_3N_4-ND1.7. (c) Charge density distribution maps for BCN and C3N4-ND1.7. (d) Calculated adsorption energy for H^+ on both BCN and C_3N_4-ND1.7. (e) The estimated H_2 generation pathway of C_3N_4-ND under visible-light irradiation. Reproduced from [32] / with permission of Elsevier.

17.3.2 Charge Carrier Separation

Once the electron–hole pairs are generated, the next critical step is their separation and migration to the respective reaction sites on the semiconductor surface. The efficiency of charge separation is vital for photocatalytic activity, as recombination of the pairs before they can participate in the water-splitting reaction can severely limit H_2 production.

Defect engineering can enhance charge carrier separation by creating favorable pathways for electron and hole transport. For instance, Yang et al. reported the fabrication of a high-performance photocatalyst, comprising mixed-phase molybdenum disulfide (MoS_2) incorporated with sulfur vacancies and sulfur-deficient zinc-cadmium sulfide ($Zn_{0.5}Cd_{0.5}S$) through a hydrothermal method using hydrazine hydrate as a reductant [36]. This innovative approach resulted in the formation of a unique nano-junction, termed 1T/2H-MoS_2@Vs-ZCS, which exhibited remarkable photocatalytic H_2 evolution capabilities under visible-light irradiation. The synthesized 1T/2H-MoS_2@Vs-ZCS demonstrated a superior H_2 production rate of 71.63 mmol g^{-1} h^{-1}, significantly surpassing the rates of pristine MoS_2 (0.47 mmol g^{-1} h^{-1}) and Vs-ZCS (16.56 mmol g^{-1} h^{-1}). This enhancement can be attributed to the synergistic effects arising from the hybrid MoS_2 phases, the presence of sulfur vacancies, and an optimized interfacial electronic structure, which collectively facilitate efficient charge separation. To elucidate the charge carrier separation mechanisms and characterization results, several figures from the study can be referenced. For instance, Figure 17.7a, b, which presents the photoluminescence (PL) and time-resolved PL spectra, highlights the efficient charge separation and suppressed recombination rates within the 1T/2H-MoS_2@Vs-ZCS structure. Additionally, Figure 17.7c–f provides insights into electrochemical performance, showcasing photocurrent response and impedance analysis that further confirm the improved charge transfer kinetics. Factors contributing to effective charge separation include the presence of sulfur vacancies within both Vs-ZCS and MoS_2, which act as electron traps that facilitate electron migration and H_2 adsorption. The incorporation of these vacancies not only enhances the electronic properties of the materials but also optimizes the interaction with photogenerated charges, promoting their migration toward the active sites for H_2 evolution reactions.

The charge separation mechanism in the 1T/2H-MoS_2@Vs-ZCS system is intricately linked to the formation of an ohmic contact heterojunction at the interface of the two materials. The DFT calculations revealed that the work function of Vs-ZCS is more negative than that of MoS_2, leading to the establishment of a built-in electric field at the interface (Figure 17.7g–i). This electric field plays a pivotal role in driving photogenerated electrons from the Vs-ZCS to the 1T-MoS_2 phase, while simultaneously allowing the holes to be consumed by sacrificial reagents. Consequently, this spatial separation of charge carriers is crucial for minimizing recombination losses, thereby enhancing the overall photocatalytic H_2 evolution efficiency (Figure 17.7j).

However, certain defects can serve as traps for charge carriers, effectively delaying recombination and allowing for longer migration times to the surface. Additionally, the spatial distribution of defects plays a crucial role in charge carrier dynamics. A well-designed defect landscape can facilitate efficient charge transport and minimize recombination losses. Researchers have explored various strategies to tailor the spatial arrangement of defects, such as creating hierarchical structures or gradient doping profiles, to optimize charge separation and transport.

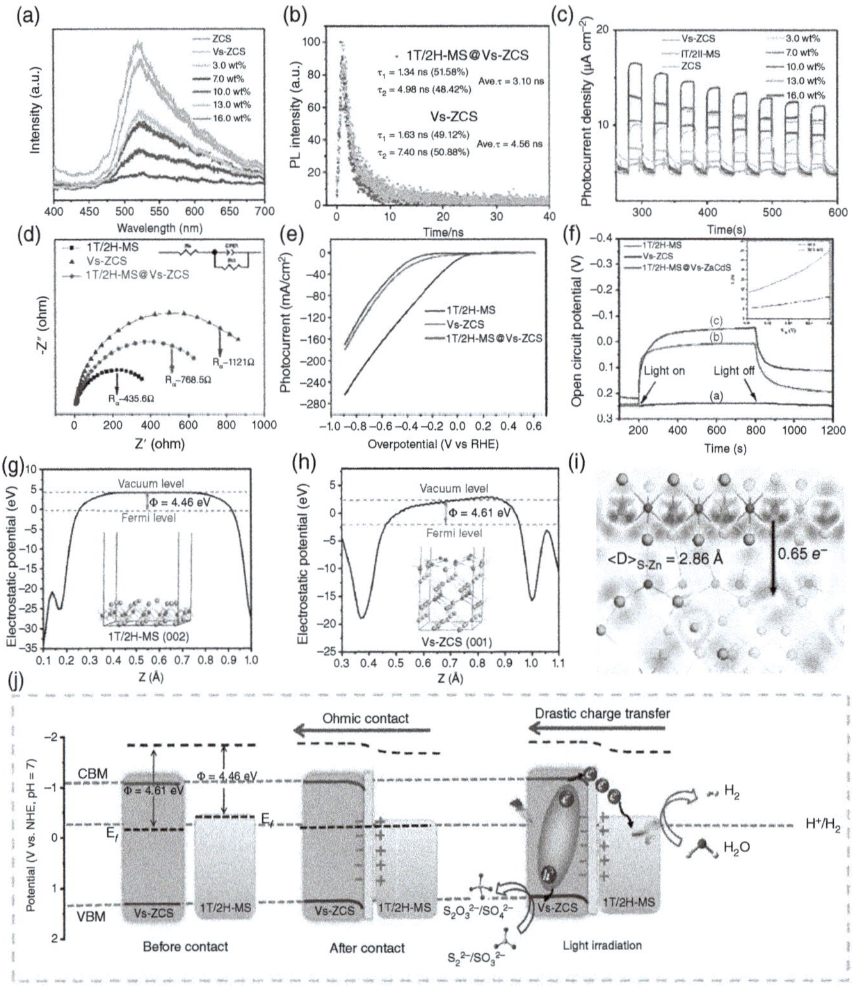

Figure 17.7 (a) PL spectra are presented, followed by (b) time-resolved photoluminescence (TRPL) decay profiles obtained under excitation at 380 nm and emission at 525 nm. (c) Transient current-time (I-t) curves and (d) electrochemical impedance spectroscopy (EIS) Nyquist plots. (e) H_2 evolution reaction polarization curves, and (f) transient photo-current response for MoS_2, Vs-ZCS, and 1T/2H-MS@Vs-ZCS samples. (g) The computed work functions for the 1T/2H-MS (002) and (h) Vs-ZCS (001) facets. (i) The interfacial chemical bonding and charge density variation of 1T/2H-MS@Vs-ZCS from the lateral perspective. (j) Proposed photocatalytic H_2 evolution mechanism over the 1T/2H-MS@Vs-ZCS system. Reproduced from [36] / with permission of Elsevier.

17.3.3 Surface Reactions and Product Formation

The final step in the photocatalytic water-splitting process involves surface reactions, where the separated electrons and holes participate in chemical reactions to produce H_2 and O_2 [37]. The surface morphology and chemical composition of the photocatalyst significantly influence these reactions [38]. Defect-engineered materials can provide abundant active sites that facilitate the adsorption of H_2O molecules and the subsequent reaction to produce H_2 [39]. The presence of defects on the surface can enhance catalytic activity by promoting the adsorption of reactants and stabilizing reaction intermediates. For example, defect-engineering-introduced OVs can serve as active sites for the adsorption of H_2O molecules, effectively lowering the energy barrier for the water-splitting reaction [40]. The enhanced surface reactivity can lead to increased rates of H_2 evolution and improved overall photocatalytic performance.

Moreover, the design of defect-engineered photocatalysts can also influence the selectivity of the reaction pathways. By controlling the type and distribution of defects, researchers can tailor the photocatalytic materials to favor the desired reaction, such as the production of H_2 over unwanted side reactions like the formation of peroxides or other oxidizing agents [41]. For instance, specific defect configurations including the incorporation of cocatalysts such as Pt [42], MoS_2 [43], CoS_2 [44], and NiS [45] can stabilize certain intermediates during the reaction, thereby steering the reaction pathway more efficiently toward H_2 evolution. Generally, the photocatalytic generation of H_2 involves two basic surface reactions: proton reduction with e^- and oxidation of electron donors with h^+. The reduction of protons in the presence of a cocatalyst can be divided into three steps: (i) the cocatalyst traps electrons from the light-harvesting semiconductor; (ii) H^+ accepts an electron and is reduced to an H atom; and (iii) two surface H atoms unite to form a H_2 molecule [46]. H* (where * denotes a site on the cocatalyst surface) as the catalytic intermediate plays an important role in the photocatalytic H_2 evolution reaction, which may occur through either the Volmer-Heyrovsky or the Volmer-Tafel mechanism. The rate of the overall reaction is largely determined by the H_2 adsorption free energy, ΔGH. If H_2 binds to the surface too weakly, the adsorption (Volmer) step will limit the overall reaction rate, whereas if the binding is too strong, the desorption (Heyrovsky/Tafel) step will limit the rate. Thus, a necessary but insufficient condition for an efficient photocatalyst is $\Delta GH \approx 0$ [47].

$$H^+ + e^- + {}^* \rightarrow H^* \text{ (Volmer step)} \tag{17.5}$$

$$H^* + H^+ + e^- \rightarrow H_2 + {}^* \text{ (Heyrovsky step)} \tag{17.6}$$

$$2H^* \rightarrow H_2 + 2^* \text{ (Tafel step)} \tag{17.7}$$

Recently, Liccardo et al. synthesized nano-Cu/TiO_2 through a microwave-assisted hydrothermal method (Figure 17.8a) [48]. This innovative synthesis involved the incorporation of varying molar ratios of oxalic acid (O_x) with commercial P25 titania, followed by the in-situ photodeposition of a 2.0 wt% Cu co-catalyst during the photocatalytic reaction aimed at enhancing H_2 production through methanol reforming under UV-light irradiation. The introduction of OVs and Ti^{3+} sites was pivotal in regulating surface reactions and promoting efficient product formation. These structural modifications not only altered the electronic properties of the TiO_2 matrix but also created favorable sites for the adsorption of reactants. The presence of these defects facilitated the formation of intra-band states between the valence and conduction bands, which allowed for improved charge carrier dynamics.

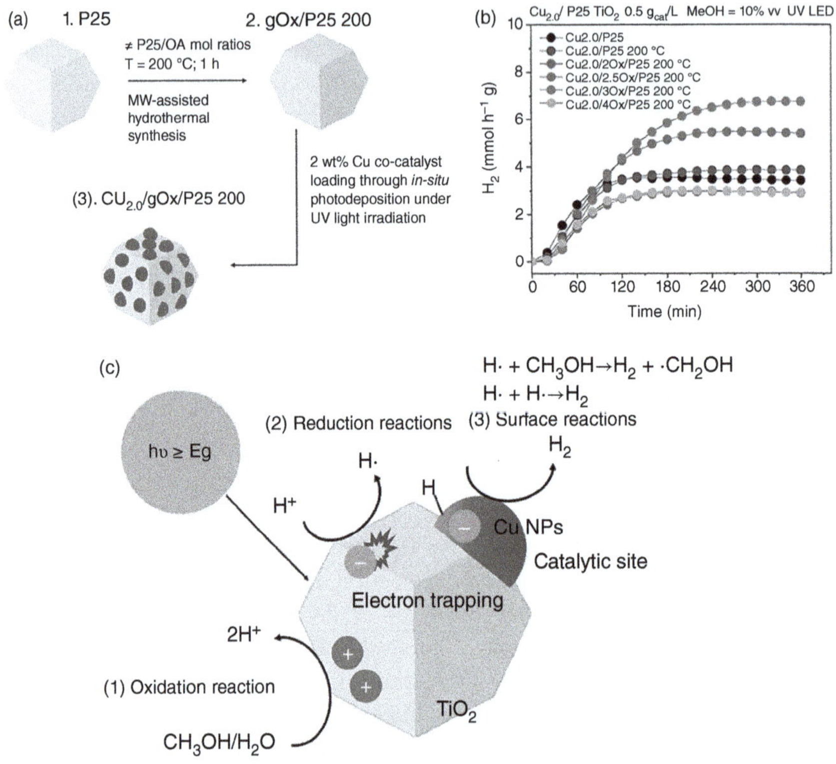

Figure 17.8 (a) A detailed procedural outline for the synthesis of $Cu_{2.0}/gO_x/P25$ 200 and (b) H_2 evolution rates for the various as-prepared photocatalysts. (c) A schematic representation illustrating the interaction mechanism in the $Cu_{2.0}/gO_x/P25$ 200 system. Reproduced from [48] / with permission of John Wiley & Sons.

This enhancement in electron mobility directly correlated with increased reactivity during the photocatalytic process. The optimized sample, named as $Cu_{2.0}/3O_x/P25$ 200, exhibited a remarkable H_2 production rate of 6.8 mmol g^{-1} h^{-1}, effectively doubling the performance of the pristine $Cu_{2.0}/P25$ system (Figure 17.8b). This significant improvement can be attributed to the synergistic effects of the Cu co-catalyst in conjunction with the engineered surface defects. The defects served as active sites that not only enhanced the adsorption of methanol and water but also facilitated the subsequent surface reactions necessary for H_2 generation. Furthermore, the structural modifications induced by defect engineering led to a more favorable reaction pathway for the conversion of methanol and H_2O into H_2 (Figure 17.8c). By optimizing the surface structure, the researchers effectively reduced barriers to reaction, allowing for a more efficient transfer of charge carriers and increased electron–hole pair separation. This resulted in a more effective catalytic cycle and ultimately enhanced H_2 production. However, the transfer path of photogenerated electrons and holes between semiconductors and cocatalysts, and the nature of active sites for H_2 evolution should be studied in depth.

17.4 Conclusion and Perspective

As the field of defect-engineered photocatalytic materials continues to evolve, a growing body of literature highlights the promising advancements and ongoing challenges. Researchers are actively investigating novel materials that exhibit superior photocatalytic performance through defect engineering, including metal oxides, sulfides, nitrides, and phosphides. Each class of materials presents unique electronic properties and defect characteristics, allowing for a diverse array of photocatalyst designs tailored for specific applications. Numerous recent studies have shown that transitioning from traditional photocatalytic materials to more complex systems, such as heterostructures or composite materials, can yield significant enhancements in H_2 evolution. For instance, coupling materials with complementary properties can facilitate charge separation and extend the light absorption range, thereby maximizing photocatalytic activity. Furthermore, the integration of co-catalysts such as noble/non-noble metals can provide additional active sites for the reduction and oxidation half-reactions, further enhancing overall efficiency.

Despite these advancements, there remain several challenges that need to be addressed in the pursuit of highly efficient defect-engineered photocatalysts. One major challenge is the reproducibility of synthesis methods, which can lead to variability in defect characteristics and, consequently, in photocatalytic performance. Standardizing synthesis protocols and characterization techniques will be crucial for comparing results across studies and establishing a clear understanding of the relationship between defect structures and photocatalytic activity. Additionally, comprehensive characterization methods must be employed to elucidate the effects of defects on photocatalytic mechanisms. Techniques such as advanced electron microscopy, X-ray diffraction, and spectroscopy can provide valuable insights into the types and distributions of defects, as well as their interactions with charge carriers and surface species. Understanding these interactions will be key to optimizing photocatalyst designs and unlocking their full potential for H_2 production.

In addition, research efforts are increasingly focused on developing multifunctional materials that can operate efficiently under realistic solar conditions, including varying light intensities and atmospheric compositions. By advancing our understanding of defect-engineering principles and integrating them with emerging technologies, it may be possible to create highly efficient, stable, and economically viable photocatalysts. Innovations in nanostructuring techniques and material composition have the potential to further enhance the performance of defect-engineered photocatalysts. For example, the development of nanostructured materials can lead to increased surface area, allowing for more active sites for photocatalytic reactions. This increased surface area can significantly enhance the interaction between the photocatalyst and reactants, thereby improving H_2 evolution rates. In addition to nanostructuring, the exploration of hybrid systems that combine various materials holds great promise. By integrating different semiconductor materials, researchers can create heterojunctions that facilitate charge transfer between the components, enhancing overall photocatalytic efficiency. Such hybrid systems can be designed to exploit the strengths of each individual component, leading to improved light absorption, charge carrier dynamics, and ultimately, H_2 production rates.

The role of computational modeling and machine learning in accelerating the discovery of new defect-engineered photocatalytic materials is also gaining traction. By employing advanced computational techniques, researchers can predict the properties and performance of novel materials before they are synthesized. This can significantly reduce the time and resources required for experimental validation, allowing for a more systematic exploration of the vast compositional space available for photocatalytic materials. Moreover, the development of in-situ characterization techniques will be crucial for gaining insights into the dynamic processes occurring during photocatalytic reactions. Monitoring changes in the electronic structure, defect states, and surface chemistry in real time can provide valuable information that can inform the design of more efficient photocatalysts. This knowledge can ultimately lead to a deeper understanding of the underlying mechanisms governing H_2 evolution and the role of defects in enhancing photocatalytic performance.

Looking toward the future, several other key areas of this research field warrant attention. First, the exploration of novel photocatalytic materials beyond traditional semiconductors is crucial. Emerging materials such as perovskites, 2D materials, and metal-organic frameworks (MOFs) offer exciting opportunities for enhanced photocatalytic activity. These materials may exhibit unique properties that can be exploited through defect engineering to achieve unprecedented efficiencies in H_2 evolution. Second, the integration of nanotechnology in photocatalyst design can lead to significant improvements in performance. By manipulating materials at the nanoscale, researchers can enhance light absorption, increase surface area, and optimize charge carrier dynamics. The incorporation of plasmonic nanoparticles, for instance, can boost the local electromagnetic field, resulting in improved light harvesting and increased reaction rates. Moreover, the development of innovative reactor designs that maximize the interaction between photocatalysts and reactants will be essential. Photocatalytic systems that can effectively harness sunlight while providing optimal conditions for H_2 evolution will be critical for practical applications. This includes the design of photoreactors that can accommodate varying light intensities and reactant concentrations, ensuring consistent performance over time. Lastly, the pursuit of cost-effective and environmentally friendly synthesis routes for defect-engineered photocatalysts is paramount. As the demand for H_2 production increases, it is vital to ensure that the materials used are produced sustainably and at scale. Researchers should prioritize green chemistry principles and economical processing techniques to reduce the ecological footprint of photocatalyst production.

Conflict of Interest

The authors declare that they have no known competing financial interests or personal relationships that could have appeared to influence the work reported in this chapter.

References

1 Hassan, Q., Viktor, P., Al-Musawi, T.J. et al. (2024). *Renew. Energy Focus* 48: 100545.

2 Sadeq, A.M., Homod, R.Z., Hussein, A.K. et al. (2024). *Sci. Total Environ.* 939: 173622.

3 Evro, S., Oni, B.A., and Tomomewo, O.S. (2024). *Int. J. Hydrog. Energy* 78: 1449–1467.

4 Teng, J., Li, W., Wei, Z. et al. (2024). *Angew. Chem. Int. Ed.* 63: e202416039.

5 Mohd, S. and Khan, A.M. (2024). *Sustainable Green Catalytic Processes*, 141–163.

6 Qi, K., Imparato, C., Almjasheva, O. et al. (2024). *J. Colloid Interface Sci.* 675: 150–191.

7 Abhishek, B., Jayarama, A., Rao, A.S. et al. (2024). *Int. J. Hydrog. Energy* 81: 1442–1466.

8 Muhammad, P., Zada, A., Rashid, J. et al. (2024). *Adv. Funct. Mater.* 34: 2314686.

9 Wang, Y., Yu, W., Wang, C. et al. (2024). *EScience* 4: 100228.

10 Zu, D., Wei, H., Lin, Z. et al. (2024). *Adv. Funct. Mater.* 34: 2408213.

11 Liu, Y., Chen, X., Kamali, M. et al. (2024). *Adv. Funct. Mater.* 34: 2405741.

12 (a) Chen, M., Sun, H., Lu, Q. et al. (2024). *Chem. Eng. J.* 497: 154235. (b) Wu, K., Lyu, C., Cheng, J. et al. (2024). *Carbon Energy* 6: e485.

13 Mohamadpour, F. and Amani, A.M. (2024). *RSC Adv.* 14: 20609–20645.

14 Mishra, R.K. and Verma, K. (2024). *Smart Mater. Manuf.* 2: 100052.

15 Chen, X., Shen, S., Guo, L., and Mao, S.S. (2010). *Chem. Rev.* 110: 6503–6570.

16 Telkhozhayeva, M. and Girshevitz, O. (2024). *Adv. Funct. Mater.* 34: 2404615.

17 Kim, J., Jung, S.-M., Im, H. et al. (2024). *Chem. Eng. J.* 501: 157385.

18 Zhu, H., Zhu, W., Kang, S.-Z. et al. (2024). *Appl. Surf. Sci.* 644: 158783.

19 Muhammad, Z., Wang, Y., Zhang, Y. et al. (2023). *Adv. Mater. Technol.* 8: 2200539.

20 Huang, J., Feng, X., Bi, F. et al. (2025). *Sep. Purif. Technol.* 359: 130804.

21 Mohapatra, L., Paramanik, L., Choi, D., and Yoo, S.H. (2024). *Appl. Surf. Sci.* 685: 161996.

22 Kogularasu, S., Lee, Y.Y., Sriram, B. et al. (2024). *Angew. Chem.* 136: e202311806.

23 Zeng, L., Lu, Z., Li, M. et al. (2016). *Appl. Catal. B Environ.* 183: 308–316.

24 Tian, Y., Zeng, D., Shen, T. et al. (2024). *Fuel* 357: 129927.

25 Zafar, Z., Yi, S., Li, J. et al. (2022). *Energy Environ. Mater.* 5: 68–114.

26 Patil, S.B., Basavarajappa, P.S., Ganganagappa, N. et al. (2019). *Int. J. Hydrog. Energy* 44: 13022–13039.

27 Reddy, N.R., Reddy, P.M., Jyothi, N. et al. (2023). *J. Alloys Compd.* 935: 167713.

28 Chen, Y., Ma, S., Yang, Y. et al. (2024). *J. Colloid Interface Sci.* 664: 640–649.

29 Khan, J., Khan, M., Uddin, W. et al. (2024). *Results Opt.* 16: 100730.

30 Xiao, F., Xiao, T., Zhao, L., and He, X. (2024). *Renew. Energy* 230: 120778.

31 Sun, T., Fan, J., Liu, E. et al. (2012). *Powder Technol.* 228: 210–218.

32 Chen, X., Guo, Y., Li, J. et al. (2024). *Chem. Eng. J.* 496: 153947.

33 Nishioka, S., Osterloh, F.E., Wang, X. et al. (2023). *Nat. Rev. Methods Primers* 3: 42.

34 Takanabe, K. (2017). *ACS Catal.* 7: 8006–8022.
35 Zheng, C., Guo, Y., Zhang, C. et al. (2025). *Applied Catalysis B: Environment and Energy* 365: 124879.
36 Yang, L., Yang, H., Tian, Q. et al. (2024). *Fuel* 360: 130548.
37 Kranz, C. and Wächtler, M. (2021). *Chem. Soc. Rev.* 50: 1407–1437.
38 Sohail, M., Rauf, S., Irfan, M. et al. (2024). *Nanoscale Adv.* 6: 1286–1330.
39 Liu, F. and Fan, Z. (2023). *Chem. Soc. Rev.* 52: 1723–1772.
40 Huang, Y., Meng, L., Xu, W., and Li, L. (2023). *Adv. Funct. Mater.* 33: 2305940.
41 Cheng, Z., Zhang, X., Bo, C. et al. (2024). *Int. J. Hydrog. Energy* 55: 542–549.
42 Tayyab, M., Liu, Y., Liu, Z. et al. (2023). *Chem. Eng. J.* 455: 140601.
43 Sun, B., Huang, C., Yang, C. et al. (2024). *Appl. Catal. B: Environ. Energy* 345: 123720.
44 Mamiyev, Z. and Balayeva, N.O. (2022). *Catalysts* 12: 1316.
45 Chen, X., Huang, T., Kuo, D.-H. et al. (2021). *Appl. Catal. B Environ.* 298: 120542.
46 Wang, X. and Li, C. (2017). *J. Photochem. Photobiol. C Photchem. Rev.* 33: 165–179.
47 Li, T., Hu, T., Dai, L., and Li, C.M. (2020). *J Mater Chem A* 8: 23674–23698.
48 Liccardo, L., Moras, P., Sheverdyaeva, P.M. et al. (2024). *Adv. Sustainable Syst.* 8: 2300418.

18

Enhanced H_2O_2 Production Over Defective Photocatalytic Materials

Mohsen Padervand[1], Alireza Bargahi[2], Mahshid Sarkheil[1], Elmuez Dawi[3], and Abdelkader Labidi[4]

[1] *University of Maragheh, Department of Chemistry, Faculty of Science, Amirkabir Highway, Maragheh, Iran*
[2] *Razi University, Department of Chemistry, Faculty of Science, Zakariya Razi Blvd, Kermanshah, Iran*
[3] *College of Humanities and Sciences, Department of Mathematics and Sciences, Ajman University, Ajman, UAE*
[4] *Shaanxi University of Science and Technology, School of Environmental Science and Engineering, Weiyang District, Xi'an, 710021, China*

18.1 Introduction

Hydrogen peroxide (H_2O_2) is a strong and eco-friendly oxidant that produces only H_2O and O_2 as by-products of the reactions. It has been applied in the paper industry, organic transformations, [1] sterilization, [2] wastewater treatment, and fuel cells [3]. H_2O_2 also serves as a clean energy alternative with an energy density comparable to compressed hydrogen (35 MPa) and offers greater convenience and safety in storage and transportation [4]. The annual demand for H_2O_2 is 2.2 million tons, which is expected to reach 5.7 million tons by 2027 [5]. However, anthraquinone oxidation, the most common production method, is limited by high energy consumption, production costs, complicated industrial processes, and harmful by-products [6].

Photocatalysis is a phenomenon in which electron–hole pairs are generated by the photoexcitation of semiconducting materials. During this process, the photocatalyst interacts repeatedly to change the chemical conversion rate of the reaction partners and regenerates itself after each cycle of interaction [7, 8]. Key requirements for efficient photocatalyst design include understanding charge transfer, elucidating reaction mechanisms, innovative syntheses and nanoarchitectures, and accessible photoactive sites for photon delivery [9]. In this regard, strategies such as doping pristine materials (e.g., TiO_2) and designing hybrid heterostructures by incorporating inorganic semiconductors bring high efficiency and improved ability to harvest solar radiation [10].

Defect engineering is a vital strategy for improving photocatalyst performance by influencing light absorption, charge separation, and interfacial reactions [11]. Defects can alter the pathway of photocatalytic reactions by acting as active sites, but do not necessarily counteract the negative effects of charge recombination, which

Defect Engineering in Photocatalysis: From Fundamentals to Applications,
First Edition. Edited by Chuanyi Wang, Abdelkader Labidi, and Qiuhui Zhu.

highlights the need for defect control [12]. The interaction of various defects, such as vacancies, dopants, crystal distortion, and surface defects, may lead to a unique positive synergistic effect on the charge carrier concentration and light harvesting [13]. They play an exceptional role in mediating rate-limiting catalytic stages and stabilizing charge separation, representing a key step toward designing high-performance photocatalysts [14, 15].

The light-based process is a promising and sustainable alternative for H_2O_2 generation. Photoredox catalysis over semiconductors can produce H_2O_2 under sunlight via a dual-pathway mechanism involving the water oxidation reaction (WOR) and oxygen reduction reaction (ORR). Effective photocatalysts require broad solar spectrum absorption, efficient electron–hole separation, and surface redox reactions. However, the cooperative optimization of these factors is challenging, and the surface interactions between O_2, protons, H_2O, and H_2O_2 are not fully understood, especially in the ORR [16]. However, a suitable photocatalyst for large-scale industrial H_2O_2 production has not yet been realized.

18.2 Defect Chemistry

18.2.1 Definition and History

Defects are defined as departures from the perfect atomic or electronic structure of a material and include several types, such as vacancies, interstitial atoms, dopants, and dislocations. Intensive research on defects can be traced back to the early 20th century, notably through landmark investigations by Frenkel and Schottky on ionic crystals, in which the existence of vacancies and interstitials provided a mechanistic understanding of the observed ionic conductivity. There was a paradigm shift in the 1970s following the discovery of the role of oxygen vacancies in titanium dioxide (TiO_2) in photocatalytic water splitting, which generated tremendous interest in defect-mediated catalysis. Contemporary defect engineering has incorporated sophisticated techniques, such as plasma treatment and chemical vapor deposition (CVD), allowing the intentional and controlled introduction of defects at the atomic level. For example, the deliberate creation of nitrogen vacancies in graphitic carbon nitride (g-C_3N_4) by thermal etching has been shown to promote charge separation via hole capture and recombination suppression, thereby exemplifying the potential of defect engineering to regulate and improve material properties.

18.2.2 Classes of Defects

Defects in photocatalysts can occur at the surface or at the interface, which are categorized based on the shape, type, and quantity of defects in the atomic structure [17]. Defects can usually be linear, [18, 19] planar, [20, 21] and point [22, 23] forms, involve crystal lattice displacement, deformation at twin boundaries and stacking faults, and cationic or ionic atom vacancies, respectively. Of course, these are point defects that are most relevant to photocatalysts, occurring in the form of vacancy, [24, 25] interstitial, [23] and antisite [26] defects as intrinsic defects, as well as doping [27]

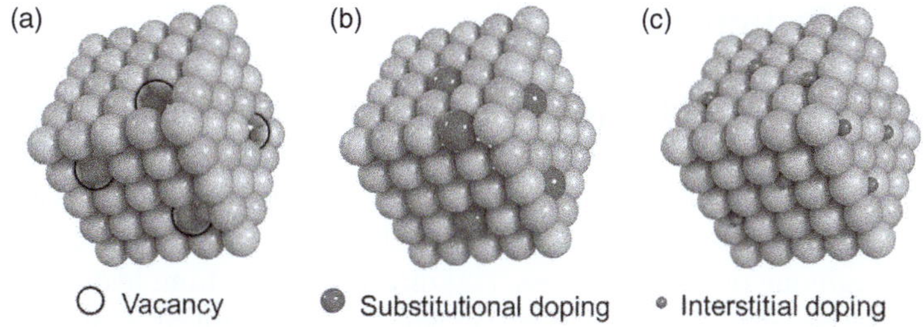

Figure 18.1 Schematic illustration of point defects:. (a) Vacancy; (b) Substitutional doping (including antisite substitution); (c) Interstitial doping in photocatalytic materials. *Source:* Reproduced from Zu et al. [23] / with permission of John Wiley & Sons / CC BY-NC-ND 4.

and impurities [28, 29] as extrinsic defects. Vacancies appear as missing metal and non-metal ions in the crystal lattice, e.g., V_{Ti} and V_o in TiO_2 [30]. Interstitials appear as atoms that occupy positions outside their normal lattice sites. Also, an antisite appears as a swapping position in the lattice, for example, a cation occupying an anionic site or vice versa. Doping occurs by introducing foreign atoms into the lattice, for example, N^{3-}, Fe^{3+}, and W^{6+} in the TiO_2 lattice [31], and impurities are created by the unintended incorporation of foreign atoms during synthesis. Examples of defects are shown schematically in Figure 18.1.

18.2.3 Role/philosophy in Photocatalysis

Defects in photocatalysis can be strategically engineered to steer material performances for specific applications [15]. By introducing defects, multiple energy levels can be created in the band gap structure, allowing for tunable light absorption and greater stability [32, 33]. Intra-band defect levels can also act as electron traps, preventing electron–hole pair recombination and prolonging the charge carrier lifetime [34]. Furthermore, they can facilitate directional charge transfer, boost the efficiency of charge carriers for specific interfacial redox reactions, and provide active sites for reactant adsorption and activation [35, 36]. Defects can be in the form of impurity atoms or vacancies, which can change the charge density distribution of the surface and change the bonding structure near the vacancies, respectively [37, 38]. However, if not properly controlled, surface and bulk defects can negatively impact photocatalyst performance by acting as recombination centers that reduce the efficiency of charge carrier utilization [15]. Inevitable interfacial defects can trap photogenerated carriers, preventing them from reaching the catalyst surface [39].

The philosophy of defect engineering in photocatalysts involves the controlled introduction and management of defects to optimize their performance. By tailoring the type, concentration, and location of defects, researchers can fine-tune material properties, such as light absorption and charge carrier dynamics. The goal is to introduce defects that enhance the desired properties while minimizing negative impacts. Defect engineering is a tool for creating more efficient photocatalytic materials, which will ultimately contribute to the development of sustainable energy and environmental solutions.

18.3 H_2O_2 Production Strategies

H_2O_2 is a valuable chemical with applications in disinfection, paper bleaching, wastewater treatment, as an oxidant in chemical synthesis, and also in fuel cells [40, 41]. Several strategies exist for H_2O_2 production, each with its own basic concepts, advantages, and disadvantages. The main methods for its production include the anthraquinone oxidation process, [42] direct synthesis from hydrogen and oxygen, as well as plasma-driven, piezocatalytic, electrochemical, [43] and photocatalytic production.

The anthraquinone oxidation process is the most widely used industrial method. It involves sequentially hydrogenation of 2-alkylanthraquinone, which is then oxidized by air in reactors to produce H_2O_2 and regenerate the anthraquinone. However, this method is costly and impractical for routine use.

H_2O_2 can be generated directly by the reaction of hydrogen (H_2) and oxygen (O_2) using an appropriate catalyst, such as palladium- or gold-based catalysts.

Enzymatic generation of H_2O_2 employs enzymes, i.e., glucose oxidase or alcohol oxidase, to catalyze the oxidation of organic substrates, i.e., glucose or methanol, in the presence of oxygen, ultimately yielding H_2O_2 as a byproduct.

The oxidative chemical conversion of alcohols, such as isopropanol, or hydrocarbons, such as methane, using oxygen or air and a catalyst, is another method for H_2O_2 production. In contrast, plasma-assisted synthesis involves the generation of H_2O_2 by exposing H_2O and O_2 to plasma, leading to the creation of reactive species that then react to produce H_2O_2.

Electrochemical generation of H_2O_2 is achieved through the cathodic reduction of molecular O_2 in an electrolytic cell, generally by a two-electron oxygen reduction process represented by the reaction ($O_2 + 2H^+ + 2e^- \rightarrow H_2O_2$).

Photocatalytic H_2O_2 synthesis employs semiconductors such as TiO_2 or g-C_3N_4 under light. H_2O and O_2 are reduced and oxidized to produce H_2O_2 via redox reactions, which is a promising route for green H_2O_2 synthesis.

Although the anthraquinone process is efficient, it is marred by its high energy intensity (~16 MJ kg^{-1}) and associated environmental risks. Electrocatalytic processes using carbon-based catalysts such as nitrogen-doped graphene to facilitate the two-electron oxygen reduction reaction (2e$^-$ ORR) are constrained by the requirement for expensive proton-exchange membranes and precise pH control of the electrolyte. Photocatalytic processes using semiconductors such as TiO_2 and g-C_3N_4 provide a more environmentally friendly option; however, their efficiency is hindered by charge recombination. Enzyme-based approaches, such as those employing glucose oxidase, are highly selective but lack scalability. Defect engineering has been proven to break these constraints, as exemplified by oxygen-deficient tungsten oxide (WO_3), which exhibits a 5-fold enhancement in H_2O_2 yield upon visible light irradiation owing to enhanced charge separation. Additionally, hybrid platforms that couple photocatalysis with electrocatalysis, including the coupling of bismuth vanadate ($BiVO_4$) photoanodes and carbon cathodes, demonstrate superior efficiency in terms of the spatial separation of oxidation and reduction reactions, thereby constituting a valuable pathway for maximizing H_2O_2 production. All the aforementioned processes have several advantages and disadvantages in terms of energy and environmental policies, which restrict their wide application in the real world. To opt for cost-effective and environmentally friendly strategies, governments should prioritize these key points in close collaboration with researchers and the academic community to minimize the adverse effects on ecosystems and human quality of life. Table 18.1 summarizes the advantages and disadvantages

Table 18.1 Comparison of the advantages and disadvantages of H_2O_2 production strategies.

Strategies	Advantages	Disadvantages
Anthraquinone oxidation process	• High yield and purity of H_2O_2. • Well-established and scalable for large-scale production.	• Complex and energy-intensive process. • Requires organic solvents and generates waste. • High capital and operational costs.
Direct synthesis from hydrogen and oxygen	• Simple and potentially more sustainable process. • No organic solvents or byproducts.	• Risk of explosion due to the mixture of H_2 and O_2. • Requires precise control of the reaction conditions (e.g., temperature, pressure, and catalyst selectivity). • Lower yield compared to the AO process.
Enzymatic production	• Mild reaction conditions (ambient temperature and pressure). • High selectivity and specificity.	• High cost of enzymes. • Limited scalability and stability of enzymes. • Requires organic substrates, which may not be sustainable.
Chemical oxidation of alcohols or hydrocarbons	• Simple and straightforward process.	• Low selectivity and yield. • Generates byproducts and waste. • Not suitable for large-scale production.
Plasma-driven synthesis	• Can operate under ambient conditions. • No catalysts or solvents required.	• High energy consumption. • Limited understanding and optimization of the process. • Low production rates.
Electrochemical synthesis	• Can be powered by renewable electricity, making it environmentally friendly. • On-site production is possible, reducing transportation costs.	• Requires efficient and stable electrocatalysts. • High energy consumption and relatively low production rates. • Expensive setup and maintenance.
Photocatalytic production	• Utilizes sunlight or artificial light as the energy source. • Environmentally friendly and sustainable.	• Low efficiency due to the rapid recombination of charge carriers. • Requires optimization of photocatalysts and reaction conditions. • Limited scalability for industrial applications.

of the extensively used methods for the generation of H_2O_2 worldwide. This helps readers easily judge the popularity of certain approaches and the limitations of others in the scientific community.

18.4 Types of Solid Photocatalysts

Photocatalysts can be classified into various types with unique properties and characteristics. Inorganic materials, such as TiO_2, exhibit ultraviolet (UV) activity and stability, and zinc oxide (ZnO) has the benefit of adjustable defects and low production cost. Organic materials such as g-C_3N_4 display visible-light activity and can be produced at a large scale. Hybrid materials, such as Au/TiO_2 and MoS_2/graphene, exploit plasmonic enhancement and improved conductivity to enhance photocatalytic activity. Moreover, supported materials such as Pt/TiO_2 and Co_3O_4/SiO_2 show improved charge transfer and high dispersion, thereby increasing the photocatalyst material pool [14, 16].

18.5 Defective Photocatalytic Materials for H_2O_2 Production

Methods to improve photocatalytic activity involve doping, defect engineering, heterojunction formation, and reactor design optimization. For example, the sulfur doping of g-C_3N_4 has been demonstrated to have 80% selectivity owing to enhanced oxygen adsorption [43]. Similarly, defect engineering, i.e., the introduction of oxygen vacancies in WO_3, can promote charge separation, thereby resulting in enhanced photocatalytic efficiency. The creation of heterojunctions, such as TiO_2/g-C_3N_4 composites, can be utilized to extend the charge carrier lifetimes and thus photocatalyst activity. Novel reactor configurations, such as microfluidic reactors, can also be utilized to optimize mass transfer and light distribution, thereby improving photocatalyst performance and system efficiency. Defective materials, such as g-C_3N_4 with three-coordinate nitrogen vacancies, have been demonstrated to exhibit high photocatalytic H_2O_2 generation rates of 1098 $\mu mol\,h^{-1}\,g^{-1}$ [43]. Similarly, oxygen-deficient TiO_2 nanowires demonstrated an impressive 10-fold enhancement in photocatalytic activity under visible light irradiation, illustrating the potential of defect engineering to enhance the performance of photocatalytic materials. This type of materials engineering is utilized in the development of photoactive solids and has gained significant attention for achieving practical photocatalysis. The authors focused on $ZnIn_2S_4$ with rich sulfur vacancies, a typical system, which was integrated with metal organic frameworks for photocatalytic H_2O_2 production from oxygen reduction under visible irradiation. However, the photocatalytic activity of the support is limited because of the agglomeration and rapid recombination of the photo-generated electrons and holes on its surface, which restricts its high efficiency. As is commonly known, safety concerns correlated with H_2O_2 instability raise the need for sustainable production methods. Inserting the solid defects in photocatalytic materials is expected to not only highlight the importance of light-triggered systems to increase the production yield, but also provide helpful evidence to resolve these obstacles.

According to the literature, O-N defects in g-C_3N_4 significantly enhance the photocatalytic generation of H_2O_2 by raising adsorption capacity and activating molecular oxygen on the surface. This implies opening new windows to make use of suitable defects to activate the targets through efficient mechanisms. In another study, it was proven that cyano defects and carbon vacancies in carbon nitride significantly improved its efficiency toward H_2O_2 production. A mechanistic description of the action is schematically represented in Figure 18.2a, b [44, 45].

In another report, in 2025, Liao et al. reported the synthesis of vacancy-defective carbon nitride nanowire clusters (MNCN and MACN) for photocatalytic H_2O_2 production utilizing a simple self-assembly/pyrolysis method and melamine-cyanuric acid supramolecular crystals (MCS) as the precursor (Figure 18.3a) [46]. In this investigation, the authors indicated that defects are typically formed within high-temperature pyrolysis or incomplete polymerization, preventing the complete conversion of complexes into the CN framework, resulting in vacancy defects in the obtained products. The existence of these defects can change the lattice structure, leading to a decrease in the intensity of the –CN stretching vibration in the MNCN and MACN materials. Among these, the clusters prepared in argon (MACN) displayed the greatest photocatalytic efficiency, with a remarkable H_2O_2 production rate of 4.11 mmol g^{-1} h^{-1}, which is 15.68 times higher than that of bulk carbon nitride (BCN). Based on the detailed mechanism (Figure 18.3b), Liao's group revealed that under visible light exposure, MACN material absorbs photons, producing electron–hole pairs, which are then transferred to these active sites. On the one hand, carbon vacancies (C vacancies) within MACN material play a crucial role by effectively adsorbing and activating O_2 molecules, generating superoxide radicals ($\cdot O_2^-$) which can water molecules to produce H_2O_2. On the other hand, the synergistic interplay between morphology tuning and defect engineering in MACN

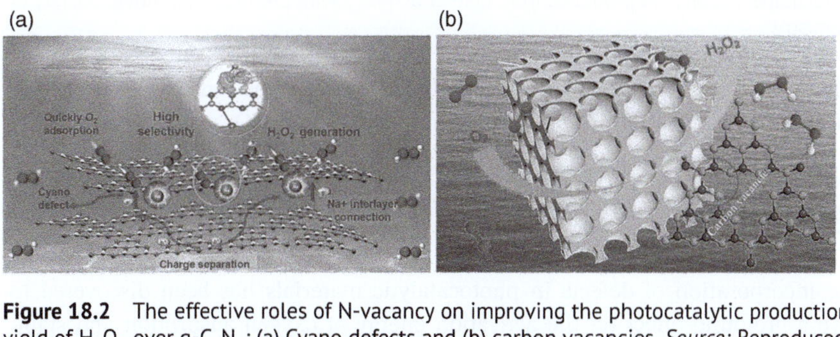

Figure 18.2 The effective roles of N-vacancy on improving the photocatalytic production yield of H_2O_2 over g-C_3N_4: (a) Cyano defects and (b) carbon vacancies. *Source:* Reproduced from Chen et al. [44] / with permission of John Wiley & Sons. *Source:* Reproduced from Lei et al. [45] / with permission of American Chemical Society.

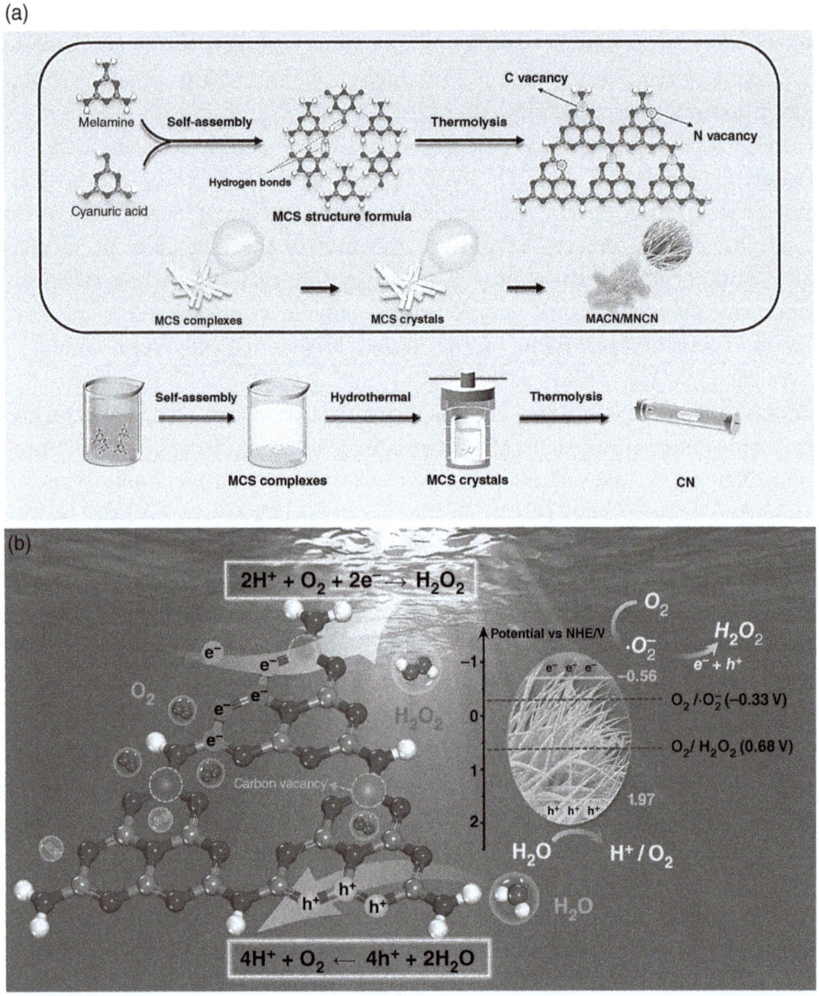

Figure 18.3 (a) Preparation protocol of defective carbon nitride carbon and nitrogen vacancies, and (b) mechanism of photocatalytic H_2O_2 evolution over the defective photocatalyst. *Source:* Reproduced from Liao et al. [46] / with permission of Royal Society of Chemistry.

material has led to an excellent selectivity for the two-electron oxygen reduction pathway ($O_2 + 2e^- + 2H^+/H_2O_2$), remarkably improving H_2O_2 production.

18.6 Mechanism of Defects

The incorporation of defects in photocatalytic materials has been discovered to have a profound effect on the activation energy of oxygen adsorption and stabilization of key intermediates, i.e., the *OOH species. In situ X-ray photoelectron spectroscopy (XPS) studies revealed that oxygen vacancies in TiO_2 effectively lower the oxygen dissociation barrier by approximately 0.5 eV, thereby improving the efficiency of the photocatalytic process. Furthermore, density functional theory (DFT) investigations revealed that the existence of carbon vacancies in g-C_3N_4 facilitates proton-coupled electron transfer processes, thus favoring the production of H_2O_2 as the dominant product. This discovery highlights the critical role of defect engineering in improving the photocatalytic performance. In general, several factors govern the mechanism of H_2O_2 production and the rate efficiency of the reaction over photoactive materials with engineered metal and non-metal defects. The existence of metal and nonmetal vacancies generates electron-rich and hole-rich sites in the crystals, which can effectively promote the oxidation and/or reduction steps of the reactions on heterogeneous catalysts. Charge transport and migration can also be affected by this factor, and therefore, the photocatalytic efficiency. The remaining electron–hole species on the surface can consequently participate in oxidation–reduction and elementary reactions involving atmospheric agents such as water molecules and molecular oxygen to promote H_2O_2 generation through facile mechanistic stages. Surface redox reactions are always in urgent need of charge carriers to retain electrochemical reactions in increasingly high production yields. This aligns with the light absorption properties of the heterogeneous catalysts, which are predominantly governed by the type of defect, determining the wide applications of photocatalysts for real-world platforms. Undoubtedly, the Earth's environment can positively benefit from such progress in photocatalysis improvement toward addressing the energy crisis and pollution concerns.

18.7 Defect Insertion in Photocatalysts

Pilot-scale reactors based on defective TiO_2 have demonstrated H_2O_2 yields of up to 50 μmol h^{-1} g^{-1}. Meanwhile, industrial partners such as Evonik are engaged in evaluating decentralized units for H_2O_2 synthesis. To enable large-scale industrial roll-out of these technologies, scalable synthesis methods, e.g., flame spray pyrolysis, that provide efficient large-scale production pathways for defective TiO_2 and other photocatalytic materials are being improved. The focus from this viewpoint is on the types of photocatalysts that can be simply prepared from inexpensive raw materials. One such material is g-C_3N_4, which can be easily fabricated from urea via pyrolysis. Researchers have revealed that the insertion of carbon vacancies makes this material an excellent photocatalyst for H_2O_2 production. The introduction of carbon vacancies from a two-step single-electron indirect reduction pathway to a one-step two-electron direct pathway. This corresponds to the presence of some amino groups around the carbon vacancies in the carbon nitride structure after modification. The activity of engineered g-C_3N_4 for H_2O_2 production was 30-fold higher than that of a pure support material under visible light. Nitrogen vacancies can also play different roles in this type of photocatalyst, which highlights the critical role of nonmetal vacancies in the structure.

18.8 Opportunities and Challenges

In the context of H_2O_2 photocatalytic synthesis based on either the 2e$^-$ ORR pathway or the 2e$^-$ WOR route, researchers have demonstrated that an ideal photocatalyst demands simultaneously satisfy good light harvesting features (such as an appropriate band gap), excellent separation performance for keeping the photogenerated carriers alive (namely electrons-holes), as well as efficient catalytic capacity for the higher H_2O_2 yield in practical applications. In fact, previously published reports on photocatalysts still suffer from the relatively low photoactivity and selectivity of single-product H_2O_2, which remains a significant challenge for real-world applications. The future prospects to produce photocatalytic H_2O_2 involve the development of solar-driven systems for water treatment and the possibility of coupling with carbon dioxide reduction reactions. Nonetheless, there are challenges that need to be tackled to fully achieve the potential of the systems, such as controlling defect healing processes, minimizing the costs associated with noble metals, and constructing scalable reactor systems that are effectively applicable within an industrial setting. In view of such concerns, macromolecular semiconductors, e.g., defective metal organic frameworks (MOFs), defective covalent organic frameworks (COFs), and defective semiconducting polymers, are considered promising candidates for addition as the co-catalysts for ameliorating the H_2O_2 yield. The principal points in this regard are the energy-band structures of the semiconductor candidates, which can be systematically tuned during the preparation steps. Importantly, their light-harvesting features, charge migration, and oxidation–reduction capacity can be optimized for reliable photocatalytic H_2O_2 generation through robust reaction systems. Based on these considerations, researchers are highly motivated to develop novel engineering strategies taking advantage of coupling cost-effective supports with novel defective semiconductors that are easy to prepare from cheap raw precursors, aimed at industrialization of H_2O_2 production platforms at the lab scale. Economic issues in this way could be an urgent factor that can drastically govern the scaling-up procedure.

18.9 Future Perspectives

In comparison with traditional fabrication strategies, such as indirect oxidation of fossil resources, direct production of H_2O_2 via photocatalytic methods alongside electrocatalytic industrial patterns appears highly promising. These systems are simple to operate, safe, environmentally friendly, clean, and sustainable, specifically when powered by solar energy and integrated with other natural sources such as wind energy. Promising directions for future research in the field of photocatalysis are likely to be dominated by innovations in new materials and systems. These include artificial intelligence-guided design of dual-defect systems, creation of bio-inspired catalysts with self-healing defects, and merging of photoelectrochemical reactors and hybrid architectures. These innovative strategies are likely to be instrumental in addressing current limitations and achieving the complete potential of photocatalytic technologies. Notably, well-established photoactive systems (such as g-C_3N_4 and TiO_2) have gained particular attention for selective H_2O_2 production under light. In fact, most of these systems make use of organic sacrificial species to enhance the efficiency of charge separation during photocatalysis in various media. This can result in the generation of polluted H_2O_2. Consequently, further costs are incurred to obtain pure H_2O_2 as the main product. Therefore, researchers should focus on developing photocatalytic systems that do not use extra scavenging compounds to increase the aim of raising photocatalytic yield. Thus, efforts should focus on using defective photoactive materials obtained from the organic precursors and engineering systems with metal and nonmetal vacancies to create electron-rich and hole-rich sites to control the mechanism of production. It is anticipated that with the growing interest of the scientific community and greater interaction among theoreticians, experimentalists, and chemical engineers, this objective will be achieved in the near future by commercializing lab-scale synthetic pathways and addressing challenges related to defective materials usage for H_2O_2 production (Figure 18.4).

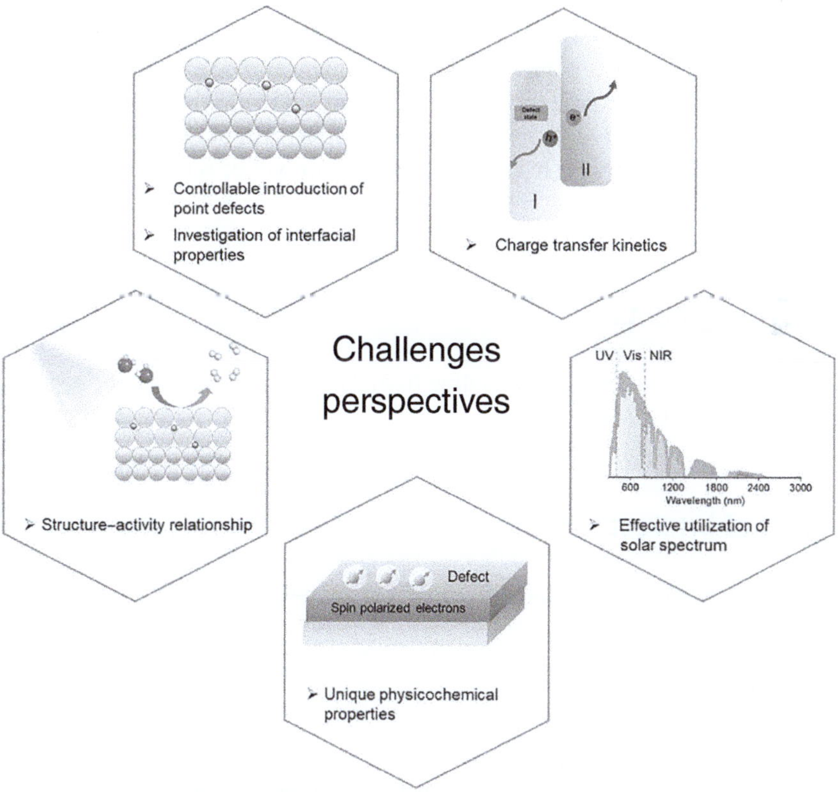

Figure 18.4 Future development paths for highly efficient point defects incorporated heterojunction photocatalysts. *Source:* Reproduced from Zu et al. [23] / with permission of John Wiley & Sons / CC BY-NC-ND 4.0.

18.10 Conclusions

Defect-engineered photocatalysts offer a sustainable and feasible pathway for H_2O_2 synthesis; however, their broader implementation will require concerted interdisciplinary efforts to resolve the significant challenges pertaining to material stability, scalability, and cost-effectiveness. Significant advances in artificial intelligence, reactor design, and defect characterization are anticipated to facilitate the industrial exploitation of these revolutionary photocatalytic systems, thereby bridging the gap between laboratory experiments and commercially relevant technologies. To date, a large number of visible-light-active photocatalysts have been reported to show the capacity for H_2O_2 generation. However, the production yield remained at a low level, which is not worthy of scaling up for real-world applications. Interestingly, engineering the crystal bulk and surface indicated the great potential of commercial photocatalysts (as cost-effective materials) for industrialization when these systems are prepared, benefiting from several types of metal and nonmetal defects in their structures. Moreover, coupling them with porous photoactive platforms originating from organic frameworks such as COFs and MOFs makes them excellent candidates for enhancing the H_2O_2 production yield. In a few instances, enriching the carbon nitride framework with carbon, nitrogen, and oxygen vacancies demonstrated that inserting such vacancies influences the selectivity of the reaction in different ways to obtain a cleaner product with the least contamination. Undoubtedly, the environmental aspects of scaling up the reaction systems could drastically hinder the development of such systems, which governments will be seriously concerned about in the future.

Acknowledgments

The authors acknowledge partial funding from Ajman University under the grant number 2025-IRG-CHS-4.

Conflict of Interest

The authors declare that they have no conflict of interest.

References

1 Targhan, H., Evans, P., and Bahrami, K. (2021). A review of the role of hydrogen peroxide in organic transformations. *J. Ind. Eng. Chem.* 104: 295–332.

2 McDonnell, G. (2014). The use of hydrogen peroxide for disinfection and sterilization applications. In: *Patai's Chemistry of Functional Groups* (ed. Z. Rappoport, J.F. Liebman, and I. Marek), 1–34. Wiley.

3 Sanli, A.E. (2013). A possible future fuel cell: the peroxide/peroxide fuel cell. *Int. J. Energy Res.* 37 (12): 1488–1497.

4 Mase, K., Yoneda, M., Yamada, Y., and Fukuzumi, S. (2016). Seawater usable for production and consumption of hydrogen peroxide as a solar fuel. *Nat. Commun.* 7 (1): 11470.

5 Zhao, W., Yan, P., Li, B. et al. (2022). Accelerated synthesis and discovery of covalent organic framework photocatalysts for hydrogen peroxide production. *J. Am. Chem. Soc.* 144 (22): 9902–9909.

6 Campos-Martin, J.M., Blanco-Brieva, G., and Fierro, J.L.G. (2006). Hydrogen peroxide synthesis: an outlook beyond the anthraquinone process. *Angew. Chem. Int. Ed.* 45 (42): 6962–6984.

7 Braslavsky, S.E. (2007). Glossary of terms used in photochemistry 3rd edition: (iupac recommendations 2006). *Pure Appl. Chem.* 79 (3): 293–465.

8 Yang, X. and Wang, D. (2018). Photocatalysis: from fundamental principles to materials and applications. *ACS Appl Energy Mater* 1 (12): 6657–6693.

9 Serpone, N. and Emeline, A.V. (2012). Semiconductor photocatalysis - past, present, and future outlook. *J. Phys. Chem. Lett.* 3 (5): 673–677.

10 Sordello, F., Calza, P., Minero, C. et al. (2022). More than one century of history for photocatalysis, from past, present and future perspectives. *Catalysts* 12 (12): 1572.

11 Maarisetty, D. and Baral, S.S. (2020). Defect engineering in photocatalysis: formation, chemistry, optoelectronics, and interface studies. *J Mater Chem A* 8 (36): 18560–18604.

12 Zhou, W. and Fu, H. (2018). Defect-mediated electron-hole separation in semiconductor photocatalysis. *Inorg. Chem. Front.* 5 (6): 1240–1254.

13 Wang, Z., Xiao, M., You, J. et al. (2022). Defect engineering in photocatalysts and photoelectrodes: from small to big. *Acc. Mater. Res.* 3 (11): 1127–1136.

14 Pastor, E., Sachs, M., Selim, S. et al. (2022). Electronic defects in metal oxide photocatalysts. *Nat. Rev. Mater.* 7 (7): 503–521.

15 Zafar, Z., Yi, S., Li, J. et al. (2022). Recent development in defects engineered photocatalysts: an overview of the experimental and theoretical strategies. *Energy Environ. Mater.* 5 (1): 68–114.

16 Zeng, X., Liu, Y., Hu, X., and Zhang, X. (2021). Photoredox catalysis over semiconductors for light-driven hydrogen peroxide production. *Green Chem.* 23 (4): 1466–1494.

17 Zhang, S., Si, Y., Li, B. et al. (2021). Atomic-level and modulated interfaces of photocatalyst heterostructure constructed by external defect-induced strategy: a critical review. *Small* 17 (1): 2004980.

18 Bai, S., Zhang, N., Gao, C., and Xiong, Y. (2018). Defect engineering in photocatalytic materials. *Nano Energy* 53: 296–336.

19 Tang, G., Hou, L., Li, J. et al. (2021). Line defects in plasmatic hollow copper ball boost excellent photocatalytic reaction with pure water under ultra-low CO_2 concentration. *J. Colloid Interface Sci.* 603: 530–538.

20 Liu, X., Wang, H., Lu, H. et al. (2023). Grain-interior planar defects induced by heteroatom monolayer. *Adv. Powder Mater.* 2 (4): 100130.

21 Nakabayashi, M., Nishiguchi, K., Liang, X. et al. (2023). Characterization of planar defect in layered perovskite photocatalyst $Y_2Ti_2O_5S_2$ by electron microscopy and first-principles calculations. *J. Phys. Chem. C* 127 (16): 7887–7893.

22 Zhang, H., Zhang, X., Yang, G., and Zhou, X. (2018). Point defect effects on photoelectronic properties of the potential metal-free C_2N photocatalysts: insight from first-principles computations. *J. Phys. Chem. C* 122 (10): 5291–5302.

23 Zu, D., Wei, H., Lin, Z. et al. (2024). The role of point defects in heterojunction photocatalysts: perspectives and outlooks. *Adv. Funct. Mater.* 34 (48): 2408213.

24 Bargahi, A., Joshaghani, M., and Rafiee, E. (2023). Brown TiO_2-based Fe-Ni thiospinel binary nanocomposite as a sustainable catalyst with synergistic improvement assisted by H_2O_2 and ascorbic acid for round-the-clock removal of rhodamine B. *Ind. Eng. Chem. Res.* 62 (43): 18091–18104.

25 Ding, W., Li, X., Su, S. et al. (2023). Cationic vacancy engineering of p-TiO_2 for enhanced photocatalytic nitrogen fixation. *Nanoscale* 15 (8): 4014–4021.

26 Liao, C., He, Z., Wang, F. et al. (2024). Anti-site defect-induced cascaded sub-band transition in $CuInS_2$ enables infrared light-driven CO_2 reduction. *ACS Nano* 18 (52): 35480–35489.

27 Takata, T. and Domen, K. (2009). Defect engineering of photocatalysts by doping of aliovalent metal cations for efficient water splitting. *J. Phys. Chem. C* 113 (45): 19386–19388.

28 Foo, C., Li, Y., Lebedev, K. et al. (2021). Characterisation of oxygen defects and nitrogen impurities in TiO_2 photocatalysts using variable-temperature X-ray powder diffraction. *Nat. Commun.* 12 (1): 661.

29 Korake, P.V., Achary, S.N., and Gupta, N.M. (2015). Role of aliovalent cation doping in the activity of nanocrystalline CdS for visible-light-driven H_2 production from water. *Int. J. Hydrog. Energy* 40 (28): 8695–8705.

30 Elrahoumi, R., Zhu, L., Wagner, E. et al. (2023). Doping-induced Ti^{3+} state and oxygen vacancies in TiO_2: a single-chip combinatorial investigation. *Mater. Chem. Phys.* 308: 128283.

31 Abraham, C. and Gomathi Devi, L. (2022). Incorporation of Fe^{3+} ions into the W^{6+} and N^{3-} doped TiO_2: exploration of crucial role of Fe^{3+} dopant ion and correlation of adsorption characteristics with reaction dynamics. *Surf. Sci.* 717: 121986.

32 Liu, Y., Yang, Y., Zhang, B. et al. (2023). Electronic structure and optical properties of CdS/BiOI heterojunction improved by oxygen vacancies. *J. Alloys Compd.* 955: 170235.

33 Miah, M.H., Khandaker, M.U., Rahman, M.B. et al. (2024). Band gap tuning of perovskite solar cells for enhancing the efficiency and stability: issues and prospects. *RSC Adv.* 14 (23): 15876–15906.

34 Xu, F., He, Y., Zhang, J. et al. (2024). Prolonging charge carrier lifetime via intraband defect levels in s-scheme heterojunctions for artificial photosynthesis. *Angew. Chem. Int. Ed.* 64 (2): e202414672.

35 Rezaei, M., Nezamzadeh-Ejhieh, A., and Massah, A.R. (2024). A comprehensive review on the boosted effects of anion vacancies in photocatalytic solar water splitting: focus on g-C_3N_4 with carbon and nitrogen vacancies. *Energy Fuel* 38 (10): 8406–8436.

36 Sharma, M., Sajwan, D., Gouda, A. et al. (2024). Recent progress in defect-engineered metal oxides for photocatalytic environmental remediation. *Photochem. Photobiol.* 100 (4): 830–896.

37 Kong, Y., Liu, L., and Xia, S. (2016). Research on electronic structure and optical properties of $Ga_{0.75}Al_{0.25}N$ with vacancy defects. *Optik (Stuttg).* 127 (20): 8662–8669.

38 Liang, J., Wen, X., Wei, S., and Zheng, S. (2021). Effect of impurity atoms on the adsorption/dissociation of hydrogen sulfide and hydrogen diffusion on the Fe(100) surface. *ACS Omega* 6 (22): 14701–14712.

39 Yang, Z., Kang, M., Chen, L. et al. (2023). Interfacial engineering over tungsten oxide by constructing z-scheme interatomic junction for efficient photocatalytic tetrachlorophenol degradation. *Appl. Surf. Sci.* 609: 155306.

40 Das, P., Roeser, J., and Thomas, A. (2023). Solar light driven H_2O_2 production and selective oxidations using a covalent organic framework photocatalyst prepared by a multicomponent reaction. *Angew. Chem. Int. Ed.* 62 (29): e202304349.

41 Mavrikis, S., Göltz, M., Perry, S.C. et al. (2021). Effective hydrogen peroxide production from electrochemical water oxidation. *ACS Energy Lett.* 6 (7): 2369–2377.

42 Chen, Q. (2008). Development of an anthraquinone process for the production of hydrogen peroxide in a trickle bed reactor-from bench scale to industrial scale. *Chem. Eng. Process. Process Intensif.* 47 (5): 787–792.

43 Xue, Y., Ma, C., Yang, Q. et al. (2023). Construction of g-C_3N_4 with three coordinated nitrogen (N_{3c}) vacancies for excellent photocatalytic activities of N_2 fixation and H_2O_2 production. *Chem. Eng. J.* 457: 141146.

44 Chen, L., Chen, C., Yang, Z. et al. (2021). Simultaneously tuning band structure and oxygen reduction pathway toward high-efficient photocatalytic hydrogen peroxide production using cyano-rich graphitic carbon nitride. *Adv. Funct. Mater.* 31 (46): 2105731.

45 Lei, J., Chen, B., Lv, W. et al. (2019). Robust photocatalytic H_2O_2 production over inverse opal g-C_3N_4 with carbon vacancy under visible light. *ACS Sustain. Chem. Eng.* 7 (19): 16467–16473.

46 Liao, Z., Hou, H., Du, J. et al. (2025). Engineering vacancy-defective carbon nitride nanowire clusters for dramatically enhanced visible-light-driven photocatalytic H_2O_2 production. *J Mater Chem A* 13: 9849–9864.

19

Photocatalytic NH_3 Synthesis by Defect-Engineered Photocatalytic Materials

Ximing Li[1], *Lin Tian*[1], *and Chuanyi Wang*[1,2]

[1] *Shaanxi University of Science and Technology, School of Environmental Science and Engineering, Weiyang University Street, Xi'an 710021, PR China*
[2] *Department of Environmental and Sustainable Engineering, Faculty of Engineering, Chulalongkorn University, 254 Phayathai Road, Pathumwan, Bangkok, 10330, Thailand*

19.1 Introduction

Ammonia is an essential chemical feedstock in modern agriculture and the chemical industry, primarily utilized in fertilizer production, and is regarded as a crucial intermediate for energy storage and low-carbon energy carriers, holding a pivotal role in the global chemical economy [1–3]. Currently, industrial-scale ammonia synthesis primarily relies on the Haber–Bosch process. However, this process requires harsh reaction conditions, including high temperatures (>300 °C) and high pressures (>15 MPa), which often lead to significant energy consumption and substantial greenhouse gas emissions [4, 5]. Therefore, improving the traditional industrial synthesis of ammonia and exploring efficient, mild, and sustainable methods for ammonia synthesis holds significant importance for scientific research and industrial applications. In recent years, there has been a growing exploration of nitrogen fixation technologies under mild conditions. Among these, photocatalytic nitrogen fixation technology has emerged as a focal point of research due to its advantages, including mild operating conditions, energy efficiency, and environmental sustainability. Photocatalytic nitrogen fixation is a green synthesis technology that utilizes solar light irradiation on the surface of semiconductor materials to generate redox-active electron–hole pairs, thereby reducing N_2 to NH_3. The photocatalytic nitrogen fixation reaction can be divided into three steps: (i) when the energy of the incident light satisfies the condition $h\nu > Eg$, the semiconductor photocatalyst is excited, resulting in the transition of electrons from the valence band to the conduction band, generating photoinduced electron–hole pairs; (ii) carrier separation, where the electron–hole pairs are separated and migrate to the surface of the catalyst under the influence of the built-in electric field; (iii) the photoinduced electrons that migrate to the active sites on the surface engage in redox reactions with the adsorbed molecules [6–8]. H_2O is oxidized to O_2 via photogenerated holes, while N_2 undergoes a series of reduction steps to combine with

Defect Engineering in Photocatalysis: From Fundamentals to Applications,
First Edition. Edited by Chuanyi Wang, Abdelkader Labidi, and Qiuhui Zhu.

photogenerated electrons and protons to form NH_3. In 1977, Schrauzer et al. [9] first reported the nitrogen fixation capability of titanium dioxide-based photocatalysts under ultraviolet light. This study utilized Fe-doped TiO_2 as the catalyst and successfully synthesized NH_3 through ultraviolet irradiation at room temperature and normal atmospheric pressure. The exploration of photocatalytic nitrogen fixation has been continuously pursued. Despite several decades of research into various photocatalysts for nitrogen reduction reactions, such as metal oxides [10–12], metal sulfides [13, 14], and g-C_3N_4 [15], the majority of these photocatalysts still face significant challenges, including low quantum efficiency, poor stability, a lack of effective active sites, rapid recombination of electron–hole pairs, and inadequate adsorption/activation capabilities. These issues limit the adsorption and activation of N_2, resulting in a solar-to-chemical conversion efficiency of merely 0.1%, [16, 17], which is far from sufficient to meet the demands of large-scale industrial production. To address this issue, researchers are actively exploring modification strategies for photocatalysts to enhance photocatalytic nitrogen fixation performance, with defect engineering being a crucial modification approach.

In a typical photocatalytic reaction, the kinetic characteristics of charge carriers generated by light excitation, including charge carrier migration rates and surface reaction activity, significantly influence the performance of photocatalysts. Defect engineering plays a significant role in optimizing the dynamical behavior of photogenerated charge carriers, which has led to increasing attention on defect engineering for catalyst modification. This approach allows for the regulation of the physicochemical properties and microscopic electronic structures of catalysts, categorized into point defects (such as vacancies and dopants), line defects (such as dislocations), planar defects (such as grain boundaries), and bulk defects (such as lattice disorder and voids), depending on their dimensions. Defects in the bulk are often regarded as limiting factors for the diffusion of photogenerated charge carriers and serve as recombination centers for photoexcited electron–hole pairs. In contrast, surface defects typically act as adsorption sites where the photogenerated charge carriers are transferred to the adsorbed species, thereby further enhancing the separation of charge carriers in photocatalysts.

Defect engineering offers significant advantages in enhancing the photocatalytic nitrogen fixation performance: (i) optimization of the bandgap structure of semiconductor photocatalysts; (ii) modulation of charge carrier transport kinetics to enhance light-induced electron transfer; and (iii) facilitation of nitrogen adsorption and activation on the surface of the semiconductor. Research has shown that morphological defects in photocatalysts can significantly affect the photophysical properties of semiconductors, the energy barriers for nitrogen fixation, and the mobility of photogenerated charge carriers. The incorporation of defects not only provides sufficient catalytic active sites for the adsorption of water and nitrogen molecules but also effectively activates N≡N, introduces defect levels within the material's energy band to broaden the spectral response range, accelerates charge carrier separation, and significantly enhances the lifetime of photogenerated charge carriers, thereby contributing positively to the performance improvement of photocatalytic ammonia synthesis reactions [18–20]. This chapter provides an overview of the types of defects in photocatalytic materials engineered for NH_3 synthesis, highlighting their advantages and challenges. It emphasizes the significance of defect-engineered photocatalytic materials in the process of NH_3 synthesis.

19.2 Principle of Photocatalytic Nitrogen Fixation Reaction

Photocatalytic nitrogen fixation is a process that utilizes photocatalytic semiconductor materials to facilitate the conversion of nitrogen gas molecules into usable ammonia. This process is based on the principles of semiconductor band theory and electron theory, and it encompasses three main steps. Step 1: in the photoexcitation phase, when photons with energy equal to or greater than the bandgap energy are absorbed by the semiconductor, electrons transition from the valence band top to the conduction band. Simultaneously, a hole is created in the valence band, resulting in the generation of electron–hole pairs. Step 2: during the electron–hole migration phase, electrons are transferred to the active sites in the conduction band (CB) of the photocatalyst through either bulk diffusion or the application of an external electric field, while holes are moved to the active sites in the valence band (VB) of the photocatalyst. Step 3: during the phases of photoreduction and photooxidation, the electrons in the conduction band undergo reduction reactions with nitrogen gas to produce ammonia, while the photogenerated holes in the valence band participate in the oxidation of water, yielding protons and molecular oxygen. This three-step process significantly influences the apparent efficiency of photocatalytic nitrogen fixation [21, 22] (Figure 19.1). Therefore, when designing photocatalysts, it is essential to comprehensively consider the redox potentials of the photoexcited conduction band electrons and valence band holes to ensure that they meet the specific requirements for nitrogen reduction reactions and water oxidation reactions.

The photocatalyst should possess a broad light absorption spectrum to generate a greater number of photoexcited electrons and holes, thereby enhancing nitrogen fixation efficiency. Furthermore, efforts should be made to minimize the recombination of photo-induced electrons and holes in the photocatalyst. After the generation of electron–hole pairs, they migrate to the surface of the catalyst. However, many charge carriers tend to recombine, resulting in significant energy dissipation that drastically reduces the utilization efficiency of the electron–hole pairs, ultimately leading to a decrease in reaction efficiency. Currently, commonly employed strategies for extending the lifetime of photogenerated charge carriers include the construction of heterojunctions, dopant incorporation of heteroatoms, and the surface modification of catalysts. In addition, enhancing the nitrogen adsorption capacity of the catalyst and reducing the activation energy for nitrogen molecules is also crucial. In catalytic reactions, the first step involves the adsorption of reactants onto the catalyst surface. In photocatalytic nitrogen fixation, the adsorption of nitrogen is one of the challenging aspects of photocatalytic synthesis. Currently, efficient nitrogen adsorption capacity is achieved primarily by modifying the catalyst surface, such

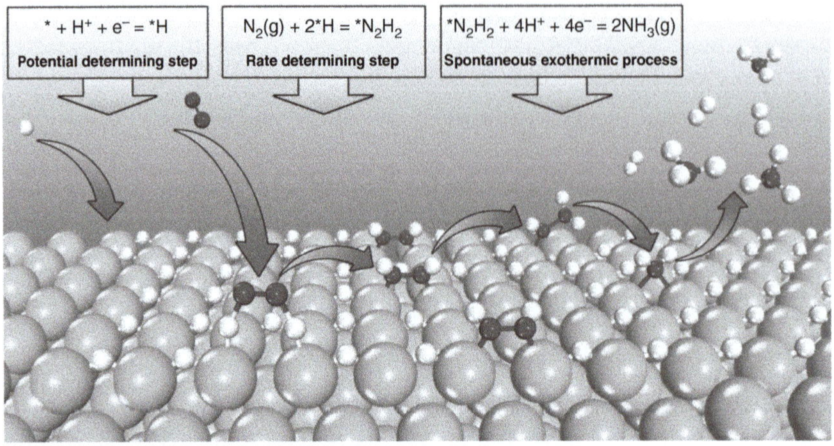

Figure 19.1 Schematic diagram of the nitrogen reduction reaction (NRR) surface hydrogenation mechanism on the catalyst. *Source:* Reproduced from Ling et al. [21] / with permission of American Chemical Society.

as creating surface defects to expose more active sites, thereby ensuring enhanced nitrogen adherence on the catalyst surface [23, 24].

19.3 Roles of Defects in Photocatalytic Nitrogen Reduction

19.3.1 Regulating the Transport Dynamics of Charge Carriers

Carriers play a crucial role in photocatalytic reactions, with defects exerting a profound influence on photocatalytic nitrogen reduction reactions by modulating the transport dynamics of carriers. Efficient carrier migration can reduce recombination losses and enhance reaction activity. Defects can act as traps for charge carriers, thereby extending the effective lifetime of electrons and holes and increasing the number of charge carriers involved in N_2 reduction. For instance, oxygen vacancies and metal doping defects can enhance charge separation. When localized states within the bandgap interact with the valence band or conduction band, they facilitate the migration of charge carriers. Zhang et al. [25] achieved efficient nitrogen fixation under visible light by incorporating oxygen vacancies (OVs) onto BiOBr nanosheets. The test results indicate that the introduction of OVs nearly doubles the average carrier lifetime of BiOBr compared to the standard BiOBr, revealing that the trapping sites induced by OVs significantly extend the carrier lifetime by serving as initial electron acceptors. Moreover, OVs not only facilitate the trapping and transfer of electrons, thereby extending charge carrier lifetimes, but also optimize the transfer efficiency of photogenerated charges to adsorbed nitrogen species, significantly enhancing the rate and selectivity of nitrogen reduction. Defect engineering can reduce the energy barrier of rate-limiting steps in the nitrogen fixation reaction process. In 2022, Di and his colleagues synthesized $Bi_2Sn_2O_7$

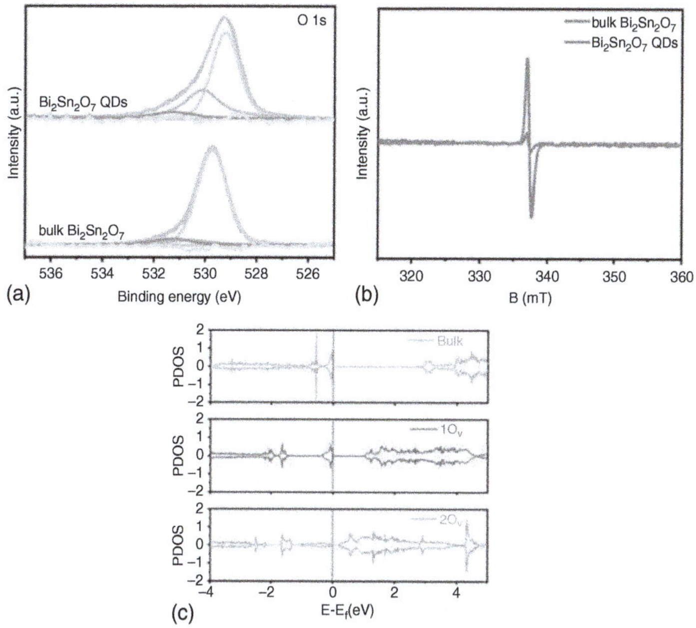

Figure 19.2 (a) Fitted XPS O-1s spectrum. (b) EPR spectrum. (c) Density of states (DOS) curve for the Bi sites in the crystal, along with models containing one and two vacancies in each lattice, denoted as −1OV and −2OV, respectively. *Source:* Reproduced from Zhang et al. [26] / with permission of Elsevier.

with quantum-sized dimensions to tailor the surface active sites, leading to an optimization of the band structure [26]. The results from X-ray photoelectron spectroscopy (XPS) and electron paramagnetic resonance (EPR) indicate the successful introduction of oxygen vacancies into the material (Figure 19.2a, b). The impact of oxygen vacancies on the electronic structure was investigated using density functional theory (DFT) (Figure 19.2c). As the concentration of oxygen vacancies in the material increases, more valence electrons from Bi atoms neighboring the oxygen vacancies are promoted to the conduction band, resulting in an upward shift of the Fermi level and a gradual reduction in the bandgap. The quantum size effect of the material and the vacancy structure facilitate charge migration and nitrogen molecule activation in $Bi_2Sn_2O_7$ quantum dots (QDs), ultimately exhibiting excellent photocatalytic nitrogen fixation performance.

However, if the distribution of defects is uneven or excessive, it can lead to the over-localization and recombination of electrons and holes, thereby reducing photocatalytic efficiency. Therefore, a rational design of defects must not only consider their quantity but also focus on their distribution and types to achieve optimum photocatalytic nitrogen fixation performance.

19.3.2 Engineering the Band Gap Structure of Catalysts

Defects in photocatalytic materials can significantly extend the range of light response by generating new energy states. The bandgap of the material determines

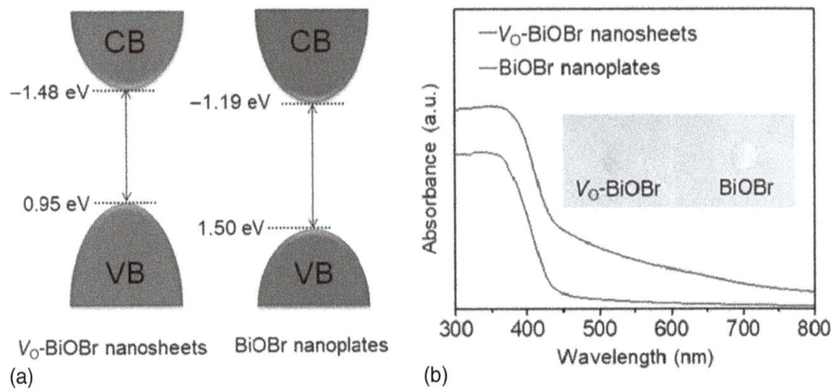

Figure 19.3 (a) Energy level alignment of VO-BiOBr nanosheets and BiOBr nanoplates. (b) UV–vis DRS spectra of VO-BiOBr nanosheets and BiOBr nanosheets. *Source:* Reproduced from Xue et al. [27] / with permission of American Chemical Society.

its ability to absorb light, and conventional photocatalysts typically only function effectively in the ultraviolet region. Defects, such as metal doping, oxygen vacancies, and lattice imperfections, can generate "electronic states" within the bandgap. This facilitates the narrowing of the bandgap and serves as mid-gap states for photoexcited electrons, thereby broadening the range of light responsiveness. In 2018, Zhong and his colleagues investigated the impact of oxygen vacancies on the photocatalytic nitrogen fixation properties of BiOBr nanosheets [27]. After the introduction of oxygen vacancies, VO-BiOBr exhibits significantly enhanced visible light absorption, reduced band gap width, and altered charge transport characteristics compared to defect-free BiOBr nanosheets. After the introduction of oxygen vacancies, the bandgap of BiOBr nanosheets decreased from 2.69 to 2.43 eV (Figure 19.3a). Additionally, the optical absorption edge redshifted to 470 nm, which is 30 nm longer than that of the BiOBr nanosheets. In addition, VO-BiOBr nanosheets exhibit enhanced light absorption capabilities within the visible light range of 450–750 nm (Figure 19.3b). The positioning of the intermediate energy states can be modulated by adjusting the type and concentration of defects. When introducing defects, it is essential to balance their corresponding reduction and oxidation capabilities to optimize catalytic performance and enhance the overall efficiency of the photocatalytic system.

19.3.3 Promote the Adsorption and Activation of N_2

The adsorption of nitrogen molecules is a crucial step in the photocatalytic nitrogen reduction reaction. Defects on the surface of the catalyst can interact strongly with N_2, thereby enhancing the adsorption strength. Defects can enhance the activation of N_2 by modifying the electronic structure of the catalyst, thereby increasing its interaction with N_2. Defects, such as metal doping, can effectively weaken the N≡N triple bond and reduce the activation energy. In 2022, Zeng et al. reported a nitrogen-doped carbon nanotube array modified with dual defects for photocatalytic

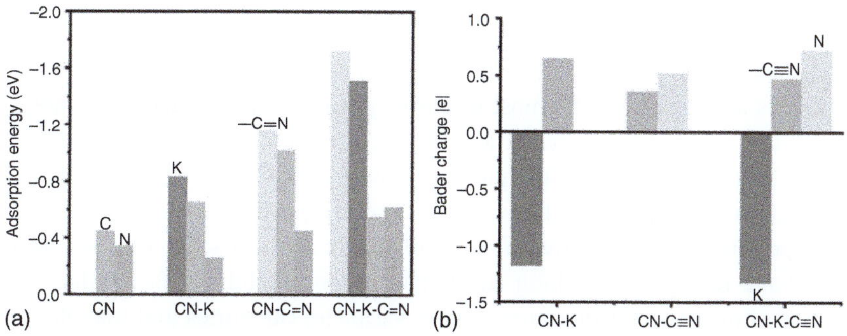

Figure 19.4 (a) Adsorption energies of different samples for N_2. (b) Charge densities at various positions in nitrogen-doped g-C_3N_4. *Source:* Reproduced from Liu et al. [28] / with permission of Elsevier.

nitrogen fixation, achieving an ammonia production rate of 23.5 mmol h^{-1} g^{-1}, which represents a two-order of magnitude enhancement compared to conventional nitrogen-doped carbon [28]. The theoretical calculation results indicate that nitrogen is more favorably adsorbed at the K^+ and —C≡N sites rather than at the C and N sites. This suggests that the introduced defects can effectively enhance the adsorption of nitrogen (Figure 19.4a). In the dual-defect structure constructed from materials, the non-metallic defect C≡N activates nitrogen molecules through an electron-donating effect, while the metallic defect K^+ attracts photogenerated electrons. The synergy between these two defects facilitates the activation of nitrogen and the separation of photogenerated charge carriers, resulting in a reduced energy barrier for nitrogen activation and an enhanced ammonia production rate (Figure 19.4b). However, excessive defects may lead to non-selective adsorption or diminish the overall stability of the catalyst. Additionally, defects can cause uneven distribution of electronic density on the catalyst surface, thereby negatively impacting the selectivity of the reaction. Therefore, it is crucial to strike a balance between the benefits derived from introducing defects and the potential decrease in catalyst performance that may result from these defects when optimizing the adsorption capacity of the catalyst.

19.4 Defects in Photocatalysts for Enhanced Nitrogen Photoreduction

Defect engineering involves the introduction of defects, such as dopant atoms and lattice vacancies, into materials to modulate the crystal and electronic structures of catalysts, enhance charge transfer, and adjust the surface adsorption properties of reaction intermediates, thereby influencing various characteristics of crystalline materials, including their optical, electrical, and magnetic properties. The intricately designed defect structures can modify the band edge and band gap energy of photocatalysts, enhancing the photogenerated electron transfer and demonstrating significant potential for promoting nitrogen gas adsorption and activation.

19.4.1 Oxygen Vacancies

Oxygen vacancies are commonly found on the surfaces of nano-sized transition metal oxides and represent the most extensively studied type of defect in photocatalysts. The semiconductor materials with oxygen vacancies possess abundant localized electrons in oxides. Consequently, in most photocatalytic reactions, semiconductor catalysts enriched with oxygen vacancies frequently exhibit exceptional catalytic performance due to the presence of highly active reactive sites. Oxygen vacancies and nitrogen gas exhibit strong adsorption capabilities, facilitating the adsorption of nitrogen. Furthermore, oxygen vacancies can attract and trap photogenerated electrons, thereby enhancing the separation rate between electrons and holes, which significantly accelerates the reduction reaction rate of nitrogen. Hence, it is evident that oxygen vacancies play a crucial role in enhancing the activity and rate of nitrogen fixation reactions in photocatalysts. In 2020, Li et al. [29] synthesized a series of Bi_5O_7Br nanostructures through a simple low-temperature thermal treatment process for photocatalytic nitrogen fixation. The synthesized Bi_5O_7Br nanotubes exhibit a high concentration of OVs. Electron spin resonance (ESR) measurements indicate that the substantial presence of OVs facilitates the separation of electron–hole pairs, which is advantageous for the adsorption and fixation of N_2. Experimental results indicate that the Bi_5O_7Br nanotube samples prepared at 40 °C exhibit the highest electron transfer rates, attributed to the presence of a high concentration of oxygen vacancies. The photocatalytic nitrogen fixation rate achieved is 12.72 mM g^{-1} h^{-1}. Simultaneously, researchers established both oxygen-rich and oxygen-deficient models to further validate the critical role of oxygen vacancy concentration variations in influencing nitrogen fixation activity. In 2023, Huang and colleagues achieved remarkable piezoelectric photocatalytic ammonia synthesis by synergizing defects with the piezoelectric field for the first time [30]. This experiment altered the local dipole states of $BaTiO_3$ by modifying the oxygen vacancies, thereby enhancing the piezoelectric polarization of $BaTiO_3$ and facilitating the separation of photogenerated charge carriers. Furthermore, the unpaired single electrons in the d orbitals of Ti^{3+} sites surrounding the oxygen vacancies facilitate the adsorption and activation of N_2 molecules through a d-π feedback mechanism, thereby enabling efficient photocatalytic nitrogen fixation. For this reason, $BaTiO_3$ with an appropriate number of oxygen vacancies can achieve a nitrogen fixation rate of up to 106.7 μmol g^{-1} h^{-1}. This study provides a new example of utilizing the synergistic effect of defects and piezoelectric fields to enhance the photocatalytic activity for ammonia synthesis.

Despite the advantages of oxygen vacancies in enhancing catalytic activity, improving light absorption, and facilitating electron transfer, semiconductor catalytic materials containing oxygen vacancies also face challenges such as stability, poor selectivity, material degradation, and synthesis difficulties. Therefore, further investigation and improvement regarding the synthesis and practical applications of oxygen-deficient semiconductor materials as photocatalysts for ammonia synthesis are warranted.

19.4.2 Nitrogen Vacancies

Nitrogen vacancies (NVs), a prevalent type of anion vacancy, not only facilitate electron capture leading to a downward shift in the conduction band and a reduction in electron–hole recombination but also exhibit a higher affinity for N_2 molecules due to their structural and size similarities with nitrogen atoms, thereby promoting selective chemical adsorption and activation of N_2. To generate nitrogen vacancies, it is essential to identify suitable base materials that contain nitrogen atoms that can be easily segregated while maintaining stable catalytic performance even in the presence of nitrogen defects. Graphitic carbon nitride (g-C_3N_4) has emerged as an excellent candidate for research, garnering extensive investigation in recent years. In 2020, the Xue research group and colleagues developed ultra-thin porous PFL-g-C_3N_4 with nitrogen vacancies using a simple molecular self-assembly method. This material was utilized in the study of photocatalytic nitrogen fixation [31]. The test results indicate that the surface N/C atomic ratio decreases from 1.33 (g-C_3N_4) to 0.87 (PFL-g-C_3N_4), suggesting the presence of nitrogen vacancies in PFL-g- C_3N_4. The existence of these nitrogen vacancies in PFL-g-C_3N_4 was further confirmed by EPR spectroscopy (Figure 19.5a). Although g-C_3N_4 and PFL-g-C_3N_4 exhibit similar single Lorentzian spectral lines, the signal intensity of PFL-g-C_3N_4 is significantly enhanced compared to pure g- C_3N_4, providing further validation of the nitrogen vacancies. The steady-state photoluminescence results indicate that

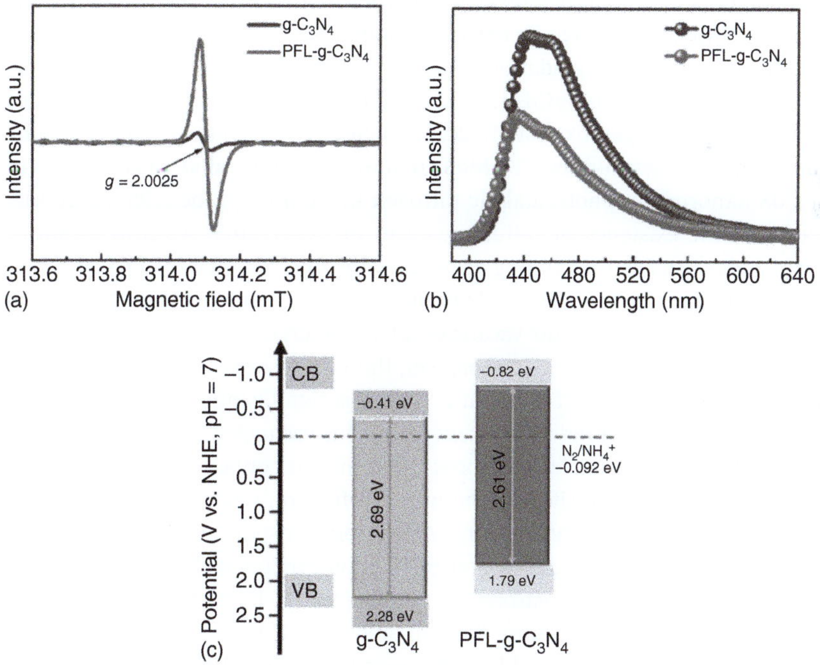

Figure 19.5 (a) The room temperature EPR spectra of g-C_3N_4 and PFL-g-C_3N_4. (b) The stable PL spectra of g-C_3N_4 and PFL-g-C_3N_4. (c) The band structure of g-C_3N_4 and PFL-g-C_3N_4. *Source:* Reproduced from Xue et al. [31] / with permission of Elsevier / CC BY-NC-ND 4.0.

the PL intensity of PFL-g- C_3N_4 is significantly reduced compared to that of g- C_3N_4 (Figure 19.5b). This suggests that PFL-g- C_3N_4 can more effectively suppress the recombination rate of photogenerated charge carriers. Furthermore, it is evident that nitrogen vacancies also play a crucial role in promoting the separation and transfer of photogenerated electrons and holes. In addition, compared to pure g-C_3N_4 the conduction band (CB) of PFL-g- C_3N_4 exhibits a more negative potential, indicating that the presence of nitrogen vacancies enables PFL-g- C_3N_4 to shift toward a more negative conduction band, thereby enhancing the efficiency of photocatalytic nitrogen fixation (Figure 19.5c). Consequently, the nitrogen fixation rate of PFL-g- C_3N_4 can reach 8.20 mM h^{-1} g^{-1}, representing an approximately 2.8-fold improvement in the photocatalytic nitrogen fixation rate compared to pure g- C_3N_4.

In summary, nitrogen vacancies selectively adsorb and activate N_2, significantly enhancing the separation efficiency of photogenerated charge carriers, resulting in increased photogenerated electrons and facilitating the transfer of photogenerated electrons from the photocatalyst to adsorbed nitrogen, thus demonstrating extensive application potential in photocatalytic nitrogen fixation catalysts.

19.4.3 Sulfur Vacancies

Sulfur vacancies (SVs), analogous to oxygen vacancies and nitrogen vacancies, can significantly enhance photoresponse characteristics, facilitate the separation of electron–hole pairs, and accelerate charge carrier migration. In recent years, their potential in NRR has garnered increasing attention, alongside their applications in HER [32] and pollutant degradation research [33, 34]. As of now, sulfides such as In_2S_3 [35], MoS_2[14] and ZnMoCds [36] have demonstrated exceptional capabilities in photocatalytic nitrogen fixation. In 2020, the Sun research group reported sulfur vacancy-rich and oxygen-doped 1T-MoS_2 nanosheets (denoted as SV-1T-MoS_2) to assist CdS nanorods in photocatalytic nitrogen fixation [37]. The XPS test results indicate that the existence of sulfur vacancies in SV-1T-MoS_2 leads to a shift of approximately 0.2 eV in the binding energies of the metallic phase (1T) S $2p_{3/2}$ and S $2p_{1/2}$, in comparison to 1T-MoS_2. This shift has been validated through calculations, demonstrating that sulfur vacancies affect the crystallographic properties of the metallic phase (Figure 19.6a). Moreover, the ESR test results indicate that the signal peak intensity of SV-1T-MoS_2 at $g = 2.004$ is significantly lower than that of 1T-MoS_2, further confirming the presence of sulfur vacancies (Figure 19.6b). The abundance of sulfur vacancies enhances the available contact area and provides a greater number of active sites for the nitrogen fixation reaction. Additionally, it can improve the ability of photogenerated electron transport and surface electron mobility, thereby facilitating the separation of photogenerated electrons and holes. Based on the aforementioned reasons, the synthesized SV-1T-MoS_2/Cd exhibited an exceptional photocatalytic nitrogen fixation efficiency of 8220.83 μmol L^{-1} h^{-1} g^{-1} (Figure 19.6c). In 2016, Hu et al. [38] synthesized a novel g-C_3N_4/ZnSnCdS hybrid heterojunction photocatalyst via a hydrothermal method. This catalyst demonstrates outstanding nitrogen fixation capabilities under nitrogen gas and exhibits excellent performance under visible light illumination. The catalyst, composed of

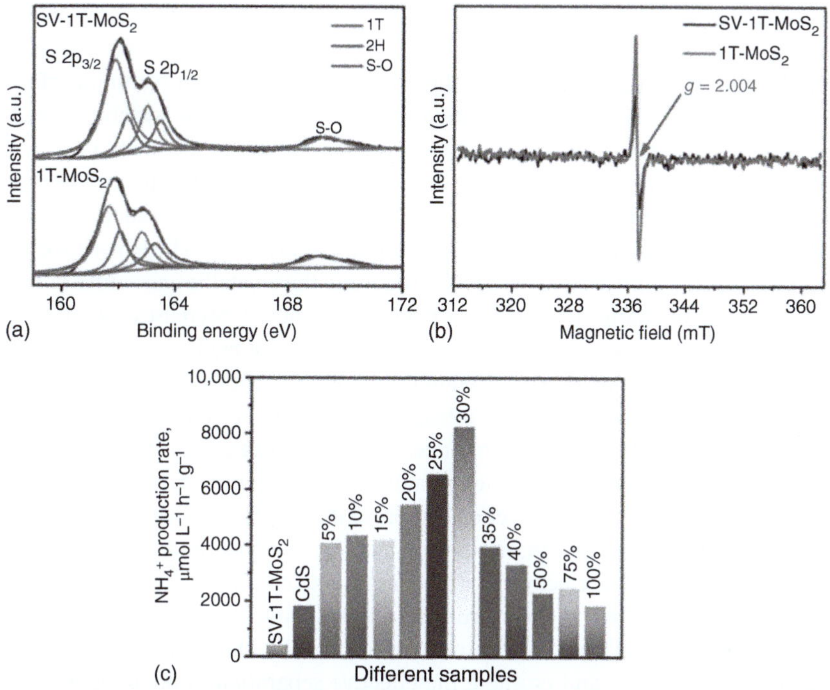

Figure 19.6 (a) The XPS spectra of SV-1T-MoS_2 and 1T-MoS_2. (b) The ESR spectra of SV-1T-MoS_2 and 1T-MoS_2. (c) Photocatalytic ammonia production rates for all catalysts. *Source:* Reproduced from Sun et al. [37] / with permission of American Chemical Society.

various metal sulfides with abundant sulfur vacancies, facilitates the adsorption and activation of nitrogen molecules while promoting interfacial charge transfer, significantly enhancing the capacity for nitrogen photofixation. Moreover, the sulfur vacancies in ZnSnCdS not only serve as active sites for the adsorption and activation of N_2 molecules but also facilitate the charge transfer at the interface between g-C_3N_4 and N_2 molecules, significantly enhancing nitrogen fixation capability. Under the synergistic effect of heterojunctions and sulfur vacancies, the prepared heterojunction photocatalyst with 80% ZnSnCdS exhibits the highest NH^{4+} yield under visible light, surpassing that of pure g-C_3N_4 and ZnSnCdS by 33.2 times and 1.6 times, respectively.

Sulfur, as a crucial non-metal element in nitrogenase enzymes, plays a significant role in biological nitrogen fixation. Additionally, sulfur vacancies serve as the primary reaction centers in semiconductor catalysts, facilitating the separation of electron–hole pairs and accelerating charge carrier migration. Thus, transforming sulfur vacancies into sulfur-containing catalysts and exploring their applications in NRR presents a promising avenue for research

19.4.4 Other Types of Defects and Synergism of Multifold Defects

In addition to lattice vacancies, doping with foreign atoms is also an effective approach for defect engineering. The introduction of external atoms into the material can significantly alter its electronic structure and surface properties. Currently, there have been reports on the incorporation of elements such as Ce, V, Co, Mo, and Ni, which can enhance the photocatalytic nitrogen fixation performance of semiconductor materials. Among these, Fe^{3+} has been widely utilized in the modification research of nitrogen-fixing photocatalysts due to its low cost and excellent effectiveness. Sun et al. [39] reported a highly efficient iron-doped photocatalytic nitrogen fixation material, Bi_2MoO_6, elucidating the crucial role of iron doping in enhancing the photocatalytic activity of Bi_2MoO_6. The experimental results indicate that with the increase in iron content, the crystalline structure of the material remains unchanged, while the surface work function decreases, facilitating electron migration and separation. The presence of iron also facilitates the effective storage and separation of photo-generated charges, significantly enhancing the photocatalytic reduction efficiency of nitrogen. Notably, 0.5% Fe-Bi_2MoO_6 exhibits the highest nitrogen fixation performance, with a nitrogen fixation rate under visible light irradiation that is 3.7 times greater than that of pure Bi_2MoO_6.

The construction of heterojunctions represents an alternative strategy for introducing defects through interface engineering. Doping and point defects can effectively adjust the electronic structure characteristics of heterojunctions, resulting in alterations to the band structure and subsequently influencing photocatalytic performance. These heterojunctions exhibit the characteristics of direct bandgap semiconductors and promote the effective separation of photo-generated electron–hole pairs by forming type-II band alignment, thereby enhancing photocatalytic performance. He et al. [40] reported a novel $KNbO_3/Bi_4O_5Br_2$ heterojunction catalyst for photocatalytic nitrogen fixation. The analysis of the energy band structure indicates that $KNbO_3$ and $Bi_4O_5Br_2$ exhibit suitable energy band potentials for the formation of a Type II heterojunction. Due to the Fermi level of $KNbO_3$ being higher than that of $Bi_4O_5Br_2$, the contact between the two materials induces electron migration, establishing a built-in electric field directed from $KNbO_3$ to $Bi_4O_5Br_2$. This electric field enhances the transfer of electrons, thereby improving the efficiency of the catalyst. Consequently, this catalyst demonstrates a high photocatalytic activity. Gao et al. synthesized a unique sandwich-structured tubular BiOBr/g-C_3N_4 photocatalyst using a simple surfactant-assisted solvothermal method. The compatible band structures of g-C_3N_4 and BiOBr form a Type II heterojunction that facilitates the separation of electrons and holes. As a result, the BiOBr/g-C_3N_4 composite exhibits high stability and excellent photocatalytic nitrogen fixation performance under visible light irradiation [41].

In addition to addressing the aforementioned defects, the utilization of the interactions and synergistic effects of multiple defects to regulate the performance of catalysts provides a novel strategy for optimizing the performance of photocatalysts in photocatalytic nitrogen fixation reactions. By meticulously designing multiple lattice defects, such as doped impurity atoms, lattice vacancies, and the construction of heterojunctions, the electronic structure and surface properties of photocatalysts can be effectively tuned to enhance the transfer of photoexcited electrons and the adsorption capacity of intermediates. This promises to achieve higher ammonia synthesis rates and improved photocatalytic efficiency, thereby advancing the application of photocatalytic nitrogen fixation technology in sustainable ammonia production.

19.5 Summary and Future Perspectives

Ammonia is a crucial chemical raw material that plays an essential role in the production of modern agricultural fertilizers, significantly contributing to the advancement of human society. Photocatalytic nitrogen fixation offers a greener pathway for the synthesis of NH_3. In the research of photocatalytic nitrogen fixation, the application and regulation of defects play a crucial role in enhancing catalytic performance. Defect engineering contributes to the expansion of the light response range, improvement of the separation efficiency and migration capacity of photogenerated electrons and holes, as well as optimization of the adsorption and activation processes of nitrogen. Despite significant advancements in nitrogen photonic catalysis technology over the past five years, its solar energy conversion efficiency has yet to reach 1%, falling far short of the minimum 10% standard required for industrial application. Therefore, the precise control of defect type, quantity, and localization is crucial for the rational design and construction of efficient defect-engineered photocatalysts. This article reviews the types of defects in defect-engineered photocatalytic materials used for NH_3 synthesis and their prospective applications in the field of photocatalysis. Although defect catalysts have demonstrated promising performance in the photocatalytic NRR, they still encounter numerous challenges. To advance the development in this field, research can be conducted in several key areas.

1) The controllability of defects.
   The controllability of defect types, concentrations, and positions is a key factor in enhancing the performance of photocatalysts. Although current technologies still face challenges in precisely introducing specific defect types and concentrations, which impose limitations on catalytic activity, the potential for fine-tuning defects can be realized through selective doping of heteroatoms or by optimizing synthetic processes. For example, the adoption of different doping strategies (such as p-type or n-type doping) can tune the electronic structure of materials, thereby optimizing catalytic performance. Additionally, investigating the interactions and synergistic effects between defects, as well as their specific contributions during nitrogen adsorption and activation processes, will provide crucial theoretical insights for catalyst design. Further research addressing these issues will enable us to design more targeted catalysts, thereby enhancing reaction efficiency.
2) Stability of Defects
   The stability of defects is a critical factor that impacts the long-term activity and performance of photocatalysts. Specifically, certain anionic defects, such as OVs, are prone to oxidation during photocatalytic processes, which can lead to a deterioration in catalytic performance. To address this issue, the development

of effective stabilization strategies becomes particularly essential. Research indicates that utilizing acceptor-type dopants can enhance defect stability by modifying the atomic environment of the material. Moreover, anchoring single atoms at defect sites and constructing heterogeneous structures can effectively suppress the oxidation of defects while also improving the overall performance of the photocatalyst during photocatalytic processes.

3) Multi-strategy Coupled Design
Integrating defect engineering with other catalyst modification techniques, such as constructing Z-scheme heterostructures, can effectively achieve spatial separation of ammonia generation and water oxidation sites, thereby reducing the occurrence of side reactions while optimizing reaction efficiency. The Z-scheme structure not only facilitates enhanced separation of photogenerated charge carriers but also ensures that ammonia and oxides are directed toward distinct reaction sites, thereby improving selectivity.

Conflict of Interest

The authors declare that they have no known competing financial interests or personal relationships that could have appeared to influence the work reported in this chapter.

References

1 Foster, S.L., Bakovic, S.I.P., Duda, R.D. et al. (2018). *Nat. Catal.* 1: 490–500.
2 Shi, L., Yin, Y., Wang, S., and Sun, H. (2020). *ACS Catal.* 10: 6870–6899.
3 McEnaney, J.M., Singh, A.R., Schwalbe, J.A. et al. (2017). *Energy Environ. Sci.* 10: 1621–1630.
4 Oshikiri, T., Ueno, K., and Misawa, H. (2016). *Angew. Chem. Int. Ed.* 55: 3942–3946.
5 van der Ham, C.J.M., Koper, M.T.M., and Hetterscheid, D.G.H. (2014). *Chem. Soc. Rev.* 43: 5183–5191.
6 Wu, H., Tan, H.L., Toe, C.Y. et al. (2020). *Adv. Mater.* 32: 32.
7 Li, X.-H., Li, H., Jiang, S.-L. et al. (2023). *ACS Catal.* 13: 7189–7198.
8 Seefeldt, L.C., Yang, Z.-Y., Lukoyanov, D.A. et al. (2020). *Chem. Rev.* 120: 5082–5106.
9 Schrauzer, G.N. and Guth, T.D. (1977). *J. Am. Chem. Soc.* 99: 7189–7193.
10 Rong, X., Chen, II., Rong, J. et al. (2019). *Chem. Eng. J.* 371: 286–293.
11 Zhang, S., Zhao, Y., Shi, R. et al. (2021). *Angew. Chem. Int. Ed.* 60: 2554–2560.
12 Huang, Y., Zhu, Y., Chen, S. et al. (2021). *Adv. Sci.* 8: 2003626.
13 Sultana, S., Mansingh, S., and Parida, K.M. (2019). *J Mater Chem A* 7: 9145–9153.
14 Sun, S., Li, X., Wang, W. et al. (2017). *Appl Catal B* 200: 323–329.
15 Liu, S., Wang, S., Jiang, Y. et al. (2019). *Chem. Eng. J.* 373: 572–579.

16 Shen, H., Yang, M., Hao, L. et al. (2021). *Nano Res* 15: 2773–2809.
17 Chen, X., Li, N., Kong, Z. et al. (2018). *Mater. Horiz.* 5: 9–27.
18 Xiong, J., Di, J., Xia, J. et al. (2018). *Adv. Funct. Mater.* 28: 1801983.
19 Xiong, J., Song, P., Di, J., and Li, H. (2020). *Chem. Eng. J.* 402: 126208.
20 Cheng, M., Xiao, C., and Xie, Y. (2019). *J Mater Chem A* 7: 19616–19633.
21 Ling, C., Zhang, Y., Li, Q. et al. (2019). *J. Am. Chem. Soc.* 141: 18264–18270.
22 Hao, D., Liu, Y., Gao, S. et al. (2021). *Mater. Today* 46: 212–233.
23 Cui, X., Tang, C., and Zhang, Q. (2018). *Adv. Energy Mater.* 8: 1800369.
24 Zhang, S., Zhao, Y., Shi, R. et al. (2019). *EnergyChem* 1: 100013.
25 Li, H., Shang, J., Ai, Z., and Zhang, L. (2015). *J. Am. Chem. Soc.* 137: 6393–6399.
26 Zhang, Y., Di, J., Qian, X. et al. (2021). *Appl Catal B* 299.
27 Xue, X., Chen, R., Chen, H. et al. (2018). *Nano Lett.* 18: 7372–7377.
28 Liu, G., Tang, Z., Gu, X. et al. (2022). *Appl Catal B* 317: 121752.
29 Li, P., Zhou, Z., Wang, Q. et al. (2020). *J. Am. Chem. Soc.* 142: 12430–12439.
30 Yuan, J., Feng, W., Zhang, Y. et al. (2024). *Adv. Mater.* 36: 36.
31 Xue, Y., Kong, X., Guo, Y. et al. (2020). *J. Mater.* 6: 128–137.
32 Fang, Z., Weng, S., Ye, X. et al. (2015). *Mater. Interfaces* 7: 13915–13924.
33 Lee, J., Ham, S., Choi, D., and Jang, D.-J. (2018). *Nanoscale* 10: 14254–14263.
34 Wang, L., Xu, X., Wang, Y. et al. (2018). *Inorg. Chem. Front.* 5: 1470–1476.
35 He, Z., Wang, Y., Dong, X. et al. (2019). *RSC Adv.* 9: 21646–21652.
36 Zhang, Q., Hu, S., Fan, Z. et al. (2016). *Dalton Trans.* 45: 3497–3505.
37 Sun, B., Liang, Z., Qian, Y. et al. (2020). *Mater. Interfaces* 12: 7257–7269.
38 Hu, S., Li, Y., Li, F. et al. (2016). *ACS Sustain. Chem. Eng.* 4: 2269–2278.
39 Meng, Q., Lv, C., Sun, J. et al. (2019). *Appl Catal B* 256: 117781.
40 Yue, L., Zeng, Z., Ren, X. et al. (2024). *Front. Chem. Sci. Eng.* 18.
41 Gao, K., Zhang, C., Zhu, H. et al. (2023). *Chem. Eur. J.* 29.

20

Summary

Atif Sial[1], Abdelkader Labidi[1], Haitao Ren[2], Ting Gao[1], and Chuanyi Wang[1,3]

[1] *Shaanxi University of Science and Technology, School of Environmental Science and Engineering, Xi'an, 710 021, PR China*
[2] *Xijing University, Xi'an Key Laboratory of Advanced Photo-Electronics Materials and Energy Conversion Device, Technological Institute of Materials & Energy Science (TIMES), Xi'an, 710 123, China*
[3] *Department of Environmental and Sustainable Engineering, Faculty of Engineering, Chulalongkorn University, 254 Phayathai Road, Pathumwan, Bangkok, 10330, Thailand*

20.1 Role of Defects in Photocatalysis

Defects, which can include vacancies, interstitials, impurities, and surface anomalies, are often viewed as detrimental in many materials. However, this book primarily highlights the beneficial potential of vacancy creation and doping with metal and nonmetal elements as forms of defect engineering when strategically introduced into photocatalytic materials. It also emphasizes their crucial role in influencing the physicochemical properties of these materials during photocatalytic reactions, particularly in the fields of energy conversion and environmental remediation applications. Throughout the book, it has been established that defects can significantly influence the generation, separation, and recombination of charge carriers, which are critical for photocatalytic activity. The incorporation of defects leads to a more favorable charge carrier dynamics, effectively enhancing the overall efficiency of photocatalytic reactions. By systematically analyzing the roles of various defects, this chapter aims to provide a holistic understanding of the current state of research, its implications, and future directions.

Photocatalysis serves as a highly effective and straightforward approach for the conversion of chemical energy and the remediation of environmental pollutants [1]. A pivotal aspect of advanced solar absorption and generation technologies lies in the innovation of a single photocatalyst capable of operating across an extensive range of radiation while maintaining suitable redox energy levels [2]. Consequently, the strategic optimization of the photocatalyst's structure for applications in energy and environmental sectors is critically significant. In this context, defects in semiconductor materials can significantly influence the performance of photocatalysts. These defects can manifest as either surface or bulk imperfections, each located within distinct regions of the material's structure. For multicomponent photocatalytic systems,

Defect Engineering in Photocatalysis: From Fundamentals to Applications,
First Edition. Edited by Chuanyi Wang, Abdelkader Labidi, and Qiuhui Zhu.

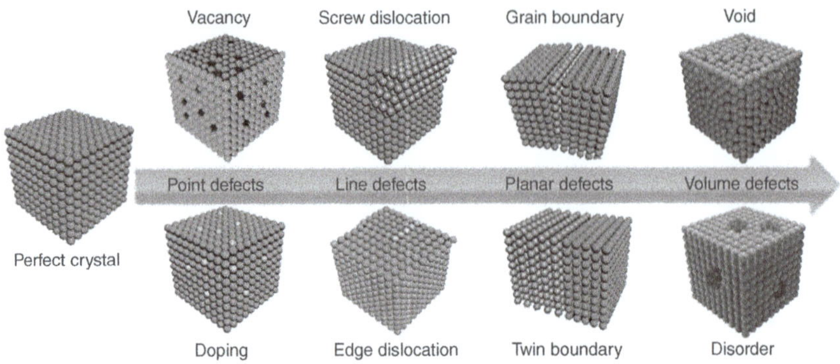

Figure 20.1 Schematic representations depicting the diverse range of defects present within photocatalytic materials. *Source*: Reproduced from Lei et al. [3] / with permission of Elsevier.

the interface formed by the face-to-face contact of different components introduces interfacial defects, which typically arise from the surface imperfections of the individual materials involved. When considering the dimensions of these defects, photocatalytic materials can be categorized into four primary types: zero-dimensional defects, commonly referred to as point defects (including vacancies and dopants); one-dimensional defects, known as line defects, which include screw dislocations and edge dislocations; two-dimensional defects, called planar defects, such as grain boundaries and twin boundaries; and finally, three-dimensional defects, characterized by volume lattice disorder and voids within the material's structure. This classification framework is illustrated in Figure 20.1, providing a comprehensive overview of how defects can be systematically organized within photocatalytic materials.

An architecture abundant in defects and cost-effective in nature emerges as a highly viable candidate for enhancing water-splitting and pollutant degradation processes. For instance, defective g-C_3N_4 has demonstrated a remarkable enhancement in H_2 production, which can be ascribed to the substitution of surface hydrogen in the amino groups with KSCN (potassium thiocyanate). This alteration facilitates the introduction of cyanamide defects, thereby creating coordinating sites that are instrumental in catalytic reactions [4]. Comparably, treating g-C_3N_4 with $NaBH_4$ (as a reducing agent) can effectively create nitrogen defects on its surface [5]. Additionally, cyanamide defects can be generated at the -C-NH_2 position of g-C_3N_4 via alkali-assisted thermal polymerization, which facilitates the necessary band position [6]. However, in acidic conditions, the hydrolysis of these cyanamide defects results in the synthesis of urea-modified g-C_3N_4. Therefore, these methodologies demand subsequent chemical treatments for the precise engineering of defects within the g-C_3N_4 structure. Likewise, pit-engineered ultrathin BiOCl nanosheets demonstrate significantly enhanced O_2 evolution activity compared to their bulk counterparts [7]. In contrast, Wang and his team synthesized a novel photocatalyst, SA-Fe@TiO_2, by anchoring single-atom Fe sites onto nitrogen-doped biochar and coupling it with anatase TiO_2, illustrating the transformative impact of defect engineering on enhancing surface reactivity and photocatalytic antibiotic degradation performance. The strategic incorporation of single-atom Fe generated shallow-state defects that facilitated effective charge carrier separation and transfer, significantly improving the photocatalytic activity. As a result of the uniform distribution of TiO_2 and the strong electronic interactions between the TiO_2 NPs and single-atom Fe sites, the SA-Fe@TiO_2 catalyst exhibited a remarkable 4.3-fold increase in the degradation rate of sulfamethoxazole, showcasing its robust ability to generate reactive species such as h^+, •OH, and $•O_2^-$ during photocatalytic processes [8]. Besides, various defect types are recognized as crucial contributors to photocatalytic reactions. For instance, oxygen vacancies (OVs) in In_2O_3, pit-engineered WO_3 nanosheets, and structural distortions in ZnSe, SnS_2, and Sn_S all play significant roles in enhancing catalytic efficiency [9]. Following the analysis previously discussed, Table 20.1 delineates the primary functions attributed to defects within photocatalytic processes.

20.2 Advantages of Defect Engineering

20.2.1 Tailored Properties

One of the primary advantages of defect engineering is the ability to tailor the physical and chemical properties of defective nanomaterials for photocatalytic reactions. By strategically introducing defects into the crystal lattice, researchers can modify the material's electronic structure, bandgap, and surface characteristics. For instance, the work by Bian and colleagues demonstrated that defect engineering, particularly through the intentional incorporation of dopants like Ca^{2+}, has emerged as a pivotal strategy for enhancing the photocatalytic efficiency of TiO_2-based materials in H_2 evolution. The introduction of Ca^{2+} facilitates the formation of OVs, which serve as active sites for electron capture, thereby mitigating the recombination of photogenerated charge carriers. These defects not only modify the electronic structure, improving light absorption properties, but also enhance the separation and mobility of charge carriers, as evidenced by increased photocurrent responses and time-resolved photoluminescence studies [43]. In another study, He and his colleagues investigated the role of defect engineering in lanthanum ferrite ($LaFeO_3$) perovskite oxides to enhance surface reactivity and photocatalytic performance for the degradation of tetracycline hydrochloride (TC) using a photo-Fenton process. By systematically reducing the La content, they created La-deficient samples, notably the 0.7LFO variant, which exhibited a significant increase in photocatalytic activity due to the introduction of La defects that facilitated OVs and improved charge transfer dynamics. Characterization techniques indicated that these defects altered the optoelectronic properties, leading to a 93% TC degradation rate, far surpassing that of pure $LaFeO_3$. The presence of reactive species such as h^+, •OH, and $•O_2^-$ was confirmed as a key contributor to the enhanced degradation process [44]. Similarly, Hoang et al. reported in their recent study that the strategic incorporation of OVs through Li-ion deposition in metal-oxide nanoparticles (MO NPs) such as TiO_2, SnO_2, and CeO_2 has been shown to significantly enhance

Table 20.1 A comprehensive overview of various defective photocatalysts, their underlying mechanisms, and respective photocatalytic applications.

Defective materials	Defect-engineering type and role	Interaction mechanism	Photocatalytic application	References
WO_3	OVs: light absorption modulation	Creation of new energy levels	CO_2 reduction	[10]
WO_3	OVs: charge transfer and separation modulation	Formation of lower energy levels to trap carriers in the Z-scheme	CO_2 reduction	[11]
g-C_3N_4	C-vacancies: charge transfer and separation modulation	Increased carrier density	CO_2 reduction	[12]
TiO_2	OVs: charge transfer and separation modulation	Prolonged lifetime of photo-generated carriers	CO_2 reduction	[13]
BiOBr	Bi-vacancies: surface reaction modulation	Enhanced CO_2 adsorption and chemical activation	CO_2 reduction	[14]
ZnS	S-vacancies/Zn-dopant: surface reaction modulation	Enhanced CO_2 adsorption and chemical activation	CO_2 reduction	[15]
SnS_2-C	C-dopants: surface reaction modulation	Enhanced CO_2 adsorption and chemical activation	CO_2 reduction	[16]
TiO_2	OVs/Mo-dopants: surface reaction modulation	Enhanced CO_2 adsorption and chemical activation	CO_2 reduction	[17]
TiO_2	Cu-dopant: surface reaction modulation	Enhanced CO_2 adsorption and chemical activation	CO_2 reduction	[18]
Fe_2O_3	V-dopant: light absorption modulation	Reduced band-gap	Water-splitting	[19]
C_3N_4	N-vacancies: light absorption modulation	Reduced band-gap	Water-splitting	[20]
TiO_2	B-dopants: light absorption modulation	Reduced band-gap	Water-splitting	[21]
ZnO	Ni/Zn dopants: charge transfer and separation modulation	p-n homojunction	Water-splitting	[22]
Ga_2O_3	Junctions: charge transfer and separation modulation	α-β phase heterojunction	Water-splitting	[23]
$ZnIn_2S_4$	S-vacancies: charge transfer and separation modulation	Increased carrier density	Water splitting	[24]

Table 20.1 (Continued)

Defective materials	Defect-engineering type and role	Interaction mechanism	Photocatalytic application	References
$Cd_{0.5}Zn_{0.5}S$	Twin-boundaries: charge transfer and separation modulation	Prolonged lifetime of both photogenerated carries	Water splitting	[25]
Co_3O_4	OVs/Mn dopant: charge transfer and separation modulation	Capturing carriers for Surface reactions	Water splitting	[26]
TiO_2	OVs: Light absorption modulation	Creation of new sub-valence energy band	Water oxidation	[27]
BiOCl	OVs/S dopant: surface reaction modulation	Enhanced O_2 adsorption and chemical activation	O_2 activation	[28]
C_3N_4	N-vacancies: light absorption modulation	Enhanced O_2 adsorption and chemical activation	O_2 activation	[29]
Bi_2O_3	Junctions: charge transfer and separation modulation	α-β phase heterojunction	RhB and MO degradation	[30]
Bi_2WO_6	OVs/Nb^{5+}-dopants: charge transfer and separation modulation	Capturing carriers for surface reactions	RhB degradation	[31]
ZnS	Fe-dopant: charge transfer and separation modulation	Capturing carriers for Surface reactions	Phenolic compound degradation	[32]
$BiOIO_3$	OVs/Co-dopant: carrier generation and separation modulation	Promoted charge generation, separation, and transfer	Hg^0 oxidation	[33]
Fe_2O_3	OVs/Mg dopant: charge transfer and separation modulation	Increased carrier density	Peroxymono-sulfate activation	[34]
ZnAl	OVs/Co-dopants: surface reactions modulation	Enhanced adsorption and chemical activation	Peroxymono-sulfate activation	[35]
$BiVO_4$	OVs: charge transfer and separation modulation	Z-scheme recombination	O_2 evolution	[36]
$SrTiO_3$	OVs/Al-dopants: light absorption modulation	Reduced band-gap	O_2 evolution	[37]
$Bi_2O_2SiO_3$	Bi-dopant: surface reaction modulation	Enhanced adsorption and chemical activation	NO removal	[38]

(Continued)

Table 20.1 (Continued)

Defective materials	Defect-engineering type and role	Interaction mechanism	Photocatalytic application	References
ZrO_2	P-dopant: charge transfer and separation modulation	Increased carrier density	H_2 evolution	[39]
ZnS	S-vacancies/ Cu-dopant: charge transfer and separation modulation	Formation of lower energy levels to trap carriers in the Z-scheme	H_2 evolution	[40]
$SrTaO_2N$	Mg-dopant: charge transfer and separation modulation	Reduced band-gap	H_2 evolution	[41]
$BiVO_4$	Fe-dopant: stability modulation	Enhanced current density	H_2 evolution	[42]

their photocatalytic and electrocatalytic properties for H_2 evolution and biomass conversion. By promoting lattice expansion and facilitating phase transitions, most notably the anatase-to-rutile transformation in TiO_2, Li-ion doping effectively tailors the electronic structure and surface characteristics of the NPs. This modification leads to a reduction in bandgap energy and an increase in charge transfer efficiency. Remarkably, the Li-TiO_2 and Li-SnO_2 variants demonstrated a substantial increase in catalytic activity, achieving up to 50% yield in the selective oxidation of 2,5-hydroxymethylfurfural to 2,5-furandicarboxylic acid, alongside lower overpotentials in H_2 evolution reactions, clearly emphasizing the critical role of defect engineering in optimizing the performance of photocatalytic materials [45].

20.2.2 Enhanced Photocatalytic Performance

Defect-engineered materials often show significantly enhanced photocatalytic performance compared to their pristine counterparts. This is largely due to improved charge carrier dynamics, which are critical for photocatalytic reactions. For instance, Feng et al. successfully tuned the OVs in Bi_5O_7I NPs to significantly enhance photocatalytic performance for H_2 evolution and water-splitting reactions. By utilizing a facile solution method to synthesize NPs with a precise ethylene glycol-to-water ratio, abundant surface OVs were created, which not only reduced the bandgap from 2.93 to 2.62 eV, thus improving visible light absorption, but also generated additional defect states within the bandgap. The concentration of defects was notably increased due to this structural modification, which facilitates effective charge carrier separation and transport. The optimized Bi_5O_7I NPs exhibited remarkable photocatalytic activity, with production rates of H_2, O_2, and H_2O_2 reaching 1.18, 106.7, and 3.05 mmol g^{-1} h^{-1}, respectively, significantly outperforming their counterparts without OVs [46]. Besides, in advancing photocatalytic efficiency, defect engineering encompasses a broader spectrum of strategies beyond merely creating OVs, including the precise control of particle size, manipulation of grain boundaries, and stabilization of different phases within ZnO through high-pressure torsion (HPT). This innovative approach facilitates the formation of a high-pressure rocksalt phase while refining the particle size from 25 to 9 nm, which in turn enhances the material's structural integrity and surface characteristics. Characterization techniques such as Raman spectroscopy and X-ray photoelectron spectroscopy (XPS) reveal a systematic increase in defect density, contributing to improved charge separation and migration. The optical bandgap reduction from 3.1 to 2.9 eV further enhances the light absorption capacity, resulting in a notable increase in H_2 production rate to 1.35 $\mu mol\,g^{-1}\,h^{-1}$ under Xe light irradiation, compared to only 0.07 $\mu mol\,g^{-1}\,h^{-1}$ for the unprocessed material [47]. Furthermore, Wang and coworkers synthesized mesoporous TiO_2@rGO nanocomposites featuring mixed-crystal structures and OVs to enhance photocatalytic performance for the degradation of 2,4,6-trichlorophenol (2,4,6-TCP). Through a hydrothermal method, they created E-TiO_2@rGO and N_2-TiO_2@rGO under different atmospheres, with the E-TiO_2@rGO demonstrating superior photocatalytic activity due to its unique mixed-crystal structure that facilitated effective charge carrier separation and transport. The presence of OVs was instrumental in improving light absorption and generating reactive oxygen species (ROS), which are critical for the degradation process. The E-TiO_2@rGO nanosheets achieved a degradation rate constant of 0.06874 min^{-1}, significantly outperforming commercial TiO_2 P25 and other variants [48].

20.2.3 Increased Surface Reactivity

Surface defects, which arise from disruptions in the periodic arrangement of atoms within the photocatalytic matrix, can be meticulously controlled and characterized to optimize their function. Among the various types of defects, the introduction of vacancies through thermal treatment or chemical processes can facilitate the generation of charge traps, enhancing charge carrier isolation and improving overall photocatalytic performance. However, an excess of these defects can lead to undesirable recombination centers that hinder efficient charge transfer. Moreover, the incorporation of aliovalent dopants can yield beneficial defect states that serve as effective traps, mitigating photocorrosion and enhancing the material's responsiveness to visible light. For instance, Lira and co-workers demonstrated the enabled O_2 absorption on reduced rutile TiO_2 through effective charge removal from Ti_3þ (deeper layers). Bulk Ti_3þ defects in a crystal that served as an e/hþ pair trapper lead to the inhibition of excited carriers reassembly and cause upward band bending [49]. Likewise, Xiang and his team conducted a compelling study on In^{3+}-doped $SrTiO_3$, highlighting the transformative role of defect engineering in augmenting the surface reactivity of photocatalysts. The introduction of In^{3+} ions not only mitigated Al^{3+}-induced lattice contraction but also facilitated the formation of shallow-state defects that significantly enhanced charge carrier dynamics at the atomic scale. By increasing the availability of reactive sites on the photocatalyst surface, these defects promote efficient electron–hole separation, thus improving the material's overall photocatalytic performance. Advanced characterization techniques revealed that the optimized In^{3+}-doped $SrTiO_3$ exhibited remarkable H_2 production rates of 1.40 mmol h^{-1} and O_2 production rates of 0.69 mmol h^{-1}, attributed to the enhanced surface properties stemming from the precise control of defect states [50]. In contrast, Lin et al. synthesized W_2N_3-NV, a two-dimensional (2D) transition metal nitride with nitrogen vacancies, utilizing a molten salt-assisted calcination method to enhance the photocatalytic degradation of the antibiotics, including penicillin and amoxicillin. By constructing the Ag_3PO_4@W_2N_3-NV composite photocatalyst, they demonstrated a remarkable increase in photocatalytic activity, achieving apparent rate constants that were 77.7 and 42.9 times greater than those of pure Ag_3PO_4, respectively. The presence of nitrogen vacancies on W_2N_3-NV facilitated the formation of defects and dangling bonds, promoting effective bonding with Ag_3PO_4 and the creation of W-O chemical bonds at the interface. This innovative coupling provided a rapid transfer channel for interfacial photogenerated charges, significantly enhancing the separation and transport of charge carriers [51]. Therefore, the collective interplay of different defect types and their precise manipulation in materials can considerably elevate several kinds of photocatalytic reactions.

20.2.4 Versatility in Material Options

Defect engineering stands at the forefront of advancing photocatalytic technologies, showcasing immense versatility across a broad spectrum of material options, particularly in enhancing the performance of various metal oxides (MOs) such as TiO_2, ZnO, and Bi_2O_3. This approach systematically modifies the geometric and electronic structures of photocatalysts, enabling tailored properties that address specific photocatalytic applications, including pollutant degradation, H_2 evolution, and CO_2 reduction. By introducing point defects in TiO_2, researchers have demonstrated significant improvements in light absorption, which directly correlates with enhanced photocatalytic activity. For example, Zhang et al. explored the role of defect engineering in a hierarchical 0D/2D TiO_2/g-C_3N_4 heterojunction designed for the remediation of combined antibiotic and heavy metal pollution. By introducing defects through separate plasma treatments for each component, the researchers observed an "anisotropic-trapping" effect in the TiO_2/D-CN configuration, which enhanced the photocatalytic degradation of TC in the presence of Cu^{2+}. Specifically, TiO_2/D-CN effectively inhibited Cu^{2+} accumulation, preventing blockage of active sites, whereas D-TiO_2/CN and D-TiO_2/D-CN configurations demonstrated reduced degradation efficiency. Additionally, the formation of Cu sites facilitated surface plasmon resonance, enhancing light absorption and further promoting the degradation process [52]. Furthermore, metals such as ZnO benefit from the introduction of metal ions (Fe and Mn) defects, which improve charge carrier separation and reduce the recombination rates, thereby enhancing their overall photocatalytic water-splitting efficiency [53].

Line defects, including edge and screw dislocations within these materials, further modify their electronic properties and facilitate improved charge transport dynamics. The incorporation of dislocations in ZnO has been shown to enhance visible light absorption and improve overall photocatalytic performance, particularly in the degradation of pharmaceutical pollutants [54]. Similarly, planar defects, such as grain boundaries and twin boundaries, play a pivotal role in facilitating the adsorption and diffusion of reactants, thus optimizing the photocatalytic output. For instance, defect-rich ZnO nanofibers with engineered grain boundaries have exhibited superior photocatalytic activity in the degradation of methylene blue, highlighting the importance of these structural modifications [55]. Volume defects, particularly voids, also significantly influence the photocatalytic performance by altering surface morphology and light–matter interactions, as demonstrated in composite materials like ZnO/Ag and ZnO/TiO_2, which have shown enhanced activity for bacterial disinfection and CO_2 reduction [56]. Ultimately, the strategic application of defect engineering not only enhances the physicochemical properties of these photocatalytic materials but also broadens the scope of their applicability in sustainable energy and environmental remediation efforts. The ability to tailor the defect landscape in materials like Bi_2O_3 [57] and $LuFe_2O_4$ [58] further emphasizes the versatility in material options, allowing researchers to optimize photocatalysts for specific tasks effectively. For a comprehensive understanding, the various types of defects in recently reported photocatalysts, along with the synthesis methods and techniques employed, are compiled in Table 20.2.

20.2.5 Cost-effectiveness

Defect engineering represents a transformative approach in enhancing the cost-effectiveness of photocatalytic materials, particularly through the targeted manipulation of defects in nanomaterials such as carbon nanotubes (CNTs) and graphene. The book elaborates on how defects, which can be categorized as point defects (including vacancies, interstitials, and substitutionals) or extended defects (such as dislocations and grain boundaries), significantly influence the material properties essential for photocatalysis. For instance, the introduction of vacancies in graphene can lead to increased surface reactivity by creating unsaturated coordination sites, which are vital for photocatalytic reactions [89]. This enhancement is crucial for applications in environmental remediation, where defect-engineered graphene photocatalysts have demonstrated remarkable efficacy in degrading pollutants under visible light, showcasing their potential for sustainable water treatment solutions [90]. Similarly, carbon nanotubes, when subjected to controlled defect engineering, exhibit improved charge carrier dynamics, which facilitate the generation of electron–hole pairs while minimizing recombination losses. This property is particularly advantageous in energy conversion applications, such as in dye-sensitized solar cells and photocatalytic water splitting, where efficiency is paramount [91]. Moreover, defect engineering allows for the tuning of the electronic band structure of these photocatalytic materials, enabling better responsiveness to visible light and thus widening their practical applications in solar energy utilization. The findings point to specific examples, such as defect-engineered transition metal dichalcogenides, which have been shown to enhance photocatalytic activity and stability, making them ideal candidates for next-generation

Table 20.2 Outline of different defective photocatalysts, highlighting defect types, fabrication methods, and characterization techniques.

Defective materials	Defect-engineering types	Synthesis method	Characterization techniques	References
TiO_2	OVs/Zr_4t-dopants	Vacuum annealing	HRTEM, XPS	[59]
TiO_2	OVs/Ti-vacancies	Heat treatment under air	STEM, EPR, XPS	[60]
TiO_2	Ti^{3+}-dopants	Solvothermal method	HRTEM, EPR, XPS	[61]
TiO_2	Edge dislocation	Hydrothermal reaction	HRTEM, Raman	[62]
TiO_2	OVs	H_2 treatment	EPR, XPS	[63]
WO_3	OVs	Vacuum annealing	HRTEM, XPS, EPR	[64]
Fe_2O_3	OVs/Mg dopant	Magnetron sputtering	XPS, Raman	[65]
ZnO	OVs/Zn vacancies	Hydrothermal reaction	XPS, EPR	[66]
$BiVO_4$	Fe-dopant	Hydrothermal reaction	EPR, XPS, TRFS	[67]
Ga_2O_3	Phase-junctions	Thermal calcination	TA, HRTEM, XPS, Raman	[68]
Bi_2O_3	Phase-junctions	Hydrothermal reaction	HRTEM, XPS	[69]
Bi_2WO_6	OVs	Hydrothermal reaction	HRTEM, XPS, Raman	[70]
$Mn_xCo_{3\_x}O_4$	Mn/Co-vacancies	Thermal calcination	EXAFS, XPS	[71]
TiO_2	Distortion	Thermal calcination	HRTEM, XPS, EXAFS	[72]
Nb_2O_5	Lattice disorder	Hydrothermal reaction	AFM, EPR, XPS, HRTEM	[73]
$Bi_7O_9I_3$	OVs	Solvothermal treatment	FESEM, EPR, XPS	[74]
$Bi_{12}O_{17}C_{l2}$	OVs	Solvothermal method	HAADF-STEM, EPR, XPS	[75]
BiOBr	OVs	Solvothermal method	HRTEM, EPR, XPS	[76]
γ-FeOOH	Fe-vacancies	Room temperature reduction	AFM, TEM-STM, EPR, XPS, EXAFS	[77]
SnCoFe hydroxide	Sn-vacancies	Ar_2 plasma	HAADF-STEM, EPR, XPS	[78]
C_3N_4	C/N-vacancies	NH_3 treatment	TEM, EPR, XPS	[79]
C_3N_4	N-vacancies	N_2 treatment	STEM, XPS	[80]
Bi_2MoO_6	S-dopants	Solvothermal method	TEM, XPS	[81]
ZnS	S-vacancies/ Cr-dopants	Hydrothermal method	HRTEM, EPR, XPS	[82]
CdS	Facet-junctions/ S-vacancies	Hydrothermal method	TEM, XPS	[83]
CdS	Dislocation	Hydrothermal method	HRTEM, XPS	[84]
CdSe	Dislocation	CVD process	HRTEM	[85]
$ZnIn_2S_4$	S-vacancies	Solvothermal method	AFM, HRTEM, EPR	[86]
CeO_2	Facet-junctions	Solvothermal reaction	HRTEM, XPS	[87]
$Cd_{0.5}Zn_{0.5}S$	Twin boundaries/ S-vacancies	hydrothermal method	HRTEM	[88]

photocatalysts [92]. Furthermore, this book underscores the importance of precise characterization techniques, including scanning electron microscopy (SEM) and transmission electron microscopy (TEM), in understanding and optimizing these defects to achieve desired photocatalytic properties. Ultimately, the integration of defect engineering not only augments the photocatalytic efficiency of materials but also contributes to their mechanical strength and durability, addressing the critical aspects of longevity and cost-effectiveness that are essential for commercial viability. Overall, the systematic exploration of defect engineering in nanomaterials like carbon nanotubes and graphene underscores its potential to revolutionize the field of photocatalysis, providing significant advancements in both environmental applications and energy conversion technologies, while simultaneously reducing costs and enhancing material performance through the strategic manipulation of defects. By harnessing the unique properties of various materials, researchers can create adaptable photocatalytic systems that not only improve efficiency but also expand the range of potential applications, ultimately contributing to a more sustainable future.

20.3 Disadvantages of Defect Engineering

While defect engineering presents numerous advantages, it is essential to recognize the challenges and limitations associated with this approach. Understanding these drawbacks is crucial for developing effective strategies to mitigate them.

20.3.1 Complexity of Defect Characterization

The complexity of defect characterization in photocatalytic materials is a considerable barrier to advancing research and application in this field. Accurately identifying and quantifying defects requires sophisticated analytical techniques that can be both resource-intensive and time-consuming. Generally, the characterization of defects in photocatalytic materials involves several advanced techniques, each with its distinctive limitations. TEM requires thin samples, which can alter defect characteristics; however, this technique provides high-resolution images that are essential for observing defect structures. XPS is sensitive to surface contamination and primarily offers limited bulk information, making it necessary for careful sample handling to ensure accuracy. Electron paramagnetic resonance (EPR) is restricted to materials with unpaired electrons and can produce overlapping signals that complicate defect identification, highlighting the need for precise calibration and interpretation. Atomic force microscopy (AFM) mainly provides topographical information and may overlook smaller defects, necessitating complementary methods to achieve a complete defect profile. SEM focuses on surface morphology and lacks atomic-level resolution, which can be a drawback when detailed defect analysis is required. Raman spectroscopy's sensitivity to sample quality can lead to misleading spectral data if impurities or surface irregularities are present, complicating the characterization of defects that are critical for photocatalytic performance. Its limited spatial resolution, typically around 1–2 μm, hampers the ability to analyze nanoscale features, such as those found in semiconductor nanostructures and graphene. This limitation can result in overlooking critical insights into how localized defects influence electronic properties and photocatalytic efficiency. Auger electron spectroscopy (AES) is surface-sensitive and requires high vacuum, which can complicate the analysis of certain materials. To address these challenges, a combination of complementary techniques, including in-situ adaptations where appropriate, can enhance the robustness of defect characterization and provide a more comprehensive understanding of their influence on photocatalytic performance.

Additionally, integrating computational modeling and machine learning techniques offers a promising avenue to predict defect behavior and properties, thereby reducing reliance on costly experimental techniques. For instance, computational simulations can provide insights into defect formation energies and migration pathways, guiding researchers in the design of materials with optimal defect structures for enhanced photocatalytic performance. Furthermore, collaborative networks that facilitate resource sharing among researchers can help mitigate the limitations posed by expensive equipment and technical expertise. Establishing shared core facilities or centralized databases for defect characterization could enable broader access to advanced analytical techniques, fostering innovation and accelerating research in defect engineering. By addressing these disadvantages, the field can progress toward harnessing the full potential of defect-engineered materials for sustainable photocatalytic applications.

20.3.2 Adverse Impacts on Photocatalytic Performance

The introduction of defects into photocatalytic materials can sometimes yield adverse effects that undermine their intended performance. For instance, while engineered OVs in complex nanomaterials such as perovskite oxides can enhance photocatalytic activity by facilitating improved charge separation, an excessive concentration of these vacancies may lead to structural instability and increased recombination rates of charge carriers. A study by Liu et al. explored the effects of La/AL co-doping in strontium titanate ($SrTiO_3$) nanostructures, revealing that while a controlled introduction of defects initially improved photocatalytic performance, an overabundance resulted in diminished stability and efficiency due to enhanced defect-induced charge carrier recombination. This finding underscores the delicate balance required in defect engineering; too few defects may fail to optimize photocatalytic properties, while too many can compromise the structural integrity and longevity of the material [93].

Moreover, the unintended consequences of defect engineering extend beyond just structural stability. Excessive defects can significantly alter the electronic properties of complex materials, leading to the formation of trap states that hinder charge carrier mobility. For example, in layered transition metal dichalcogenides (TMDs) such as MoS_2 [94], an overabundance of defects can introduce localized states within the band gap, which trap charge carriers and limit their mobility, thereby reducing photocatalytic efficiency. This is particularly critical for applications that rely on rapid electron transfer, such as H_2 evolution and CO_2 reduction, where efficient charge transport is essential for optimal performance [95]. The challenge lies in meticulously controlling the type, quantity, and distribution of defects introduced into these advanced photocatalytic materials to ensure that their beneficial effects are maximized while minimizing any detrimental impacts.

20.3.3 Scalability and Reproducibility Challenges

A significant drawback of defect engineering lies in the challenge of scaling up the production of defect-engineered materials while maintaining consistent quality and performance. Laboratory-scale synthesis methods often yield promising results, but translating these findings to industrial-scale production presents substantial hurdles. Techniques such as sol–gel synthesis, hydrothermal synthesis, and chemical vapor deposition (CVD) can effectively introduce defects at a small scale; however, replicating these processes on a larger scale can lead to inconsistencies in defect formation and variations in material properties. A study by Li et al. investigated the scalability issues of nitrogen-doped graphene quantum dots (GQDs), revealing that variations in synthesis parameters during scale-up resulted in inconsistent defect densities and distributions, adversely impacting their photoluminescent and photocatalytic properties. Moreover, materials such as metal–organic frameworks (MOFs) and 2D materials pose additional challenges in achieving uniform defect characteristics across larger batches. For instance, the introduction of defects in materials such as black phosphorus (BP) can significantly affect their electronic and optical properties, which are critical for applications in photodetectors and photonic devices. Defects in BP, such as vacancies or interstitials [96], can alter its band structure and increase its susceptibility to oxidation, which is a critical concern for stability in ambient conditions. If defect levels are not properly controlled during scaling, it can lead to performance variability and instability, undermining the reliability necessary for commercial applications and large-scale deployment. The optimization of defect concentrations in BP can enhance its light absorption and carrier mobility [97], but achieving this balance during large-scale synthesis remains a significant challenge [98].

20.3.4 Limited Understanding of Defect Interactions

The interactions between defects and other material properties remain an area of active research, with a limited understanding of how these interactions influence photocatalytic performance. The complexity of defect behavior in various environments can lead to unpredictable outcomes, complicating the design process for new materials. For example, the presence of moisture, temperature fluctuations, and the nature of the catalytic environment can all significantly affect the stability and reactivity of defects within photocatalytic materials. A study by Wang et al. highlighted that in dual-defective layered materials such as tin disulfide (SnS_2), while certain defects may enhance photocatalytic activity in controlled laboratory settings, their behavior can be adversely affected by environmental factors, leading to reduced effectiveness in real-world applications [99]. Specifically, moisture can lead to the hydrolysis of defects, altering their electronic properties and diminishing their catalytic efficiency. Additionally, temperature variations can induce thermal fluctuations that either stabilize or destabilize defect configurations, directly impacting charge carrier dynamics and photocatalytic performance. These challenges underscore the need for a comprehensive understanding of defect interactions within the context of the operational environment. Without this knowledge, the optimization of materials for practical photocatalytic applications, such as environmental and energy, remains hindered, as the designed defect characteristics may not translate effectively to real-world conditions. To address these limitations, researchers must focus on investigating defect behavior in diverse environments and develop predictive models that encompass environmental variables. By gaining insights into how defects interact with their surroundings, it will be possible to design more robust photocatalytic materials that maintain high efficiency under realistic operational conditions.

20.4 Summary and Outlooks

The field of defect engineering in photocatalysis presents a wealth of opportunities for further exploration and innovation. This book summarizes recent developments in defect-engineered photocatalysts, emphasizing the significance of various defect types in tailoring electronic and microstructural properties to enhance conductivity and charge carrier concentrations for improved photocatalytic efficiency. Effective strategies for defect-controlled synthesis have been discussed in detail, alongside advanced characterization techniques for defect identification and quantification. The strategic incorporation of defects in photocatalytic materials has proven to significantly enhance charge carrier dynamics, light absorption, and surface reactivity, leading to superior performance in applications such as water splitting, CO_2 photoreduction, nitrogen fixation, organic synthesis, biomedical fields, and pollutant degradation. However, the complexities and challenges inherent to defect engineering must not be overlooked. Issues such as unintended recombination centers due to excessive defects, the stability of anion defects in ambient conditions, and the random nature of metal defect formation necessitate rigorous mechanistic studies and advanced fabrication techniques to fully understand and optimize the relationship between defects and photocatalytic properties.

1) One significant trend is the exploration of novel defect types and their synergistic effects. For example, the combination of different types of defects, such as OVs and dopants, can lead to unique electronic structures that facilitate improved charge separation and transport. This is particularly evident in MOFs and perovskite materials, where the tunability of defects allows for the fine-tuning of band structures to cater to specific photocatalytic reactions, such as nitrogen fixation or CO_2 reduction.
2) Moreover, the use of machine learning and artificial intelligence to design defect-engineered materials is gaining traction. By analyzing vast datasets of

material properties and performance metrics, researchers can identify optimal defect configurations and predict their impact on photocatalytic activity. This approach can significantly accelerate the discovery of new materials and the optimization of existing systems, paving the way for breakthroughs in photocatalytic efficiency.

3) Recent developments have also highlighted the importance of environmental stability in defect-engineered photocatalysts. Understanding how defects interact with environmental factors, such as humidity and temperature, is crucial for ensuring the long-term performance of these materials in real-world applications. For instance, the stability of OVs in TiO_2 can be compromised in humid conditions, leading to decreased photocatalytic efficiency [9]. Strategies to enhance defect stability, such as surface passivation techniques or the development of robust composite materials, are vital for advancing the practical deployment of defect-engineered photocatalysts.
4) The potential for scaling up the production of defect-engineered materials is another area of focus. Advances in synthesis techniques, such as chemical vapor deposition (CVD), sol–gel methods, and hydrothermal synthesis, allow for the controlled introduction of defects on a larger scale, which is essential for industrial applications. Ensuring that these methods are cost-effective and reproducible will be critical for the commercialization of defect-engineered photocatalysts.

To summarize, the future of defect engineering in photocatalysis is bright, driven by continuous innovation and interdisciplinary collaboration. By leveraging advanced characterization techniques, computational modeling, and machine learning, researchers can gain deeper insights into the mechanisms of defect formation and their effects on photocatalytic performance. The ongoing exploration of novel materials and defect configurations will undoubtedly lead to significant advancements in sustainable energy technologies and environmental remediation strategies. As the scientific community continues to address the challenges and limitations associated with defect engineering, the potential for real-world applications grows, promising a more sustainable and efficient future in photocatalysis.

Conflict of Interest

The authors declare that they have no known competing financial interests or personal relationships that could have appeared to influence the work reported in this chapter.

References

1 Brédas, J.-L., Sargent, E.H., and Scholes, G.D. (2017). *Nat. Mater.* 16: 35–44.
2 Wang, X., Hisatomi, T., Liang, J. et al. (2020). *J Mater Chem A* 8: 11743–11751.

3 Lei, W., Yu, Y., Zhang, H. et al. (2022). *Mater. Today* 52: 133–160.
4 Lau, V.W.-h., Moudrakovski, I., Botari, T. et al. (2016). *Nat. Commun.* 7: 12165.
5 Saka, C. (2022). *Fuel* 322: 124196.
6 Chen, M., Sun, M., Cao, X. et al. (2024). *Coord. Chem. Rev.* 510: 215849.
7 Chen, S., Huang, D., Cheng, M. et al. (2021). *J Mater Chem A* 9: 196–233.
8 Wang, C., Zhan, Z., Liu, H. et al. (2024). *Chem. Eng. J.* 482: 148906.
9 Zafar, Z., Yi, S., Li, J. et al. (2022). *Energy Environ. Mater.* 5: 68–114.
10 Zhong, K., Sun, P., and Xu, H. (2024). *Small* 21: 2310677.
11 Shandilya, P., Sambyal, S., Sharma, R. et al. (2022). *J. Hazard. Mater.* 428: 128218.
12 Li, Y., He, Z., Liu, L. et al. (2023). *Nano Energy* 105: 108032.
13 Lettieri, S., Pavone, M., Fioravanti, A. et al. (2021). *Materials* 14: 1645.
14 Di, J., Chen, C., Zhu, C. et al. (2019). *ACS Appl. Mater. Interfaces* 11: 30786–30792.
15 Shen, H., Gao, H., Zhang, S. et al. (2025). *J Mater Chem A* 13: 2987–2996.
16 Yao, Y., Yin, W., Tang, Z. et al. (2022). *J. Phys. Chem. C* 126: 1271–1280.
17 Feng, S., Zhao, J., Bai, Y. et al. (2020). *J CO_2 Util* 38: 1–9.
18 Bai, J., Shi, Y., Ma, A. et al. (2024). *Chem. Eng. J.* 499: 156064.
19 Shuwanto, H., Abdullah, H., Ku, Y., and Lie, J. (2024). *Mater. Sci. Eng. B Solid State Mater.* 299: 117027.
20 Guo, M., Chen, M., Xu, J. et al. (2023). *Chem. Eng. J.* 461: 142046.
21 Ma, K., Zhang, M., Sun, W. et al. (2022). *Appl Catal B* 315: 121570.
22 Neaţu, F., Abramiuc, L.E., Trandafir, M.M. et al. (2020). *ChemCatChem* 12: 4642–4651.
23 Wang, Y., Zhuang, P., Yue, H. et al. (2019). *J. Phys. Chem. C* 123: 8990–9000.
24 Chen, J., Li, K., Cai, X. et al. (2022). *Mater. Sci. Semicond. Process.* 143: 106547.
25 Liu, F., Xue, F., Si, Y. et al. (2021). *ACS Appl. Nano Mater.* 4: 759–768.
26 Zhang, L., Wei, F., and Liang, H. (2024). *Mater Today Commun* 40: 109887.
27 Zhang, X., Tang, P., Zhai, G. et al. (2022). *Chem. Res. Chin. Univ.* 38: 1292–1300.
28 Zhang, C., Deng, Y., Wan, Q. et al. (2024). *Appl Catal B* 343: 123557.
29 Xie, Z., Lin, J., Zhang, W. et al. (2024). *Diam. Relat. Mater.* 141: 110658.
30 Bera, K.K., Chakraborty, M., Mondal, M. et al. (2020). *Ceram. Int.* 46: 7667–7680.
31 Chen, H., Zhang, C., Pang, Y. et al. (2019). *RSC Adv.* 9: 22559–22566.
32 Li, B., Amin, A.H., Ali, A.M. et al. (2023). *Chemosphere* 336: 139208.
33 Lin, M., Liu, H., Wang, H. et al. (2024). *Chem. Eng. J.* 484: 149472.
34 Guo, S., Liu, M., You, L. et al. (2021). *Chemosphere* 279: 130482.
35 Ma, S., Chen, D., Zhong, Y. et al. (2023). *Chem. Eng. J.* 467: 143385.
36 Yang, B., Xu, J., Bin, D. et al. (2021). *Appl. Catal. B: Environ. Energy* 283: 119583.
37 Kosaka, T., Teduka, Y., Ogura, T. et al. (2020). *ACS Catal.* 10: 13159–13164.
38 Li, N., Zhao, W., Zhang, J. et al. (2024). *Catalysts* 14: 275.
39 Dong, S., Li, Y., Zhang, X. et al. (2023). *Mater. Des.* 231: 112068.
40 Bao, L., Ali, S., Dai, C. et al. (2024). *ACS Nano* 18: 5878–5889.
41 Xu, J., Luo, Y., Guo, Q. et al. (2024). *Chin. J. Catal.* 65: 70–78.
42 Sun, Y., Suriyaprakash, J., Shan, L. et al. (2024). *Appl. Catal. B: Environ. Energy* 355: 124209.

43 Bian, R., An, S., Wang, X. et al. (2024). *Int. J. Hydrog. Energy* 51: 787–795.
44 He, N., Yu, Z., Yang, G. et al. (2024). *Chem. Eng. J.* 484: 149613.
45 Hoang, D.T., Lim, D., Kang, M.J. et al. (2024). *Adv. Sustainable Syst.* 8: 2400094.
46 Feng, J., Ran, X., Wang, Y. et al. (2024). *Appl. Surf. Sci.* 655: 159517.
47 Shundo, Y., Nguyen, T.T., Akrami, S. et al. (2024). *J. Colloid Interface Sci.* 666: 22–34.
48 Wang, F., Zhao, Q., Li, H. et al. (2024). *Chem. Eng. J.* 481: 148855.
49 Chen, H., Zhang, F., Zhang, W. et al. (2018). *Appl. Phys. Lett.* 112: 112.
50 Xiang, L., Shi, L., Jiang, J. et al. (2024). *Inorg. Chem.* 64: 351–360.
51 Lin, Y., Yang, C., Niu, Q., and Luo, S. (2022). *Adv. Funct. Mater.* 32: 2108814.
52 Zhang, J., Wang, X., Dai, J. et al. (2024). *Colloids Surf A Physicochem. Eng. Asp.* 680: 132624.
53 Mishra, R.K., Choi, G.J., Ryu, J.W. et al. (2024). *Energy Fuel* 38: 6300–6313.
54 Mirzaeifard, Z., Shariatinia, Z., Jourshabani, M., and Rezaei Darvishi, S.M. (2020). *Ind. Eng. Chem. Res.* 59: 15894–15911.
55 Pantò, F., Dahrouch, Z., Saha, A. et al. (2021). *Appl. Surf. Sci.* 557: 149830.
56 Padmanabhan, N.T., Thomas, N., Louis, J. et al. (2021). *Chemosphere* 271: 129506.
57 Huang, R., Wei, Y., Gao, T. et al. (2021). *Ceram. Int.* 47: 205–213.
58 Hunnestad, K.A., Das, H., Hatzoglou, C. et al. (2024). *Nat. Commun.* 15: 5400.
59 Shaddad, M.N., Cardenas-Morcoso, D., García-Tecedor, M. et al. (2019). *ACS Omega* 4: 16095–16102.
60 Wang, Q., Zhang, S., He, H. et al. (2021). *Chem. Asian J.* 16: 3–19.
61 Wang, K., Yang, N., Xiao, B. et al. (2024). *ACS Appl. Nano Mater.* 7: 21842–21851.
62 Sarkar, D., Goswami, S., Sarngan, P.P. et al. (2024). *Surf Interfaces* 54: 105247.
63 Zhang, Q., Yao, C., Zhang, F. et al. (2025). *J Mater Chem A* 13: 5180–5190.
64 Li, R., Wang, Q., Wang, Y. et al. (2024). *ACS Appl. Mater. Interfaces* 16: 51738–51747.
65 Saritaş, S., Muglu, G., Turgut, E. et al. (2024). *Phys. Rev. B-Condens. Matter.* 677: 415704.
66 Wei, Y., Li, Y., Xu, Y. et al. (2025). *Appl. Catal. B: Environ. Energy* 365: 124878.
67 Wang, J., Guo, L., Xu, L. et al. (2021). *Nano Res* 14: 1294–1304.
68 Khan, K., Rehman, Z.U., Yao, S. et al. (2024). *Nanotechnol. Rev.* 13: 20240104.
69 Yu, X., Huang, X., Huang, Y. et al. (2021). *Colloids Surf. A Physicochem. Eng. Asp.* 626: 127068.
70 Xiong, J., Zeng, H.-Y., Yang, Z.-L. et al. (2023). *Ceram. Int.* 49: 35927–35938.
71 Li, K., Zhang, R., Gao, R. et al. (2019). *Appl Catal B* 244: 536–545.
72 Zhang, T., Zheng, P., Gao, J. et al. (2024). *Nat. Commun.* 15: 6827.
73 Trench, A.B., Moura, J.P.C., Antonin, V.S. et al. (2024). *Adv. Powder Technol.* 35: 104404.
74 Liang, R., Wu, H., Hu, Z. et al. (2024). *Chin. Chem. Lett.* 36: 110136.
75 Guan, M., Lu, N., Zhang, X. et al. (2024). *Carbon Energy* 6: e420.
76 Li, Z., Hou, Y., Shen, Y. et al. (2024). *Water Res.* 267: 122524.
77 Bora, D.K., Ghosh, D., Jana, A. et al. (2024). *ACS Appl. Eng. Mater.* 2: 975–987.

78 Wu, X., Zhu, Q., Cong, Y. et al. (2025). *Mater. Chem. Front.* 9: 445.
79 Liang, J., Fang, N., Liu, C. et al. (2024). *Chem. Eng. J.* 486: 150155.
80 Xia, Y., Yang, H., Ho, W. et al. (2024). *Appl Catal B* 344: 123604.
81 Liu, Z., Luo, M., Cao, Y. et al. (2024). *J. Catal.* 430: 115347.
82 Bao, L., Jia, Y., Ren, X. et al. (2024). *J. Mater. Sci. Technol.* 199: 75–85.
83 Fu, H., Wang, J., Zhang, J. et al. (2024). *Int. J. Hydrog. Energy* 73: 665–672.
84 Kumari, D., Phogat, P., Singh, S., and Jha, R. (2024). *Mater. Sci. Eng. B* 301: 117212.
85 Das, S., Tripathi, A., Parida, A. et al. (2024). *ACS Appl. Electron. Mater.* 6: 6522–6533.
86 He, J., Yang, Z., Wang, Z. et al. (2025). *Chem. Eng. J.* 503: 158433.
87 Xu, B., Jia, L., Yang, H. et al. (2024). *Rare Metals* 43: 2026–2038.
88 Hamid, M.A.S.H., Zengin, Y., and Boz, I. (2024). *Int. J. Hydrog. Energy* 52: 103–114.
89 Zhang, L., Jia, Y., Gao, G. et al. (2018). *Forensic Chem.* 4: 285–297.
90 Chen, S., Li, J., Zhou, W. et al. (2024). *Coord. Chem. Rev.* 507: 215749.
91 Deshmukh, M.A., Park, S.-J., Hedau, B.S., and Ha, T.-J. (2021). *Sol. Energy* 220: 953–990.
92 Tan, Z.H., Kong, X.Y., Ng, B.-J. et al. (2023). *ACS Omega* 8: 1851–1863.
93 Qin, Y., Fang, F., Xie, Z. et al. (2021). *ACS Catal.* 11: 11429–11439.
94 Panchal, D., Sharma, A., and Pal, S. (2023). *Mater Today Sustain* 21: 100264.
95 Liu, C., Kong, C., Zhang, F.-J. et al. (2021). *J. Korean Ceram. Soc.* 58: 135–147.
96 Mu, H., Yu, W., Yuan, J. et al. (2022). *Mater. Futures* 1: 012301.
97 Fan, S., Li, J., Cao, H.-Q. et al. (2022). *J. Mater. Chem. C* 10: 14053–14079.
98 He, G., Dong, T., Yang, Z., and Ohlckers, P. (2019). *Chem. Mater.* 31: 9917–9938.
99 Sainbileg, B., Lai, Y.-R., Chen, L.-C., and Hayashi, M. (2019). *Phys. Chem. Chem. Phys.* 21: 26292–26300.

Index

Defect Engineering in Photocatalysis: From Fundamentals to Applications,
First Edition. Edited by Chuanyi Wang, Abdelkader Labidi, and Qiuhui Zhu.

e

f

g

p